Vegetation Ecology

A companion website with additional resources is available at
www.wiley.com/go/vandermaarelfranklin/vegetationecology
with Figures and Tables from the book

Vegetation Ecology

Second Edition

Eddy van der Maarel & Janet Franklin
University of Groningen, The Netherlands
Arizona State University, USA

Ⓦ**WILEY-BLACKWELL**

A John Wiley & Sons, Ltd., Publication

This edition first published 2013 © 2013 by John Wiley & Sons, Ltd.

Blackwell Publishing was acquired by John Wiley & Sons in February 2007. Blackwell's publishing program has been merged with Wiley's global Scientific, Technical and Medical business to form Wiley-Blackwell.

Registered office: John Wiley & Sons, Ltd, The Atrium, Southern Gate, Chichester, West Sussex, PO19 8SQ, UK

Editorial offices: 9600 Garsington Road, Oxford, OX4 2DQ, UK
The Atrium, Southern Gate, Chichester, West Sussex, PO19 8SQ, UK
111 River Street, Hoboken, NJ 07030-5774, USA

For details of our global editorial offices, for customer services and for information about how to apply for permission to reuse the copyright material in this book please see our website at www.wiley.com/wiley-blackwell.

The right of the author to be identified as the author of this work has been asserted in accordance with the UK Copyright, Designs and Patents Act 1988.

Library of Congress Cataloging-in-Publication Data

Vegetation ecology / [edited by] Eddy van der Maarel & Janet Franklin. – 2nd ed.
p. cm.
Includes bibliographical references and index.
ISBN 978-1-4443-3888-1 (cloth) – ISBN 978-1-4443-3889-8 (pbk.)   1. Plant ecology.
2. Plant communities.   I. Maarel, E. van der.   II. Franklin, Janet, 1959–
QK901.V35 2013
581.7–dc23

2012018035

A catalogue record for this book is available from the British Library.

Wiley also publishes its books in a variety of electronic formats. Some content that appears in print may not be available in electronic books.

Front cover image: Vegetation mosaic in the calcium-poor coastal dunes in North Holland, with dune heathland, scrub, a dune lake and parabolic dunes in the background; photo Eddy van der Maarel (March 2005).
Back cover image: Sonoran Desert scrub vegetation of the Arizona Uplands, also known as 'saguaro-palo verde forest', shown here in the Rincon Mountains, Saguaro National Park, east of Tucson, Arizona; photo Janet Franklin (April 2012).
Cover Design by Design Deluxe

Set in 10.5/12 pt Classical Garamond by Toppan Best-set Premedia Limited
Printed and bound in Malaysia by Vivar Printing Sdn Bhd

1   2013

Contents

10 Vegetation and Ecosystem   285
Christoph Leuschner

11 Diversity and Ecosystem Function   308
Jan Lepš

12 Plant Functional Types and Traits at the Community, Ecosystem and World Level   347
Andrew N. Gillison

The color plate section can be found between pp. 272–273.

A companion website with additional resources is available at
www.wiley.com/go/vandermaarelfranklin/vegetationecology
with Figures and Tables from the book

Contributors

Mehdi Abedi, Institute of Botany, University of Regensburg, D-93040 Regensburg, Germany
mehdi.abedi@biologie.uni-regensburg.de

Dr Mike P. Austin, CSIRO Ecosystem Sciences, GPO Box 1700, Canberra ACT 2601, Australia
mike.austin@csiro.au

Professor Dr Jan P. Bakker, Community and Conservation Ecology Group, University of Groningen, P.B. 14, NL-9750 AA Haren, The Netherlands
j.p.bakker@rug.nl

Dr Maik Bartelheimer, Institute of Botany, University of Regensburg, D-93040 Regensburg, Germany
maik.bartelheimer@biologie.uni-regensburg.de

Dr Robert Baxter, School of Biological and Biomedical Sciences, University of Durham, South Road, Durham DH1 3LE, UK
robert.baxter@durham.ac.uk

Professor Elgene O. Box, Geography Department, University of Georgia, Athens, Georgia 30602-2502, USA
boxeo@uga.edu

Professor Mary L. Cadenasso, Department of Plant Sciences, University of California Davis, Mail Stop 1 1210 PES, One Shields Avenue, Davis, CA 95616-8780, USA
mlcadenasso@ucdavis.edu

Dr Bengt Å. Carlsson, Department of Ecology and Evolution, Uppsala University, Norbyvägen 18D, SE-752 36 Uppsala, Sweden
bengt.carlsson@ebc.uu.se

Dr Ron G.M. de Goede, Department of Soil Quality, Wageningen University, P.O. Box 47, 6700 AA Wageningen, The Netherlands
ron.degoede@wur.nl

Dr Juliane Drobnik, Institute of Botany, University of Regensburg, D-93040 Regensburg, Germany
juliane.drobnik@biologie.uni-regensburg.de

Professor Janet Franklin, School of Geographical Sciences and Urban Planning, Coor Hall, 975 S. Myrtle Ave., Fifth Floor, Arizona State University, P.O. Box 875302, Tempe AZ 85287-5302, USA
janet.franklin@asu.edu

Professor Kazue Fujiwara, Laboratory of Vegetation Science, Yokohama National University, Tokiwadai 79-7, Hodogaya-ku, Yokohama 240-8501, Japan
kazue@ynu.ac.jp

Dr Andrew N. Gillison, Center for Biodiversity Management, P.O. Box 120, Yungaburra 4884 QLD, Australia
andygillison@gmail.com

Professor Brian Huntley, School of Biological and Biomedical Sciences, University of Durham, South Road, Durham DH1 3LE, UK
brian.huntley@durham.ac.uk

Professor Dr Thomas W. Kuyper, Department of Soil Quality, Wageningen University, P.O. Box 47, 6700 AA Wageningen, The Netherlands
thom.kuyper@wur.nl

Professor Jan Lepš, Department of Botany, Faculty of Biological Sciences, University of South Bohemia, Branišovská 31, CZ-370 05 České Budějovice, Czech Republic
suspa@bf.jcu.cz

Professor Dr Christoph Leuschner, Plant Ecology, Albrecht-von-Haller-Institute for Plant Sciences, University of Göttingen, Untere Karspüle 2, D-37073 Göttingen, Germany
cleusch@gwdg.de

Professor Scott J. Meiners, Department of Biological Sciences, Eastern Illinois University, 600 Lincoln Avenue, Charleston, IL 61920-3099, USA
sjmeiners@eiu.edu

Professor Samuel J. McNaughton, Department of Biology, Syracuse University, 114 Life Sciences Complex, Syracuse, NY 13244-1220, USA
sjmcnaug@mailbox.syr.edu

Professor Robert K. Peet, Biology Department, University of North Carolina, 413 Coker Hall, Chapel Hill, NC 27599-3280, USA
peet@unc.edu

Dr Steward T.A. Pickett, Cary Institute of Ecosystem Studies, Box AB, Millbrook, NY 12545-0129, USA
picketts@ecostudies.org

Professor Dr Peter Poschlod, Institute of Botany, University of Regensburg, D-93040 Regensburg, Germany
peter.poschlod@biologie.uni-regensburg.de

Dr Petr Pyšek, Institute of Botany, Academy of Sciences of the Czech Republic, CZ-252 43 Průhonice, Czech Republic
pysek@ibot.cas.cz

Professor Marcel Rejmánek, Department of Evolution & Ecology, University of California Davis, 5337 Storer Hall, Davis, CA 95616, USA
mrejmanek@ucdavis.edu

Professor David M. Richardson, DST-NRF Centre for Invasion Biology, Stellenbosch University, Private Bag X1, Matieland 7602, South Africa
rich@sun.ac.za

Professor David W. Roberts, Department of Ecology, Montana State University, Bozeman, MT 59717-3460, USA
droberts@montana.edu

Sergey Rosbakh, Institute of Botany, University of Regensburg, D-93040 Regensburg, Germany
sergey.rosbakh@biologie.uni-regensburg.de

Professor Håkan Rydin, Department of Ecology and Evolution, Uppsala University, Norbyvägen 18D, SE-752 36 Uppsala, Sweden
hakan.rydin@ebc.uu.se

Dr Arne Saatkamp, Aix-Marseille Université – IMBE, Faculté des Sciences de St Jérôme, F-13397 Marseille cedex 20, France
arnesaatkamp@gmx.de

Dr Mahesh Sankaran, National Centre for Biological Sciences, Tata Institute of Fundamental Research, GKVK, Bellary Road, Bangalore, India
mahesh@ncbs.in; and

Institute of Integrative & Comparative Biology, Faculty of Biological Sciences, 9.18 LC Miall Building, University of Leeds, Leeds LS2 9JT, UK
m.sankaran@leeds.ac.uk

Professor Brita M. Svensson, Department of Ecology and Evolution, Uppsala University, Norbyvägen 18D, SE-752 36 Uppsala, Sweden
brita.svensson@ebc.uu.se

Professor Dr Jelte van Andel, Community and Conservation Ecology Group, University of Groningen, Centre for Life Sciences, P.O. Box 11103, 9700 CC Groningen, The Netherlands
j.van.andel@biol.rug.nl

Professor Dr Eddy van der Maarel, Community and Conservation Ecology Group, University of Groningen, Centre for Life Sciences, P.O. Box 11103, 9700 CC Groningen, The Netherlands
Home address: De Stelling 6, 8391 MD Noordwolde fr, The Netherlands
eddy.arteco@planet.nl

Preface

This book started as a multi-authored account on the many-sided topic of Vegetation Ecology (more commonly called plant community ecology in North America) because this modern field of science can hardly been treated by one or a few authors. In this second edition still more topics have been treated in separate chapters. As editors we have certainly had some influence on the choice and contents of the various chapters, but nevertheless the chapters are independent essays on important aspects of vegetation ecology.

This edition consists of 16 chapters following an introductory chapter, three more than in the first edition. In addition to the 13 original chapters, which are all updated and adapted to the new structure (described below), we were able to include three new topical chapters. In connection with this new structure, the introduction no longer contains the mini-essays that were in the first chapter of the first edition. Instead we will refer to that chapter (van der Maarel 2005), as several authors in this book do, and present a simplified introduction in this edition.

We have modified the sequence of topics by starting the this edition with the chapters (2–4) that deal mainly with the concept, structure, environmental relations and dynamics of plant communities. The second group of chapters (5–9) continue on the internal organization of plant communities. Subsequent chapters (10–12) deal with the structural and functional aspects and processes in plant communities as part of ecosystems. Here the emphasis is on the organization of plant communities in relation to the ecosystem of which they form a part. Chapters 13 and 14 deal with human impacts on plant communities in their ecosystem and landscape setting. The final chapters (15–17) address communities and geographically larger units in their distribution over regions and continents.

The editors, having been ecological pen friends for over 25 years, have thoroughly enjoyed the correspondence and their first face-to-face meeting in 2011. It was very rewarding to learn about each other's specialisations and favourite vegetation types – which we were also allowed to show on the cover of our book. We also enjoyed the vivid exchange of views with the chapter authors and we hope that their chapters will be appreciated both as essays in their own right and as intrinsic parts of this book.

We should like to thank several members of the editorial and production staff of Wiley-Blackwell Science at Oxford for their help. Particular thanks go to Aileen Castell, Mitch Fitton, Kelvin Matthews and Senior Commissioning Editor Ward Cooper for getting us started and keeping an eye on the work.

Finally we hope that this book will find its way across the world of vegetation scientists and plant ecologists.

Eddy van der Maarel
Janet Franklin

1

Vegetation Ecology: Historical Notes and Outline

Eddy van der Maarel[1] and Janet Franklin[2]
[1]University of Groningen, The Netherlands
[2]Arizona State University, USA

1.1 Vegetation ecology at the community level

1.1.1 Vegetation and plant community

Vegetation ecology, the study of the plant cover and its relationships with the environment, is a complex scientific undertaking, regarding the overwhelming variation of its object of study, both in space and in time, as well as its intricate interactions with abiotic and biotic factors. It is also a very modern science with important applications in well-known socio-economic activities, notably nature management, in particular the preservation of biodiversity, sustainable use of natural resources and detecting 'global change' in the plant cover of the earth.

Vegetation, the central object of study in vegetation ecology, can be loosely defined as a system of largely spontaneously growing plants. Not all growing plants form vegetation, for instance, a sown corn field or a flowerbed in a garden do not. But the weeds surrounding such plants do form vegetation. A pine plantation will become vegetation after some years of spontaneous growth of the pine trees and the subsequent development of an understorey.

From the early 19th century onwards, vegetation scientists have studied stands (small areas) of vegetation, which they considered samples of a plant community (see Mueller-Dombois & Ellenberg 1974; Allen & Hoekstra 1992). Intuitively, and later on explicitly, such stands were selected on the basis of uniformity and discreteness. The vegetation included in the sample should look uniform and should be discernable from surrounding vegetation. From early on, plant communities have been discussed as possibly or certainly integrated units which can be studied as such and classified. Most early European and American vegetation scientists did not explicitly make a distinction between actual stands of vegetation and the abstract concept of the plant community. This distinction was more

Vegetation Ecology, Second Edition. Eddy van der Maarel and Janet Franklin.
© 2013 John Wiley & Sons, Ltd. Published 2013 by John Wiley & Sons, Ltd.

important in the 'Braun-Blanquet approach' (Westhoff & van der Maarel 1978). This approach, usually called **phytosociology**, was developed in Central Europe in the early decades of the 20th century, notably by J. Braun-Blanquet from Zürich, and later from Montpellier. The Braun-Blanquet approach, also known as the Zürich–Montpellier school, became the leading approach in vegetation science. It has a strong emphasis on the typology of plant communities based on descriptions of stands, called relevés. This can be understood because of its practical use (see also Chapter 2). However, Braun-Blanquet (1932, 1964) paid much attention to the relations of plant communities with the environment and the interactions within communities (see Section 1.1.2), which is now incorporated in the concept of ecosystem.

A **plant community** can be conveniently studied while separated from its abiotic and biotic environment with which it forms an ecosystem, even if this separation is artificial. In a similar way, a community of birds, insects, molluscs or any other taxonomic group under study, including mosses and lichens, can be studied separately as well (see Barkman 1978). One can also describe a biotic community, i.e. the combination of a plant community and several animal groups (Westhoff & van der Maarel 1978).

Uniformity and distinctiveness. As mentioned above, the delimitation of stands of vegetation in the field is based on an internal characteristic, i.e. **uniformity**, and an external one, i.e. **distinctiveness**. Distinctiveness of a stand has been much discussed and interpreted. Distinctiveness implies discontinuity with surrounding vegetation. This is sometimes very obviously environmentally determined, for example in the case of a depression in a dry area, or the roadside vegetation between the road and a ditch in an artificial landscape. However, more usually the distribution of the local plant populations is decisive. This has been the case since H.A. Gleason (e.g. 1926) observed that species are 'individualistically' distributed along omnipresent environmental gradients and thus cannot form bounded communities. Note that this observation referred to stands of vegetation, even if the word community was used! The wealth of literature on ordination (see also Chapter 3) offers ample evidence of the 'continuum concept of vegetation' (McIntosh 1986).

Gleason and many of his adherers criticized the community concept of F.E. Clements (e.g. 1916), the pioneer in succession theory, who compared the community with an organism and, apparently, recognized plant community units in the field. However, this 'holistic approach' to the plant community had little to do with the recognition of plant communities in the field.

Shipley & Keddy (1987) simplified the controversy by reducing it to the recognition of different boundary patterns in the field. They devised a field method to test the 'individualistic and community-unit concepts as falsifiable hypotheses'. They detected the concentration of species distribution boundaries at certain points along environmental gradients. In their study – as in other studies – boundary clusters are found in some cases and not in others. Coincidence of distribution boundaries occur at a steep part of an environmental gradient, and at places with a sharp spatial boundary or strong fluctuations in environmental conditions (see also Chapter 3).

The occurrence of different boundary situations as such is of theoretical importance. They can be linked to the two types of boundary distinguished by C.G. van Leeuwen and put in a vegetation ecological framework (see Westhoff & van der Maarel 1978; van der Maarel 1990). The first type is the *limes convergens* which can be identified with an **ecotone** *sensu stricto* or tension zone. Here species boundaries can be determined strictly by abiotic conditions, which shift abruptly, in space and/or in time, although interference between species may play a part (e.g. Shipley & Keddy 1987); the ecotone may also be caused or sharpened by plants, the so-called vegetation switch (Wilson & Agnew 1992). The opposite type of boundary, *limes divergens* or **ecocline**, is typically what we now call a gradient where species reach local distribution boundaries in an 'individualistic' way along gradually changing environmental conditions (van der Maarel 1990).

Despite the general appreciation of the individualistic character of species distributions, it has been recognized that 'there is a certain pattern to the vegetation with more or less similar groups of species recurring from place to place' (Curtis 1959). This was further elucidated by R.H. Whittaker (e.g. 1978). Indeed, the individualistic and community concepts are now generally integrated (e.g. van der Maarel 2005).

1.1.2   Plant communities: integrated, discrete units or a convenient tool

Concepts. Within the neutral definitions of plant community quite different ideas and opinions on the nature of the plant community have been expressed since the early 20th century and the discussion is still going on. The controversy between Clements and Gleason has been an important element in this discussion. Allen & Hoekstra (1992) posited that the contrasting viewpoints of the two masters were influenced by the differences in the landscapes where they grew up. Clements was brought up in the prairie landscape of Nebraska and viewed plant communities as units from horseback, while Gleason walked through the forest, from tree to tree, aware of the small-scale differences within the community. Thus, the different environments may have had a decisive influence on their 'perspective'.

However, two outstanding European contemporaries of Clements and Gleason do not fit this interpretation. The Russian plant ecologist G.I. Ramenskiy, who is generally considered the father of ordination and who was a Gleasonian avant la lettre, demonstrated the individuality of species distributions along gradients with meadow vegetation. On the other hand, the Finnish forest ecologist A.K. Cajander developed an authoritative typology of Finnish forests (e.g. Trass & Malmer 1978). Apparently, emphasizing that continuities, or rather discontinuities, can be done in any plant community type and this has to do with intellectual attitude rather than upbringing and field experience. Westhoff & van der Maarel (1978) considered that the 'organismal concept' of Clements versus the 'individualistic concept' of Gleason, can rather be interpreted as the 'social structure' concept and the 'population structure' concept, respectively (see van der Maarel 2005).

Definitions. One or more of these different plant community concepts are reflected in the many plant community definitions available. The definition by Westhoff & van der Maarel (1978) is representative of phytosociology as it was developed in Central Europe, notably by J. Braun-Blanquet, and in Northern Europe by G.E. Du Rietz. However, it also reflects ideas from early Anglo-American plant ecology, both in Great Britain (A.G. Tansley) and the USA (F.E. Clements), notably the emphasis on the interrelations between community and environment and on species interactions: 'a part of a vegetation consisting of interacting populations growing in a uniform environment and showing a floristic composition and structure that is relatively uniform and distinct from the surrounding vegetation'.

Several later definitions of the plant community reflected the outcome of the more recent debates on the holistic and individualistic concepts, and on the reality of emergent properties. They may emphasize the co-occurrence of populations (Looijen & van Andel 1999), interactions between individuals (Parker 2001), or the 'phenomenological' coincidence (Grootjans *et al.* 1996). '**Emergent properties**' are causing the whole to be more than the sum of its parts, such as dominance–diversity relations (Whittaker 1965; Wilson *et al.* 1998). Weiher & Keddy (1999) proposed the term 'assembly rules'. Grime (2001) paid attention to the mechanisms of plant community assembly. Details and more literature on aspects of integration are found in van der Maarel (2005).

In conclusion, a plant community is generally recognized as a relatively uniform piece of vegetation in a uniform environment, with a recognizable floristic composition and structure, that is relatively distinct from the surrounding vegetation. Even if the populations of the participating species are usually distributed individualistically in the landscape, they may well interact within the community and build up an integrated unit with emergent properties. At the same time, plant communities can be convenient units for conveying information about vegetation and its environment.

1.1.3 Vegetation survey and sampling

Whatever our aim, approach and scale of observation, vegetation – whether loosely defined or approached as a plant community, or as a unit in a higher level of integration – should be described and analysed. Vegetation characteristics are either derived from plant morphological characters, usually called *structure*, or from the plant species recognized, the *floristic composition*. In Chapter 2, R.K. Peet & D.W. Roberts present a detailed account of community description. Amongst the many different objectives, there are four common ones:

1 phytosociological: community classification and survey, dealt with in Chapter 2;
2 ecological: correlation of the variation in vegetation composition with variation in environmental factors, dealt within Chapter 3;
3 dynamical: study of vegetation changes; see Chapter 4;
4 applied: nature conservation and management, the subject of Chapter 14.

Size of the sample plot; minimal area. A contemporary approach to the selection of plot size and shape for vegetation sampling is discussed in Chapter 2, while only a brief history of the development of the minimal area concept is provided here. The size of a sample plot will depend on the type of vegetation and may vary from a few square metres to several hectares. **Minimal area** is defined here (in line with Mueller-Dombois & Ellenberg 1974 and Westhoff & van der Maarel 1978) as a 'representative area on which the species of regular occurrence are found'. In various schools (Braun-Blanquet 1932; Cain & Castro 1959) determination of a species–area relationship has been recommended as a way to identify minimum area, on the assumption that the curve would reach an asymptote at which a '*saturated community*' (Tüxen 1970) would be reached. However, in practice this occurs only in species-poor communities whereas in communities richer in species a semi-logarithmical or a log–log function is found; see Chapter 11 on Diversity for more on functions.

In conclusion, a 'minimal area' to be sampled, related to species richness, canopy height and species dominance relations, remains difficult to determine. Instead a 'representative' sampling area should be selected, the size of which can be chosen on the basis of field experience with different vegetation types as represented in various textbooks. For further information see van der Maarel (2005), who has also summarized minimal area data for 38 community types. These data are summarized in Chapter 2.

Vegetation characteristics. Vegetation structure and floristic composition are usually measured or estimated on a plant community basis. *Structure* includes: **stratification**, the arrangement of phytomass in layers; **cover**, as percentage of the surface area of the sample plot; **phytomass**, expressed as dry-weight g/m^2, kg/m^2 or t/ha (1 t/ha = 10 kg/m^2), or as productivity in g/m^2/yr; and **leaf area index**, LAI, and its derivate specific leaf area. These elements appear particularly in Chapters 10–12, and see, for example, Mueller-Dombois & Ellenberg (1974). The description of the characteristics and spatial position of organs, as in textural descriptions, including drawings of vegetation profiles, has not become a standard procedure. Structural research rather proceeds via the species composition combined with the allocation of species to life-form or other categories (see also Chapter 12). Structural analysis of above-ground plant parts should be (but is seldom) completed with an analysis of the below-ground parts, as stimulated by Braun-Blanquet (1932, 1964; Dierschke 1994). Species data should not only be collected above-ground but also below-ground. Titlyanova *et al.* (1999) showed how in steppes the below-ground phytomass (which can store 70% of the net primary production) is more homogeneously distributed, both over the area and over the species. The **dominance–diversity curves** for 19 species in steppe vegetation based on percentage dry weight contributions of species to green phytomass and below-ground organs are quite different.

Species composition includes a list of species for the sample plot (usually vascular plants only), with expressions of their quantitative occurrence, usually broadly called **abundance**. This comprises: (1) *abundance* proper, the number of individuals on the sample plot – because individuality in many (clonal) plant species is difficult to determine (see Chapter 5), the concept of **plant unit,** a plant or part of a plant (notably a shoot) behaving like an individual, is needed; (2) *frequency*, the number of times a species occurs in subplots within the sample; (3) *cover* of individual species is usually estimated along a cover scale – many scales have been proposed, the most current of which are described in Chapter 2; (4) *cover–abundance* is a combined parameter of cover – in case the cover exceeds a certain level, e.g. 5% – and abundance. This 'total estimate' (Braun-Blanquet 1932) has been both criticized as a wrong combination of two independently varying parameters and praised as a brilliant integrative approach. It is analogous to the *importance value* developed by Curtis (1959) – the product of density, frequency and cover – which has been popular in the USA for some decades. Several proponents of a combined cover–abundance estimation have nevertheless found it necessary to convert the abundance categories from the combined scale into approximate cover values. Two combined scales still in use are the Domin or Domin-Krajina scale (see Chapter 2) and the most frequently used Braun-Blanquet scale which, in several variants, has been in use since the 1920s. Van der Maarel (1979) suggested an 'ordinal transform' scale replacing the modern nine-point Braun-Blanquet scale by the values 1–9. This scale was also included in Westhoff & van der Maarel (1978) and has found wide acceptance. Van der Maarel (2007) also suggested a cover-based interpretation of this scale by transforming the abundance categories so that they approximate a ratio scale, where the means of the cover classes form a geometrical (×2) series (see Table 1.1). Peet & Roberts (Chapter 2) concentrate on cover values, but emphasize that cover intervals should confirm to the Braun-Blanquet scale, which the geometrical-ordinal scale does.

1.1.4 Plant communities and plant community types

Typology and syntaxonomy. When plant communities are described in the field by means of relevés (or other types of analysis), they can be compared with each other and an abstract typology can be developed. Plant community types must

Table 1.1 Extended Braun-Blanquet cover-abundance scale and ordinal transform values (OTV) according to van der Maarel (1979) with interpreted cover value intervals for low cover values. See also van der Maarel (2007).

Braun-Blanquet	Abundance category	Cover: interpreted interval	OTV cover interval	OTV
r	1–3 individuals	$c \leq 5\%$		1
+	few individuals	$c \leq 5\%$	$0.5 < c \leq 1.5\%$	2
1	abundant	$c \leq 5\%$	$1.5 < c \leq 3\%$	3
2m	very abundant	$c \leq 5\%$	$3 < c \leq 5\%$	4
2a	irrelevant	$5 < c \leq 12.5\%$		5
2b	'	$12.5 < c \leq 25\%$		6
3	'	$25 < c \leq 50\%$		7
4	'	$50 < c \leq 75\%$		8
5	'	$c > 75\%$		9

be based on characteristics analysed in the field. Originally, the decisive characteristic was the physiognomy, i.e. the dominance of certain growth-forms such as trees, shrubs and grasses. The different physiognomic types were called formations and were usually described for large areas by plant geographers, such as E. Warming (see Mueller-Dombois & Ellenberg 1974 and Chapter 15). Later the floristic composition became decisive. For this community type the term association became standard under the definition adopted at the 1910 Botanical Congress (see also Chapter 2).

R. Tüxen considered a type as an ideal concept – in line with German philosophers – which could empirically be recognized as a 'correlation concentrate'. Tüxen's idea was elaborated by H. von Glahn who distinguished three steps in classification: (1) identification, through reconnaissance and comparison; (2) elaboration of a maximal correlative concentration, i.e. first of vegetation, second of environmental characteristics, through tabular treatment (and nowadays multivariate methods); (3) systematic categorization, i.e. arranging the type in a system of plant communities (Westhoff & van der Maarel 1978).

The Braun-Blanquet approach developed a hierarchical system of plant community types which resembles the taxonomy of organisms. Each syntaxon is defined by a characteristic species combination, a group of **diagnostic taxa** which may include **character** ('faithful') **taxa**, **differential taxa** and **companions**. The confinement of taxa to syntaxa is seldom absolute and degrees of **fidelity** have been recognized. The distribution area of characteristic species seldom coincide with that of their syntaxon: they can be much wider, but also smaller, or overlap only partly. This has been elucidated by Westhoff & van der Maarel (1978) and particularly Dierschke (1994). Other challenges arise. At what level in the syntaxonomical hierarchy should a newly described syntaxon be placed? Syntaxa of a lower rank often show floristic similarities to syntaxa from different classes. These and other problems were discussed by Westhoff & van der Maarel (1978); see also van der Maarel (2005) and Chapter 2. After this system has long been distrusted and left aside in Anglo-American ecology, the concise description of vegetation classification by Robert H. Whittaker (e.g. 1978) came close to the European approach and stimulated worldwide interest.

Numerical classification. The development of numerical methods for the classification – as well as the ordination – of plant community samples started after the Second World War in various countries, e.g. Th. Sørensen in Denmark, D.M. de Vries in the Netherlands, J.T. Curtis in the USA and W.T. Williams in the UK (see Westhoff & van der Maarel 1978). Application of these methods on a larger scale was initiated in 1969 by the Working Group for Data-Processing of the International Association of Vegetation Science. The aim of this group was first of all to build up a database of phytosociological relevés. This implied the unification of the identity and nomenclature of the plant species involved and the development of a coding system. Numerical clustering and table arrangement programmes were developed, two of which received much attention and application.

TABORD (van der Maarel *et al.* 1978) is an agglomerative method based on a similarity analysis and subsequent fusion of relevés and clusters and a subsequent arrangement of clusters in an ordered phytosociological table. A chi-square analysis was implemented to indicate the fidelity of species to clusters. The elaborated version FLEXCLUS by O. van Tongeren (in Jongman *et al.* 1995) is searching for a cluster structure on an optimal level of similarity and an ordination, so that the structure is reticulate rather than hierarchical.

TWINSPAN (Hill 1979), a divisive method on the basis of the position of relevés along axes of a correspondence analysis ordination and a subsequent tabular ordering, is by far the most popular method and its popularity has grown since it was incorporated in the program TurboVeg for phytosociological classification of very large data sets (Hennekens & Schaminée 2001). Attractions of the latter programs are the capacity and speed and the relatively low number of options one has to consider, but this has distracted the attention from their weaknesses: the strictly hierarchical approach and the problems with correspondence analysis, which are discussed in Chapter 3. Numerical classification is treated extensively in Chapter 2.

Classification of natural and semi-natural vegetation. Under this denominator, R.K. Peet and D.W. Roberts in Chapter 2 present a comprehensive and sophisticated guide to conceptual and methodological issues in the development, interpretation and use of modern vegetation classifications based on large-scale surveys. Vegetation description and classification are integral to contemporary planning, management and monitoring for conservation of natural communities. Chapter 2 examines several large-scale national and multinational classification systems and finds that standardization of methods and nomenclature are attributes of successful classification systems. Peet and Roberts outline all components of vegetation classification: planning and data acquisition; numerical classification or other approaches to creating vegetation classes or entities (entitation); community characterization, determination (assigning new observations to classes), integration and documentation. Numerical classification typically involves calculating distance or similarity measures from community composition data and then applying some sort of clustering or partitioning algorithm. Chapter 2 outlines the variety of methods currently applied to the vegetation classification problem and their relative merits for use with ecological community data.

1.1.5 Vegetation and environment: discontinuities and continuities

M.P. Austin, in Chapter 3, treats vegetation and environment in a coherent way, indeed as **vegetation ecology**. This term was coined by Mueller-Dombois & Ellenberg (1974), both of whom were educated in Germany in the tradition of continental-European phytosociology. Anglo-American vegetation ecology has its roots in plant ecology – and is usually called so. However, the study of plant communities in the UK with A.G. Tansley, in the USA with Cowles, F.E. Clements and later R.H. Whittaker, and in continental Europe with J. Braun-Blanquet and H. Ellenberg, has always been an ecological rather than a botanical undertaking, despite the differences in approach (McIntosh 1986).

Community and continuum. Austin (Chapter 3) makes clear that both vegetation and environment are characterized by discontinuities and continuities and that their interrelationships should be described by multivariate methods of ordination and classification. He shows how three key *paradigms* have emerged during the history of vegetation ecology, which we can conveniently label 'association', 'indirect gradient' and 'direct gradient'; the differences between the paradigms are smaller than is often believed and vegetation ecology can further develop when a synthesis of the three paradigms is developed.

Measuring the environment. Austin (Chapter 3) emphasizes the importance of a framework of environmental factors which should be developed for any study of vegetation and environment. The special attention paid to climatic and derivate microclimatic factors leads to the notion of the 'hierarchy of spheres' influencing vegetation in an order of impact (van der Maarel 2005; see also Chapter 14).

A useful distinction within the environmental factors is between (i) indirect, distal factors, notably altitude, topography and landform, and (ii) direct factors such as temperature, groundwater level and pH – which are determined by indirect factors, and resource factors such as water availability and nutrients. Generally, vegetation ecology is more meaningful when the environmental factors available for vegetation–environment studies are more physiologically relevant. Austin also re-introduces the concept of scalars, major integrated environmental complexes, once introduced by Loucks (1962) in an ordination study of forests, but largely neglected afterwards.

An additional way of characterizing the environment of a plant community is to use indicator values assigned to the participating plant species. The best known system of values is that of H. Ellenberg (Ellenberg *et al.* 1992), with indicator values for most of the Central and West European vascular plant species regarding moisture, soil nitrogen status, soil reaction (acidity/lime content), soil chloride concentration, light regime, temperature and continentality. The system is also mentioned in Chapter 12. The values generally follow a (typically ordinal) 9- or 10-point scale, based on field experience and some measurements. They are used to calculate (weighted) mean values for plots and communities, which is a calibration problem, discussed by ter Braak (in Jongman *et al.* 1995).

Indirect ordination, direct ordination. Austin (Chapter 3) explains how **indirect ordination** determines environmental gradients on the basis of the variation in the vegetation data, while **direct ordination** starts from the variation in environmental factors and then determines the distribution of plant species along these environmental gradients. Indirect ordination is numerically developed in many different methods, of which correspondence analysis, and its derivate canonical correspondence analysis and non-metric multidimensional scaling are treated in detail by Austin, while relating the appropriateness of these methods to the character of the distribution of species along environmental gradients.

Classification and ordination as complementary approaches. From Chapters 2 and 3 it becomes clear that classification and ordination are both useful and can usually be profitably integrated in plant community studies. In this connection, an old approach may be mentioned, based on the observation that in coarse-grained relatively dynamic and homogeneous ecotone environments, plant communities are relatively poor in species and simply structured, whereas fine-grained relative constant and divergent ecocline environments, plant communities are richer in species, more structured and integrated. 'Ecotone communities' can be more easily classified and be included in a hierarchy, while 'ecocline communities' cannot be easily classified and are more liable to be ordinated together with related communities. A framework for combining both numerical approaches is presented in Fig. 1.1. As a 'golden mean' it was recommended to apply both approaches, with an optimally effective syntaxonomy on the alliance/order level (van der Maarel unpublished).

1.1.6 Vegetation dynamics

In Chapter 4, S.T.A. Pickett, M.L. Cadenasso and S.J. Meiners adopt the vision that vegetation dynamics is governed by three general processes: differential **site**

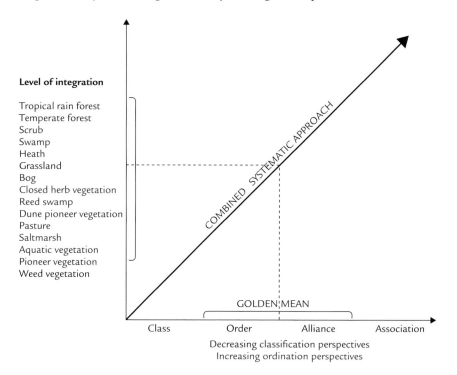

Fig. 1.1 Relation between the level-of-integration in vegetation and the relative success of classification vs ordination in a 'combined systematic approach'. (Based on a figure designed by E. van der Maarel in consultation with V. Westhoff & C.G. van Leeuwen, and presented in a lecture at the International Botanical Congress in Edinburgh, 1964.)

availability, differential **species availability** and differential **species performance**. If a site becomes differentially available, species are differentially available at that site, and/or species perform differentially at that site. As a result the composition and/or structure of vegetation will change.

Analytical methods. The two main methods for analysing vegetation dynamics are the repeated description of **permanent plots** and the description of sites of different ages, forming a **chronosequence** ('space-for-time substitution'). There is a long tradition of permanent plot studies in Europe, starting in 1856 with the Park Grass Experiment at Rothamsted near London (mentioned in Bakker *et al.* 1996). Nowadays thousands of such plots are under regular survey, many surveyed initially to help solve management problems (Chapter 12).

Types of disturbance and types of vegetation dynamics. As Pickett *et al.* (1987) explained and Chapter 4 discusses further, site availability is largely the result of a disturbance; differential species availability is a matter of dispersal (Chapter 6); and differential species performance is based on the differences in ecophysiology and life history (Chapter 12), which is the outcome of species interactions (Chapters 7 and 9) and herbivory (Chapter 8). Chapter 4 also elucidates how vegetation dynamics are increasingly affected by human activities (see also Chapter 14).

One of the interesting consequences of the primate of disturbance is that primary sites are more carefully analysed and mostly seem to have at least some legacy. So, the classical distinction between primary and secondary succession is replaced by a gradient between two extremes. After a disturbance, the time needed for the vegetation to reach a new stable state will vary. Fig. 1.2 indicates how we can distinguish between fluctuation (on the population level), patch dynamics, secondary succession, primary succession, secular succession and

	Fluctuation	Gap, patch dynamics	Cyclic succession	Secondary succession	Primary succession	Secular succession
Organism–environment	10^{-1}–1 yr 10^{-2}–10 m	1–10 yr 10^{-2}–10 m				
Population–environment	1 yr 1–10 m	1–10 yr 10–10^2 m	1–10^2 yr 10–10^2 m	10–10^2 yr 10–10^2 m	10–10^3 yr 10–10^2 m	
Microcommunity–environment	1 yr 1–10 m	1–10 yr 10–10^2 m				
Phytocoenosis–environment	1 yr 1–10 m	1–10 yr 10–10^2 m	1–10^2 yr 10–10^2 m	10–10^2 yr 10–10^2 m	10–10^3 yr 10–10^2 m	
Regional landscape				10^0–10^2 yr 10^2–10^4 m	10^2–10^3 yr 10^2–10^4 m	10^2–10^4 yr 10^2–10^4 m
Biome					10^3–10^6 yr 10^4–10^6 m	
Biosphere						10^6–10^7 yr

Fig. 1.2 Spatial scales (m) and temporal scales (yr) of studies of ecological objects and their dynamics. (Based on similar schemes in van der Maarel 1988 and Gurevitch *et al.* 2002.)

long-term vegetation change in response to (global) changes in climate (see Chapter 17), and how the time scale varies from less than a year to thousands of years. Dynamic studies of plant populations, especially clonal plants, may vary from 10 to 10^3 yr (examples in White 1985). Cyclic successions may take only a few years in grasslands rich in short-lived species (e.g. van der Maarel & Sykes 1993), 30–40 yr in heathlands (e.g. Gimingham 1988) and 50–500 yr in forests (e.g. Veblen 1992). The duration of successional stages at the plant community level ranges from less than a year in early secondary stages in the tropics to up to 1000 yr in late temperate forest stages. Finally, long-term succession in relation to global climate change may take a hundred to a million years (e.g. Prentice 1992).

Development of vegetation and soil. In Chapter 4, Pickett *et al.* point to the fact that in between disturbances biomass will accumulate. More generally, succession is a process of building up biomass and structure, both above ground in the form of vegetation development, and below ground in the form of soil building. Odum (1969), in his classical paper on ecosystem development, was one of the first to present an overall scheme of gradual asymptotic biomass accumulation and a peak in gross production in the 'building phase' of a succession.

   The contribution to these developments by individual species varies with the type of succession and the successional phase. The old phytosociological literature already described different types of species while emphasizing the 'constructive species', i.e. the species with a high biomass production which build up the vegetation (Braun-Blanquet 1932). Russian ecologists have used the term **edificator** for this type of species (see e.g. White 1985). Usually these species are dominants. Grime (2001) summarized the conditions for the development of dominance and mentioned maximum plant height, plant morphology, relative growth rate and accumulation of litter as important traits for dominants.

1.1.7   Pattern and process in the plant community

The phrase '**pattern and process**' has become a standard feature of community ecology since A.S. Watt published his seminal paper (Watt 1947). The basic idea is that within a plant community, which is in a steady state, changes may occur patchwise as a result of local disturbance (exogenous factors) or plant senescence (endogenous factors); in the gaps formed, regeneration will occur that will initially lead to a patch of vegetation which is different from its surroundings. These processes are 'fine-scale vegetation dynamics' (Chapter 4) within a community, rather than of the community as a whole or of larger units.

Spatial pattern analysis. Spatial patterns of plant units of particular species comprise the development of patches, that may form a clumped distribution, regular (overdispersed) dispersion and homogeneous (random) distribution. The statistical analysis of these patterns was introduced in plant ecology by Greig-Smith (1957) and Kershaw (1964), who were particularly interested in the causes of patch formation. Kershaw distinguished between morphological, environmental and sociological patterns. Morphological patterns arise from the growth-form of plants, in particular clonal plants (see Chapter 5). Environmental patterns are related to spatial variation in environmental factors (see Chapter 3), for instance soil depth. Sociological patterns result from species interactions (see Chapter 7) and temporal changes in the behaviour of plants.

The development of analytical methods has proceeded and has been regularly reviewed (e.g. Dale 1999; Fortin & Dale 2005; Franklin 2010), but the ecological application of these methods has remained limited and will not be treated further in this book.

Patch dynamics. On the other hand, the study of patch dynamics in relation to internal environmental dynamics has continued and has found a place in Chapter 4 by Pickett *et al.* Within-community patch development as linked to disturbance, particularly gap formation, started in the 1920s in forests by R. Sernander in Sweden and A.S. Watt in Great Britain (Hytteborn & Verwijst 2011). When the investigated forest plots and the gaps are large, the dynamics are considered a regeneration succession (see Chapter 4) and the succession stages have been described as their own plant communities. In the European syntaxonomical system, these stages have remote positions, being different classes (e.g. Rodwell *et al.* 2002).

Watt (1947) described similar patch dynamics in bogs (where he had studied the work of the Swede H. Osvald from 1923), heathlands and grasslands (see also van der Maarel 1996). In bogs the well-known mosaic of hollows and hummocks appeared to be dynamically related and was described as a '**regeneration complex**'. Watt considered the different stages as seral and also as separate communities, involved in a **cyclic succession**. Whether or not to call these cyclical processes 'succession' is a matter of definition and of scale (e.g. Glenn-Lewin & van der Maarel 1992). An alternative term *Mosaik-Zyklus* has been proposed by the German animal ecologist H. Remmert (Remmert 1991: '*mosaic-cycle*'). A mosaic-cycle is a special case of patch dynamics where the changes are triggered largely by endogenous factors, in particular plant senescence. Exogenous factors generally also play some part (Burrows 1990).

Regeneration niche and the carousel model. The work by Watt on grasslands inspired P.J. Grubb, one of his pupils, to elaborate the concept of regeneration niche in a paper as influential as Watt's (Grubb 1977). The essence of this concept is that gaps arise everywhere, through the death or partial destruction of plant units, the natural death of short-lived species and all sorts of animal activities, and the open space can be occupied by a germinating seed or by a runner of a clonal plant. In grazed grasslands, local removal of plant parts, trampling and deposition of dung are additional causes of gaps, often large ones.

Where gaps arise more or less continuously in grasslands and plant species become not only locally extirpated because of disturbances and/or death but also have continuous opportunities to re-establish, species may show a high fine-scale mobility. At the same time, patch dynamics can contribute considerably to the co-occurrence of many plant species on small areas of grassland. The limestone grassland on the alvars of southern Öland (Sweden), which is rich in annuals, as a whole appeared to be remarkably constant in floristic composition, while

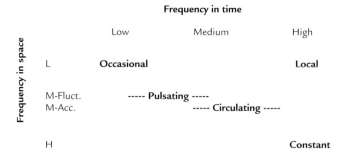

Fig. 1.3   Types of within-community plant species mobility based on frequency in space and time in 10 × 10 cm subplots in limestone grassland during 1986–1994. Mean spatial frequency values divided into high, >75% (H), medium, 35–75% (M; M-Fluc., with large between-year differences; M-Acc, accumulating frequency) and low, <25% (L). Temporal frequency values divided into H (occurring in >66% of the years), M (33–66%) and L (<33%). (After van der Maarel 1996.)

the species composition on subplots from 10 to 100 cm² changed from year to year. Van der Maarel & Sykes (1993) quantified this mobility as (1) cumulative frequency, i.e. the cumulative number of subplots a species is observed in over the years and (2) cumulative species richness, i.e. the mean number of species that is observed in a subplot over the years. A 'carousel model' was suggested to characterize this 'merry-go-round' of most species. In this short, open grassland on summer-dry soil, many short-lived species are involved and germination is a main process in (re-)establishment of species. Several types of mobility could be distinguished, mainly based on mean frequency and mean cumulative frequency (Fig. 1.3). Lepš (Chapter 11) discusses these aspects of regeneration as contribution to the species richness in communities.

1.2   Internal organization of plant communities

1.2.1   Clonality in the plant community

In Chapter 5, B. Svensson, H. Rydin & B. Carlsson give an account of the processes and ecological significance of vegetative spread by clonal plants. They make clear that clonal spread is a form of dispersal – even if (diaspore) dispersal as discussed in Chapter 6 will be seen as dispersal proper. Clonality is largely an internal community process, but it may link a community to neighbouring communities, or still further away, as the chapter describes. Neighbour effects have long been recognized in phytosociology as vicinism (van der Maarel 1995). See also Section 1.3.2. Important sources of clonal variation include the length of the ramets formed (notably rhizomes, stolons and runners) and the speed with which these are formed.

Svensson *et al.* pay attention to the distinction between 'phalanx' and 'guerilla' forms of vegetative reproduction of species, which they consider as endpoints on a continuum variation. Ecologists may resist the metaphor of plants as warriors, and are confused about the spelling of guerilla (the correct spelling of the originally Spanish word being guerrilla). Moreover, the two strategies do not seem to even resemble the two types of warfare involved. Nevertheless the distinction between the two types is useful.

Of special interest for vegetation ecology is the characterization of vegetation types regarding the relative importance of clonal species and their role in patch dynamics (Section 1.1.7; Chapter 4), the relation between clonality and competition and co-existence (Chapters 7 and 11) and the relation between clonality and herbivory (Chapter 8).

1.2.2   Seed ecology and assembly rules in plant communities

This title for Chapter 6 by P. Poschlod *et al.* suggests that the original focus on diaspore dispersal in the first edition of this book has now been broadened towards the ecology of diaspores and their dispersal and germination, in relation to community assembly. The following Chapters 7 on species interactions and 11 on diversity further discuss assembly rules while Chapter 6 is now also linked to Chapter 4 on vegetation dynamics. As to vegetation succession, the availability of diaspores is one of the major characteristics of secondary (post-agricultural and post-disturbance) succession, versus the lack of diaspores on the virginal substrates of a primary succession. On a smaller temporal and spatial scale, the mobility of plants through clonal and diaspore dispersal is a driving force in 'pattern and process' in the plant community. Fine-scale mobility of plants as described in the carousel model and similar contexts is very much a matter of dispersal to open space as it becomes available.

Poschlod *et al.* make clear that dispersal is one of the essential factors which determine the composition of the species pool of a plant community (Zobel *et al.* 1998, who, incidentally, consider species reservoir a better – i.e. a more appropriate – term than species pool). The community reservoir is supplied through dispersion from the local reservoir around the community, which in its turn is supplied by the regional reservoir through migration and speciation. This chapter is also a natural place to treat the soil seed bank, which – as Poschlod *et al.* state – is in fact rather a diaspore bank. Zobel *et al.* (1998) suggested including the diaspore bank in the community pool, thus including the so-called persistent diaspores. It is debatable whether species that never germinate should also be included in the target community – because the environment may not be suitable for them. However, there are many examples of species apparently not being suitable for an environment and nevertheless occurring there, if only ephemerally. This is usually a matter of 'mass effect', the availability of numerous diaspores meeting favourable conditions for germination just outside the mother community, also known as **vicinism** (van der Maarel 1995).

1.2.3   Species interactions structuring plant communities

The concise chapter on species interactions by J. van Andel, Chapter 7, gives a survey of the different types of species interaction and then pays attention to the following types of interaction: competition, allelopathy, parasitism, facilitation and mutualism. The focus on competition, the classical main type of interaction, is no longer predominant in this edition, even though competition as a mechanism to arrange species packing along gradients (see Chapter 2) remains important in vegetation ecology. The typically community-structuring force of facilitation is now a more fascinating topic in vegetation ecology. Another important community-structuring interaction type with a rapidly growing body of literature devoted to it is mycorrhiza. Van Andel treats it as an important aspect of mutualism, while it also forms part of the topic 'interactions between higher plants and soil-dwelling organisms', elaborated in Chapter 9.

Van Andel's chapter is one of the few where bryophytes are treated in some detail. In addition, the review paper by Rydin (1997) and the detailed competition study by Zamfir & Goldberg (2000) can be mentioned.

1.2.4   Terrestrial plant–herbivore interactions

In Chapter 8, M. Sankaran & S.J. McNaughton present an integrative account on herbivory, with links to Chapters 4, 6, 7, 11 and 14. The idea of co-evolution comes to mind (e.g. Howe & Westley 1988) in view of the broad spectrum of plant types and plant parts being eaten and the equally broad spectrum of herbivores, as well as the often intricate mutual adaptations between plants and animals in each type of interaction.

Plants deal with herbivory by avoidance or tolerance (i.e. compensation for damage), and a range of compensatory responses is discussed. There is a range from symbiotic to parasitic aspects of grazing. Finally, herbivores and herbivore diversity have major effects on plant diversity and pattern formation.

1.2.5   Interaction between higher plants and soil-dwelling organisms

In Chapter 9, T.W. Kuyper & R.G.M. de Goede concentrate on the interactions between plants and soil organisms that occur around and in roots. The three major processes described are N-fixation by bacteria, mycorrhiza with fungi and root-feeding by invertebrates. The gradual transition and alteration between symbiotic and antagonistic aspects is related to the ranges of interactions described in the two preceding chapters.

A link to Chapter 13 on plant invasions is the often noticed difference in behaviour of invasive plants in their new regions compared to their old, which is related to the difference in accompanying soil-dwelling organisms. A link to Chapter 4 follows from the elucidation of the two hypotheses about the driving force of succession. If mycorrhizal fungi are causes of plant dynamics (driver hypothesis), the presence of specific mycorrhizal fungi is required for the growth of specific plants. If soil organisms are merely passive followers of plant species dynamics (passenger hypothesis), specific plants are required to stimulate the growth of specific mycorrhizal fungi.

1.3   Structure and function in plant communities and ecosystems

1.3.1   Vegetation and ecosystem

The plant community together with the animals within, the soil underneath and the environment around is now generally considered an integrated unit, the ecosystem. Nevertheless, most vegetation studies are restricted to the above-ground plants, even if it is long since known (e.g. Braun-Blanquet 1932) that the below-ground components are of decisive importance for the anchoring of plants, the uptake of water and nutrients and the storage of photosynthates. Most of the large biomass is made up of roots and seeds.

Root-related phenomena such as nitrogen-fixation and mycorrhiza are now being included in vegetation studies (Chapter 9). Evidently, the dense contacts between roots, biological turnover (through biomass consumption and de-composition, humus formation and partial re-use of mineralized components) and nutrient cycling are convincing contributions to the notion of integrated ecosystems.

Chr. Leuschner, in Chapter 10, focuses on trophic levels between which matter and energy are exchanged. An important part of the primary production ends up in the below-ground plant parts. Here decomposition and humus forma-tion take place. In an ecosystem in steady state there is a balance between net primary production and organic matter decomposition. This balance is reached in later stages of succession. As Leuschner states, after perturbation an ecosystem can often rapidly regain certain structural properties. As an example, Titlyanova & Mironycheva-Tokareva (1990) described the building up of the below-ground structure during secondary succession in just a few years. On the other hand, the recovery to steady state in steppe grassland may take 200 yr. This also relates to the actual discussion on the relation between diversity and ecosystem function (Chapter 11).

Ecosystem ecologists have no doubt about the reality of emergent properties. It is as if these properties appear clearer, the higher the level of integration is at which we are looking at ecosystems. Ultimately we are facing clear aspects of regulation at the 'gaia' level of the global ecosystem. Leuschner finishes his chapter with a treatment of four biogeochemical cycles: carbon, nitrogen, phos-phorus and water. These cycles are studied on the global level and these processes at this level return in Chapter 17.

1.3.2   Diversity and ecosystem function

Chapter 11 by J. Lepš is on diversity or **biodiversity** as it is called nowadays. It starts with a brief treatment of some diversity indices: α or **within-community**, β or **between-community** and γ or **within-landscape** diversity, basically a pro-duct of α and β. These are all concerned with species diversity, or rather taxon diversity, the variation in taxa. In addition, within-taxon or (phylo-)genetic diversity is receiving increasing attention. Relatively new aspects of biodiversity are **phylogenetic distinctiveness**, based on taxonomic distinctiveness, **numerical**

distinctiveness, based on the rarity of occurrence, and **distributional distinctiveness**, i.e. endemism of taxa (van der Maarel 2005). Lepš makes the point that the diversity of a community is largely a function of the species pool and the forms of distinctiveness can indeed be determined in the species pool.

As Lepš confirms, diversity has both an aspect of species *richness*, i.e. the number of species, and of *evenness*, the way species quantities are distributed. These two aspects are more related than is generally recognized by users of diversity indices. According to the relation between the various diversity indices described by M.O. Hill, the well-known indices of Simpson and Shannon are similar in that the most abundant species to some extent determines the diversity, but Simpson does this more than Shannon.

Chapter 11 emphasizes the relation between diversity and ecosystem function. Much research has been triggered by the symposium volume by that name edited by Schulze & Mooney (1994). As Lepš elucidates, biotic diversity can be better understood if it can be divided into functional components. If we manage to distinguish such types and allocate each species to a type, diversity – i.e. species richness – can then be approached as the number of functional types multiplied by the mean number of species per type. Important contemporary studies of biodiversity are concerned with productivity, disturbance, co-existence and stability in the plant community.

1.3.3 Plant functional types and traits at the community, ecosystem and world level

Chapter 12 by A.N. Gillison treats the characteristics and function of life-forms and growth-forms in a contemporary fashion under the denominator of plant functional types. As in the previous chapter, such a treatment has to exceed the level of integration of the plant community, and is indeed relevant up to the global level, where it relates to Chapter 15. A **plant functional type** (PFT) is a group of plant species sharing certain morphological–functional characteristics. The notion of plant function seems to go back to Knight & Loucks (1969) – who related plant function and morphology to environmental gradients – and Box (1981) – who correlated 'ecophysiognomic' plant types with climatic factors, and used climatic envelopes for selected sites to predict the combination of forms (see also Chapter 15). Peters (1991) mentioned this study with its validated global model as a good example (one of the few) of predictive ecology.

In a way the abundant use of PFTs is a revival of the attention paid to *life-forms* during the period 1900–1930. **Life-forms** were seen as types of adaptation to environmental conditions, first of all by E. Warming who spoke of epharmonic convergence after the term *epharmony* – 'the state of the adapted plant' – coined as early as 1882 by J. Vesque. Life-form systems from this early period include those of E. Warming from 1895, C. Raunkiær from 1907, G.E. Du Rietz from 1931 and J. Iversen from 1936 (Table 1.2; see also Table 15.2). Environmental adaptation is most obvious in the life-form system of Raunkiær.

Table 1.2 Some classical life-form systems of vascular plants. **A** Main life-form groups according to Du Rietz (1931). **B** Growth-forms according to Warming (1909); only main groups distinguished. **C** Main terrestrial life-forms according to Raunkiær (1934), largely following Braun-Blanquet (1964). **D** Hydrotype groups acording to Iversen (1936).

A

Physiognomic forms	Based on general appearance at full development
Growth-forms	Largely based on shoot formation (*sensu* Warming)
Periodicity-based life-forms	Based on seasonal physiognomic differences
Bud height-based life-forms	Based on height of buds in the unfavourable season (*sensu* Raunkiær)
Bud type-based life-forms	Based on differences in type and structure of buds
Leave-based life-forms	Based on form, size, duration of the leaves

B

Hapaxanthic (monocarpic) plants	Plants which reproduce only once and then die; including annuals, biennials and certain perennials, e.g. *Agave*
Pollakanthic (polycarpic) plants	Plants which reproduce repeatedly
Sedentary generative	Primary root or corm long-lived, with only generative reproduction
Sedentary vegetative	Primary root short-lived, with both generative and some vegetative reproduction
Mobile stoloniferous	Creeping above-ground with stolons which develop rootlets
Mobile rhizomatous	Extending below-ground with rhizomes
Mobile aquatic	Free-floating aquatic plants

C

Phanerophytes (P)	Perennial plants with perennating organs (buds) at heights > 50 cm Tree P; Shrub P; Tall herb P; Tall stem succulent P.
Chamaephytes (Ch)	Perennial plants with perennating organs at heights < 50 cm Woody (frutescent) dwarf-shrub Ch; Semi-woody (suffrutescent) dwarf-shrub Ch; herbaceous Ch., low succulent Ch., pulvinate Ch.
Hemicryptophytes (H)	Perennial plants with periodically dying shoots and perennating organs near the ground Rosette H; Caespitose H; Reptant H.
Geophytes (Cryptophytes) (G)	Perennials loosing above-ground parts and surviving below-ground during the unfavourable period Root-budding G; Bulbous G; Rhizome G; Helophyte G.
Therophytes (T)	Annuals, completing their life-cycle within one favourable growing period, surviving during the unfavourable period as seed or young plant near the ground Ephemeral T (completing cycle several times per growing period; Spring-green T; Summer-green T; Rain-green T; Hibernating green T (green almost all year)

(Continued)

Table 1.2 *(Continued)*

D	
Terriphytes	Terrestrial plants without aerenchyma
Seasonal xerophytes	
Euxerophytes	
Hemixerophytes	
Mesophytes	
Hygrophytes	
Telmatophytes	Paludal plants (growing in swamps and marshes) with aerenchyma
Amphiphytes	Aquatic plants with both aquatic and terrestrial growth-forms
Limnophytes	Aquatic plants in a strict sense

Plant strategy is a concept more recent than life-form that is also closely related to PFT. The best known system of plant strategies is that by Grime (2001; earlier publications cited there), with competitors (C) adapted to environments with low levels of stress and disturbance, *stress-tolerators* (S) to high stress and low disturbance and *ruderals* (R) to low stress and high disturbance. Strategies are 'groupings of similar or analogous genetic characteristics which recur widely among species or populations and cause them to exhibit similarities in ecology'. Such characteristics have also been called *attributes* (e.g. the '*vital attributes*' of Noble & Slatyer 1980), used in relation to community changes caused by disturbances. However, nowadays the term *trait* (probably borrowed from genetics) is predominantly used. These concepts and their use are all discussed by Gillison in Chapter 12.

The three strategy types proposed by Grime have been maintained virtually unchanged, even if the system has been regularly criticized. CSR theory has some predecessors, mentioned by Grime (2001). The most interesting is the theory of L.G. Ramenskiy, who distinguished three types of life history strategies (Rabotnov 1975), which are astonishingly similar to the CSR types (Onipchenko *et al.* 1998; Grime 2001). Onipchenko used Ramenskiy's ideas in combination with ideas by Yu.E. Romanovskiy on two ways a population can succeed in the competition for limiting resources, i.e. reducing the equilibrium resource requirement R* (Tilman 1982) and developing a high resource capture capacity and a high population growth rate when the resource is available. Onipchenko *et al.* (1998) elucidated the 'RRR' – Ramenskiy/Rabotnov/Romanovskiy – typology.

1.4   Human impact on plant communities

This section comprises two topics which are almost entirely concerned with human impact on plant communities. Several other chapters also provide information on human impacts, notably Chapter 4 on disturbance, Chapter 6 on diversity, Chapter 8 on grazing, Chapter 11 on diversity and Chapter 17 on global change.

1.4.1 Plant invasions and invasibility of plant communities

In the new edition of Chapter 13, M. Rejmánek, D.M. Richardson and P. Pyšek consider the burgeoning literature on biological invasions, a research focus motivated by the need to understand why a small percentage of introduced plants become invasive with significant environmental and economic impacts. Chapter 13 presents the characteristics of invasive species, the pathways of migration of invasive species, the characteristics of environments and plant communities open to invasion and the main impacts of invaders. Of special interest are the relations between invasive and local native species and the often different behaviour of invasive species in their new, alien environment. An interesting suggestion is that invasibility of plant communities by exotics is mainly caused by fluctuations in resource availability (*cf.* Grime 2001). Other factors affecting community invasibility, reviewed in Chapter 13, include functional type diversity, spatial heterogeneity of the environment and the disturbance responses and life-history traits of resident species. A very interesting and important conclusion which is emerging is that stable environments with little anthropogenic disturbance tend to be less open to invasive species.

Invasion is a function of the interaction of a compatible habitat for invaders with propagule pressure. Only few invasive species become dominant in new environments and act as a 'transformer species'. They have major effects on the biodiversity of the local native community. They all transform the environment and different ways of transformation are treated. Useful information is provided on the perspectives of eradication of invasive species. As a rule of thumb, species which have invaded an alien area for more than 1 ha, can hardly be eradicated. As the authors conclude, plant invasions as 'natural' community experiments actually provide important opportunities to study basic ecological and evolutionary processes as well as address important applied research problems.

1.4.2 Vegetation conservation, management and restoration

Chapter 14 by J.P. Bakker is ample proof of the profit made by conservation, management and restoration ecology of the development of vegetation ecology. Phytosociological classification facilitates communication over national boundaries on target plant communities and vegetation mapping can be used for land use planning. Still more importantly, ecological theory regarding the behaviour of plant species along gradients (Chapter 3), succession (Chapter 4), diaspore dispersal, species pool and seed bank dynamics (Chapter 6) and diversity (Chapter 11) has been developed and applied in these chapters. The development of ecohydrology as a basis for the restoration of nutrient-poor wetlands is particularly impressive.

Many of Bakker's examples of successful management projects are from Western Europe where, indeed, both theory and practice have been developed constantly. For a world perspective, see also Perrow & Davy (2002).

1.5   Vegetation ecology at regional to global scales

1.5.1   Vegetation observed at different spatial and temporal scales and levels of integration

We introduce this group of chapters by considering scales of organization from the plant community level upward. The plant community as defined in Section 1.1 is a realistic concept only at a certain scale of observation, i.e. the scale at which it is possible to judge the relative uniformity and distinctness. This 'community scale' will vary with the structure of the community. On the next higher ecological level plant communities are part of ecosystems, while geographically they are part of community complexes. Mueller-Dombois & Ellenberg (1974) distinguish four types of community complex:

1   *mosaic complex*, such as the hummock–hollow complex in bogs;
2   *zonation complex* along a local gradient, e.g. a lake shore;
3   *vegetation region*, roughly equivalent to a formation;
4   *vegetation belt*, a zonation complex along an elevational gradient, i.e. a mountain.

In practice, ecological and geographical criteria are mixed in obtaining 'levels of organization', for instance Allen & Hoekstra (1992): 1. Cell < 2. Organism < 3. Population < 4. Community < 5. Ecosystem < 6. Landscape < 7. Biome < 8. Biosphere.

Each discipline or approach involved in the study of plants and ecosystems, respectively, usually extends beyond its 'central' level of organization. The intricate relations between organization and scale are extended by including temporal scales. A summary of these considerations is presented in Fig. 1.2, which combines a scheme relating levels of organization to temporal scales of vegetation dynamics with a scheme relating spatial to temporal research scales. Essential elements in the hierarchical approach to organization levels and scales are the recognition of (1) mosaic structures, with elements of a mosaic of a smaller grain size being mosaics of their own at a larger grain size; (2) different processes governing patterns at different scales; and (3) different degrees of correlation between vegetational and environmental variables at different grain sizes.

1.5.2   Vegetation types and their broad-scale distribution

In Chapter 15, E.O. Box & K. Fujiwara treat vegetation typology mainly in relation to the broadscale distribution of vegetation types. On a world scale, vegetation types have largely been defined physiognomically, in the beginning (early 19th century) by plant geographers, including A. Grisebach, who coined the term formation as early as 1838. Several readers will share the first author's memory of the famous world map of formations by H. Brockmann-Jerosch and E. Rübel decorating the main lecture hall of many botanical institutes. Box and Fujiwara emphasize the ecological context in which these physiognomic systems were developed. In fact, the English term plant ecology was coined in the translation of the book on ecological plant geography by Warming (1909).

It is clear that there is a growing interest in subordinating floristic units to physiognomic ones. This is also directly relevant for vegetation mapping (Chapter 16). The integrated physiognomic–floristic approach has indeed been proven to be effective since its apparently first study and vegetation map by van der Maarel & Westhoff in the 1960s (see van Dorp *et al.* 1985).

Chapter 16 also pays attention to the problems of modelling and mapping larger areas of vegetation which have lost most of their original vegetation as a result of human land use, and to the development of the concept of *potential natural vegetation* for large-scale vegetation mapping. Reconstruction of vegetation types developing after human impact would have stopped is of course difficult.

Global vegetation distribution patterns can be better understood using a plant functional approach securely rooted in ecophysiology – an approach Box has been instrumental in developing. Chapter 15 traces the development of climate-based global vegetation models from simple but powerful mechanistic rule-based models through contemporary dynamic simulation models of the vegetation, land surface and ocean–atmosphere system. The authors emphasize the importance of understanding the role of global vegetation in the earth system for studies of global change, including – but not limited to – climate change and land use change.

1.5.3   Mapping vegetation from landscape to regional scales

Chapter 16 by J. Franklin was developed for the second edition to address recent developments in vegetation mapping at the local to regional scale that combine traditional elements of photointerpretation and field mapping with powerful new data products and tools from remote sensing and geographic information science (GISci). While contemporary vegetation mapping at landscape to regional extents shares principles and techniques with global vegetation modelling and mapping (Chapter 15), it is typically carried out at the categorical resolution of the plant community. Therefore mapping must capture community attributes of structure and composition. The availability of high-resolution digital aerial imagery has allowed image processing algorithms and geographical models to be effectively married with the expert abilities of a photointerpreter. The result is multi-attribute vegetation databases replacing conventional vegetation maps that were constrained in their information content by the limits of traditional cartography. This new generation of vegetation maps depict extents ranging from local landscapes to subcontinents, at spatial resolutions ranging from 1 km down to extremely fine. Vegetation and land cover maps are being used for purposes ranging from monitoring land use change to environmental planning and management.

1.5.4   Vegetation ecology and global change

Chapter 17 by B. Huntley and R. Baxter deals with global pollution problems including deposition of N compounds and increasing tropospheric concentrations of various pollutants, increasing UV-B and increasing CO_2 concentration, but with particular focus on global warming (climate change). Of interest in this connection are models to help understand and predict future changes of broad ecosystem types, and problems of species to cope with changes and of dispersing to newly available suitable environments.

Studies on the effects of global changes, and especially climate, on vegetation at the broad scale rely heavily on palaeo-ecological studies. In a way these studies are extrapolations into the future of the processes of secular succession. Secular succession, also called vegetation history (Huntley & Webb 1988), was already recognized in early phytosociology – e.g. by Braun-Blanquet (1932) under the name synchronology – as the ultimate vegetation succession.

Models that simulate ecosystem processes and vegetation dynamics in response to climatic drivers, and with feedbacks to the atmosphere, may suffer from uncertainty regarding estimation of crucial parameters, leading to an often broad range of the parameter predicted. Moreover, it may appear that essential parameters have been overlooked. Nevertheless, the further development of predictive models, from the scale of species ranges to that of global vegetation, must be encouraged.

1.6   Epilogue

Vegetation ecology has grown tremendously since its first textbook appeared (Mueller-Dombois & Ellenberg 1974). Ever since, many thousands of papers have been published in international journals. Although only a small minority of them have been cited in this book, it is hoped that the growth of the science, both in depth and in breadth, will become clear from the 16 chapters that follow. The growing breadth is also expressed in the involvement of scientists from other disciplines in vegetation ecology, notably population ecology, ecophysiology, microbiology, soil biology, entomology, animal ecology, landscape ecology, physical geography, geology and climatology. The updated and new chapters in this second edition highlight developments in the field during that past 5–10 years, but retain their firm grounding in the deeper history of the development of key concepts in the classic literature.

It is encouraging that international cooperation between plant ecologists all over the world has also grown impressively. The authorship of this book includes colleagues from Africa, Asia, Australia, Europe and the USA. Several chapters conclude with a summary of achievements, others offer perspectives for the future of our science. Let us hope that the book will indeed contribute to the further development of vegetation ecology.

References

Allen, T.F.H. & Hoekstra, T.W. (1992) *Toward a Unified Ecology*. Columbia University Press, New York, NY.

Bakker, J.P., Olff, H., Willems, J.H. & Zobel, M. (1996) Why do we need permanent plots in the study of long-term vegetation dynamics? *Journal of Vegetation Science* 7, 147–156.

Barkman, J.J. (1978) Synusial approaches to classification. In: *Classification of Plant Communities*. 2nd edn (ed. R.H. Whittaker), pp. 111–165. Junk, The Hague.

Box, E.O. (1981) *Macroclimate and Plant Forms: an Introduction to Predictive Modeling in Phytogeography*. Junk, The Hague.

Braun-Blanquet, J. (1932) *Plant Sociology. The Study of Plant Communities*. Authorized English translation of 'Pflanzensoziologie' by G.D. Fuller & H.S. Conard. McGraw-Hill Book Company, New York, NY.

Braun-Blanquet, J. (1964) *Pflanzensoziologie*. 3. Auflage. Springer-Verlag, Wien.

Burrows, C.J. (1990) *Processes of Vegetation Change*. Unwin Hyman, London.

Cain, S.A. & Castro, G.M. de Oliveira (1959) *Manual of Vegetation Analysis*. Harper & Brothers, New York, NY.

Clements, F.E. (1916) *Plant Succession. An Analysis of the Development of Vegetation*. Carnegie Institution, Washington, DC.

Curtis, J.T. (1959) *The Vegetation of Wisconsin*. University of Wisconsin Press, Madison, WI.

Dale, M.T.R. (1999) *Spatial Pattern Analysis in Plant Ecology*. Cambridge University Press, Cambridge.

Dierschke, H. (1994) *Pflanzensoziologie*. Verlag Eugen Ulmer, Stuttgart.

Du Rietz, G.E. (1931) Life-forms of terrestrial flowering plants. *Acta Phytogeographica Suecica* 3, 1–95.

Ellenberg, H., Weber, H.E., Düll, R., Wirth, V., Werner, W. & Paulißen, D. (1992) Zeigerwerte von Pflanzen in Mitteleuropa, 2nd ed. *Scripta Geobotanica* 18, 1–258.

Fortin, M.-J. & Dale, M.R.T. (2005) *Spatial Analysis: a Guide for Ecologists*. Cambridge University Press, Cambridge.

Franklin, J. (2010) Spatial point pattern analysis of plants. In: *Perspectives on Spatial Data Analysis* (eds S.J. Rey & L. Anselin), pp 113–123. Springer, New York, NY.

Gimingham, C.H. (1988) A reappraisal of cyclical processes in *Calluna* heath. *Vegetatio* 77, 61–64.

Gleason, H.A. (1926) The individualistic concept of the plant association. *Bulletin of the Torrey Botanical Club* 53, 1–20.

Glenn-Lewin, D.C. & van der Maarel, E. (1992) Patterns and processes of vegetation dynamics. In: *Plant Succession – Theory and Prediction* (eds D.C. Glenn-Lewin, R.K. Peet & T.T. Veblen), pp. 11–59. Chapman & Hall, London.

Greig-Smith, P. (1957) *Quantitative Plant Ecology*. Butterworths, London.

Grime, J.P. (2001) *Plant Strategies, Vegetation Processes, and Ecosystem Properties*. 2nd edn. John Wiley & Sons, Chichester.

Grootjans, A.P., Fresco, L.F.M., de Leeuw, C.C. & Schipper, P.C. (1996) Degeneration of species-rich *Calthion palustris* hay meadows; some considerations on the community concept. *Journal of Vegetation Science* 7, 185–194.

Grubb, P.J. (1977) The maintenance of species-richness in plant communities: the importance of the regeneration niche. *Biological Reviews of the Cambridge Philosophical Society* 52, 107–145.

Gurevitch, J., Scheiner, S.M. & Fox, G.A. (2002) *The Ecology of Plants*. Sinauer Associates, Sunderland, MA.

Hennekens, S.M. & Schaminée, J.H.J. (2001) TURBOVEG, a comprehensive data base management system for vegetation data. *Journal of Vegetation Science* 12, 589–591.

Hill, M.O. (1979) TWINSPAN – A FORTRAN program for arranging multivariate data in an ordered two-way table by classification of the individuals and attributes. Cornell University, Ithaca, NY.

Howe, H.F. & Westley, L.C. (1988) *Ecological Relationships of Plants and Animals*. Oxford University Press, New York, NY.

Huntley, B. & Webb III, T. (1988) *Handbook of Vegetation Science, part 7, Vegetation History*. Kluwer Academic Publishers, Dordrecht.

Hytteborn, H & Verwijst, Th. (2011) The importance of gaps and dwarf trees in the regeneration of Swedish spruce forests: the origin and content of Sernander's (1936) gap dynamics theory. *Scandinavian Journal of Forest Research* 26, 3–16.

Iversen, J. (1936) Biologische Pflanzentypen als Hilfsmittel in der Vegetationsforschung. Ein Beitrag zur ökologischen Charakterisierung und Anordnung der Pflanzengesellschaften. *Meddelelser fra Skalling-Laboratoriet* 4, 1–224.

Jongman, R.H.G., ter Braak, C.J.F. & van Tongeren, O.F.R. (1995) *Data Analysis in Community and Landscape Ecology*. Cambridge University Press, Cambridge.

Kershaw, K.A. (1964) *Quantitative and Dynamic Plant Ecology*. Edward Arnold, London.

Knight, D.H. & Loucks, O.L. (1969) A quantitative analysis of Wisconsin forest vegetation on the basis of plant function and gross morphology. *Ecology* **50**, 219–234.

Looijen, R.C. & van Andel, J. (1999) Ecological communities: conceptual problems and definitions. *Perspectives in Plant Ecology, Evolution and Systematics* **2**, 210–222.

Loucks, O. (1962) Ordinating forest communities by means of environmental scalars and phytosociological indices. *Ecological Monographs* **32**, 137–166.

McIntosh, R.P. (1986) *The Background of Ecology: Concept and Theory*. Cambridge University Press, Cambridge.

Mueller-Dombois, D. & Ellenberg, H. (1974) *Aims and Methods of Vegetation Ecology*. John Wiley and Sons, New York. Reprint published in 2002 by Blackburn, Caldwell, NJ.

Noble, I.R. & Slatyer, R.O. (1980) The use of vital attributes to predict successional sequences in plant communities subject to recurrent disturbance. *Vegetatio* **43**, 5–21.

Odum, E.P. (1969) The strategy of ecosystem development. *Science* **164**, 262–270.

Onipchenko, V.G., Semenova, G.V. & van der Maarel, E. (1998) Population strategies in severe environments: alpine plants in the northwestern Caucasus. *Journal of Vegetation Science* **12**, 305–318.

Parker, V.T. (2001) Conceptual problems and scale limitations of defining ecological communities: a critique of the CI concept (Community of Individuals). *Perspectives in Plant Ecology, Evolution and Systematics* **4**, 80–96.

Perrow, M.R. & Davy, A.J. (2002) *Handbook of Ecological Restoration. Vol. 1 Principles of Restoration. Vol. 2 Restoration in Practice*. Cambridge University Press, Cambridge.

Peters, R.H. (1991) *A Critique for Ecology*. Cambridge University Press, Cambridge.

Pickett, S.T.A., Collins, S.L. & Armesto, J.J. (1987) Models, mechanisms and pathways of succession. *Botanical Review* **53**, 335–371.

Prentice, I.C. (1992) Climate change and long-term vegetation dynamics. In: *Plant Succession – Theory and Prediction* (eds D.C. Glenn-Lewin, R.K. Peet & T.T. Veblen), pp. 293–339. Chapman & Hall, London.

Rabotnov, T.A. (1975) On phytocoenotypes. *Phytocoenologia* **2**, 66–72.

Raunkiær, C. (1934) *The Life Forms of Plants and Statistical Plant Geography*. Clarendon Press, Oxford.

Remmert, H. (1991) *The Mosaic Cycle Concept of Ecosystems*. Ecological Studies 85. Springer-Verlag, Berlin.

Rodwell, J.S., Schaminée, J.H.J., Mucina, L., Pignatti, S., Dring, J. & Moss, D. (2002) The Diversity of European Vegetation. An overview of phytosociological alliances and their relationships to EUNIS habitats. *Report EC-LNV 2002/054*, Wageningen.

Rydin, H. (1997) Competition among bryophytes. *Advances in Bryology* **6**, 135–168.

Schulze, E.-D. & Mooney, H.A. (eds) (1994) *Biodiversity and Ecosystem Function*. Springer-Verlag, Berlin.

Shipley, B. & Keddy, P.A. (1987) The individualistic and community-unit concepts as falsifiable hypotheses. *Vegetatio* **69**, 47–55.

Tilman, D. (1982) *Resource Competition and Community Structure*. Princeton University Press, Princeton, NJ.

Titlyanova, A.A. & Mironycheva-Tokareva, N.P. (1990) Vegetation succession and biological turnover on coal-mining spoils. *Journal of Vegetation Science* **1**, 643–652.

Titlyanova, A.A., Romanova, I.P., Kosykh, N.P. & Mironycheva-Tokareva, N.P. (1999) Pattern and process in above-ground and below-ground components of grassland ecosystems. *Journal of Vegetation Science* **10**, 307–320.

Trass, H. & Malmer, N. (1978) North European approaches to classification. In: *Classification of Plant Communities*, 2nd edn (ed. R.H. Whittaker), pp. 201–245. Junk, The Hague.

Tüxen, R. (1970) Einige Bestandes- und Typenmerkmale in der Struktur der Pflanzengesellschaften. In: *Gesellschaftsmorphologie* (ed. R. Tüxen), pp. 76–98. Junk, The Hague.

van der Maarel, E. (1979) Transformation of cover-abundance values in phytosociology and its effects on community similarity. *Vegetatio* **39**, 97–114

van der Maarel, E. (1988) Vegetation dynamics: patterns in time and space. *Vegetatio* **77**, 7–19.

van der Maarel, E. (1990) Ecotones and ecoclines are different. *Journal of Vegetation Science* **1**, 135–138.

van der Maarel, E. (1995) Vicinism and mass effect in a historical perspective. *Journal of Vegetation Science* **6**, 445–446.

van der Maarel, E. (1996) Pattern and process in the plant community: fifty years after A.S. Watt. *Journal of Vegetation Science* **7**, 19–28.

van der Maarel, E. (2005) Vegetation ecology – an overview. In: *Vegetation Ecology* (ed. E. van der Maarel), pp. 1–51. Blackwell Publishing, Oxford.

van der Maarel, E. (2007) Transformation of cover-abundance values for appropriate numerical treatment –alternatives to the proposals by Podani. *Journal of Vegetation Science* **18**, 767–770.

van der Maarel, E. & Sykes, M.T. (1993) Small-scale plant species turnover in a limestone grassland: the carousel model and some comments on the niche concept. *Journal of Vegetation Science* **4**, 179–188.

van der Maarel, E., Janssen, J.G.M. & Louppen, J.M.W. (1978) TABORD, a program for structuring phytosociological tables. *Vegetatio* **38**, 143–156.

van Dorp, D., Boot, R. & van der Maarel, E. (1985) Vegetation succession in the dunes near Oostvoorne, The Netherlands, since 1934, interpreted from air photographs and vegetation maps. *Vegetatio* **58**, 123–136.

Veblen, T.T. (1992) Regeneration dynamics. In: *Plant Succession – Theory and Prediction* (eds D.C. Glenn-Lewin, R.K. Peet & T.T. Veblen), pp. 152–187. Chapman & Hall, London.

Warming, E. (1909) *Oecology of Plants: An Introduction to the Study of Plant Communities*. English edition of the Danish textbook *Plantesamfund. Grundtræk af den økologiske Plantegeografi.* (1895) by M. Vahl, P. Groom & B. Balfour. Oxford University Press, Oxford.

Watt, A.S. (1947) Pattern and process in the plant community. *Journal of Ecology* **35**, 1–22.

Weiher, E. & Keddy, P. (eds) (1999) *Ecological Assembly Rules. Perspectives, Advances, Retreats*, pp. 251–271. Cambridge University Press, Cambridge.

Westhoff, V. & van der Maarel, E. (1978) The Braun-Blanquet approach. In: *Classification of Plant Communities*, 2nd edn (ed. R.H. Whittaker), pp. 287–297. Junk, The Hague.

White, J. (ed) (1985) *Handbook of Vegetation Science, Part 3, The Population Structure of Vegetation*. Junk, Dordrecht.

Whittaker, R.H. (1965) Dominance and diversity in land plant communities. *Science* **147**, 250–260.

Whittaker, R.H. (1978) Approaches to classifying vegetation. In: *Classification of Plant Communities*, 2nd edn (ed. R.H. Whittaker), pp. 1–31. Junk, The Hague.

Wilson, J.B. & Agnew, A.D.Q. (1992) Positive-feedback switches in plant communities. *Advances in Ecological Research* **23**, 263–336.

Wilson, J.B., Gitay, H., Steel, J.B. & King, W. McG. (1998) Relative abundance distributions in plant communities: effects of species richness and of spatial scale. *Journal of Vegetation Science* **9**, 213–220.

Zamfir, M. & Goldberg, D.E. (2000) The effect of density on interactions between bryophytes at the individual and community levels. *Journal of Ecology* **88**, 243–255.

Zobel, M., van der Maarel, E. & Dupré, C. 1998. Species pool: the concept, its determination and significance for community restoration. *Applied Vegetation Science* **1**, 55–66.

2

Classification of Natural and Semi-natural Vegetation

Robert K. Peet[1] and David W. Roberts[2]

[1]University of North Carolina, USA
[2]Montana State University, USA

2.1 Introduction

Vegetation classification has been an active field of scientific research since well before the origin of the word ecology and has remained so through to the present day. As with any field active for such a long period, the conceptual underpinnings as well as the methods employed, the products generated and the applications expected have evolved considerably. Our goal in this chapter is to provide an introductory guide to participation in the modern vegetation classification enterprise, as well as suggestions on how to use and interpret modern vegetation classifications. Some notes on the historical development of classification and the associated evolution of community concepts are provided by van der Maarel & Franklin in Chapter 1, and Austin describes numerical methods for community analysis in Chapter 3. While we present some historical and conceptual context, our goal in this chapter is to help the reader learn how to create, interpret and use modern vegetation classifications, particularly those based on large-scale surveys.

2.1.1 Why classify?

Early vegetation classification efforts were driven largely by a desire to understand the natural diversity of vegetation and the factors that create and sustain it. Vegetation classification is critical to basic scientific research as a tool for organizing and interpreting information and placing that information in context. To conduct or publish ecological research without reference to the type of community the work was conducted in is very much like depositing a specimen in a museum without providing a label. Documenting ecological context can range

Vegetation Ecology, Second Edition. Eddy van der Maarel and Janet Franklin.
© 2013 John Wiley & Sons, Ltd. Published 2013 by John Wiley & Sons, Ltd.

from a simple determination of the local community to a detailed map showing a complicated spatial arrangement of vegetation types as mapping units. This need for documenting ecological context is also scale transgressive with vegetation classification schemes contributing equally to research from small populations of rare species to that involving global projection of human impacts (Jennings *et al.* 2009). Frameworks other than vegetation classification are conceivable for documenting ecological context. For example, environmental gradients and soil classifications have often been used to define site conditions. However, these require *a priori* knowledge of factors important at a site while vegetation classification, in contrast, lets the assemblage of species and their importance serve as a bioassay.

Use of vegetation classification has increased over the past few decades. Vegetation description and classification provides units critical for inventory and monitoring of natural communities, planning and managing conservation programmes, documenting the requirements of individual species, monitoring the use of natural resources such as forest and range lands, and providing targets for restoration. Vegetation types are even achieving legal status where they are used to define endangered habitats and where their protection is mandated (see Waterton 2002). For example, the European Union has created lists of protected vegetation types, and vegetation types are being used to develop global red lists of threatened ecosystems (e.g. Rodríguez *et al.* 2011).

2.1.2 The challenge

The goal of vegetation classification is to identify, describe and interrelate relatively discrete, homogeneous and recurrent assemblages of co-occurring plant species. Vegetation presents special challenges to classification as it varies more or less continuously along environmental gradients and exhibits patterns that result from historical contingencies and chance events (Gleason 1926, 1939). Not surprisingly, multiple solutions are possible and as Mucina (1997) and Ewald (2003) have explained, adopting one approach over another should be based on practical considerations.

Although there is considerable variability in approaches taken to vegetation classification, most initiatives embrace some basic assumptions about vegetation and its classification. Four such widely adopted assumptions were articulated by Mueller-Dombois & Ellenberg in their classic 1974 textbook.

1   Similar combinations of species recur from stand to stand under similar habitat conditions, though similarity declines with geographic distance.
2   No two stands (or sampling units) are exactly alike, owing to chance events of dispersal, disturbance, extinction, and history.
3   Species assemblages change more or less continuously if one samples a geographically widespread community throughout its range.
4   Stand similarity varies with the spatial and temporal scale of analysis.

These underlying assumptions have led to the wide adoption of a practical approach wherein community types are characterized by attributes of vegetation records that document similar plant composition and physiognomy with the vegetation classification relying on representative field records (plots) to define the central concept of the type. Subsequent observations of vegetation are determined as belonging to a unit through their similarity to the type records for the individual communities.

Another challenge that increasingly confronts the vegetation classification enterprise is that with the widespread adoption of classification systems for inventory, monitoring, management and even legal status, classification systems need to have comprehensive coverage, stability in the classification units and a transparent process for revising those units. This new and broader set of applications suggests that we need to move toward consensus classifications that combine the inquiry of many persons into a unified whole, and that the rules for participation be open and well defined. However, we must also recognize that as the applications of vegetation classification migrate from the pure scientific arena to one of management and policy, the categories are likely to evolve in ways that find their origin not just in science but also in policy and public opinion (Waterton 2002).

2.2   Classification frameworks: history and function

Vegetation classifications systems can vary from local to global and from fine-scale to coarse-scale, and the approaches to vegetation classification used tend to reflect the scale of the initiative. Classification schemes used at the global scale tend to focus on growth-forms or physiognomic types that reflect broad-scale climatic variation rather than species composition (discussed by Box & Fujiwara in Chapter 15). In the present chapter our focus is on actual or realized natural and semi-natural vegetation. These are generally bottom-up classifications where units are defined by sets of field observations where species occurrences and/or abundances were recorded. Vegetation classification has a rich history (discussed in Chapter 1) with the many and varied approaches reviewed in detail by Whittaker (1962, 1973), Shimwell (1971) and Mueller-Dombois & Ellenberg (1974). Subsequent synthetic overviews by Kent (2012), McCune & Grace (2002), and Wildi (2010) summarize, compare and evaluate commonly used methods.

Although local-scale projects can use any classification criteria that provide a convenient conceptual framework for the project at hand, such local and idiosyncratic classifications do not allow the work to be readily placed in a larger context. The growing recognition of the need for vegetation classification research to place new results in context means that a consistent conceptual framework is needed for all components of the classification process (De Cáceres & Wiser 2012). Below we summarize key components of two such frameworks: European phytosociology as it has evolved from the school of Braun-Blanquet, and the more recently developed US National Vegetation Classification. We then summarize the differences and compare these classifications to those encountered in other national-level initiatives.

2.2.1 The Braun-Blanquet approach and contemporary European phytosociology

By far the most widely applied approach to vegetation classification is that developed by Josias Braun-Blanquet. The method centres on recording fine-scale vegetation composition. The basic unit of observation is the plot (or relevé) within which all species are recorded by vertical stratum and the abundance of each is estimated, usually using an index of cover/abundance. Related plots are combined in tabular form and groups of similar plots are defined as communities based on consistency of composition. The basic unit, adopted at the International Botanical Congress in 1910, is the association, which is defined as having 'definite floristic composition, presenting a uniform physiognomy, and growing in uniform habitat conditions.' The community is then characterized by the constancy of shared taxa and specific diagnostic species that provide coherency to the group and set it off from other groups. Historically, table sorting was done by hand, while today computer-aided sorting is the rule with numerous algorithms available to automate the process (see Section 2.6.1 for more detail, or consult Braun-Blanquet 1964 or Westhoff & van der Maarel 1973). Similar associations that share particular diagnostic species are combined into higher-level assemblages, there being five primary levels (Association, Alliance, Order, Class and Formation).

Once an author has developed one or more new or revised associations, that author reviews past published work, designates the critical diagnostic species, assigns a unique name following the International Code of Phytosociological Nomenclature (Weber *et al.* 2000), places it within the hierarchy and submits the work for publication. The process is similar to that required to establish a new species. In both cases the author examines documented occurrences, writes a monograph wherein the examined occurrences are typically reported, and specifies plots or a type specimen that serve to define the type. In the Braun-Blanquet system one plot is designated the nomenclatural type for each association, the nomenclature follows a formal code that gives priority to the first use of a name, and the resultant associations are then available in the literature for scientists to discover and accept or not.

The strongest attributes of the Braun-Blanquet system are the consistency of the approach, the enormous number of plots that have been recorded (with an estimated total for Europe alone of 4.3 million; Schaminée *et al.* 2009), and the large number of published descriptions of vegetation types. Weaknesses include a seeming arbitrary definition of units, the lack of requirement that new units be integrated with established units, and the lack of any formal registry of published units. Some potential users find the naming system awkward, which is why the recent vegetation classification of Great Britain divorced itself from the traditional nomenclature, despite the fundamental units otherwise closely approximating the associations of the Braun-Blanquet system (Waterton 2002; Rodwell 2006).

The literature on European vegetation is so enormous that summarizing it has proven extremely difficult. Community types have been synthesized for quite a few countries and other geographic units, but these efforts have not yet been integrated. In 1992 The European Vegetation Survey was established with the goal of fostering collaboration and synthesis (Mucina *et al.* 1993; Rodwell *et al.* 1995). One direct result has been a number of trans-national overviews of thematic types and a summary of types at the alliance level and above by Rodwell *et al.* (2002). In addition, there has been movement toward standards for collecting plot data (Mucina *et al.* 2000), and the development of the software program TurboVeg (Hennekens & Schaminée 2001) for managing plot data has led to considerable standardization in data content and format.

2.2.2 The United States National Vegetation Classification

The development of the US National Vegetation Classification (USNVC) provides clear contrasts with the European classification enterprise, although both have roughly equivalent primary units (in both cases called associations), and both are based on vegetation plot records. Historically, when vegetation classification was undertaken in North America by academic ecologists, the approaches tended to be idiosyncratic and specific to the particular project. In the absence of leadership from the academic community, various federal land management and environmental regulatory agencies in the USA created classification systems for their own purposes, such as for wetlands (Cowardin *et al.* 1979), land-cover (Bailey 1976), and forest management (Pfister & Arno 1980).

Vegetation classification in the USA has matured considerably over the past few decades in response to three initiatives. First, starting in the 1970s, The Nature Conservancy, a non-profit organization, encouraged the development of state programmes to inventory the status of biodiversity for conservation planning. The lack of consistency in inventory units between states ultimately led to a national vegetation classification system based on types provided by state programmes, published literature and expert opinion (Anderson *et al.* 1998). Although at first largely subjective, the units were defined to be non-overlapping and to constitute a formal list of recognized types. This effort grew into an international classification (see Grossman *et al.* 1998; Anderson *et al.* 1998; Jennings *et al.* 2009). As this system has matured, emphasis has been placed on both providing linkage to original data and describing the variation in each type across its geographic range. Second, growing recognition of the need for common standards for geospatial data across government agencies led to the establishment of the US Federal Geographic Data Committee (USFGDC), including a subcommittee for standardizing vegetation classification activities across government agencies. Although this standard is formally recognized only for cross-tabulating classifications, it is beginning to have broad application in its own right. Third, members of the Ecological Society of America (ESA) recognized the diversity of approaches and standards in use across the country, the need to allow broad participation by interested parties and the importance of peer review of proposed changes in the classification. ESA established a Panel on Vegetation Classification in 1994 that subsequently proposed standards for vegetation classification (Jennings *et al.* 2009).

These three independent initiatives formed a formal partnership to advance the USNVC that led to adoption of a new USFGDC standard in 2008, including rules for documentation and peer review of proposed new and revised types. As a consequence of this partnership, the USA has a national classification with a definitive set of associations (*c.* 6200 at this writing), and mechanisms for modification of this list by interested parties are being developed. By requiring that accepted types span the known range of variation, that they not overlap, and that they be based on vegetation records in public archives, the system is more forward looking than current European initiatives. However, at this time the US community types have only limited linkage to archived data, and the formal descriptions of types are not always sufficiently detailed to allow creation of keys or expert systems for determination of vegetation occurrences. Thus, while the US infrastructure is very progressive, the content will require considerably more development to catch up with the established European initiatives.

The USNVC formal hierarchy differs from that of the Braun-Blanquet system in that it is not derived entirely from lumping smaller units into larger ones. Instead it has three upper levels that provide a top-down, physiognomic hierarchy with units that are global in conception (Formation Class, Formation Subclass and Formation). Nested below these are three middle levels based on biogeographic and regional environmental factors (Division, Macrogroup and Group). At the base are Associations, which are combined into Alliances that nest into the middle-level Groups. This three-tier, eight-level hierarchy is intended to provide interpretable and widely applicable units across all spatial scales. The nesting is not always as seamless as it is in the Braun-Blanquet approach, but is intended to facilitate a broader range of applications.

2.2.3 Attributes of successful classification systems

The recent British National Vegetation Classification (NVC) programme is a model for standardized data collection in a vegetation classification system. This programme was led by John Rodwell who described the methodology in a user's handbook (Rodwell 2006). There are standard rules for placement of plots, size of plots and data to be collected. Standard forms were used to minimize drift in field methods. Any large new initiative would be well advised to adopt the level of standardization employed in the UK NVC, and small programmes should adopt methods and goals consistent with well-established programmes in order to maximize compatibility. The Braun-Blanquet, British and US initiatives all have their own standard nomenclatures, although the formats and rules vary considerably between the systems. Finally, the US system remains unique in requiring public archiving of supporting plot data and providing systems for interested stakeholders to formally propose changes, both of which are likely to lead to more rigorously defined types.

2.3 Components of vegetation classification

There are ten primary components to vegetation classification, their complexity depending on the situation, but all of them being important. We define those components here, and starting in Section 2.4 we address each in some detail.

Project planning. Delimiting the geographic and ecological extent or range of the study allows data needs to be defined and existing data to be identified and evaluated. Often this will involve extensive preliminary work to aid in the selection of field sites, perhaps through stratification relative to composition or environment or successional development, or in more human-dominated systems through locating the remaining examples of natural and semi-natural vegetation.

Data acquisition. Once the objective of the study is defined, quantitative data characterizing vegetation composition must be acquired as new records or from databases of previously collected vegetation records. At a minimum, each record should contain the date and location of observation, some attributes of the site, a list of plant taxa and some measure of importance for each taxon.

Data preparation. Before the vegetation composition data can be analysed, the observation records need to be combined into a single data set wherein inconsistences in field methods, scales of observation, measures of abundance, units of environment, resolution of species identifications and inconsistent taxonomic authorities have all been resolved. Although the goal is straight forward, complete integration without loss of information is often impossible and this component often involves a number of difficult and subjective decisions.

Community entitation. This is the most essential step in classification as it is the creation of the entities that constitute the classification units. A broad range of methods can be employed, often iteratively and in combination, to define the classification units or 'types'. As vegetation often varies continuously in time and space, there is nothing conceptually as solid as a species and different investigators following different rules and protocols often come up with different classification units.

Cluster assessment. Once entities have been defined, it is important to critically analyse the results to determine that the types are relatively homogeneous and distinct from other types (Lepš & Šmilauer 2003), and to assure that distributions of species within types exhibit high fidelity and ecologically interpretable patterns. The criteria often involve formal assessment of the quantitative similarity (or dissimilarity) of vegetation composition within versus between types and the calculation of quantitative indices of species fidelity to types.

Community characterization. Entities must be characterized in a way that allows additional occurrences to be recognized with less than a full-scale reanalysis, and also allows placement in a larger system of community types. Traditionally, this has included assessment of the typical abundance and frequency of taxa, and in many cases identification of indicator species and the typical range of environmental conditions.

Community determination. Users need to be able to determine to which classification unit an instance of vegetation should be assigned, be it a published or archived record of vegetation or a new field observation. Tools range from dichotomous keys, to methods that use mathematical similarity, to expert systems. Determinations range from binary (yes/no), to assignment to multiple types with various designated degrees of fit.

Classification integration. Vegetation classification is often intended to expand or revise an established vegetation classification system. Often this involves changes in established units, or replacement of previously published units. This, in turn, requires that levels of resolution (e.g. fineness of splitting), criteria for peer review and the importance of stability in classification systems be addressed systematically, more so than has historically been the case. For effective communication, community types need names, and the names need to be compliant with the current standards of the classification system (e.g. Weber *et al.* 2000; USFGDC 2008).

Classification documentation. The results of vegetation classification initiatives need to be documented, both as to the units recognized and the data analysed. Different classification systems have different requirements, formats and protocols. Publication with tables summarizing composition is always important, and vegetation records used in the analysis should be deposited in a public database.

2.4   Project planning and data acquisition

The fundamental unit for recording vegetation is the plot (or relevé). Associated with the plot are records of its location, size, physical setting and vegetation composition. The distribution and placement of plots, their size and shape, and the attributes to be recorded vary among recognized protocols and are important decisions to make when initiating a new project, or to recognize when using existing plot data.

2.4.1   Plot distribution and location

The first step in a vegetation classification project is to define the geographic and compositional variation in vegetation to be classified as this will determine the number of plots needed and the difficulty of acquiring them. This step can be accomplished by literature review, consulting with regional experts and preliminary field reconnaissance. Next, existing relevant vegetation plot data should be identified. This is not always straightforward for while some plot data are available in public archives (e.g. VegBank; see www.vegbank.org, Peet *et al.* 2012) and many data sets are described in indices of plot databases (e.g. Global Index of Vegetation-Plot Databases (GIVD); see www.givd.info, Dengler *et al.* 2011), many other data sets are not widely known and must be discovered by contacting likely sources. Once the availability of extant plot data is assessed and the need for new plots has been ascertained, the next step is to estimate the effort required to obtain those new plots.

The physical distribution of plots across the study area can be determined in a number of ways and these will reflect the objectives of the project. Traditionally, plots have been placed using preferential sampling where the investigator subjectively locates them to cover the range of variation needed for the project. The potential for bias in this method is obvious, so sometimes field plots are randomly located, or the landscape is stratified and plots are placed randomly within the strata. An alternative form of stratification often employed is the gradsect method where vegetation samples are stratified along known gradients of compositional variation (see Gillison & Brewer 1985; Austin & Heyligers 1989, 1992). As random and stratified sampling might undersample rare or unknown types, it is not uncommon for a probability designed sample to be supplemented with preferential plots on types poorly represented in the sample. Also, as the spatial extent of the project increases, the need for both stratification and some component of preferential sampling increases. For example, if sampling the range of variation in riparian vegetation across a moderate-sized European country or American state, there would inevitably be preferential selection of regions within which the sampling would occur. In contrast, if the objective of the project were an inventory of the area of each vegetation type, or of standing timber, objective sampling methods would be more critical. An example of this is the Forest Inventory and Analysis plot system of the United States Forest Service designed to monitor the timber supply of the nation. This system uses a base grid of sample points with one plot located randomly in each of the 125,100 2430-ha hexagonal cells (Bechtold & Patterson 2004; Gray *et al.* 2012).

The potential for bias in preferentially located plots has led to considerable introspection and some critical analysis. Preferential sampling is often favoured in human-manipulated landscapes where patches of natural and semi-natural vegetation tend to be small and influenced by recent land use. Roleček *et al.* (2007) explain that while probability designed sampling schemes better meet certain statistical assumptions, preferential sampling yields data sets that cover a broader range of vegetation variability including rare types that might otherwise have been missed. Random sampling is required when the sample units must represent a single statistical population. In vegetation sampling generally the intention is to distinguish types that are not necessarily members of the same statistical population.

Michalcová *et al.* (2011) further considered the problems inherent in using large plot databases wherein many of the plots are likely to represent preferential sampling. They found that sets of preferential samples contained more endangered species and had higher beta diversity, whereas estimates of alpha diversity and representation of alien species were not consistently different between preferentially and stratified-randomly sampled data. Thus, if the goal is to characterize the range of compositional variation or maximize species coverage, then at least some element of preferential sampling can be important.

2.4.2   Plot size and shape

Choice of plot size and shape can significantly influence perception of vegetation for a number of reasons. First, vegetation is spatially variable at nearly all scales.

This variation can be driven by underlying environmental variation, biological interactions, or historical events (Nekola & White 1999). Secondly, species number increases with plot size, the logarithm of species richness usually varying directly with the logarithm of plot area (Fridley *et al.* 2005). Plot shape has a similar trade-off in that plots with low perimeter to area ratios (squares and circles) tend to minimize spatial pattern (within-plot heterogeneity) and thus species number, whereas plots with high edge-area ratios (e.g. long, thin plots) maximize representation of the range of patch types and species.

Historically, the solution to the trade-off between homogeneity and completeness was to create a species–area curve to assess the 'minimum area' needed to represent a particular type of vegetation. Unfortunately there is no objective stopping rule for plot area. In addition, plots were generally located preferentially in homogenous vegetation, but again this was subjective as some pattern can nearly always be found within a plot. Plot size also traditionally varied with vegetation height so as to capture a snapshot of the total community, and Dengler *et al.* (2008) observe that as a rule of thumb, plots are roughly as large in square metres as vegetation is high in decimetres.

In excess of four million vegetation plots are available in various archives (see Schaminée *et al.* 2009; Dengler *et al.* 2011). Integrating subsets of these plots for various analyses is complicated by the diversity of plot sizes and shapes. In addition, metrics such as species constancy and plot similarity can vary with plot size (Dengler *et al.* 2009). Collectively, these considerations have led a series of authors to propose that a standard set of plot sizes be adopted to facilitate future data integration and analysis. For example, Chytrý & Otýpková (2003) proposed plot sizes of $4\,m^2$ for sampling aquatic vegetation and low-grown herbaceous vegetation, $16\,m^2$ for grassland, heathland and other herbaceous or low-scrub vegetation types, $50\,m^2$ for scrub, and $200\,m^2$ for woodlands. In contrast, many North American ecologists have followed a tradition established by Whittaker (1960) of recording forest vegetation in $1000\,m^2$ plots, reflecting the generally higher tree species richness of North American forests as compared to European forests.

Peet *et al.* (1998) proposed that because there is no one correct scale for observing vegetation and because different factors influence composition at different scales, vegetation should be recorded at multiple scales, both to facilitate data integration across projects and to allow investigation of processes working at different scales. They proposed a specific protocol with plots on a nearly log scale of 0.01, 0.1, 1, 10, 100, 400 and $1000\,m^2$. For their study they suggested $100\,m^2$ as the smallest acceptable total plot size, calling smaller-scale pattern 'within-community variation'. Such nested designs largely originated with Whittaker *et al.* (1979), with alternative protocols subsequently proposed (e.g. Stohlgren *et al.* 1995; Dengler 2009). All these protocols note increased variance in composition among subsamples at smaller scales and recommend that there be multiple small plots within each large plot to increase the range of this variance documented. Although there is no consensus as to the optimal size or arrangement of nested plots, some form of nested sampling is highly desirable, if for no other reason than to maximize the potential for aggregating the plots with those from other studies. Moreover, with careful plot design, relatively little extra effort is required to include nested plots within the largest plot. Users of nested protocols should, however, be cautious not to aggregate dispersed sub-plots into larger subsamples as this will inflate species numbers owing to the subplots spanning an artificially high range of within-community variation.

2.4.3   Plot records

In its simplest form, a plot record contains information about the observation event, the site and the plants observed at the site. Lists of required and recom-mended plot attributes have been codified for numerous plot protocols and with remarkably similar prescriptions (e.g. Mucina *et al.* 2000; Rodwell 2006; Jennings *et al.* 2009). These prescriptions often recognize two kinds of plots; occurrence plots are those used to determine the vegetation type at a site or document its presence, whereas classification plots are those intended for devel-opment or improvement of a classification. Occurrence plots require only a subset of the observations required of a classification plot, reducing the time needed for data collection.

Information about the observation of the plot that describes the event – such as the date, the persons involved, the geo-coordinates (including the datum and the precision of the record), the unique identifier of the observation – and the physical layout of the plot should be recorded as metadata. If the plot is observed more than once, it is important to separate data that are constant between meas-urements, such as geo-coordinates, from information particular to the observa-tion event, such as date. A text description of the location is encouraged. The second group of observations contains facts about the site and its overall vegeta-tion. Basic topographic information such as slope, aspect and elevation are nearly always collected. Most other environmental data are difficult to standardize, so these are usually tailored to the project or its larger context. For example, soil chemistry data can be very helpful for interpreting plots in a project, but results can vary greatly with protocol, and even between labs using a consistent proto-col; consequently, combining soils data from plots collected in separate projects must be done with caution. Finally, summary records about the physical structure of the vegetation are often required, such as height and cover in different vertical strata. These seemingly simple measurements also vary significantly with proto-col, so care must be taken to retain consistency in data collection across a project and when integrating data from multiple projects.

Taxon identification and documentation present several challenges. Inevitably some taxa observed in the field will be unknown. As multiple taxa are often unknown, it is best to link a collection to a specific line number on the field data page so that future ambiguities are minimized. The taxon list should have each taxon recorded to the highest resolution possible, be it variety or family. Recog-nition of infraspecific types can prove invaluable during future data integration as varieties and subspecies often migrate to full species status, and future splitting would not be possible without special information being recorded, such as variety or subspecies name. Care should be taken to follow standard authorities for the taxa recognized and to record that authority (as opposed to the authority for creation of the name) so that in the future the meaning of the name can be

Table 2.1   A comparison of several cover scales used for recording vegetation plots including the traditional Braun-Blanquet scale (1928), the original Domin scale (1928), a variant of the Domin scale by Krajina (1933), and the scales of the Carolina (Peet *et al.* 1998) and New Zealand vegetation surveys (Allen 1992). The shading indicates how the newer indices nest into the Braun-Blanquet scheme.

Range of cover	Braun-Blanquet	Domin	Krajina	Carolina	New Zealand
Single individual	r	+	+	1	1
Sporadic or few	+	1	1	1	1
0–1%	1	2	1	2	1
1–2%	1	3	1	3	2
2–3%	1	3	1	4	2
3–5%	1	4	1	4	2
5–10%	2	4	4	5	3
10–25%	2	5	5	6	3
25–33%	3	6	6	7	4
33–50%	3	7	7	7	4
50–75%	4	8	8	8	5
75–90%	5	9	9	9	6
90–95%	5	10	9	9	6
95–100%	5	10	10	10	6

evaluated. This step is necessary because the meaning of a taxonomic name can vary among treatments, and a taxon can have different names in different treatments owing to multiple, contrasting circumscriptions (see Franz *et al.* 2008; Jansen & Dengler 2010).

Each species in a plot is typically assigned a cover class value, and in many cases a cover class value is assigned specific to each stratum in which it occurs. Cover is the percentage of the earth surface covered by a vertical projection of the leaves, though typically small holes within a single individual's crown are ignored. Cover class is an ordinal variable, typically with 5–10 possible values. Numerous scales have been proposed (summarized in van der Maarel 1979, Dengler *et al.* 2008 and Jennings *et al.* 2009; see also Table 2.1). The most frequently employed cover index is the 1–5 scale of Braun-Blanquet (1928) or some variant of it. Almost as common are variants of the 1–10 Domin scale (1928), such as that of Krajina (1933), the UK National Vegetation Classification (Rodwell 2006), the New Zealand Survey (Allen 1992) and the Carolina Vegetation Survey (Peet *et al.* 1998). In selecting a cover scale, there are three important guidelines. First, it should be approximately logarithmic until at least 50% cover. This is because the human mind perceives cover in roughly a logarithmic way; we can perceive the difference between 1 and 2%, but not 51 and 52%, as the first pair represents a doubling while the second is a small relative increase. Second, the index should be replicable between observers to the level that almost always two observers will be within one value of each other. Third, it is highly desirable that the scale can be directly mapped onto the numeric units of the Braun-Blanquet scale to assure that data sets from diverse times and places can be integrated for at least some purposes.

2.5   Data preparation and integration

Once plot data have been collected, either in the field or from plot archives, it is necessary to integrate and standardize the data for analysis. This requires that inconsistencies in plot method, size and taxonomy be addressed in a consistent and well-reasoned manner, with each step recorded for archiving with the database when the project is completed.

2.5.1   Taxonomic integration

Construction of a taxonomically homogeneous data set can be challenging and typically requires investigator judgments on numerous inconsistencies. Because taxonomic adjustments will differ in their implications for data analysis, researchers should typically develop two data sets, one designed to address questions of species richness (species richness data set) and one designed to address questions where between-plot similarity must be assessed (analysis data set). In the species richness data set, all entities recorded as different species in a plot should be retained as distinct, regardless of the taxonomic resolution. In the analysis data set there should be a standard set of taxa used across all plots, and where taxa are inconsistently resolved they should usually be lumped together. If a small percentage of occurrences are reported only to the genus level, these taxon occurrences should be discarded from the analysis data set; if most occurrences are unknown one should lump them to the genus. Taxa not resolved to at least the genus level should be dropped from the analysis data set as such groups usually have little commonality in traits or distribution. If many taxa in a plot are not known to the species level, the plot should be dropped from the data set. The trickiest cases are where many observations are known to species and still a significant number are known only to genus. What if in a data set 70% of *Carex* occurrences are known to species and they span 20 taxa? Perhaps the numerous occurrences of *Carex* species should be dropped because they contain much less information than those identified to species, but the price is that there are missing records of shared taxa. Moreover, if one data source were consistently of lower resolution, there would be a signal attributable to a specific study.

Integrating taxon occurrences across data sets of mixed provenance presents greater uncertainty as to synonymy than does a single survey. Even when two occurrences are unambiguously assigned the same taxonomic name, it is still necessary to verify that the taxonomic concepts are equivalent. This is because one taxonomic name can have many meanings in terms of specific sets of specimens, and a certain set of specimens could have many different names

(Berendsohn *et al.* 2003; Jansen & Dengler 2010). Franz *et al.* (2008) describe the situation with the grass known as *Andropogon virginicus* in the *Flora of the Carolinas* (Radford *et al.* 1968), which when examined across eight taxonomic treatments reveals nine distinct sets of specimens variously arranged into 17 taxonomic concepts (combinations of the nine sets of specimens) and labelled with 27 scientific names. Thus, when plots from multiple sources report the presence of *Andropogon virginicus*, it is impossible to know how to combine them without knowledge of the taxonomic treatments the original authors followed, and even then there could easily be ambiguities requiring lumping to obtain unambiguous bins of taxon occurrences. The current situation in Europe serves to illustrate the mind-numbing complexity of integrating accurately and precisely across data sets. Schaminée *et al.* (2009) stated that in order to establish the TurboVeg-based joint European vegetation database SynBioSys Europe (Schaminée *et al.* 2007), 30 national species lists with 300,000 names had to be mapped against each other. This mapping is strictly one of synonymy and different applications of names are in many cases not accounted for, leaving many potential traps for the unwary data aggregator.

2.5.2 Plot data integration across data sets

Compared to taxonomic integration, merging other aspects of plot records is relatively straightforward, even if somewhat arbitrary. For the most part there are only three major impediments: inconsistencies in cover scales, plot size and definition of vertical strata.

To the extent that cover scales nest into a small number of bins, such as those of the Braun-Blanquet scale, it is easiest to simply condense the number of bins. Where this nesting approach is not possible, one can convert the cover scale value to an absolute cover value and then back to a new cover scale value. In doing so the reader is advised to convert to the geometric mean of the range rather than the arithmetic mean as species occurrences tend to occur disproportionately in the lower portion of each cover class. Where no such conversions seem reasonable, analyses should be conducted with simple presence-absence data. In fact, some authors have argued that there is more interpretable information in presence-absence data than cover data because the degree of absence of a species cannot be known or readily estimated (Lambert & Dale 1964; Smartt *et al.* 1976; Wilson 2012, but see Beals 1984; McCune 1994).

Combining plots of different sizes in a single database is at best problematic. The variable most sensitive to plot size is species richness, but a rough correction can be achieved by adjusting the species richness of plots that are at most within a doubling of the target size by use of the species–area relationship (see Fridley *et al.* 2005). More problematic and uncertain are the implications of plot size differences for calculation of similarity and designation of species constancy and indicator species. The reduction in species number with decreasing plot size of necessity decreases similarity to larger plots with more species, and constancy is similarly sensitive to plot size (Dengler *et al.* 2009). As a rule of thumb, all comparisons of richness should be made with plots of identical size, and all studies based on species similarity or constancy should be based on plots that do not range more than perhaps four-fold in area.

Most plot protocols call for recognition of vertical strata within a community, for which separate cover values are assigned for species. Unfortunately, these classes are not consistent between protocols. For example, the height cutoffs for strata can vary, and the actual definition of a stratum can vary from being based on the height of the individual plants (e.g. Mucina *et al.* 2000) to simple vertical bands of leaf area (e.g. Allen 1992). Vertical strata can be combined for purposes of data integration and Jennings *et al.* (2009) suggest that a simple probabilistic calculation of species total cover across strata (C_i) can be calculated as

$$C_i = \left[1 - \prod_{j=1}^{n} \left(1 - \frac{\%cov\ j}{100} \right) \right] \times 100$$

assuming the leaf area in each stratum (*%cov j*) is statistically independent of the other strata.

2.5.3 Sampling intensity

The distribution of plot frequencies in phytosociological databases is far from even. Some types of vegetation have hundreds or even thousands of plots, whereas others may be represented by only a small number. In some forms of analysis, the plots from the abundant types would dominate. Consequently, if we want an analysis to span the range of vegetation variation in the database, it may prove necessary to sample from the database in a stratified random fashion. Knollová *et al.* (2005) proposed several methods for stratifying phytosociological databases related to distribution along environmental or geographical axes, or relative to between-plot variation in species composition. Subsequently, a resampling method based on between-plot dissimilarity in species composition was proposed by De Cáceres *et al.* (2008). Lengyel *et al.* (2011) proposed a resampling method based on species composition where subsets of the database are selected randomly and the subsets with the lowest mean dissimilarity and lowest variance in similarity are retained for purposes of stratification.

2.6 Community entitation

In Section 2.3 we distinguished between classification as the creation of classes *versus* the assignment of objects to classes. In this section we address the creation of classes or types from an undifferentiated data set of vegetation plots or relevés, which we will refer to as entitation – the creating of entities. Assignment of new vegetation plots to existing classifications is discussed as determination in Section 2.9 below.

Vegetation scientists may have a broad range of ultimate objectives for classifying vegetation (see Section 2.1.1). From an operational perspective, however, the objective of vegetation classification is fairly simple – to create a set of vegetation types or syntaxa where (1) the types are mutually exclusive (no vegetation plot belongs to more than one type) and (2) the types are exhaustive (all vegetation plots are assigned to a type). Mathematicians call such a set of classes a 'partition': every object is a member of strictly one set, and every set has at least one member. Perhaps not surprisingly, there is an extremely large number of ways to produce such a partition. In general, methods of vegetation classification can be characterized as expert-based *versus* algorithmic, with the algorithmic methods divided into numerical *versus* combinatorial.

2.6.1   Classification by table sorting

Vegetation classification by sorting of phytosociological tables has a long history in vegetation ecology, with methodological monographs from Braun-Blanquet (1928), Ellenberg (1956) and Becking (1957), and with subsequent reviews by Westhoff & van der Maarel (1973) and Mueller-Dombois & Ellenberg (1974, chapter 9).

In table sorting methods, the data on species occurrence or abundance by plot are organized in a rectangular matrix with species as rows and plots as columns. The objective is to order the rows and columns of the tables to create a block-structured table where abundances for individual species are concentrated in adjacent columns of a row, and species with similar distributions are concentrated in adjacent rows so that plots of similar composition occur in proximity in the table. Based on successive re-ordering of the rows and columns, the table can be divided into sections or blocks of co-occurring species with the blocks arranged in a diagonal down the table. Vegetation plots that include one (or more) of these blocks are assigned to the same syntaxon, and species that compose a given block are considered diagnostic of the syntaxon in which they occur. The specific meaning of diagnostic has been the subject of considerable scientific development. Szafer & Pawlowski (1927), Becking (1957), Whittaker (1962), Westhoff & van der Maarel (1973) and Mueller-Dombois & Ellenberg (1974) distinguish 'character species' based on fidelity of occurrence within classes and 'differentiating (or differential) species' that are diagnostic in differentiating one class from another class while not necessarily being restricted to the focal class.

Table sorting by inspection was superseded many years ago by computer-aided approaches. The direct optimization of structured tables by iterative algorithms is difficult due to the extremely large number of possible solutions. The number of distinct table orderings is $n! \times m!$ where n is the number of plots and m is the number of species; even a simple table of 10 plots and 20 species has $10! \times 20! > 8.8 \times 10^{24}$ possible orderings. Developing efficient numerical approaches to producing sorted tables thus became an area of active research (Westhoff & van der Maarel 1973).

In the past decade the development of computer-based or computer-aided table sorting has received renewed attention motivated in part by the need to manage tables of truly enormous size, such as when combining multiple national classifications into European-wide classifications (Bruelheide & Chytrý 2000). Given the difficulty of direct optimization of large tables, most approaches have centred on statistical characterization of diagnostic species (see Section 2.8.2). Among the more notable advancements was the COCKTAIL algorithm for defining species groups developed by Bruelheide (2000). COCKTAIL starts with a preselected group of relevés or species and employs an iterative membership algorithm to refine the list of member species in each species group. Once no further candidate species are identified for membership in the type, a new type is begun from an alternative initial relevé or species group. Species fidelity to a type is based on the *u* statistic (see Section 2.8.3).

2.6.2   Numerical classification

The most common approach to vegetation classification is by numerical means. Typically this requires defining a similarity or dissimilarity matrix among all the vegetation plots, and then clustering the plots into types. In operation, it is a three-step process of (1) defining (dis)similarity, (2) choosing a clustering algorithm, and (3) choosing the number of clusters revealed or desired. All three decisions strongly affect the results and have to be made in concert.

Dissimilarity and distance. There is an extraordinary number of dissimilarity/ distance indices proposed or employed in vegetation ecology. Goodall (1973), Orlóci (1978), Hubálek (1982) and Legendre & Legendre (1998) all present comprehensive descriptions of indices that have been employed in community ecology; Mueller-Dombois & Ellenberg (1974), Kent (2012) and Ludwig & Reynolds (1988) emphasize shorter lists of commonly used indices. Confusingly, many indices have been independently derived and given more than one name. Other indices have a different name for the similarity index and its complement, the dissimilarity index, but vegetation ecologists often ignore the distinction and use the same name for both. Important distinctions among the indices concern the distinction between dissimilarity and distance and the use of presence/ absence versus abundance data.

   Dissimilarity and distance are similar concepts that characterize, on a quantitative scale, how different vegetation sample plots are from each other, but the mathematical bases of dissimilarity and distance are different. Dissimilarity is based on set theory and represents the ratio of the disjunct elements of two sets (belonging to one or the other but not both) to the union of the two sets. Plots that have no species in common have a dissimilarity of 1, and plots that are identical have a dissimilarity of 0, with all other possibilities scaled [0,1]. Distance is a geometric concept and represents the sum of all the pairwise differences in abundance for species which occur in one or both plots. Identical plots have a distance of 0. Plots with no species in common have a distance determined by species richness (for presence/absence indices) or standing crop (for quantitative indices) of the two plots; there is no upper bound. In practice a matrix is constructed with *n* rows and *n* columns for *n* vegetation plots where

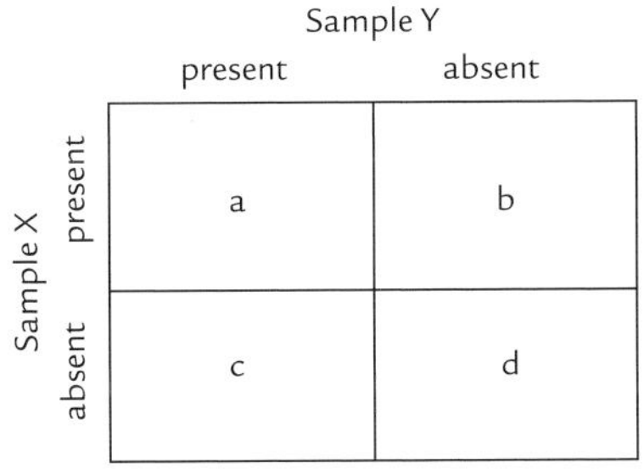

Sample Y

present   absent

Fig. 2.1 Contingency table notation for presence/absence dissimilarity indices.

each row or column in the matrix expresses the dissimilarity or distance of one vegetation plot to all the other plots. Both dissimilarity and distance follow a set of axioms:

1   $d_{ii} = 0$, reflexive property; the dissimilarity or distance from a plot to itself is zero;
2   $d_{ij} = d_{ji}$, symmetric property; dissimilarity is independent of direction.

These two axioms are generally true of all dissimilarities or distances employed in vegetation ecology. Some, but not all, indices meet a third axiom:

3   $d_{ik} \leq d_{ij} + d_{jk}$, i.e. the dissimilarity or distance of a plot to another plot is less than or equal to the sum of the distances involving any third plot.

The third axiom is called the triangle inequality property and does not hold for many dissimilarity indices. Indices that meet all three axioms are 'metric' and play a key role in analyses based in linear algebra.

Dissimilarity indices for presence/absence data often employ a 2×2 contingency table notation (Fig. 2.1). One of the earliest commonly used indices is the Steinhaus index (the complement of the Jaccard index): $(b + c)/(a + b + c)$; see Fig 2.1. This index can be viewed as the ratio of the number of species in one but not both plots to the pooled species list of the two plots. A commonly used alternative is the Marczewski index (the complement of the Sørensen index): $(b + c)/(2a + b + c)$, the ratio of the species in one but not both plots to the average number of species in the two plots. Both indices ignore d, the number of species in the data set that don't occur in either plot. Goodall (1973) and Legendre & Legendre (1998) argue strongly that ecologists should only use presence/absence dissimilarity indices that ignore joint absence (d).

For quantitative dissimilarity/distance indices, the abundance scale used can have a profound effect on the results. In vegetation ecology the scale is often not purely numeric (e.g. the Braun-Blanquet cover/abundance scale), and a lively debate has developed concerning the appropriate use of such data in quantitative analyses (Podani 2005; van der Maarel 2007). Nonetheless, most vegetation ecologists have adopted a pragmatic approach, and transform such scales to a numeric scale (see van der Maarel 1979; Noest *et al.* 1989). For scales with discrete classes of abundance, the widths of the intervals and the values chosen to represent each interval (often the interval midpoint but preferably the geometric mean of the endpoints) strongly affect the results. In general, linear abundance scales should be transformed to a convex scale (e.g. square root or log) that emphasizes differences for low values in the scale.

In addition to transformation, standardization of data can have a strong effect on results. Three standardizations are in common use in vegetation ecology: species maximum standardization, sample total standardization and Wisconsin double standardization. Species maximum standardization divides the abundance of each species in each plot by the maximum value observed for that species in all plots in the data set, giving all species an equal voice in the calculation of dissimilarity/distance, and thus strongly de-emphasizing differences in dominance among sample plots. This can be useful where indicator species (Section 2.8.2) exhibit low abundances and need increased weighting relevant to the dominants. However, this transformation can also increase the noise associated with rare species in the data and may work best where rare species are removed or down-weighted. In a comprehensive analysis of the performance of different dissimilarity indices on simulated data, Faith *et al.* (1987) found that a species maximum standardization improved the performance of most indices; however, their simulated data may have contained disproportionately few rare species.

Sample total standardization divides the abundance of each species in a plot by the sum of abundances for all species in that plot. This transformation treats total abundance for each plot as equal, eliminating differences in productivity or standing crop among samples. This standardization can be effective when data were collected in different years or seasons, by different parties, or measured on different scales. Sample total standardization plays an important role when using geometric distances, such as Euclidean or Manhattan distance (see Table 2.2). Geometric distances quantify the differences between plots without accounting for what the plots may have in common and can give a distorted perspective. A sample total standardization scales the differences relative to the total abundance and eliminates such problems. Some dissimilarity/distance indices (e.g. chord distance described later) have an inherent sample total standardization.

In Wisconsin double standardization, named for its use by Bray & Curtis (1957), data are first standardized by species maximum standardization and then by sample total standardization. Bray & Curtis's rationale for this sequence was that different life-forms (trees versus non-trees) were measured on different scales and the species maximum standardization achieved a common scale. The subsequent sample total standardization corrected for the fact that not all plots had the same number of measurements.

Commonly used quantitative dissimilarity indices in vegetation ecology include the Bray–Curtis index (Table 2.2). The Bray–Curtis index has been criticized for not being a true metric (Orlóci 1978). However, in comparative tests it has often performed extremely well (Faith *et al.* 1987). Alternatively, the

Table 2.2   Definitions of dissimilarity and distance; $d_{i,j}$ is the dissimilarity of plot i to plot j, $x_{i,k}$ is the abundance of species k in plot i for p species, $x_{i,+}$ is the sum of abundances for all species in plot i.

Index	Equation
Bray-Curtis	$d_{ij} = \sum_{k=1}^{p} \lvert x_{ik} - x_{jk} \rvert \Big/ \sum_{k=1}^{p} x_{ik} + x_{jk}$
Marczewski-Steinhaus	$d_{ij} = \sum_{k=1}^{p} \lvert x_{ik} - x_{jk} \rvert \Big/ \sum_{k=1}^{p} max(x_{ik}, x_{jk})$
Euclidean distance	$d_{ij} = \sqrt{\sum_{k=1}^{p} (x_{ik} - x_{jk})^2}$
Manhattan distance	$d_{ij} = \sum_{k=1}^{p} \lvert x_{ik} - x_{jk} \rvert$
Hellinger distance	$d_{ij} = \sqrt{\sum_{k=1}^{p} \left(\sqrt{\dfrac{x_{ik}}{x_{i+}}} - \sqrt{\dfrac{x_{jk}}{x_{j+}}} \right)^2}$
Chord distance	$d_{ij} = \sqrt{\sum_{k=1}^{p} \left(\dfrac{x_{ik}}{\sqrt{x_{i+}^2}} - \dfrac{x_{jk}}{\sqrt{x_{j+}^2}} \right)^2}$

Marczewski–Steinhaus index (Table 2.2) is similar to the Bray–Curtis index, but it is a true metric.

Geometric distances employed in vegetation ecology include Euclidean and Manhattan (or city-block) distance (Table 2.2). Euclidean distance is the common distance we use to measure the distance between objects in our three-dimensional world and seems quite intuitive. Due to its use of squared abundances, however, it is quite sensitive to the range of abundances in the data. Manhattan distance is named for its similarity to walking distances in a city where all distances occur along the principal axes and travel along the diagonals is not possible. Both Euclidean and Manhattan distance benefit from a sample total standardization. Legendre & Gallagher (2001) examined the behaviour of a number of dissimilarities and distances on artificial data and observed that Hellinger distance (Table 2.2) performed well at recovering ecological gradients. Hellinger distance can be viewed as the Euclidean distance of square root transformed sample total standardized data. Orlóci (1967, 1978) has demonstrated good results with chord distance (Table 2.2). Both Hellinger and chord distance use inherent sample total standardization.

Hierarchical agglomerative clustering. Hierarchical agglomerative clustering algorithms begin with each vegetation sample in its own 'cluster' and then iteratively fuse the least dissimilar clusters at each step. Ultimately, after $n - 1$ fusions (for n vegetation plots), all the plots are in a single cluster. The algorithms differ in how they define 'least dissimilar' for clusters with more than one member (Table 2.3). Over the years many algorithms have been proposed

Table 2.3 Hierarchical agglomerative clustering criteria; $d_{A,B}$ is the dissimilarity between cluster A and B, $d_{i,j}$ is the dissimilarity between plots i and j, $i \in A$ indicates plot i is a member of set A, $|A|$ is the number of members of cluster A, $\overline{d_A}$ is the mean coordinate on axis d for plots in cluster A, and Var d_k is the variance of dissimilarities formed in fusing cluster A with B.

Linkage	Equation
Single	$d_{A,B} = min\{d_{ij} : i \in A, j \in B\}$
Complete	$d_{A,B} = max\{d_{ij} : i \in A, j \in B\}$
Average	$d_{A,B} = \dfrac{1}{|A| \times |B|} \sum_{i \in A} \sum_{j \in B} d_{ij}$
Centroid	$d_{A,B} = \sqrt{\sum_{d=1}^{D} \left(\overline{d_A} - \overline{d_B}\right)^2}$
Ward's	$d_{A,B} = Var\, d_k : k \in A \cup B$

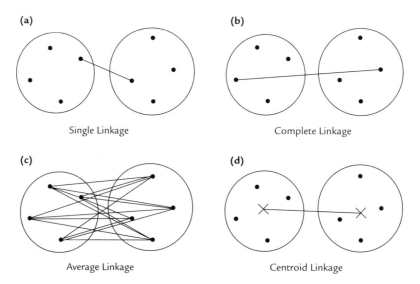

(a)

Single Linkage

(b)

Complete Linkage

(c)

Average Linkage

(d)

Centroid Linkage

Fig. 2.2 Hierarchical agglomerative algorithms differ specifically in how they define dissimilarity between clusters with more than one member.

and tested based in multidimensional geometry, graph theory, and information theory. We restrict our discussion to algorithms commonly used in vegetation ecology.

In single linkage (nearest neighbour) clustering, the dissimilarity of two clusters is the dissimilarity between the two least dissimilar members of the respective clusters (Fig. 2.2a). As clusters get larger, there are more members you could be least dissimilar to, and existing clusters have a tendency to grow at the expense of starting new clusters. This leads to the phenomenon of 'chaining' (Williams *et al.* 1966) where new vegetation plots are continually added to one large existing cluster. Due to the tendency to exhibit strong chaining, single linkage clustering is now rarely employed (Legendre & Legendre 1998; Podani 2000; McCune & Grace 2002).

In complete linkage clustering, the dissimilarity of two clusters is the dissimilarity between the two most dissimilar vegetation plots of the respective clusters (Fig. 2.2b). This approach emphasizes maximum rather than minimum dissimilarity among clusters. As clusters get larger, there are more members to be potentially maximally dissimilar to, and joining existing clusters gets harder. This leads to more numerous, equally-sized often spherical clusters. In both the single linkage and complete linkage algorithms, the dissimilarity between clusters is decided by a single dissimilarity and the algorithms operate at the plot-level rather than the cluster-level (Williams *et al.* 1966). The algorithms are, therefore, sensitive to unusual plots or outliers.

In the average linkage method (also called UPGMA or Unweighted Paired Group using Averages; Sokal & Sneath 1963), the dissimilarity is the average dissimilarity of each plot in each cluster to all the plots on the other cluster (Fig. 2.2c). Average linkage performs intermediate to single linkage and complete linkage, i.e. it is less prone to chaining than single linkage, but may form irregularly shaped clusters of varying size.

Ward's algorithm attempts to minimize the sums of squared distances from each plot to the centroid of its cluster (Fig. 2.2d), equivalent to variance minimization (Legendre & Legendre 1998). Beginning with every plot in its own cluster it fuses those clusters that result in the minimum increase in the sum of squared distances. Because it is based on a sum-of-squares criterion, the algorithm is most appropriately applied to a Euclidean distance matrix of plot dissimilarities (Legendre & Legendre 1998). However, many vegetation ecologists have been successful in applying Ward's algorithm to other dissimilarities such as Sørensen's (e.g. Wesche & von Wehrden 2011). Ward's algorithm tends to create compact spherical clusters where much of the variability in the dendrogram is compressed in the smaller clusters. This makes choosing relatively few large clusters rather easy, but sometimes hides considerable variability among the more numerous smaller clusters.

Lance & Williams (1967) realized that many of the existing hierarchical agglomerative algorithms could be generalized to a single algorithm with specific coefficients in the among-cluster distance equation. This algorithm is mostly known today as 'flexible-β' after one of the coefficients in the algorithm. Fig. 2.3d shows a flexible-β dendrogram with β set at the commonly employed value of −0.25. With this value (and suitable constraints on the other coefficients), flexible-β is intermediate to average linkage and complete linkage, and is generally recognized as a good compromise. By assigning increasingly negative values (e.g. −0.5) to β, the flexible-β algorithm more nearly approximates Ward's algorithm and provides an alternative that alleviates the concerns over requiring Euclidean distance.

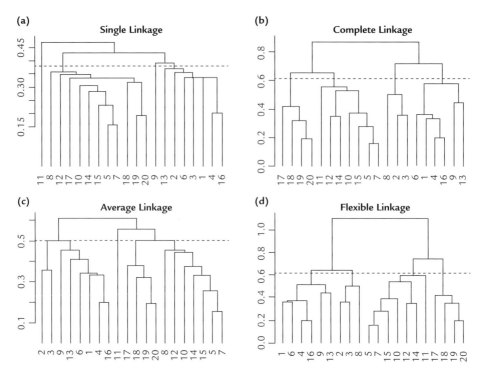

Fig. 2.3 Dendrograms for hierarchical agglomerative clustering algorithms based on the same dissimilarity matrix but using different linkages.

All the hierarchical agglomerative algorithms initially produce a dendrogram that portrays the sequence of fusions into clusters of the sample plots. Dendrograms are aggregated from the bottom up. Early fusions of clusters in the algorithm constrain later fusions, and in hierarchical clustering the assignment of plots to clusters is never re-evaluated. Consequently, the relatively few clusters produced near the top of the dendrogram may show considerable artefact in plot assignment. While highly informative, dendrograms can be visually misleading as plots that are adjacent to each other but attached to different 'branches' higher up may be quite dissimilar. An example is shown in Fig. 2.3 where the complete linkage and flexible-β algorithms produce what seem to be quite different dendrograms; re-ordering the plots along the horizontal axis would show that the solutions are very similar and the four cluster solutions are identical.

Dendrograms must be 'sliced' to generate clusters of plots on the same 'branch' and the question of where to slice is a critical issue. Given an *a priori* desired number of clusters you can solve for the height at which to slice. Fig. 2.3 shows all four dendrograms sliced to produce four clusters. Often, however, the correct or desired number of clusters is not known, and we are interested in finding natural breaks in the dendrogram where the results are relatively insensitive to the precise height at which we slice. In the example shown, natural breaks result in two or four clusters for complete linkage (Fig. 2.3b), two, three or five clusters for average linkage clustering (Fig. 2.3c) and two or three clusters for flexible-β (Fig. 2.3d). Further down in the dendrogram it is much more difficult to visually identify natural breaks and algorithmic approaches may be required.

Hierarchical divisive clustering. Hierarchical divisive clustering begins will all plots in a single cluster, which is then divided into two subclusters recursively until the clusters get too small or too homogeneous to subdivide according to criteria established by the user. Hierarchical divisive clustering algorithms are combinatorial, as opposed to numerical, and computing optimal results may be impossible. Accordingly, most divisive algorithms do not examine all possible solutions.

   Two divisive algorithms are currently used in vegetation ecology: Two Way Indicator Species Analysis (TWINSPAN) and Divisive Analysis Clustering (DIANA). TWINSPAN (Hill 1979) iteratively partitions the first axis of a correspondence analysis ordination (see Chapter 3). In practice the algorithm makes a number of *ad hoc* adjustments in choosing the exact point at which to partition at each step. Because TWINSPAN bifurcates each branch, the original algorithm always produces classifications where the number of classes is a power of two. Roleček *et al.* (2009) recently proposed a modification of the algorithm that employs a measure of cluster heterogeneity to determine which branches to split further. The result is a more natural classification with similar levels of cluster heterogeneity.

   Kaufman & Rousseeuw (1990) introduced a hierarchical divisive algorithm, DIANA, that operates on a dissimilarity matrix. At each iteration DIANA identifies the cluster with the largest diameter (maximum within-cluster dissimilarity, equivalent to the complete linkage criterion). Within that cluster the plot with the greatest average dissimilarity to all other plots in that cluster is identified and set aside as the seed for a 'splinter group'. All plots in the cluster that are more similar to the splinter group than the original cluster are then assigned to the splinter group, which forms a new cluster. Because DIANA is numerical, rather than combinatorial, it is fairly rapid but somewhat sensitive to outliers. Like hierarchical agglomerative algorithms, DIANA produces a dendrogram rather than clusters, and must be sliced to generate clusters. Because of the maximum diameter criterion, DIANA produces results most similar to complete linkage hierarchical agglomerative clustering.

Non-hierarchical partitioning algorithms. Non-hierarchical partitioning algorithms attempt to derive clusters from an undifferentiated set of vegetation plots directly without a hierarchical dendrogram. In contrast to hierarchical approaches the number of clusters must be specified in advance. Non-hierarchical partitioning of objects into types is mathematically difficult due to the extraordinary number of possible solutions. For example, to classify only ten vegetation plots into non-overlapping types there are 118515 possible solutions. To simplify finding good solutions to this problem, many non-hierarchical algorithms search for suitable 'seeds' to start each cluster and then assign each vegetation plot to the nearest seed. The original approach was called the k-means algorithm (Hartigan & Wong 1979), which minimized the sum of squared distances between points and the centroid of the cluster to which they were assigned. Modifications of the algorithm generally involve methods to choose the initial seeds and iteratively re-designate seeds. The k-means algorithm is strongly biased to create circular clusters of equal size rather than identifying natural discontinuities in the data. In addition, the algorithm is sensitive to the initial choice of seeds, and often requires multiple independent starts to ensure a good (although not necessarily optimal) solution.

Kaufman & Rousseeuw (1990) introduced a variation on k-means clustering called Partitioning Around Medoids (PAM). In the PAM algorithm, the seed for cluster formation (the medoid) represents an actual plot, called the representative object, rather than a geometric centroid. A deterministic algorithm selects the initial medoids, and because PAM does not require calculating centroids, it can operate on a broad range of dissimilarity indices other than Euclidean distance. Roberts (2010) defined two iterative non-hierarchical partitioning algorithms called OPTPART and OPTSIL. OPTPART iteratively reassigns plots to clusters to maximize the ratio of within-cluster similarity to among-cluster similarity. OPTSIL iteratively reassigns plots to clusters to maximize the similarity of a plot to its assigned cluster compared to the next most similar cluster (see Section 2.7.2 for more detail). Fuzzy clustering algorithms have also been proposed as an alternative to non-hierarchical algorithms wherein plots can have partial membership in multiple types (Equihua 1990; Podani 1990; De Cáceres *et al.* 2010a). These approaches recognize that not all plots are representative of a single type and sometimes are intermediate to clearly recognized types, but the resulting classification structure is more complex.

Non-hierarchical partitioning methods are subject to the requirement that the number of clusters to be solved must be specified in advance. They can also be slow to converge to a solution for some data sets. In practice, it is generally necessary to try multiple starts for a variety of cluster numbers and to compare the results to identify the best solution based on cluster validity statistics (Section 2.7), cluster characterization based on ancillary data (Section 2.8), or synthesis tables of the clusters. On the other hand, non-hierarchical partitioning algorithms generally are not subject to the artefact of fusion sequences constraining results because all plots are re-examined for best fit at each iteration.

2.7 Cluster assessment

The two objectives of assessing vegetation classes derived from any clustering method are to assure that (1) types are relatively homogeneous and distinct from other types, and (2) distributions of species within types exhibit high fidelity and ecologically interpretable patterns. Assessing the goodness of clustering ('cluster validity') is a vast field with a voluminous literature. Aho *et al.* (2008) present a recent review of cluster assessment methods for vegetation classifications. These authors distinguish geometric evaluators based on dissimilarity matrices versus non-geometric evaluators based on species distributions within clusters, often with a view to identifying diagnostic species. Some methods attempt to measure structure in a vegetation table directly.

2.7.1 Table-based methods

Feoli & Orlóci (1979) proposed a method termed Analysis of Concentration (AOC) to assess the structure of vegetation tables based on the density of non-zero values within species and sample blocks recognized by the vegetation ecologist. Blocks with high density (dominated by the presence of species in plots within the block) and blocks of low density (dominated by the absence of species in plots within the block) are compared to a random expectation by χ^2 analysis. Deviation from expectation is a direct measure of the degree of structuring of the table, and it is possible to scale the divergence to a relative scale of [0,1]. Many of the optimization criteria from iterative table-sorting algorithms (e.g. Podani & Feoli 1991; Bruelheide & Flintrop 1994) can be used to measure the quality of the final results even when that algorithm was not employed to define the classes. Generally these statistics are insensitive to the ordering of species or plots within blocks, but still measure cluster structure from the table.

2.7.2 Dissimilarity-based methods

Dissimilarity-based methods of cluster assessment operate on dissimilarity matrices, and can be applied whether numerical clustering was employed in defining the types or not. Aho *et al.* (2008) refer to these approaches as geometric evaluators and list five statistics useful in assessing goodness of clustering: Average Silhouette Width (Rousseeuw 1987), C-Index (Hubert & Levin 1976), Gamma (Goodman & Kruskal 1954), the PARTANA (PARtition ANAlysis) ratio (Roberts 2010, Aho *et al.* 2008), and Point Biserial Correlation (Brogden 1949). Two of these indices are highlighted below.

Rousseeuw (1987) defined silhouette width as a measure of the degree to which plots are more similar (less dissimilar) to the type to which they are assigned than to the most similar alternative type. Positive values indicate a good fit, and negative values indicate samples more similar to another cluster than to the cluster to which they are assigned. Thus, the quality of each cluster can be assessed by the mean silhouette widths of all plots assigned to that cluster and the number of negative silhouette widths, and the overall quality of the classification can be assessed by the global mean silhouette width and the number of negative silhouette widths. Silhouette width is a 'local' evaluator in the sense that each plot is only compared to the single other cluster to which it is least dissimilar regardless of the number of clusters. That comparative cluster may be different for every plot within a cluster. The PARTANA ratio, the dissimilarity-based statistic defined by Roberts (Roberts 2010; Aho *et al.* 2008), calculates the ratio of the mean similarity of plots within types to the mean similarity of plots among types. Good clusters have a high within-cluster similarity and low among-cluster similarities, and plots that fit well within their cluster have a higher mean similarity to their cluster than to other clusters. In contrast to sil-houette width, PARTANA is a global statistic that compares every cluster to every other cluster.

2.7.3 Indicator species methods

Statistical analysis of diagnostic or indicator species is often used as an evaluator of clustering effectiveness. The IndVal statistic (Dufrêne & Legendre 1997, and as modified by Podani & Csányi 2010; see Section 2.8.4) and the OptimClass approach of Tichý *et al.* (2010) have both been effectively used in selecting 'optimal' solutions from competing alternative classifications. However, as the identification of diagnostic and indicator species is of significant interest in com-munity characterization, it is treated in Section 2.8.

2.8 Community characterization

Once a set of types has been developed, it is desirable to develop concise rep-resentations of the compositional and ecological characteristics of the types. The data often represent a large number of species and plots, as well as possible environmental attributes, and efficient summaries are required for effective communication.

2.8.1 Synoptic tables

One common and simple approach is to produce a synoptic table for the types recognized with species as rows, types as columns, and values of frequency, mean abundance, or preferably both, for each species in each type entered into the table. In US vegetation classifications such tables are often called constancy/abundance tables. Similarly to the more expansive structured tables described in Section 2.7.1, the species (table rows) are often ordered to highlight the diag-nostic species of the types. In large data sets with numerous types, even the synoptic tables can get quite large and unwieldy.

2.8.2 Diagnostic and indicator species

Deriving statistical indices of diagnostic or indicator species has been an area of significant activity in the past decade. Here we distinguish two groups of ap-proaches: probabilistic versus composite. In general the probabilistic approaches calculate the 'fidelity' of species to types or clusters based on presence/absence data and evaluate the deviation of species occurrence within types from a random distribution of taxa. Generally, each type or class is considered indi-vidually against all the other types pooled. The composite approach combines fidelity and the distribution of a species' abundance across types to create a single index. Because the null distribution of this index is not known, the deviation from expectation for the index values has to be estimated by permutation techniques.

Juhász-Nagy (1964) in De Cáceres *et al.* (2008) described three aspects of species fidelity that influence the indices in use today: Type I – the occurrence of a species typically only within a vegetation type, although it may not occur in all (or even most) plots within the type; Type II – the commonness or ubiquity of a species within a type although the species may be widespread outside the type; Type III – joint fidelity where a species occurs primarily within a single type and occurs in all (or most) plots within that type. The first case we might call 'sufficient' in that the occurrence of that species is sufficient to indicate the type, the second case we might call 'necessary' in that if you are in that type you should see that species, and the third case we might call necessary and sufficient.

2.8.3 Probabilistic indices of species fidelity

The general approach to probabilistic identification of diagnostic species is to calculate an index of concentration, and then the probability of obtaining as high or higher a concentration of a species within a given type as is observed. For simplicity, these indices are generally calculated on presence/absence data and concentration is calculated as number of occurrences (though see Willner *et al.* 2009). The most common approach is to produce a 2×2 contingency table of occurrences of a species in a type and calculate the Φ index (Sokal & Rohlf 1995: 741, 743). Following notation established by Bruelheide (2000), the analysis is as follows:

N = total number of sites
N_p = number of sites in type of interest
n = number of occurrences of species of interest
n_p = number of occurrences of species in type

$$\Phi = \frac{N \times n_p - n \times N_p}{\sqrt{n \times N_p \times (N - n) \times (N - N_p)}}$$

Φ takes values in [−1, 1], reflecting perfect avoidance to perfect concordance of the species in the type. The statistical significance of the index can be calculated from Fisher's exact test. Bruelheide (2000) proposed that for species that occurred more than ten times a normal distribution approximation could be used, calculating

$$u = \frac{n_p - \mu}{\sqrt{n \times N_p / N \times (1 - N_p / N)}}$$

dividing the observed number of occurrences n_p minus the expected number of occurrences ($\mu = n \times N_p / N$) by the standard deviation of the binomial, preferably after applying a continuity correction to the numerator. Chytrý *et al.* (2002)

preferred to divide by the standard deviation of a hypergeometric random variable and called the resulting value u_{hyp}.

$$u_{hyp} = \frac{n_p - \mu}{\sqrt{n \times N_p \times (N - n) \times (N - N_p)/(N^2 \times (N - 1))}}$$

In either case the index of fidelity is scaled in units of standard deviation from expectation, rather than [−1, 1]. As we are primarily interested in positive values of the index, a one-tailed test of significance can be performed on the index.

Chytrý *et al.* (2002) compared a range of statistical indices (including Φ, u and u_{hyp}) for use in identifying diagnostic species on a classified data set from dry grasslands in the Czech Republic. Rankings achieved by the probabilistic indices were very similar, although correction for continuity tended to reduce the values for rare species. Tichý & Chytrý (2006) argued that fidelity indices such as Φ are sensitive to variability in the size of types or clusters, and proposed a modification of the Φ coefficient that normalizes cluster size. The number of occurrences for a species and the number of occurrences within the type of interest are rescaled to a constant cluster size while maintaining the ratio of within-type to out-of-type occurrences. The new equalized Φ values are comparable across clusters of different sizes. By adjusting the size of the normalized cluster relative to the total number of plots, the index can be made more or less sensitive to rare species relative to more common species. Normalized Φ values are not appropriate for testing statistical significance, so significance testing should occur before normalizing. Alternatively, the data can be subsampled to equal cluster sizes before the analysis (see Section 2.5.3).

De Cáceres *et al.* (2008, 2009) present a detailed discussion of the importance of context in identifying diagnostic and differential species. Willner *et al.* (2009) studied a range of fidelity indices on real data and found that differences in context were more important than the use of different indices of fidelity in identifying diagnostic species. The range of other vegetation types considered strongly influences the determination of species values. Approaches that compare the presence of species within types to outside the type can find character species with high fidelity, but miss many differential species that are not globally differential. A solution proposed long ago by Goodall (1953) is to compare the distribution of species within types to the type where the species is next most common.

Most of the numeric approaches to identifying indicator species focus on species with high fidelity as opposed to differential species. Tsiripidis *et al.* (2009) developed a method based on taxon relative constancy within types to identify differential species directly. While the algorithm is somewhat *ad hoc*, it proved successful when applied to both simulated and actual data, and is logically related to thresholds used in more classical phytosociology. Alternatively, a statistical numeric approach to identifying differential species is to use classification trees (Breiman *et al.* 1984) or random forest classifiers (Breiman 2001) on the plot-level compositional data to identify species useful in predicting the membership of plots in types (see Section 2.9.2).

2.8.4 Composite indices

The most widely employed statistic for identifying diagnostic species in a classification is Dufrêne & Legendre's (1997) IndVal statistic. Using the notation introduced by Bruelheide (2000; see Section 2.8.3)

$$IV_{ip} = A_{ip} \times B_{ip} \times 100 \quad \text{where} \quad A_{ip} = \frac{\sum_{j \in p} a_{ij}/N_p}{\sum_c^C \sum_{j \in c} a_{ij}/N_c}; B_{ip} = \frac{n_p}{N_p}$$

where IV_{ip} is the indicator value of species i to cluster p, a_{ij} is the abundance of species i in plot j, c is a cluster from one to C clusters, and N_c is the number of plots in cluster c.

The first term (A_{ip}) is the average abundance of the species in plots in the cluster of interest divided by the sum of the average abundances in all clusters. Calculating the sum of averages is an unusual calculation, but in this case it makes the relative abundances independent of cluster size. The second term (B_{ip}) is simply the relative frequency of the species in the cluster (Type II fidelity, as given earlier).

To achieve a maximum indicator value a species must occur in every plot assigned to that type and no plots outside the type. Species that are restricted to a single type, but which occur in only a subset of the plots assigned to that type, are given an indicator value equal to their frequency; species that occur in every plot of the type, but which also occur in other types, are assigned an indicator value proportional to their relative average abundance within the type. The values are tested for statistical significance by permutation. The IndVal statistic attempts to find species that are both necessary and sufficient (i.e. if you see the species you should be in the indicated type, and if you are in the indicated type the species should be present). As a comparative metric of overall classification efficacy, Dufrêne & Legendre (1997) proposed summing the statistically significant indicator values across species, or alternatively counting the number of significant indicator species and choosing the partition that maximizes the statistic.

The dual requirements that indicator species have high frequency in the indicated type and low abundance outside the type bias the IndVal statistic in favour of species that occur in the data at a frequency approximately equal to the mean cluster size. However, widespread species can have compact, ecologically informative distributions occurring with high fidelity in pooled types that are adjacent along gradients. De Cáceres *et al.* (2010b) developed a modified IndVal statistic that pools types into all possible larger groups and calculates the IndVal statistic (as well as the point biserial correlation) for those groups. Species with wider niche breadths could, thus, be recognized as indicative of a union of possibly several types.

Podani & Csányi (2010) noted that the first term of IndVal (A_{ip}) is independent of the number of types being considered and represents concentration as opposed to specificity. They argued that specificity should consider how many types are in the data set and proposed a modification comparing the difference of the average abundance of a species in the type minus its average abundance in all other types, normalized by the maximum average abundance for the species in any type. This has the effect of changing the scale of indicator value from [0,1] to [−1,1], where species have negative specificity to types where their average abundance is less than their average abundance in all types. Ecologists have argued for years about whether or not the lack of species can be diagnostic, but Podani & Csányi note their proposed index is consistent with the position of Juhász-Nagy (1964) that the absence of a ubiquitous species can be indicative.

2.9   Community determination

Determination is the assignment of a plot to an existing type based on comparison with the typical composition of candidate types. Determination may be absolute (or crisp) where the plot is assigned to only a single type, or fuzzy where the plot is given grades of membership in multiple types (De Cáceres *et al.* 2009, 2010a). The USNVC and VegBank allow five possible levels of determination: Absolutely Wrong, Understandable but Wrong, Reasonable or Acceptable Answer, Good Answer, and Absolutely Right (Gopal & Woodcock 1994). Alternatively, fuzzy set theory can be employed, where plots are assigned memberships in types in the range [0,1], typically where the sum of all memberships must equal one. Van Tongeren *et al.* (2008) rank the potential types in order of fit from 1 to 10 whereas De Cáceres *et al.* (2009) noted first and second best fit. In a manner similar to entitation, determination can be based on either actual compositional data or on (dis)similarities calculated among plots, or both.

Developing numerical or combinatorial approaches to correct plot assignment is exceedingly difficult. For large data sets, the number of plots and the number of types is large and the dimensionality of the problem is typically very high. However, given the importance of developing comprehensive vegetation classifications, efforts to perfect such algorithms will certainly be given high priority by vegetation and computer scientists.

2.9.1   Expert-based approaches

Type membership for plots is often determined by expert opinion. Experienced vegetation ecologists employ an understanding of data context and intuitive species weighting in selecting the appropriate type for a plot. Often when numeric approaches are used, the results are validated using determinations by experts (treated as 'truth'). However, as noted by van Tongeren *et al.* (2008), mistaken determination by experts is a source of error unaccounted for in tests of numeric methods. Perhaps more importantly, as noted by Gégout & Coudun (2012), given the size of the task of producing national or regional classifications, there simply aren't enough experts to accomplish the task.

2.9.2 Dichotomous keys

Dichotomous keys are extremely useful tools for field determination of new plots or relevés, as long as the list of possible types is not too long. Automated procedures for generating dichotomous keys are available using classification trees (Breiman *et al.* 1984) or random forest classifiers (Breiman 2001) on plot-level compositional data. However, given the stochastic nature of species distributions, dichotomous keys are limited by using the abundance of a single species (or a few pooled species) at each decision point, rather than a more synthetic perspective. In addition, dichotomous keys (and the differential species identified by them) are limited by context. If a type is widespread, then the differential species may vary by region, and application of a dichotomous key outside the region where the calibration plots were collected may prove highly error-prone. Keys must be recognized as useful but fallible tools for narrowing down the list of candidate types (Pfister *et al.* 1977; Rodwell 2006). Users must still compare the composition and environmental attributes of the indicated type and similar types to make a clear determination.

2.9.3 Numeric approaches

Ĉerná & Chytrý (2005) employed the Φ index (see Section 2.8.3) in an application of neural nets (multilayer perceptron) to predict plot membership in 11 *a priori* alliances for 4186 relevés of Czech grasslands. The neural net was fit to a subset of the relevés (the training set), limited from over-fitting by another subset of relevés (the selection set) and tested on a third set of relevés (the test set). When the training data set was randomly selected from the pool of relevés, the neural net obtained from 80.1 to 83.0% correct assignment of the test data. Surprisingly, when the training data were selected by emphasizing relevés with high numbers of diagnostic species, the accuracy declined to 77.0–79.6%. Ĉerná & Chytrý regard the use of neural nets for plot assignment as promising, but note that the model is essentially a black box and does not produce keys useful for field application.

Gégout & Coudun (2012) also employed the Φ index (see Section 2.8.3) to develop a model for assigning plots to pre-existing types. Φ was calculated for every species in every type using the data from the original (calibration) plots. Then the fidelity of a plot to a type (F_{ij}) was calculated as the mean Φ for all species in the plot to that type.

$$F_{ij} = \sum_{k=1}^{n} \Phi_{kj} / \mathrm{n}$$

This fidelity was compared to the mean fidelity of all plots used to define that type.

$$A_{ij} = \left(F_{ij} - \bar{F}_j \right) / s\left(F_j \right)$$

where A_{ij} = the affinity of plot i to type j, $\bar{F}_j$ is the mean fidelity for all plots in type j, and $s(F_j)$ is the standard deviation of F_j. Plots were assigned to the type for which they had the highest affinity. There was a 60% agreement of assignment to type compared to assignment by phytosociological experts on the calibration plots. For 800 plots independent of those used to define the types, agreement with expert assignment dropped to 47%.

Van Tongeren *et al.* (2008) developed a numerical determination approach called ASSOCIA based on a composite index combining presence/absence data and abundance data using weighted averaging. For the presence/absence data the deviance (–2 ln(likelihood)) associated with plot membership of a plot to a type is calculated for all possible types. For the abundance data a modified Euclidean distance is calculated from the plot to the centroid of all types. This approach has the significant advantage that it can employ synoptic tables, as opposed to full plot-level data, thus allowing comparisons to published classifications where the raw data are not available.

De Cáceres *et al.* (2009, 2010a) explored fuzzy approaches to determination. While the fuzzy classifiers performed well in general, they proved susceptible to poorly defined types in the set of possible choices, and differed in their response to outliers as opposed to intermediate plots.

2.10 Classification integration

With the growing importance of large, comprehensive classification systems such as that of the Braun-Blanquet system and the USNVC, it is critical that new classification work be integrated into a broader framework. Additionally, existing classifications need to be reconciled to achieve a consistent, comprehensive system (Bruelheide & Chytrý 2000; De Cáceres & Wiser 2012). There are significant challenges to achieving such integration. There are four components to managing classifications (De Cáceres *et al.* 2010a): (1) assigning new relevé data into existing types; (2) updating the types to reflect the additional data; (3) defining new types for plots that don't fit the current classification; (4) reconciling and validating the modified classifications. De Cáceres & Wiser (2012) provide guidelines to ensure that the products of vegetation classification efforts can be integrated into broader classification frameworks, modified and extended in the future, and can be used to communicate information about vegetation stands beyond those included in the original analysis.

Here is a simple overview of the problem of integration. If all of the vegetation plots that are characteristic of a classification unit under one system (say A) would be assigned to a single classification unit of another system (say B), we will say that the relationship (or mapping) is one-to-one. If, however, plots that define a type in classification A would be assigned to more than one type in classification B, we will say the mapping is one-to-many. If the mapping is one-to-many in both directions, then the classifications are significantly different and reconciliation will be difficult for the same reasons as mapping of taxonomic concepts for plants is challenging.

2.10.1   Classification resolution

Most vegetation classifications are hierarchical with lower levels nested into broader types. It makes sense to begin the discussion of classification integration at the lowest practical level, the association, as upper levels are often defined in terms of their component lower units (but see the USNVC for a combined bottom-up and top-down system). We refer to the heterogeneity of vegetation within an association (how finely divided into types the vegetation is) as classification resolution. If classifications to be reconciled differ significantly in resolution, then a one-to-one correspondence cannot be established. The best case is that in one direction the mapping is one-to-many and in the reverse direction it is one-to-one; in this case one-to-one mapping may still be achieved by lumping the more finely resolved types or splitting the more coarsely resolved types. Given a standard definition for intra-association heterogeneity, this would be a simple decision. However, no standard currently exists (although Mueller-Dombois & Ellenberg 1974 suggested that all plots within an association should have a Jaccard's similarity index of at least 25% to the typal plot). The variability in association resolution across classifications could be used to guide this decision.

2.10.2   Classification alignment, precedence and continuity

Even given similar levels of classification resolution between two adjoining classifications, it is likely that the classifications will still exhibit one-to-many relationships in both directions. Recurrent patterns of vegetation composition (associations) are determined in part by the pattern of landscapes acting on the regional species pool (Austin & Smith 1989). In an adjoining region, differences in these landscape patterns may create different recurring community patterns from the same species pool. In these cases it may be necessary to pool the plot data from both areas and seek new associations that better represent the larger-scaled pattern of community composition and distribution. Similarly, a detailed study of a narrowly circumscribed geographic region (perhaps a national park) may yield an intuitively very satisfying classification that does not map well onto a geographically broad classification (say one for all of Europe or the USA). In these cases is will be necessary to be cautious in proposing changes in the larger-scale classifications so as to avoid disharmonies in application of the classification in other regions.

Vegetation classifications represent significant scientific achievements often accomplished by a large number of people over a long period of time. Much of the utility of the classification, however, is tied to the information content of the classes. Often important ancillary information on productivity, animal habitat suitability, conservation priority and hazards are associated with each unit in the classification by accumulated experience or specific monitoring or research programmes. Maps of classification unit distribution may feature prominently in land-management activities. Significant revisions of existing classifications run the risk of making such information obsolete. Accordingly, while new methods or new data or the desire to reconcile with adjacent areas sometimes lead to revised classifications, this should be done sparingly. At a minimum, considerable effort should be given to documenting the mapping from old types to new (Section 2.10.3).

2.10.3   Cross-referencing classifications

An alternative approach to aggregating classifications into new systems is to develop a formal cross-referencing system that identifies synonymy among classifications. One approach is a set theoretic system that follows the international standard for taxonomic mapping (TDWG 2005) in defining the relationship of each type in one classification with each type in another as: (1) is congruent, (2) is contained in, (3) contains, (4) intersects with, or (5) is disjunct from, as is implemented for community classification in the VegBank archive (Peet *et al.* 2012). By knowing the relationship of a type in one classification to all types in another classification it is possible to erect higher-order relationships by network algorithms. Such an approach preserves the ancillary information associated with types in legacy classifications and minimizes unnecessary dynamics in the larger classification enterprise. On the other hand, it imposes additional complexity on regional efforts.

2.10.4   Nomenclature

Each of the major vegetation classification systems has its own nomenclatural rules. The best established and most detailed is the International Code of Phytosociological Nomenclature, which applies to units in the Braun-Blanquet system (Weber *et al.* 2000) and is maintained by the International Association for Vegetation Science. This system is modelled after the nomenclature rules for plant taxa (Dengler *et al.* 2008). Among several names for a syntaxon, the oldest validly published name has priority, and each syntaxon name is connected to a nomenclatural type (a single plot for associations, or a validly described lower-rank syntaxon in the case of a higher syntaxa), which determines the usage of the name. Syntaxon names are based on the scientific names of one or two plant species or infraspecific taxa that usually are characteristic in the particular vegetation type. An 'author citation' (i.e. the author(s) and year of the first valid publication) also forms part of the complete syntaxon name.

   The USNVC (http://usnvc.org) has less formal naming rules. Each association and alliance is assigned a scientific name based on the names of plant species that occur in the type (Jennings *et al.* 2009). Dominant and diagnostic taxa are used in naming a type and are derived from the tabular summaries of the type. The number of species names in the name can vary from one to five, with those predominantly in the same stratum separated by a hyphen (-), and those predominantly in different strata separated by a slash (/). Association or alliance names include the term Association or Alliance as part of the name to indicate the level of the type in the hierarchy, as well as a descriptive physiognomic term, such as forest or grassland.

2.11   Documentation

2.11.1   Publication

Publication is critical for disseminating the results of vegetation classification research, though it plays different roles in different classification systems. In the Braun-Blanquet system, vegetation types are defined in publications, much as species are. Typically, these publications contain synoptic tables with species as rows and communities as columns. For classification publications constructed outside the framework of the Braun-Blanquet system, tabular summaries are still important, but less emphasis is placed on sorting or identification of diagnostic species. More typically, the most characteristic species are indicated. One effective manner of doing this is by including only the prevalent species, defined as the '*n*' most frequent species, where '*n*' is the average number of species per plot (Curtis 1959). In addition, it is common to flag the species with high indicator value as defined by some standard metric, such as that of Dufrêne & Legendre (1997).

2.11.2   Plot archives

With the advent of inexpensive digital archiving of data and widespread access to digital archives over the web, there is a growing expectation that key original data will be made available in permanent public archives (Jones *et al.* 2006; Vision 2010). As a consequence, analyses can now be redone with expanded data sets or with different methodologies, and new questions can be asked through use of large quantities of available data. This trend toward archiving original data is particularly important for vegetation classification initiatives. Large national and multinational classifications need to evolve, and this is only possible if plots records are permanently archived, much like systematics depends on museum collections that have been examined and determined by a series of monographers. The USNVC now requires that plot data used to advance the classification be made available in public archives. Already in excess of 2.4 million vegetation plots are reported in the Global Index of Vegetation Databases (GIVD; Dengler *et al.* 2011), a significant proportion of which is publicly available.

   Key to efficient reuse of data is that the records conform to some standardized format. The widespread use of TurboVeg (Hennekens & Schaminée 2001) as a database for plots consistent with the Braun-Blanquet approach has meant that millions of plots can be exchanged in an efficient manner. However, TurboVeg supports only a limited range of plot types and formats. To solve this problem, Veg-X has been proposed as an international data exchange standard for vegetation plots of nearly all formats (Wiser *et al.* 2011). Widespread application of the Veg-X format would greatly simplify both sharing of data and ease of application of software tools.

2.12 Future directions and challenges

Given the pressing need for documenting and monitoring the Earth's biodiversity and for providing context for broader ecological research, vegetation classification has received increasing attention in recent decades in both academic ecology and across a broad range of user communities. This new and broader set of applications also suggests that we need to move beyond individual and idiosyncratic classifications toward large, consensus classifications that combine the effort of many persons to produce and maintain a unified and comprehensive whole, subject to revision in an open and transparent manner. Toward this end, individual workers should conform to established standards for collecting and archiving plot data. Not only will this significantly advance vegetation classification, but it will also facilitate future international collaboration and synthesis.

Computer databases and numerical approaches will become increasingly important for developing large consensus classifications. While a single preferred protocol is unlikely to emerge, increased testing of competing approaches on large regional or national classifications should provide insights into the task-specific utility of each approach. Transparent algorithms should be strongly preferred, although the specific nature of vegetation research means that special-purpose software may still be required. As emphasized by De Cáceres & Wiser (2012), formal rules for assigning plot data to specific types will play an increasingly important role.

Vegetation scientists need access to the data used in vegetation classifications. Numerous plot databases currently exist (Dengler *et al.* 2011), and progress is being made on data transfer protocols that will facilitate access to and utility of such data (Wiser *et al.* 2011). The development of better tools for managing and analysing the massive vegetation data sets anticipated in future classification efforts is an area of active research and development.

The greatest future challenge may be integrating the numerous existing classifications into a comprehensive system. The USNVC includes a peer review protocol for modifying the classification. Ironically, the USA may benefit in this effort from the historical lack of emphasis on vegetation classification in North America, beginning from almost a clean slate. The long legacy of vegetation classification in Europe means that many more vegetation types are formally recognized. Thus, reconciliation of existing classifications will play a much larger role in Europe than in the USA.

Vegetation is complex and dynamic and efforts to characterize it in a formal structure are inherently problematic. Nonetheless, identifying those problem areas focuses the efforts of vegetation science into new research areas of interest to a broad range of scientists in complexity science, database design, multivariate analysis, expert systems and many other fields.

References

Aho, K., Roberts, D.W. & Weaver, T. (2008) Using geometric and non-geometric internal evaluators to compare eight vegetation classification methods. *Journal of Vegetation Science* **19**, 549–562.

Allen, R.B. (1992) RECCE: an inventory method for describing New Zealand's vegetation cover. *Forestry Research Institute Bulletin* 176. FRI, Christchurch, New Zealand.

Anderson, M., Bourgeron, P., Bryer, M.T. *et al.* (1998) *International Classification of Ecological Communities: Terrestrial Vegetation of the United States. Volume II. The National Vegetation Classification System: List of Types.* The Nature Conservancy, Arlington, VA.

Austin, M.P. & Heyligers, P.C. (1989) Vegetation survey design for conservation: Gradsect sampling of forests in Northeastern New South Wales. *Biological Conservation* **50**, 13–32.

Austin, M.P. & Heyligers, P.C. (1992) New approaches to vegetation survey design: Gradsect sampling. In: *Nature Conservation: Cost Effective Biological Surveys and Data Analysis*, (eds C.R. Margules & M.P. Austin), pp. 31–37. CSIRO, Melbourne, VIC.

Austin, M.P. & Smith, T.M. (1989) A new model for the continuum concept. *Vegetatio* **83**, 35–47.

Bailey, R.G. (1976) *Ecoregions of the United States* (map). U.S. Forest Service, Intermountain Region, Ogden, UT.

Beals, E.W. (1984) Bray-Curtis ordination: an effective strategy for analysis of multivariate ecological data. *Advances in Ecological Research* **14**, 1–55.

Bechtold, W.A. & Patterson, P.L. (2004) The enhanced Forest Inventory and Analysis Program – national sampling design and estimation procedures. *General Technical Report SRS-80*. U.S. Department of Agriculture Forest Service, Southern Research Station, Asheville, NC.

Becking, R.W. (1957) The Zürich-Montpellier school of phytosociology. *Botanical Reviews* **23**, 411–488.

Berendsohn, W.G., Döring, M., Geoffroy, M. *et al.* (2003) The Berlin Model: a concept-based taxonomic information model. In: *MoReTax – Handling Factual Information Linked to Taxonomic Concepts in Biology* [*Schriftenreihe für Vegetationskunde* 39] (ed. W.G. Berendsohn), pp. 15–42. Federal Agency for Nature Conservation, Bonn, DE.

Braun-Blanquet, J. (1928) *Pflanzensoziologie: Gründzuge der Vegetationskunde.* Springer-Verlag, Berlin.

Braun-Blanquet, J. (1964) *Pflanzensoziologie.* 3rd ed. Springer-Verlag, Berlin.

Bray, J.R. & Curtis, J.T. (1957) An ordination of the upland forest communities of southern Wisconsin. *Ecological Monographs* **27**, 326–349.

Breiman, L. (2001) Random forests. *Machine Learning* **45**, 532.

Breiman, L., Friedman, J., Olshen, R.A. & Stone, C.J. (1984) *Classification and Regression Trees.* Wadsworth & Brooks, Monterey, CA.

Brogden, H.E. (1949) A new coefficient: application to biserial correlation and to estimation of selective efficiency. *Psychometrica* **14**, 169–182.

Bruelheide, H. (2000) A new measure of fidelity and its application to defining species groups. *Journal of Vegetation Science* **11**, 167–178.

Bruelheide, H. & Chytrý, M. (2000) Towards unification of national vegetation classifications: A comparison of two methods for analysis of large data sets. *Journal of Vegetation Science* **11**, 295–306.

Bruelheide, H. & Flintrop, T. (1994) Arranging phytosociological tables by species-relevé groups. *Journal of Vegetation Science* **5**, 311–316.

Černá, L. & Chytrý, M. (2005) Supervised classification of plant communities with artificial neural networks. *Journal of Vegetation Science* **16**, 407–414.

Chytrý M. & Otýpková Z. (2003) Plot sizes used for phytosociological sampling of European vegetation. *Journal of Vegetation Science* **14**, 563–570.

Chytrý, M., Tichý, M., Holt, J. & Botta-Dukát, Z. (2002) Determination of diagnostic species with statistical fidelity measures. *Journal of Vegetation Science* **13**, 79–90.

Cowardin, L.M., Carter, V., Golet, F.C. & LaRoe, E.T. (1979) *Classification of the Wetlands and Deepwater Habitats of the United States.* U.S. Fish and Wildlife Service, Washington, DC.

Curtis, J.T. (1959) *Vegetation of Wisconsin.* University of Wisconsin Press, Madison, WI.

De Cáceres, M., Font, X. & Oliva, F. (2008) Assessing species diagnostic value in large data sets: a comparison between phi coefficient and Ochiai index. *Journal of Vegetation Science* **19**, 779–788.

De Cáceres, M., Font, X., Vicente, P. & Oliva, F. (2009) Numerical reproduction of traditional classifications and automated vegetation identification. *Journal of Vegetation Science* **20**, 620–628.

De Cáceres, M., Font, X. & Oliva, F. (2010a) The management of vegetation classifications with fuzzy clustering. *Journal of Vegetation Science* **21**, 1138–1151.

De Cáceres, M., Legendre, P. & Moretti, M. (2010b) Improving indicator species analysis by combining groups of sites. *Oikos* **119**, 1674–1684.

De Cáceres, M. & Wiser, S. (2012) Towards consistency in vegetation classification. *Journal of Vegetation Science* **23**, 387–393.

Dengler, J. (2009) A flexible multi-scale approach for standardised recording of plant species richness patterns. *Ecological Indicators* **9**, 1169–1178.

Dengler, J., Chytrý, M. & Ewald, J. (2008) *Phytosociology*. In: *Encyclopedia of Ecology* (eds S.E. Jørgensen & B.D. Fath), pp. 2767–2779. Elsevier, Oxford.

Dengler, J., Löbel, S. & Dolnik, C. (2009) Species constancy depends on plot size – a problem for vegetation classification and how it can be solved. *Journal of Vegetation Science* **20**, 754–766.

Dengler, J., Jansen, F., Glöckler, F. *et al.* (2011) The Global Index of Vegetation-Plot Databases: a new resource for vegetation science. *Journal of Vegetation Science* **22**, 582–597.

Domin, K. (1928) The relations of the Tatra mountain vegetation to the edaphic factors of the habitat: a synecological study. *Acta Botanica Bohemica* **6/7**, 133–164.

Dufrêne, M. & Legendre, P. (1997) Species assemblages and indicator species: the need for a flexible asymmetrical approach. *Ecological Monographs* **67**, 345–366.

Ellenberg, H. (1956) *Grundlagen der Vegetationsgliederung. 1. Teil: Aufgaben und Methoden der Vegetationskunde*. Ulmer, Stuttgart.

Equihua, M. (1990) Fuzzy clustering of ecological data. *Journal of Ecology* **78**, 519–534.

Ewald, J. (2003) A critique for phytosociology. *Journal of Vegetation Science* **14**, 291–296.

Faith.D.P., Minchin, P.R. & Belbin, L. (1987) Compositional dissimilarity as a robust measure of ecological distance. *Vegetatio* **69**, 57–68.

Feoli, E. & Orlóci, L. (1979) Analysis of concentration and detection of underlying factors in structured tables. *Vegetatio* **40**, 49–54.

Franz, N.M., Peet, R.K. & Weakley, A.S. (2008) On the use of taxonomic concepts in support of biodiversity research and taxonomy. Symposium Proceedings, In: *The New Taxonomy* (ed. Q.D. Wheeler). *Systematics Association Special Volume* **74**, 63–86. Taylor & Francis, Boca Raton, FL.

Fridley, J.D., Peet, R.K. Wentworth, T.R. & White, P.S. (2005) Connecting fine- and broad-scale patterns of species diversity: species-area relationships of Southeastern U.S. flora. *Ecology* **86**, 1172–1177.

Gégout, J.-C. & Coudun, C. (2012) The right relevé in the right vegetation unit: a new typicality index to reproduce expert judgment with an automatic classification programme. *Journal of Vegetation Science* **23**, 24–32.

Gillison, A.N. & Brewer, K.R.W. (1985) The use of gradient directed transects or gradsects in natural resource survey. *Journal of Environmental Management* **20**, 103–117.

Gleason, H.A. (1926) The individualistic concept of the plant association. *Bulletin of the Torrey Botanical Club* **53**, 7–26.

Gleason, H.A. (1939) The individualistic concept of the plant association. *American Midland Naturalist* **21**, 92–110.

Goodall, D.W. (1953) Objective methods for the classification of vegetation. II. Fidelity and indicator value. *Australian Journal of Botany* **1**, 434–456.

Goodall, D.W. (1973) Sample similarity and species correlation. In: *Ordination and Classification of Communities* [*Handbook of Vegetation Science* V] (ed. R.H. Whittaker), pp. 105–156. Dr. W. Junk, The Hague.

Goodman L. & Kruskal, W. (1954) Measures of association for cross-validations. *Journal of the American Statistical Association* **49**, 732–764.

Gopal, S. & Woodcock, C. (1994) Theory and methods for accuracy assessment of thematic maps using fuzzy sets. *Photogrammetric Engineering and Remote Sensing* **60**, 181–188.

Gray, A.N., Brandeis, T.J., Shaw, J.D. & McWilliams, W.H. (2012) Forest inventory vegetation database of the United States of America. *Biodiversity and Ecology* **4** (in press).

Grossman, D.H., Faber-Langendoen, D., Weakley, A.S. *et al.* (1998) *International Classification of Ecological communities: Terrestrial Vegetation of the United States. Volume I. The National Vegetation Classification System: Development, Status, and Applications.* The Nature Conservancy, Arlington, VA.

Hartigan, J.A & Wong, M.A. (1979) Algorithm AS 136: A K-Means Clustering Algorithm. *Journal of the Royal Statistical Society, Series C (Applied Statistics)* **28**, 100–108.

Hennekens, S.M. & Schaminée, J.H.J. (2001) TURBOVEG, a comprehensive data base management system for vegetation data. *Journal of Vegetation Science* **12**, 589–591.

Hill, M.O. (1979) TWINSPAN – A FORTRAN Program for Arranging Multivariate Data in an Ordered Two-way Table by Classification of the Individuals and Attributes. Cornell University, Ithaca, NY.

Hubálek, Z. (1982) Coefficients of association and similarity, based on binary (presence–absence) data: an evaluation. *Biological Reviews of the Cambridge Philosophical Society* **57**, 669–689.

Hubert, L.J. & Levin, J.R. (1976) A general framework for assessing categorical clustering in free recall. *Psychology Bulletin* **83**, 1072–1080.

Jansen, F. & Dengler, J. (2010) Plant names in vegetation databases – a neglected source of bias. *Journal of Vegetation Science* **21**, 1179–1186.

Jennings, M.D., Faber-Langendoen, D., Loucks, O.L., Peet, R.K. & Roberts, D. (2009) Characterizing Associations and Alliances of the U.S. National Vegetation Classification. *Ecological Monographs* **79**, 173–199.

Jones, M.B., Schildhauer, M.P., Reichman, O.J. & Bowers, S. (2006) The new bioinformatics: integrating ecological data from the gene to the biosphere. *Annual Review of Ecology, Evolution and Systematics* **37**, 519–544.

Juhász-Nagy, P. (1964) Some theoretical models of cenological fidelity I. *Acta Botanica Debrecina* **3**, 33–43.

Kaufman, L. & Rousseeuw, P.J. (1990) *Finding Groups in Data.* John Wiley and Sons, New York, NY.

Kent, M. (2012) *Vegetation Description and Data Analysis: A Practical Approach.* 2nd ed. Wiley-Blackwell, Oxford, UK.

Knollová, I., Chytrý, M., Tichý, L. & Hájek, O. (2005) Stratified resampling of phytosociological databases: some strategies for obtaining more representative data sets for classification studies. *Journal of Vegetation Science* **16**, 479–486.

Krajina, V.J. (1933) Die Pflanzengesellschaften des Mlynica-Tales in den Vysoke Tatry (Hohe Tatra). Mit besonderer Berücksichtigung der ökologischen Verhältnisse. *Beihefte zum Botanischen Centralblatt* **50**, 774–957; **51**, 1–224.

Lambert, J.M. & Dale, M.B. (1964) The use of statistics in phytosociology. *Advances in Ecological Research* **2**, 59–99.

Lance, G.N. & Williams, W.T. (1967) A general theory of classificatory sorting strategies. I. Hierarchical systems. *Computer Journal* **9**, 373–380.

Legendre, P. & Gallagher, E.D. (2001) Ecologically meaningful transformations for ordination of species data. *Oecologia* **129**, 271–280.

Legendre, P. & Legendre, L. (1998) *Numerical ecology*, 2nd ed. Developments in Environmental Modelling 20. Elsevier, Amsterdam.

Lengyel, A., Chytrý, M. & Tichý, L. (2011) Heterogeneity-constrained random resampling of phytosociological databases. *Journal of Vegetation Science* **22**, 175–183.

Lepš, J. & Šmilauer, P. (2003) *Multivariate Analysis of Ecological Data Using CANOCO.* Cambridge University Press, Cambridge.

Ludwig, J.A. & Reynolds, J.F. (1988) *Statistical Ecology: A Primer on Methods and Computing.* John Wiley and Sons, New York, NY.

McCune, B. (1994) Improving community analysis with the Beals smoothing function. *Ecoscience* **1**, 82–86

McCune, B. & Grace, J.B. (2002) *Analysis of Ecological Communities.* MjM Software Design, Gleneden Beach, OR.

Michalcová, D., Lvončík, S., Chytrý, M. & Hájek, O. (2011) Bias in vegetation databases? A comparison of stratified-random and preferential sampling. *Journal of Vegetation Science* **22**, 281–291.

Mucina, L. (1997) Classification of vegetation: past, present and future. *Journal of Vegetation Science* **8**, 751–760.

Mucina, L., Rodwell, J.S., Schaminée, J.H.J. & Dierschke, H. (1993) European vegetation survey: Current state of some national programs. *Journal of Vegetation Science* **4**, 429–438.

Mucina, L., Schaminée, J.H.J & Rodwell, J.S. (2000) Common data standards for recording relevés in field survey for vegetation classification. *Journal of Vegetation Science* **11**, 769–772.

Mueller-Dombois, D. & Ellenberg, H. (1974) *Aims and Methods of Vegetation Ecology*. John Wiley & Sons, Ltd, New York, NY.

Nekola, J.C. & White, P.S. (1999) The distance decay of similarity in biogeography and ecology. *Journal of Biogeography* **26**, 867–878.

Noest, V., van der Maarel, E. & van der Meulen, F. (1989) Optimum transformation of plant species cover-abundance values. *Vegetatio* **83**, 167–178.

Orlóci, L. (1967) An agglomerative method for classification of plant communities. *Journal of Ecology* **55**, 193–206.

Orlóci, L. (1978) *Multivariate Analysis in Vegetation Research*, 2nd edn. Dr. W. Junk, The Hague.

Peet, R.K., Lee, M.T., Jennings, M.D. & Faber-Langendoen, D. (2012) VegBank: a permanent, open-access archive for vegetation plot data. *Biodiversity and Ecology* **4** (in press).

Peet, R.K., Wentworth, T.R. & White, P.S. (1998) The North Carolina Vegetation Survey protocol: a flexible, multipurpose method for recording vegetation composition and structure. *Castanea* **63**, 262–274.

Pfister, R.D. & Arno, S.F. (1980) Classifying forest habitat types based on potential climax vegetation. *Forest Science* **26**, 52–70.

Pfister, R.D., Kovalchik, B.L., Arno, S.F. & Presby, R.C. (1977) *Forest Habitat Types of Montana*. USDA Forest Service General Technical Report INT-34.

Podani, J. (1990) Comparison of fuzzy classifications. *Coenoses* **5**, 17–21.

Podani, J. (2000) Simulation of random dendrograms and comparison tests: some comments. *Journal of Classification* **17**, 123–142.

Podani, J. (2005) Multivariate exploratory analysis of ordinal data in ecology: pitfalls, problems and solutions. *Journal of Vegetation Science* **16**, 497–510.

Podani, J. & Csányi, B. (2010) Detecting indicator species: some extensions of the IndVal measure. *Ecological Indicators* **10**, 1119–1124.

Podani, J. & Feoli, E. (1991) A general strategy for the simultaneous classification of variables and objects in ecological data tables. *Journal of Vegetation Science* **2**, 435–444.

Radford, A.E., Ahles, H.E. & Bell, C.R. (1968) *Manual of the Vascular Flora of the Carolinas*. University of North Carolina Press, Chapel Hill, NC.

Roberts, D.W. (2010) *OPTPART: Optimal partitioning of similarity relations*. R package version 2.0-1, http://CRAN.R-project.org/package=optpart (accessed 25 May 2012).

Rodríguez, J.P., Rodríguez-Clark, K.M., Baille, J.E.M. *et al.* (2011) Establishing IUCN redlist criteria for threatened ecosystems. *Conservation Biology* **25**, 21–29.

Rodwell, J.S. (2006) *National Vegetation Classification: User's Handbook*. Joint Nature Conservation Committee, Peterborough.

Rodwell, J.S., Pignatti, S., Mucina, L. & Schaminée, J.H.J. (1995) European Vegetation Survey: update on progress. *Journal of Vegetation Science* **6**, 759–762.

Rodwell, J.S., Schaminée, J.H.J., Mucina, L., Pignatti, S., Dring, J. & Moss, D. (2002) *The Diversity of European Vegetation. An Overview of Phytosociological Alliances and Their Relationships to EUNIS Habitats*. National Centre for Agriculture, Nature Management and Fisheries, Wageningen.

Roleček, J., Chytrý, M., Háyek, M., Lvončik, S. & Tichý, L. (2007) Sampling in large-scale vegetation studies: Do not sacrifice ecological thinking to statistical puritanism. *Folia Geobotanica* **42**, 199–208.

Roleček, J., Tichý, L., Zleny, D. & Chytrý, M. (2009) Modified TWINSPAN classification in which the hierarchy respects cluster heterogeneity. *Journal of Vegetation Science* **20**, 596–602.

Rousseeuw, P.J. (1987) Silhouettes: A graphical aid to the interpretation and validation of cluster analysis. *Journal of Computation and Applied Mathematics* **20**, 53–65.

Schaminée, J.H.J., Hennekens, S.M., Chytrý, M. & Rodwell, J.S. (2009) Vegetation-plot data and data-bases in Europe: an overview. *Preslia* **81**, 173–185.

Schaminée, J.H.J., Hennekens, S.M. & Ozinga, W.A. (2007) Use of the ecological information system SynBioSys for the analysis of large datasets. *Journal of Vegetation Science* **18**, 463–470.

Shimwell, D.W. (1971) *Description and Classification of Vegetation*. Sidgwick and Jackson, London.

Smartt, P.F.M., Meacock, S.E. & Lambert, J.M. (1976) Investigations into the proper ties of quantitative vegetational data. II. Further data type comparisons. *Journal of Ecology* **64**, 41–78.

Sokal, R.R. & Rohlf, F.J. (1995) *Biometry. The principles and Practice of Statistics in Biological Research*, 3rd edN. Freeman, New York, NY.

Sokal, R.R. & Sneath, P.H.A. (1963) *Principles of Numerical Taxonomy*. Freeman, San Francisco and London.

Stohlgren, T.J., Falkner, M.B. & Schell, L.D. (1995) A modified-Whittaker nested vegetation sampling method. *Vegetatio* **117**, 113–121.

Szafer, W. & Pawlowski, B. (1927) Die Pflanzenassoziationen des Tatra-Gebirges. A. Bemerkungen über die angewandte Arbeitsmethodik. *Bulletin International de l' Académie Polonaise des Sciences et Lettres* **B3** Suppl. 2, 1–12.

TDWG (2005) *Taxonomic Concept Transfer Schema*. Biodiversity Information Standards. http://www.tdwg.org/standards/117/ (accessed 5 July 2012).

Tichý, L. & Chytrý, M. (2006) Statistical determination of diagnostic species for site groups of unequal size. *Journal of Vegetation Science* **17**, 809–818.

Tichý, L., Chytrý, M., Hájek, M., Talbot, S.S. & Botta-Dukát, Z. (2010) OptimClass: using species-to-cluster fidelity to determine the optimal partition in classification of ecological communities. *Journal of Vegetation Science* **21**, 287–299.

Tsiripidis, I., Bergmeier, E., Fotiadis, G. & Dimopoulos, P. (2009) A new algorithm for the determination of differential taxa. *Journal of Vegetation Science* **20**, 233–240.

USFGDC (United States Federal Geographic Data Committee) (2008) *National Vegetation Classification Standard, Version 2 FGDC-STD-005-2008*. Vegetation Subcommittee, Federal Geographic Data Committee, FGDC Secretariat, US Geological Survey, Reston, VA.

van der Maarel, E. (1979) Transformation of cover-abundance values in phytosociology and its effects on community similarity. *Vegetatio* **39**, 97–144.

van der Maarel, E. (2007) Transformation of cover-abundance values for appropriate numerical treatment: Alternatives to the proposals by Podani. *Journal of Vegetation Science* **18**, 767–770.

van Tongeren, O., Gremmen, N. & Hennekens, S. (2008) Assignment of relevés by supervised clustering of plant communities using a new composite index. *Journal of Vegetation Science* **19**, 525–536.

Vision, T.J. (2010) Open data and the social contract of scientific publishing. *BioScience* **60**, 330–331.

Waterton, C. (2002) From field to fantasy: classifying nature, constructing Europe. *Social Studies of Science* **32**, 177–204.

Weber, H.E., Moravec, J. & Theurillat, J.-P. (2000) *International Code of Phytosociological Nomenclature*. 3rd edn. *Journal of Vegetation Science* **11**, 739–768.

Wesche, K. & von Wehrden, H. (2011) Surveying southern Mongolia: application of multivariate classification methods in drylands with low diversity and long floristic gradients. *Journal of Vegetation Science* **14**, 561–570.

Westhoff, V. & van der Maarel, E. (1973) The Braun-Blanquet approach. In: *Ordination and Classification of Communities* [*Handbook of Vegetation Science V*] (ed. R.H. Whittaker), pp. 617–726. Junk, The Hague.

Whittaker, R.H. (1960) Vegetation of the Siskiyou Mountains, Oregon and California. *Ecological Monographs* **30**, 279–338.

Whittaker, R.H. (1962) Classification of natural communities. *Botanical Review* **28**, 1–239.

Whittaker, R.H. (ed.) (1973) *Ordination and Classification of Communities*. [*Handbook of Vegetation Science – Part V*.]. Junk, The Hague.

Whittaker, R.H., Niering, W.A. & Crisp, M.D. (1979) Structure, pattern, and diversity of a mallee community in New South Wales. *Vegetatio* **39**, 65–76.

Wildi, O. (2010) *Data analysis in vegetation ecology*. Wiley-Blackwell, Oxford, UK.

Williams, W.T., Lambert, J.M & Lance, G.N. (1966) Multivariate methods in plant ecology. V. Similarity analyses and information-analysis. *Journal of Ecology* **54**, 427–445.

Willner, W., Tichý, L. & Chýtrý, M. (2009) Effects of different fidelity measures and contexts on the determination of diagnostic species. *Journal of Vegetation Science* **20**, 10–137.

Wilson, J.B. (2012) Species presence/absence sometimes represents a plant community as well as species abundances do, or better. *Journal of Vegetation Science* **23** (DOI: 10.1111/j.1654-1103.2012.01430.x).

Wiser, S., Spencer, N., De Cáceres, M., Kleikamp, M., Boyle, B. & Peet, R.K. (2011) Veg-X – an exchange standard for plot-based vegetation data. *Journal of Vegetation Science* **22**, 598–609.

3

Vegetation and Environment: Discontinuities and Continuities

Mike P. Austin

CSIRO Ecosystem Sciences, Canberra ACT, Australia

3.1 Introduction

The pattern of variation shown by the distribution of species among quadrats of the earth's surface chosen at random hovers in a tantalizing manner between the continuous and the discontinuous.

(Webb 1954)

The issue can be expressed as: is vegetation organized into discrete recognizable communities or as a continuum of gradually changing composition? Answering this question has a history of confused debate between conflicting schools of research, tedious descriptive accounts and a lack of hypothesis testing. McIntosh (1985) demonstrated the important role it has played in all ecological disciplines, not just plant community ecology. Lack of resolution of the issue has led some ecologists to conclude that it is irrelevant to the advancement of ecological science. However, a study of a forest, grassland or the population of a species can have little practical value without an adequate description of the associated vegetation and its correlation with environment.

3.1.1 Vegetation concepts

Two terms are used extensively in vegetation ecology, **community** and **continuum.** Definitions of the term community vary (see Chapter 1). A plant community can be broadly defined as (1) having a consistent floristic composition, (2) having uniform physiognomy, (3) occurring in a particular environment and (4) usually occurring at several locations. Implicit in the definition is the assumption that the consistent composition and uniform physiognomy is the result of

Vegetation Ecology, Second Edition. Eddy van der Maarel and Janet Franklin.
© 2013 John Wiley & Sons, Ltd. Published 2013 by John Wiley & Sons, Ltd.

biotic interactions between the species, particularly competition. The individualistic continuum concept differs in that each species is considered to have an
individualistic response to both abiotic and biotic factors such that when vegetation is viewed in relation to an environmental variable, variation in floristic
composition and structure is continuous.

3.1.2   Relationship between vegetation and environment

Detailed analysis of the relationship between vegetation and environment
requires a detailed understanding of the environmental processes that influence
vegetation, for example, knowledge of the processes that link rainfall to the
availability of water to plants and of the physiological processes that govern its
use by different species is essential.

In ecology there is often a dichotomy between experimental and observational
studies. Very few studies combine rigorous observational analysis with detailed
manipulative experiments on vegetation composition. Grime (2001) provided
an exception with experiments based on extensive surveys of grasslands in a
local area together with examples from other regions. The focus in this chapter
is on what observational studies can tell us about vegetation/environment relationships. There is an intimate dependence between developments in vegetation
concepts, mathematical methods of analysis and knowledge of environmental
processes. An account of these aspects is presented in the context of three
questions:

1   Is vegetation pattern continuous or discontinuous and how is this pattern
    related to environment?
2   What theory and methods are most appropriate for investigating such
    pattern?
3   What is the relative importance of environment and factors intrinsic to the
    vegetation in determining the observed patterns?

3.2   Early history

In order to evaluate the relative merits of different approaches to vegetation/
environment patterns it is important to know the history that has led to the
current research paradigms when assessing alternative approaches and methods
(Kuhn 1970). A Kuhnian paradigm consists of an agreed collection of facts, a
conceptual framework concerning those facts, a restricted set of problems
selected from within the framework and studied with an accepted array of
methods (Austin 1999b). Over the past 50 years, developments in three areas
have contributed to the development of current paradigms in observational plant
community ecology. These were recognition of (1) alternative theoretical frameworks, (2) the need for rigorous quantitative methods of sampling and analysing
vegetation, and (3) the need to measure more precisely the potential causal
environmental variables.

3.2.1   The continuum versus community controversy

Two conceptual approaches dominated plant community ecology prior to the 1950s, the climax community as a super-organism associated with its proponent F.E. Clements, and the association as a vegetation unit that could be classified in a similar way to a species often associated with name of J. Braun-Blanquet. The first paradigm predominated in North America and Britain, the second as phytosociology in continental Europe (see Section 1.1.1). Both 'schools' accepted that vegetation could be classified into units (communities) though their assumptions and methods differed (Whittaker 1962; Mueller-Dombois & Ellenberg 1974, Westhoff & van der Maarel 1978; McIntosh 1985). Gleason (1926) had advanced an alternative conceptual framework: 'the individualistic concept of vegetation.' This framework attracted intense opposition and subsequent neglect until the 1950s. Subsequently, it was found that other European ecologists, particularly Ramenskiy in Russia had put forward similar ideas and received similar negative responses (Whittaker 1967). Whittaker (1967) restated Gleason's ideas as two principles:

'(1) The principle of species individuality – each species is distributed in relation to the total range of environmental factors (including effects of other species) it encounters according to its own genetic structure, physiological characteristics and population dynamics. No two species are alike in these characteristics, consequently, with few exceptions, no two species have the same distributions. (2) The principle of community continuity – communities which occur along continuous environmental gradients usually intergrade continuously, with gradual changes in population levels of species along the gradient.'

In the late 1940s and 1950s, Whittaker (1956), identified with gradient analysis, and Curtis (1959), identified with the continuum concept, began to examine patterns of vegetation composition using explicit though different numerical methods. They and their students concluded that vegetation patterns were better explained by the continuum concept of continuous variation in relation to environmental gradients. Their studies generated further controversy. Three issues were often confused in the debate: (1) the discrete community versus continuum issue, (2) the use of objective numerical methods to analyse vegetation data on composition as opposed to subjective methods and (3) whether disturbed or heterogeneous stands of vegetation had been included in the sampling.

By 1970 it was recognized that quantitative methods could be applied to either continuum or community approaches and that the two approaches were not necessarily incompatible (for further commentary see Mueller-Dombois & Ellenberg 1974; Whittaker 1978a, b). A variety of different research paradigms continue to be used today (see Section 2.4).

3.3 Development of numerical methods

3.3.1 Indirect ordination

This numerical approach determines the major gradients of variation to be found in the vegetation data itself. A graphical representation of the variation in vegetation across all sites can be constructed by measuring the similarity between each site based on the species composition. The earliest method, the continuum (or compositional) index took account of only a single dimension. Methods were quickly recognized or developed which would allow several dimensions to be estimated and displayed. These were pioneered by the Wisconsin school (Bray & Curtis 1957; Curtis 1959; see Greig-Smith 1983; Kent & Coker 1992; Jongman *et al.* 1995 for details). The gradients estimated in this way need not represent environmental gradients but may represent successional changes or variation in grazing regimes. The indirect methods do not make the assumption that all major variations are due to environment as direct methods often do. An early example is the investigation of the variation in a small limestone grassland area in Wales (Gittins 1965). The example shows how the method could display patterns of variation in an individual species and may detect discontinuities in vegetation composition where they existed (Fig. 3.1). The two plots in the bottom left of the figure are very different in composition from the rest and hence disjunct. These plots were from a sheep night camp, which had become enriched with nutrient and therefore supported a flora distinct from the surrounding nutrient-poor limestone grassland.

Such an ordination diagram summarizes the major axes of variation in the vegetation data matrix. It was soon recognized that adoption of particular methods also implied ecological assumptions about the response of species. For

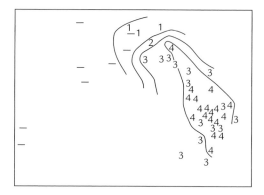

Fig. 3.1 An early example of indirect ordination analysis with two axes from a Welsh limestone grassland showing the distribution of *Helianthemum chamaecistus* with four levels of abundance plus absence. Note the two outliers in the bottom left-hand corner, which were from a sheep camp. (After Gittins 1965.)

example, assuming species response was a symmetric bell-shaped curve to the underlying gradient could result in severe so-called 'horseshoe' distortion (Swan 1970). Numerous different methods have now been developed and are widely used in ecology (for further details see texts by Legendre & Legendre 1998; Lepš & Šmilauer 2003; McCune & Grace 2002).

3.3.2 Numerical classification

Early classification of vegetation was subjective. Numerical methods were developed to provide objective procedures. They were based on the use of a similarity or association measure between plots of vegetation, grouping together those plots which were most similar. Numerous methods of classification were developed with various similarity measures and different strategies for grouping plots together (see Kent & Coker 1992; Greig-Smith 1983; Jongman *et al.* 1995). Classification is treated in Chapter 2.

Initially ordination and numerical classification were contrasted as supporting the different concepts of vegetation organization, continuum or community respectively. However, these methods can be applied without regard for the different concepts. The two methods provide complementary information about the composition of the vegetation and its relationship to environment. Ordination displays the major axes of variation while classification identifies clusters of sites and outliers. The objectivity of the methods was also seen to be illusory. Each method was explicit, consistent and repeatable but the choice of method was a highly subjective decision. Results, particularly the detection of outliers and discontinuities, are highly sensitive to data standardizations, dissimilarity measures and the statistical method used (Section 3.5).

3.3.3 Direct ordination

Direct ordination, originally termed direct gradient analysis by Whittaker (1956), is the analysis of species distributions (presence/absence or abundance data) and collective properties (e.g. species richness) in relation to environmental variables conventionally referred to as environmental gradients. Initially, the methods used were graphical and the environmental measures were crude, often simply subjective estimates of moisture (Fig. 3.2). Relatively independent developments of this graphic analysis seem to have occurred in America (Whittaker 1956), England (Perring 1959) and Europe (Ellenberg 1988, first German edition 1963; sixth edition 2010 revised and extended by Chr. Leuschner; see Chapter 10). Fig. 3.2 shows two examples. No species were found to have similar patterns of distribution. The evidence is not presented, only the interpretation which would not satisfy modern standards of statistical rigour. The evidence of dissimilar patterns of distribution among species as opposed to the long-held assumption of coincident distributions of species was, however. clear. There has been a progressive improvement in the statistical methods used since this early work (Section 3.6).

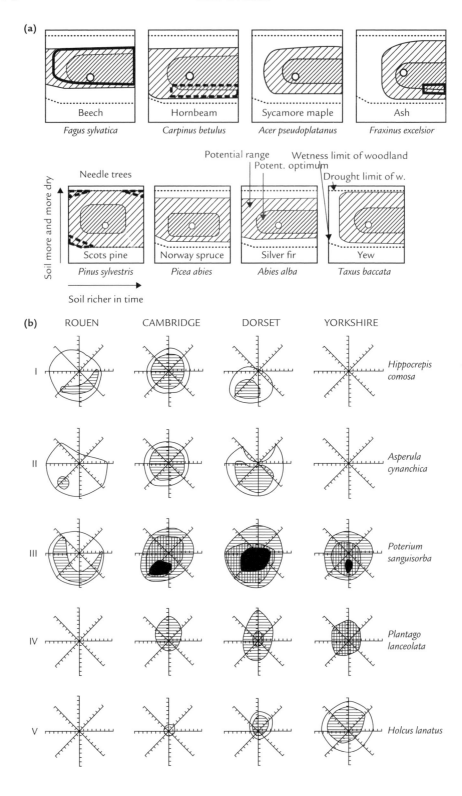

Fig. 3.2 Two early examples of direct gradient analyses from Britain and Central Europe. (a) Distribution of tree species along gradients of acidity and moisture in Central European forests (Ellenberg 1988). Thick black border encloses zone where species is dominant. Broken border encloses zone where species is co-dominant. These borders define zones where species have ecological optima as opposed to hatching zones which indicate species' estimated physiological optima. (b) Distribution of five chalk grassland species in relation to slope and aspect as represented by a diagrammatic hemispherical hill in four regions of north-west Europe. The spokes represent the eight cardinal points of the compass and slope increases in steps of five degrees from the centre (Perring 1960). Note for species I and II: within the contour the species is present while hatched area has >5% cover. For other species: within the contour the species is present; simple hatching >10% cover; cross hatch >20%; black >25%.

3.3.4 Environmental measurement

Early ecology texts emphasized the multitude of environmental factors and the complexity of their effects on different species. In contrast, Jenny (1941) had presented a simple conceptual framework for soils. The equation for soil formation put forward by Jenny is a list of factors that should be taken into account when examining how soils develop. As modified for vegetation it is:

$$V = f(cl, p, r, o, t,)$$

where V is some property of vegetation, which is a function (f) of cl = climate, p = parent material, r = topography, o = organisms, and t = time.

Each of these factors may influence plants in numerous complex ways. No mathematical expression can summarize the processes involved. As a minimum it provides a checklist of broad environmental factors to be considered (see discussion in Mueller-Dombois & Ellenberg 1974). Maximally, it can provide a conceptual framework for both survey design and environmental analysis. Comprehensive use of this framework was made by Perring (1958, 1959, 1960) to design a survey of chalk grasslands in England and northern France, analysing the vegetation data graphically (Fig. 3.2b). He restricted the study by parent material (p), and then stratified sampling by climate and topography. Topography was idealized as a hemispherical hill (an inverted pudding basin) and a stratified sample taken by slope and aspect. The results show individualistic responses by species to climate, slope and aspect and complex interactions between these variables.

Few studies since have used such an explicit approach to the analysis of vegetation/environment relationships. One contributing factor is the variety of ways environmental variables can be expressed and measured. Different types of environmental variable can be recognized, e.g. abiotic and biotic. Abiotic variables such as rainfall and soil nitrogen content directly determine plant growth and success. Biotic variables such as competition from other plants (Chapter 7), pathogens, herbivores (Chapter 8) and mycorrhiza (Chapter 9), may destroy plants (pathogens), enhance growth (mycorrhiza) or have complex effects contingent on abiotic variables (Chapter 11). Environmental variables may be considered to be either distal or proximal. Proximal and distal refer to the position of the predictor in the chain of processes that link the predictor to its impact on the plant. The most proximal gradient will be the causal variable determining the plant response (Austin 2002). Distal variables such as rainfall influence plant growth through various intermediate variables for example soil permeability and soil water holding capacity, while the equivalent proximal variable would be water availability at the root hair.

Another alternative classification of environmental variables or gradients is into indirect, direct and resource gradients (Austin & Smith 1989). Indirect variables or gradients are those that have no direct influence on plant growth. An example is altitude, a variable often correlated with vegetation composition. Altitude can only have an influence via some correlated variable, which has a direct influence (e.g. temperature or rainfall). However, these variables have correlations with altitude that are specific to a locality. Correlations based on indirect variables can be used for local prediction but cannot provide explanation in terms of ecological process. Direct environmental gradients are those where the variable has a direct influence on plant growth. Examples are pH and temperature. Resource variables are those which are consumed by plants in the course of growth (e.g. phosphorus or nitrogen). There is no absolute set of categories for these variables. Water is both a consumable resource and a direct variable when excess creates anaerobic conditions. Scalars can be developed to combine distal variables based on environmental process knowledge to give estimates of proximal, direct variables which may establish relationships with vegetation that are more robust and less dependent on location-specific correlations. Improvements in measuring and estimating environmental variables continue to be made (Section 3.6.3).

3.4 Current theory: continuum and community

3.4.1 Introduction

Progress in vegetation science depends on the development of explicit theory and numerical methods capable of discriminating between rival theories. At the present time there is little consensus on even rival theories and little agreement as to what constitute suitable methods for discrimination. Here a conceptual framework for both community and continuum concepts is presented with a suggestion of how a synthesis can be achieved. In a subsequent section, the current complex relationship between indirect ordination methods, the phenomenological models they assume and the use of artificial data to evaluate them is briefly examined (Section 3.5). Finally, the potential of direct ordination methods, now widely referred to as species distribution modelling (SDM), for resolving many of these issues is discussed (Section 3.6).

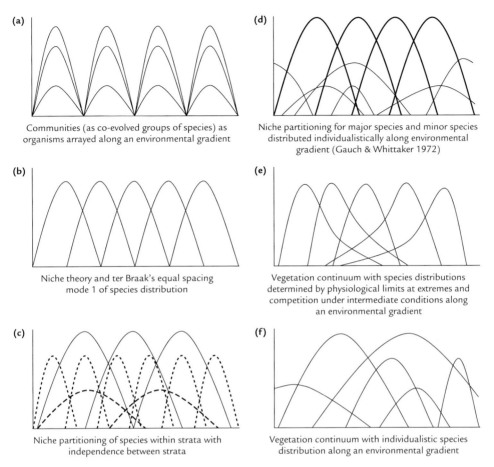

(a)

Communities (as co-evolved groups of species) as organisms arrayed along an environmental gradient

(b)

Niche theory and ter Braak's equal spacing mode 1 of species distribution

(c)

Niche partitioning of species within strata with independence between strata

(d)

Niche partitioning for major species and minor species distributed individualistically along environmental gradient (Gauch & Whittaker 1972)

(e)

Vegetation continuum with species distributions determined by physiological limits at extremes and competition under intermediate conditions along an environmental gradient

(f)

Vegetation continuum with individualistic species distribution along an environmental gradient

Fig. 3.3 Six hypothetical patterns of vegetation composition along an environmental gradient corresponding to different theories. See text for details.

3.4.2   Continuum

Alternative realizations. Fig. 3.3 shows a spectrum of possibilities from the superorganism concept of a community (Fig. 3.3a) to a totally individualistic organization (Fig. 3.3f). The second realization (Fig. 3.3b) is based on the niche concept of species partitioning a resource gradient. Species have equal ranges and amplitudes and are equally spaced along the gradient. This is the model explicitly underlying Correspondence Analysis (CA) and Canonical Correspondence Analysis (CCA) (ter Braak 1986; Jongman *et al.* 1995; see Section 3.6.2). This niche representation can be combined with the idea that each stratum (trees, shrubs etc.) partitions the gradient independently of the other strata (Fig. 3.3c). The result is a continuum with each species showing a response partially determined by growth-form.

Gauch & Whittaker (1972) put forward a detailed set of hypotheses about the patterns of species response observed along a gradient. These included equal spacing of the dominants (trees) equivalent to resource partitioning and individualistic patterns for understorey species (Fig. 3.3d). Austin (1999a) summarized results for eucalypt species along a mean annual temperature gradient. These suggest that patterns of species response change depending on position on the gradient (Fig. 3.3e). It was hypothesized that the physiology of individual species determined limits towards the extremes of the gradient while competition determined species occurrence and shape of response in the central mesic portion of the gradient. This hypothesis applies only to the tree stratum (see Section 3.6.6 for further discussion). The individualistic continuum (Fig. 3.3f) shows no patterns of species' behaviour along the gradient.

It is possible to represent phytosociological associations, as they might exist along an environmental gradient (Fig. 3.4; see also Westhoff & van der Maarel 1978). Identification of each association depends on recognition of the constant species with a wide environmental range, and differential species with narrower ranges that distinguish each association. For example, association 1 is characterized by constant species A and differential species C and D with association 2 having constant species A and B and differential species E and F. The presence of indifferent and rare species would result in a diagram that would not easily be distinguished from the continuum models presented in Fig. 3.3.

This series of hypothetical vegetation patterns demonstrates that (1) the differences between phytosociological concepts and continuum concepts may be smaller than sometimes imagined, and (2) discriminating between these

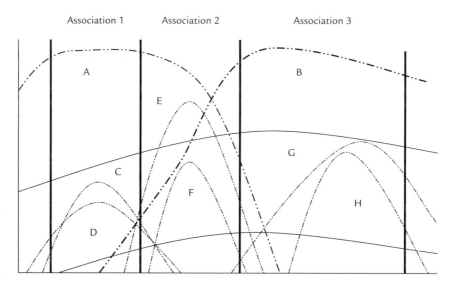

Fig. 3.4 A possible representation of phytosociological associations along an environmental gradient showing constant species (heavy broken line), differential species (light broken line C–H) and indifferent species (light solid line). Associations are distinguished by different combinations of constant and differential species.

hypotheses will require detailed data and rigorous statistical methods. No comprehensive tests have been published. There is a complication. The species responses shown in Fig. 3.3 are, with one exception, presented as symmetric bell-shaped curves. If species responses are not bell-shaped and symmetric, what implications does this have for theories of vegetation composition?

Niche theory and continuum concepts. Niche theory assumes each species has a **fundamental niche** (in the absence of competitors) in relation to some resource gradient. Species niche response is usually assumed to be a symmetric bell-shaped curve (Fig. 3.3b). Each species is usually shown as having the same response with equal width and amplitude. In the presence of competitors, the species is restricted to a **realized niche**. The optima for both the fundamental and realized niches are co-incident as in example 1a of Fig. 3.5. This is a special case of a more general theory advanced by Ellenberg (see Mueller-Dombois & Ellenberg 1974). A species' realized niche (ecological response) may be displaced from its physiological (fundamental niche) by a superior competitor. This can result in bimodal curves. Each species may have different shaped responses to different environmental gradients (Fig. 3.5). Ellenberg's ideas of the niche shapes of plant species have received little recognition in the general ecological literature (Austin 1999b). Neither niche theorists nor plant community ecologists have considered in detail the patterns suggested in Figs 3.3, 3.4, 3.5.

The various continuum and community concepts are basically phenomenological; they are descriptive without an explicit mechanistic basis. Ellenberg's hypothesis introduces species-specific physiological limits and competition as organizing processes to produce the observed patterns. Some numerical methods

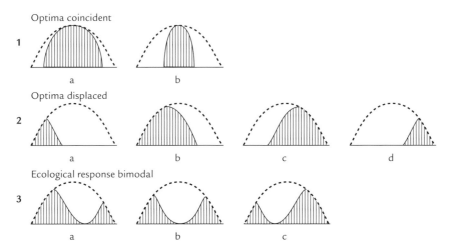

Fig. 3.5 Ellenberg's theory of species response patterns. Example 1a corresponds to classical niche theory. Other examples show interpretations of the possible responses of various species in relation to different environmental gradients. Competition from a superior competitor results in different shapes of ecological response for the species displaced from their physiological optima. (From Mueller-Dombois & Ellenberg 1974.)

Table 3.1 Frequency distributions of different shapes of ecological response surfaces for 100 common species in Tasmanian montane vegetation.[a]

	Response surface shape		
Structural group	Symmetric	Skewed	Complex
Trees	3	4	1
Shrubs	24	13	7
Herbs	8	11	7
Graminoids	8	5	6
Pteridophytes	2	0	4
All species	45	33	22

[a]Data for the 100 species occurring in at least 20 quadrats. A Monte-Carlo test showed no difference between structural groups in the relative frequencies of the shape categories. From Minchin (1989).

of vegetation analysis are explicitly based on symmetric bell-shaped curves and equal partitioning of the gradient without considering the ecological processes involved (Jongman *et al.* 1995). They have a restrictive theoretical basis that needs to be tested.

Evidence from ordinations. Ordinations can only provide evidence that a particular pattern of vegetation exists along a gradient. Indirect ordinations may display sharp discontinuities in the ordination space (Fig. 3.1). These have usually been ascribed to major differences in environmental conditions. Indirect ordinations cannot be used as proof of the existence of a continuum. The mathematical methods employed have implicit ecological assumptions about how species respond to gradients, which may determine the outcome. Most early analyses using direct ordination indicated the existence of varied species response shapes. No obvious co-incidences of species limits were observed which would support the community concept. Symmetric bell-shaped curves were no more abundant than other shapes. Minchin (1989) found only 45% of species had response surfaces which appeared unimodal and symmetric, the so-called Gaussian responses (Table 3.1).

Relatively few studies have directly tested continuum concepts or attempted to discriminate between the two concepts of community or continuum. Austin (1987) examined the continuum propositions put forward by Gauch & Whittaker (1972) using tree species in south-east Australia. A marked preponderance of skewed curves was found for the major species along a mean annual temperature gradient. Statistical modelling supported this conclusion (Austin 1999a). A test of Gauch & Whittaker's second proposition 'the modes of major species are evenly distributed along environmental gradients while those of minor species tend to be randomly distributed' rejected the proposition for major species. Other propositions could not be tested due to confounding with species richness, which increased steadily with temperature from one species at the tree line.

Minchin (1989) undertook a fuller analysis with 100 species in relation to two indirect environmental gradients – altitude and soil drainage – in montane

Tasmania. A test for the even distribution of modes indicated that species modes were clumped (all species), random (major species) or varied with structural group (growth-form). Herbs had clumped species modes while other growth-forms were random; alpha diversity (species number per unit area) was examined. Unimodal species richness patterns were evident for the different growth-forms. The modes of richness for each growth-form occupied different positions in the environmental space.

Shipley & Keddy (1987) attempted to distinguish between the continuum and community concepts using species limits. They examined species limits along transects following a water table gradient. If the community concept holds, then there should be more limits in some intervals than others along the gradient and species limits – both upper and lower – should coincide, i.e. cluster (Fig. 3.3a). If the individualistic continuum concept holds, then the average number of limits per interval along the gradient should be equal apart from random effects (Fig. 3.3f). In addition, for the continuum concept to hold the number of upper limits of species should be independent of the number of lower limits in each interval. Both upper and lower limits were found to be clustered. The individualistic continuum is rejected. No correlation between the number of upper and lower limits per interval was found. The community concept is also rejected. The results are equivocal and address only two of the possibilities represented in Figs 3.3, 3.4, 3.5. The transects ran from the edge of an *Acer saccharinum* forest into a marsh as far as the edge of the zone of aquatic species with floating leaves. This is a steep gradient from a terrestrial to an aquatic environment. Only six of the 43 species have both upper and lower limits recorded within the gradient. Most species have either an upper limit (aquatics) or a lower limit (terrestrial plants). A gradient length with less extreme moisture conditions or a less steep gradient might yield a different result.

Shipley & Keddy (1987) pointed out that they used an indirect gradient or factor-complex gradient in Whittaker's terms, namely water table depth. A clustering of limits in one interval might then indicate a discontinuity in one of the many environmental variables correlated with the factor-complex represented by water table depth. For example, anaerobic soil conditions may occur as a step function at a particular water depth in the marsh. The sharp increase in anaerobic conditions might appear to limit species at the same water level when in fact they are actually limited by different degrees of anaerobic conditions. Distance along a transect cannot be equated directly with changes in an environmental variable. The correlation of species patterns with an indirect distal variable may yield very different results from those using a direct proximal variable. Plant community ecologists have yet to specify the properties of either the community or continuum concepts in sufficient detail for any variant to be statistically distinguished from another.

3.4.3 Community

The term community is used with various meanings in the ecological literature (see McIntosh 1985; Chapter 1 and Section 3.1.1). Drake (1991) applied it to an experimental food web involving algae, bacteria, protozoans and cladocerans, plants and animals. He explored the mechanics of community assembly. The composition of invasion-resistant communities was found to depend on the order of invasion by species, as was the food web structure. The results demonstrate a number of important features involving the invasibility of some communities and the predictability of the outcome of competition among the primary producers in multiple trophic level experiments. The conclusions are relevant to vegetation ecology.

The definition of an 'ecological community' used by Drake (1991) is 'an ensemble of individuals representing numerous species which coexist and interact in an area or habitat.' This and other definitions could apply to almost any combination of species under any circumstances. They are non-operational for any form of comparative analysis of observations or experiments. To compare the results of food web or other experiments we need to know how different each 'ecological community' is from another. This leads naturally to the use of multivariate methods such as ordination and classification to measure the difference.

3.4.4   Possible synthesis

The controversy between the community-unit or association concept and the continuum concept arises because the former is an abstraction based on geographical space and the latter an abstraction based on environmental space (Austin & Smith 1989). Fig. 3.6a represents a hypothetical transect up a mountain in an area with four species showing an altitudinal zonation. Five communities can be distinguished: A, AB, B, C, and D if species associations are recognized based on their frequency of occurrence with the combinations BC and CD as ecotones (Fig. 3.6b). The communities are a result of the frequency of different altitudes along the transect, particularly the community AB on the bench at 200 m and community B on the bench at 400 m. The distribution of the four species in relation to altitude, however, is a continuum (Fig. 3.6a). Each species is spaced along the gradient approximating Gauch & Whittaker's (1972) conception of the continuum, yet communities are clearly recognizable along the transect.

Another hypothetical transect in an adjacent area where the two benches were at altitudes of 170 m and 430 m instead of 200 m and 400 m gives a different result. Here the communities would be A, B, BC, C, D, with ecotones AB and CD (Fig. 3.6c). The frequency of altitudinal classes has changed and hence the most frequent combinations of species (communities) are different. So, communities are a function of the frequency of different environments in the landscape examined. However, the altitudinal continuum would be unchanged: continua are a function of the environmental space measured. Note that the environmental gradient used here is an indirect (factor-complex) gradient. The continuum pattern observed only applies where the correlations between altitude and direct or resource gradients remain constant.

Mueller-Dombois & Ellenberg (1974, p. 205) discussed the definition of the phytosociological association and the choice of characteristic species. They point out that when an investigator is concerned with a small geographical area, many

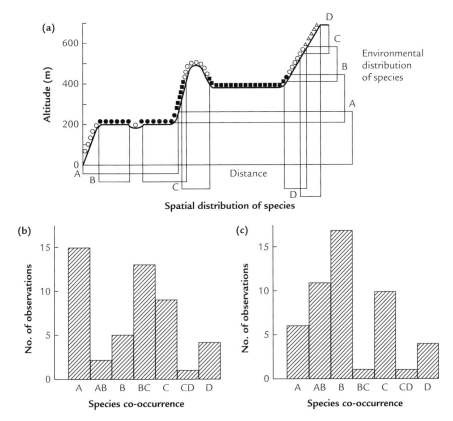

Fig. 3.6 (a) A hypothetical transect up an altitudinal gradient showing the spatial extent of the possible combinations of species. Each species has a distinct but overlapping niche with respect to the indirect environmental gradient of altitude (Austin & Smith 1989). (b) A histogram of the frequency of species combinations from the transect shown in Fig. 2.6a (Austin 1991). (c) A histogram of the frequency of the same species combinations but from a different transect where the benches occur at different altitudes (Austin 1991).

characteristic and differential species can be identified for each association. If the geographic range is increased, more and more species, which locally had a strong correlation with one association, are now found in other associations. Enlarging a study region will result in the inclusion of entirely new environments. The difficulties of identifying diagnostic species are consequences that follow naturally from the ideas presented in Fig. 3.6. However, where a single gradient is studied, the phytosociological model and the continuum model may seem very similar (Figs 3.3 and 3.4). Austin & Smith (1989) concluded:

1 The continuum concept applies to an abstract environmental space, not necessarily to any geographical distance on the ground or to any indirect environmental gradient.

2   The abstract concept of a community of co-occurring species can only be
    relevant to a particular landscape and its pattern of environmental variables;
    community is a property of the landscape.

Such a community concept is compatible with the different concepts of a continuum (Fig. 3.3). For communication and ecological management, the community will be the preferred concept to use, provided the applicable region is
clearly defined. For investigation of vegetation/environment relationships, the
continuum concept is preferable. However, this framework does not resolve
the alternative realizations of the continuum outlined in Figs 3.3, 3.4, 3.5. The
developing contribution of SDM (Section 3.6) to resolving some of these issues
is reviewed in Section 3.6.6.

3.5   Current indirect ordination methods

3.5.1   Introduction

With the availability of textbooks and software packages there has been a massive
expansion in the use of multivariate methods for indirect ordination in vegetation studies and in other areas of ecology (Kent & Coker 1992; Jongman *et al.*
1995; Legendre & Legendre 1998; McCune & Grace 2002; Lepš & Šmilauer
2003; Clarke & Gorley 2006; Zuur *et al.* 2007; Wildi 2010). Indirect ordination
can be used in two ways: (1) as a hypothesis-generating tool answering the question 'what are the major gradients of vegetation in my sample?' or (2) for testing
whether the major gradients are correlated with particular biotic (e.g. grazing)
or environmental (e.g. pH) variation in the sample. There is, however, no consensus on the most appropriate method or conceptual framework. Three major
research paradigms can be recognized associated with different methods and
assumptions about species responses to ecological gradients (Austin *et al.* 2006).
Attempts have been made to evaluate the performance of the different approaches
using real data and artificial simulated data but these are confounded by differences in data standardizations, choice of dissimilarity measure and multivariate
method adopted by different authors (e.g. Faith *et al.* 1987; Palmer 1993; Legendre & Gallagher 2001).
   Numerous indirect ordination techniques have been proposed using different
similarity measures. Many have now been shown to be effective only with certain
limited types of data sets. For example, principal components analysis (PCA)
ordination, although still widely used, will give distorted results when any species
shows unimodal response to the underlying gradient.
   Three methods define different research paradigms, correspondence analysis
(CA), principal co-ordinates analysis (PCoA) and non-metric multidimensional
scaling (NMDS). CA is a method that uses a χ^2-distance as the similarity measure.
PCoA provides an ordination which is a Euclidean representation of plots based
on any similarity measure chosen by the user (Legendre & Legendre 1998).
NMDS constructs an ordination where the distance between plots has maximum
rank order agreement with the similarity measures between plots. In theory,

NMDS can accommodate any similarity measure provided the resulting relationship between similarity and distance in ordination space remains monotonic (Minchin 1987).

3.5.2 Correspondence analysis

The most used method is correspondence analysis (CA). However, it is often used in the form of canonical correspondence analysis (CCA) where the axes of the ordination are constrained to maximize their relationship with a nominated set of environmental variables. CCA is a hybrid ordination method that combines features of direct and indirect ordination but note that any assumptions or limitations which apply to CA will apply to CCA.

The choice of an ordination method requires a suitable evaluation method. One cannot use real data for evaluation. The true gradients underlying an observed vegetation pattern can never be unequivocally known. A comparison of two methods on real data may give two different answers, both partially correct. Artificial data where the true gradients are known are necessary to evaluate methods. However, this requires that the model used to generate the data reflects a realistic theory of how vegetation varies in relation to environment.

Ter Braak (1986) in developing his CCA approach is very explicit in the mathematical assumptions implicit in the method:

1   The species' tolerances (niche widths) are equal;
2   The species' maxima are equal;
3   The species' optima are homogeneously distributed over a length of the gradient (A) that is large compared to individual species' tolerances;
4   The site scores are distributed over a length of the gradient that is large but contained within A.

The words 'homogeneously distributed' mean either that the optima or scores are equally spaced along the gradient or that they are randomly distributed according to a uniform distribution. Assumption 4 assumes a particularly sampling strategy for vegetation sampling. Assumptions 1 to 3 assume a particular species-packing model for the environmental gradient; the method is attempting to estimate the one represented in Fig. 3.3b. Ter Braak (1986) acknowledges that assumptions 1 and 2 'are not likely to hold in most natural communities'. He then claims that the usefulness of the method 'in practice relies on its robustness against violations of these conditions'. The robustness of the method has been examined with artificial data generated with different assumptions from those above (ter Braak *et al.* 1993) and considered to be satisfactory. No comparison with the performance of alternative ordination methods was made.

The claim of robustness does not accord with the work of Faith *et al.* (1987). They showed with artificial data that χ^2-distance as used by CA is unsatisfactory for estimating the true ecological distance. Minchin (1987) using similar artificial data sets showed fairly conclusively that local non-metric multidimensional scaling (LNMDS) outperformed detrended correspondence analysis (DCA) a form of CA, in recovering two-dimensional gradients. Økland (1999) examined the impact of horseshoe distortions and noise on the performance of various ordination methods including CA, DCA and CCA using artificial data. He showed that there are significant problems in distinguishing ecological signal from random noise and distortion due to the inappropriate choice of the theoretical model. In many cases the importance of the ecological signal is underestimated.

Other authors who have criticized the use of CA-based methods include Legendre & Gallagher (2001), McCune & Grace (2002) and Clarke *et al.* (2006). The balance of evidence suggests that these CA methods are not robust to departures from the species response model (Fig. 2.3b). Their use has been supported by others (e.g. Lepš & Šmilauer 2003; Kenkel 2006). Current best practice would be to compare CA with methods known to be more robust to departures from the assumed response models.

3.5.3   Principal coordinates analysis

Legendre & Legendre (1998) in their authoritative text *Numerical Ecology* provide a review of principal coordinate analysis (PCoA) based on the work of Gower (1966). There are a number of technical advantages associated with PCoA as compared to CA such as the use of a variety of dissimilarity measures, choice of data standardizations plus the possibility for constrained ordination, which is not available for NMDS (Legendre & Gallagher 2001). Legendre & Legendre (1998) and Legendre & Gallagher (2001) presented comparative studies of PCoA and CA showing the superior performance of PCoA. This was based on single gradient artificial data. See Wildi (2010) for recent examples of its use and Podani & Miklos (2002) for its application in the evaluation of the differences in dissimilarity measures and gradient properties on the 'horseshoe distortion'. Comparative studies of PCoA and NMDS using the Bray–Curtis coefficient as the dissimilarity measure on data for two-dimensional environmental gradients are needed to assess these methods.

3.5.4   Non-metric multidimensional scaling

Legendre & Legendre (1998) and McCune & Grace (2002) described NMDS. In a paper responsible for introducing this approach widely in ecology, Clarke (1993) outlined a strategy for ordination and classification of communities using NMDS with the Bray–Curtis coefficient as the similarity measure. Faith *et al.* (1987) used a form of hybrid multidimensional scaling which was further extended by Belbin (1991) and De'ath (1999). In general, NMDS has been shown to recover the ecological distance between plots and the patterns of plots along two-dimensional gradients better than PCoA and CA methods (Faith *et al.* 1987; De'ath 1999; Clarke *et al.* 2006) but there is no general consensus as to the best ordination approach in ecology (Lepš & Šmilauer 2003; Kenkel 2006; Wildi 2010).

Three major decisions define the ordination paradigm to be used: (1) data standardization to be applied to species/stand matrix; (2) similarity measure to be used; (3) ordination method.

3.5.5   Data standardization

Numerous standardizations and transformations of the data have been suggested (Greig-Smith 1983), but no general agreement has been reached on those most appropriate for vegetation data (Faith *et al.* 1987; Legendre & Legendre 1998). Two approaches are often used. Transformations, e.g. square root or log, are independent of other values in the data matrix or standardizations, for example species values are expressed as a proportion of the total stand abundance equalizing the contribution of species where values are dependent on the properties of the species/stand matrix (Lepš & Šmilauer 2003). Note that transformation will alter the relative contribution of species and stand abundance. Faith *et al.* (1987) showed that standardization improved recovery of ecological distance by dissimilarity measures for a wide variety of simulated data structures. They compared 29 combinations of dissimilarity measures and standardization; the most successful used species standardization. Equalizing species contributions implies that the presence of rare species is relatively more important than that of abundant ones and interest in stands is proportional to their richness in rare species.

In effect, differences in stand abundance, stand species richness, species abundance and dominance by individual species are regarded as unimportant to the investigation depending on the standardization chosen. Yet, these properties are well known to be influenced by the same ecological gradients as are individual species (Austin & Smith 1989; Margules *et al.* 1987; Minchin 1989). The decision to treat these collective properties as unimportant by standardizing them depends on the assumptions the researcher is making, which may not be immediately obvious. The relationship between ordination models and theoretical models of the composition of vegetation remains an area of research with many unanswered technical questions.

3.5.6   Similarity measures

A similarity measure (*S*) is calculated between every plot and every other plot. The resulting similarity matrix is used to produce an ordination (for details see Kent & Coker 1992; Legendre & Legendre 1998; McCune & Grace 2002). The results are critically dependent on the similarity measure chosen. Numerous similarity measures have been proposed (for examples see Greig-Smith 1983; Faith *et al.* 1987; Legendre & Legendre 1998; also Chapter 2). Choice depends on the assumptions the researcher is prepared to make. The common assumption is that species' responses to an environmental gradient take the form of a bell-shaped curve (Fig. 3.3). The choice of a similarity measure is often incompatible with this ecological concept (Faith *et al.* 1987). This has important consequences for the performance of similarity measures in their ability to recover information about the underlying environmental gradients.

When comparing two plots from different positions along an environmental gradient, the number of species that are absent from both plots ('double-zero matches') is critical. If two plots have no species in common it implies they are so far away from each other in environmental space that no single species can tolerate both environments. No simple measure of similarity between the two plots can measure how far apart the two plots are. When plots are closer together in environmental space then some species will occur in both plots. These species contribute information about the distance between the plots in environmental space. The number of zeros in common provides no additional information except that they are distant. Similarity measures which incorporate double-zero matching information, distort the ecological relationships.

Similarity measures summarize information in species space, which is intended to be used to construct species patterns in environmental space (Fig. 3.7). If we assume each species has a linear or unimodal response shape along an

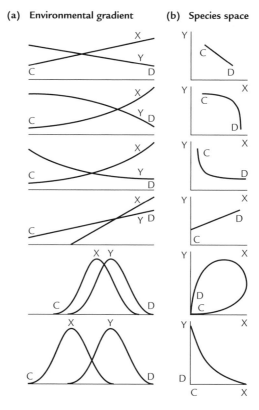

Fig. 3.7 Possible relationships between species when plotted in environmental and species space. (a) Different performances of two species X and Y along a single environmental gradient. (b) Relationships between X and Y when the performance of Y is plotted against X. It is the information in (b) which contributes to the similarity measures which ordination methods use to recreate the patterns of (a). (From Greig-Smith 1983.)

environmental gradient CD (Fig. 3.7a), then the information available in species space depends on the shape of the species response and degree of overlap between species. When the plots along the gradient CD are graphed in species space, i.e. with species abundance as axes, then the gradient becomes twisted in a complex fashion. Only in the top row is there a simple relationship between gradient CD and the equivalent line CD in species space (Fig. 3.7b). It is only linear under the circumstances represented by the first row. The relationship between plot composition in species space and the environmental gradient is rarely linear. Similarity measures estimate distance in species space (Fig. 3.7b) not in environmental space (Fig. 3.7a). If the similarity measure is based on simple linear or Euclidean concepts of distance and a PCA ordination is applied, then severe distortions of the ordinations may result, including the 'horseshoe effect'. When a series of plots from a sequence of unimodal species along an environmental gradient (e.g. as in Fig. 3.3) are ordinated, most ordination techniques represent the plots as a horseshoe or arch in two dimensions (e.g. CA). There are incompatibilities between the data analysis models used by different authors and theories of species responses. If skewed and bimodal species responses occur (Fig. 3.5), the problem becomes even more complicated.

Faith *et al.* (1987) examined the behaviour of 29 similarity measures and standardizations using artificial data sets; see also De'ath (1999). A large number of different data sets were constructed based on different assumptions about the nature of species response curves to environmental gradients. The true ecological distance between plots along the gradient was known as a consequence of using artificial data. This could be compared with the dissimilarity estimated from the compositional data. The results for three similarity measures (Fig. 3.8) show how different outcomes can occur if the data model is not equivalent to the theoretical model of species response. The Manhattan measure (Fig. 3.8a) shows the impact of total plot abundances on compositional distance when the ecological distance is such that there are no species in common. Plots with no species in common appear similar if they both have low total abundances. The Kendall measure (Fig. 3.8b) reaches a limiting dissimilarity when there are no species in common but is sensitive to plot total abundances when there are many species in common. The symmetric quantitative Kulczynski (Fig. 3.8c) provides a more balanced representation of the ecological distance. The χ^2-measure of distance used in correspondence analysis was found to perform badly relative to the Kulczynski measure, which performed best when used with data standardized to species maxima (Faith *et al.* 1987).

3.5.7 Current position

Each of the three paradigms regarding ordination research adopts a different approach to analysis. Those using CA assume that most applications involve linear or bell-shaped species responses, the standardizations implicit in use of χ^2 dissimilarity are robust to deviations from the response model and horseshoe distortions can easily be recognized (e.g. Lepš & Šmilauer 2003; Kenkel 2006). Legendre & Gallagher (2001) adopting the PCoA approach suggest that the conclusions of Faith *et al.* (1987) regarding standardizations and dissimilarity

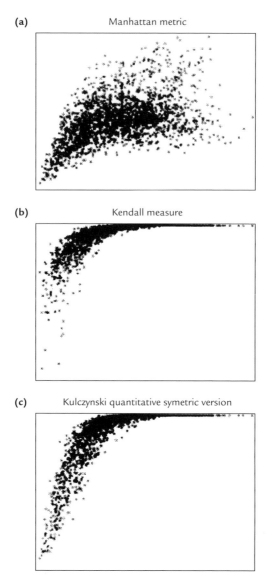

(a)   Manhattan metric

(b)   Kendall measure

(c)   Kulczynski quantitative symetric version

Fig. 3.8 The relationship between compositional dissimilarity (y axis) and the 'true' ecological distance (x axis) for artificial data calculated from known species response shapes for three different measures: (a) Manhatten metric; (b) Kendall measure; (c) Kulczynski quantitative symmetric version. (After Faith *et al.* 1987.)

measures could be combined with PCoA and reject the CA method but consider only one-dimensional gradients. The NMDS paradigm exemplified by the papers of Faith *et al.* (1987), Clarke (1993), De'ath (1999) and Clarke *et al.* (2006) may be the most robust approach at present. A comparative standard would be to use Bray–Curtis or symmetric quantitative Kulczynski coefficients after species standardization modified by De'ath's (1999) extended dissimilarity and apply NMDS.

Ordination methods are necessary to investigate the patterns found in vegetation data. Their performance depends critically on whether the assumptions they make about vegetation patterns are realistic. Unfortunately there is no census about vegetation patterns, nor has there been sufficient consistency in the comparative studies of different indirect ordination methods to reach any definitive conclusions.

3.6 Species distribution modelling or direct gradient analysis

3.6.1 Introduction

Statistical models in which species abundance or vegetation properties such as species richness are related to environmental or biotic variables by regression based methods have expanded greatly in the first decade of this century (Guisan & Zimmerman 2000; Austin 2002; Elith *et al.* 2006; Elith & Leathwick 2009a; Franklin 2009) because of the development of new methods and associated software packages. These methods allow the actual shape of species responses to ecological gradients to be investigated as opposed to indirect ordination methods where their shapes are assumed. Ordination explores the question of what are the major gradients in vegetation composition, and SDM explores the response of particular species to postulated environmental variables. These are complementary approaches to understanding vegetation variation.

The framework for any analysis of vegetation/environment relationships has three components (Austin 2002). The first component is an ecological model incorporating the ecological theory to be used or tested, for example the likely shape of the species response (Gaussian, skewed or bimodal) to an environmental variable or examining which environmental variables are most important in predicting species distributions. The second component is a data model. This concerns how the data were collected and measured. Were the data collected using a statistically designed survey procedure or is it an *ad hoc* compilation of published data? Often data are presence/absence or even presence only with no knowledge of actual absences. Whether the environmental variables selected as predictors are simply correlative or causal is an issue to be considered (Austin & van Niel 2011). The third component of the framework is a statistical model. The choice of statistical method and of the error function to be used for the species data is part of the statistical model. Assumptions made in one of the model components can confound those of another component (Austin 2002). These SDM methods offer the means to test the various models discussed in Section 3.5 and shown in Figs 3.3, 3.4, 3.5, though most recent uses have been to investigate climate change impacts.

3.6.2 Methods

A recent introductory account of these methods is provided by Elith & Leathwick (2009b), but for authoritative accounts see Elith *et al.* (2006), Elith

& Leathwick (2009a) and Franklin (2009). Numerous methods exist including machine-learning techniques (e.g. neural nets), but no consensus yet exists on either which method or which combination of methods to use (Elith & Leathwick 2009a).

Currently three methods are frequently used by plant ecologists (Elith & Leathwick 2009a; Franklin 2009) in addition to CCA.

1   Generalized linear modelling (GLM). This method is a generalization of normal least-squares analysis using maximum likelihood. It allows the analysis of various types of data and error functions, in particular presence/absence data with its binomial error function (see McCullagh & Nelder 1989).
2   Generalized additive modelling (GAM). This method is a non-parametric extension of GLM. It uses a data smoothing procedure, which has the great advantage that the exact shape of the species response does not have to be specified by a mathematical function prior to the analysis but the number of inflexions needs to specified. Yee & Mitchell (1991) introduced GAM into plant ecology. See Hastie & Tibshirani (1990).
3   Maximum entropy modelling (MaxEnt). This method is specifically designed to provide the best possible predictive models when used with presence only data (Phillips *et al*. 2006). This is a major advantage as other statistical models require at least presence/absence data or problematic pseudo-absences (Elith *et al*. 2011).

CCA (ter Braak 1986; Jongman *et al*. 1995), a combination of indirect ordination with environmental regression, is often used in vegetation science at the present time. A key step in the method constrains the ordination axes to be maximally correlated with the environmental variables included in the analysis. The assumption is that the major variation in vegetation composition is environmental and not due to succession or other historical influences, an assumption that should always be tested. In practice, most CCA applications assume that the environmental variables are linearly correlated with the ordination axes regardless of whether they are indirect or direct variables (Austin 2002). CCA does not actually require this assumption.

3.6.3   Environmental measurements

Traditionally the environmental data collected consist of variables such as altitude, slope and aspect. These variables are typically indirect distal variables and little thought is given to the processes that may result in the variable being correlated with vegetation. (In contrast, rainfall would be an example of a direct distal variable; it is known to have a direct physiological effect on plants but the proximal variable would be moisture supply at the root hair.)

Slope and aspect are examples of variables where much is known about the environmental processes that are likely to be responsible for any correlation. Re-expression of these indirect variables as direct or resource variables should improve and clarify any observed correlation. Aspect is the compass direction in which a plot on a sloping surface may be facing. A compass bearing is a circular measure where 2° and 358° are closer to each other than either is to 340°. Various data transformations have been used to correct this problem without reference to the environmental processes involved.

A major difference between north and south facing slopes is the amount of solar radiation each receives. The potential solar radiation at a point can be calculated as a complex trigonometric function of the aspect and slope of the site, depending on the position of the sun, which varies with the time of year and latitude. No simple data transformation of aspect will capture this information about radiation, a variable that has numerous direct effects on the physiology of plants. Many different combinations of aspect and slope have equivalent radiation climates. See Austin & van Niel (2011) for further discussion in relation to modelling for assessing climate change impacts.

Ecologists have long recognized the influence of aspect. Whittaker (1956) used subjective estimates to take account of high hills cutting off the direct rays of the sun. Radiation on protected north-facing valleys can be 10% of that on exposed south facing slopes in winter at northern temperate latitudes. Today, algorithms exist for calculating radiation, which may include horizon effects, direct and diffuse radiation components, sunshine hours and atmospheric properties. Dubayah & Rich (1995) provide an account of the equations involved (for further developments see Kumar *et al.* 1997). Radiation integrates many features of the plant's environment in terms of explicit physical processes, and hence it is a more relevant variable than aspect.

This process of deriving more environmental variables that are physiologically relevant can be taken further. Actual evapotranspiration, the amount of water transpired in a given time, is a crucial indicator of the drought stress undergone by a plant. A simple model of the physical process can estimate actual evapotranspiration. Rainfall is a measure of water supply in a given period. Current storage is estimated from the available soil water capacity. Potential evapotranspiration (Ep) is a measure of demand and can be derived from weather records. Water available for transpiration is given by the sum of rainfall and the amount in the soil store. If the total of these is greater than the demand, actual transpiration is equal to potential. If the total is less, then actual transpiration is equal to the total available. A measure of moisture stress is the ratio of actual to potential evapotranspiration. This measure estimates the extent to which supply satisfies demand. At the local scale, radiation is the dominant term in the equation estimating potential evapotranspiration. Differences in moisture relations of plants on different aspects depend on the relative amounts of radiation they receive. The Ep on different aspects will vary proportional to the radiation received relative to the radiation received on a flat surface.

These processes can be incorporated into a simple water balance model to estimate moisture stress (MI) scalar for different aspects:

$$Ea_t = Ep_t \text{ if } (R_t + S_{t-1}) > Ep_t \text{ where } Ep_t = Ep.RI$$

$$Ea_t = (R_t + S_{t-1}) \text{ if } (R_t + S_{t-1}) < Ep_t$$

$$S_t = R_t + S_{t-1} - Ea_t \text{ and } MI = Ea_t/Ep_t$$

Ea_t is the actual evapotranspiration for time step t; Ep_t is the potential evapotranspiration for time step t adjusted by the relative radiation index RI; R_t is rainfall for time step t; S_{t-1} is the soil moisture remaining from the last time step $t–1$.

The water balance model can use monthly average values for Ep and rainfall or actual data. When the average values reach equilibrium, the moisture stress index would then be estimated for a particular season, annually or for a specific weather sequence. Environmental scalars such as that described here for moisture stress can be made more elaborate. They may then appear more precise than those based on field observations. *Poterium sanguisorba* (now *Sanguisorba minor*), which shows a distribution centred on south-west slopes in Fig. 3.2 provides an example of the problems of scalars. The species is typical of chalk grassland species that show this distribution in England (Perring 1959). It is well known to field ecologists that south-west aspects with slopes above 15° are the most drought-prone in the northern hemisphere and characterized by species with more southern distributions. This is inconsistent with the moisture stress scalar and radiation model described earlier. The radiation model predicts that radiation is greatest on slopes facing directly south, i.e. aspect 180°. Radiation is symmetric about south, with south-east and south-west slopes receiving the same amount of radiation. The model is inadequate. Differences in potential evapotranspiration between aspects arise from differences in radiation and air temperatures. Temperatures in the afternoon are higher when radiation is falling on south-west aspects than when the same amount of radiation falls on south-east aspects in the morning. The physical model is incomplete; a significant component has been omitted. Analysis of vegetation/environment correlation is an iterative process requiring constant testing of the model with field observations. Development of environmental scalars for use in environmental modelling with GIS is a very active area of research and improvements are constantly being made. Environmental scalars integrating our physical and physiological knowledge can be generated from many environmental factors. Guisan & Zimmerman (2000) provide a useful review of current ideas. Fig. 3.9 shows the many different connections that can exist between distal variables and the more proximal direct variables. To understand vegetation patterns, we need to understand the environmental processes that are responsible.

3.6.4 Applications

CCA is usually used to determine the environmental correlates of the variation in vegetation composition while accepting the assumptions of the underlying ecological, data and statistical models. The three other methods (GLM, GAM, and MaxEnt) are frequently used to predict species distributions in a region from survey data in conjunction with a geographic information system (GIS).

The work of Leathwick in New Zealand exemplifies the use of GLM and GAM to investigate species/environment relationships taking careful account of ecological history. The forests of New Zealand are composed of three groups of tree species, broad-leaved evergreen species, Gondwanan conifers and *Nothofagus* species. Composition varies across a wide range of climatic conditions (mean annual temperature from *c.* 5.0 to 16.0°C, mean annual rainfall from 400 to

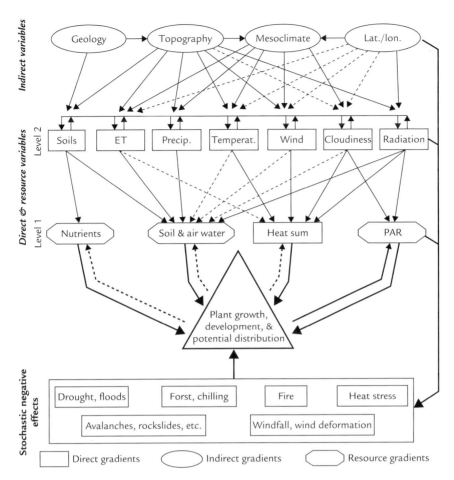

Fig. 3.9 Relationship of indirect and direct environmental variables and their possible combination into scalars. (From Guisan & Zimmerman 2000.)

>10 000 mm) and in response to historical disturbances (volcanic eruptions and earthquakes). Existing extensive plot survey data on species tree density for stems >30 cm diameter have been coupled with a GIS for New Zealand, which provides information on climate, and biophysical variables for use as environmental predictors for each plot.

Leathwick & Mitchell (1992) examined data from the central North Island of New Zealand. They modelled 11 tree species using presence/absence data and GLM with the predictors mean annual temperature, solar radiation difference, mean annual rainfall, and depth of Taupo pumice as continuous variables. Topography and drainage were treated as categorical variables. Quadratic terms for the continuous predictors were used to test for curvilinear responses. Mean annual temperature was a predictor in all models with a quadratic term significant for 10 species confirming the importance of unimodal species responses. It was the most important predictor for nine species. The solar radiation variable was the second most important predictor overall. Depth of pumice included in six models is a surrogate variable representing both a physical substrate predictor and the distance from a major historical disturbance, the Taupo volcanic eruption of 130 AD. Both environment and succession since the volcanic eruption influence current species distribution. The statistical models demonstrated the relative importance of climatic and volcanic variables in determining forest composition. Leathwick (1995) extended the analysis to the whole of New Zealand to examine the climatic relationships of 33 tree species using GAM. Many species responses to environmental variables were shown to depart from the symmetric unimodal curves often assumed.

Leathwick (1998) explored whether the distribution of the *Nothofagus* species were due to environmental variables or due to slow dispersal after postglacial climate changes and volcanic catastrophes. A proximity factor (presence of *Nothofagus* on other plots within 5 km of the plot) was used after fitting the environmental models to demonstrate significant spatial autocorrelation in the distribution of the *Nothofagus* species. This supports the non-equilibrium explanation of slow dispersal for their observed distribution. Regression models can test hypotheses about the non-equilibrium nature of forest composition.

Nothofagus species are frequently the dominant species in the communities where they occur. Many species that occur in association with *Nothofagus* also occur in identical environments without *Nothofagus* as a dominant. It is possible, therefore, to model the impact of competition from the dominant *Nothofagus* species on the co-occurring species (Leathwick & Austin 2001). The results show that density of *Nothofagus* species has significant effects on the species composition of forests. Introduction of significant interaction terms between *Nothofagus* density and temperature and moisture suggested that competition effects vary with the position of the plots on the environmental gradients of temperature and moisture. Competitive effects of dominant species conditional on environment have been demonstrated with broad-scale survey data. The development of these regression models including environment, competition and historical limitations on the dispersal of dominant species has allowed the prediction of New Zealand's potential forest composition and pattern across the whole country (Leathwick 2001).

An example of a skewed response surface found using GAM and the geographical prediction possible when a GIS with suitable layers is shown in Fig. 3.10. *Prumnopitys taxifolia* is a Gondwanan conifer with a distinct dry east-coast distribution in New Zealand. The methods used were based on those in Leathwick (2001).

The progressive development of realistic environmental processes is well demonstrated by these New Zealand studies in the case of the moisture stress indices discussed earlier. The initial environmental predictor for the moisture component was mean annual rainfall (Leathwick & Mitchell 1992), then the ratio of summer rainfall to summer potential evapotranspiration was used (Leathwick 1995). Leathwick *et al.* (1996) developed a soil water balance model to estimate an annual integral of water deficit based on a 1 in 10 year drought rainfall. In addition monthly relative humidity was found to be a significant predictor for many species (Leathwick 1998).

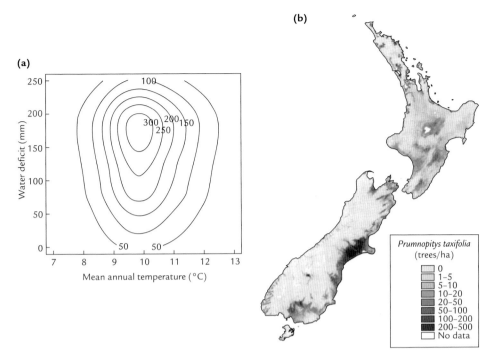

Fig. 3.10   Example of the environmental GAM model and the geographical distribution map generated from it. (a) Part of GAM model predicting response of *Prumnopitys taxifolia* density (trees/ha) to annual water deficit and mean annual temperature. Note skewed response to annual water deficit. (b) Predicted geographical distribution of *P. taxifolia* density for New Zealand. Note abundance on the dry east coasts. (Figure kindly supplied by J.L. Leathwick, Landcare New Zealand.)

The four important developments in SDM to note are:

1   progressive incorporation of better statistical methods which are more consistent with current ecological concepts;
2   increasing realism of the ecological concepts incorporated into the models;
3   improved representation of environmental processes;
4   development of specific methods for use with the abundant presence data found in herbarium and museum records (Phillips *et al.* 2006; Elith *et al.* 2006, 2011).

3.6.5   Limitations

Observational analysis of the kind described in Section 3.6.4 is often denigrated as 'mere correlation' and not causation. Shipley (2000) reminded us that correlation is merely 'unresolved' causation. Resolving an observed correlation may result in a causal explanation of no ecological interest, or it may yield a detailed set of relevant hypotheses. The New Zealand studies have provided a set of hypotheses and estimates of the relative importance of different environmental variables. This needs to be contrasted with the intuitive correlations on which many ecological hypotheses and subsequent experiments are based. One frequent limitation of species distributional modelling is the lack of a dynamic component. Vegetation is often assumed to be in equilibrium with the environment. Repeated measurements on the same plots can be used to study the successional changes in vegetation by means of a trajectory analysis using ordination (e.g. Greig-Smith 1983, p. 309). The impact of historical events can be incorporated into regression studies (e.g. time since fire). Leathwick (2001) takes an alternative approach where competition from dominants and spatial autocorrelation are introduced into the predictors to account for non-equilibrium effects due to historical events and slow dispersal of certain species.

A common limitation is the mismatch between ecological assumptions and the statistical methods used. Studies with both CCA and GLM often assume that vegetation variables have a straight-line relationship with environmental variables. There is no ecological or statistical reason to impose this limitation. Theory suggests a curvilinear unimodal response with a maximum occurring between upper and lower limits beyond which the species does not occur (Austin 2002). GLM models are often fitted with polynomial functions which have undesirable properties (Austin *et al.* 1990), although any parametric function can be used. There is a growing literature on discussions of what methods should be used and why and on comparative evaluations of different methods (Elith & Leathwick 2009a; Franklin 2009; Elith *et al.* 2011).

3.6.6 Potential of SDM for analysing the continuum concept

Species distribution modelling can investigate the patterns of hypothetical responses put forward in Figs 3.3, 3.4, 3.5 by providing descriptions of the response shape of species to an environmental gradient while allowing for the influence of other environmental predictors. Four recent papers have examined aspects of those responses.

Peppler-Lisbach & Kleyer (2009) tested species richness patterns and the continuum hypothesis using composition turnover rates based on HOF (Huisman–Olff–Fresco) models of species response curves (Huisman *et al.* 1993). The responses of 119 understorey species along a pH gradient were examined in German deciduous hardwood forests. The response of the collective property species richness approximated a hyperbola. There were positions on the pH gradient where species responses change in concert with each other, possibly as a result of a threshold effect due to toxic ions (e.g. aluminium). The majority of species responses which were not truncated were skewed or plateau responses. This may be a result of fitting the restricted HOF set of curves and the short length of the gradient (pH 2.5–6.1). The authors concluded that the hypothesis 'that skewed response curves are characteristic for the extremes of a gradient (Austin 1990) is confirmed for a specific group of species, but not as a general pattern.' However the concept that species responses were independent and hence species turnover rates were constant along the gradient was rejected

(Peppler-Lisbach & Kleyer 2009). Peper *et al.* (2011) undertook a similar approach but in relation to an extreme grazing gradient where the endpoint was sites with no plants. There was as a consequence a monotonic decline in species richness and the majority of species showed sigmoidal negative responses.

Heikkinen & Mäkipää (2010) examined three different questions: '(1) are species optima uniformly distributed along a soil fertility (C/N ratio) gradient, (2) is niche width dependent on the location of a species optimum, and (3) does skewness of the response curves depend on the location of the optimum?' They concluded that (1) the density of optima peaked at a relatively low C/N where 'optima of . . . species were highly packed,' (2) 'niche width was negatively correlated with density of optima,' and (3) 'skewness of the response curves was positively correlated with location of optima' but non-significant. Patterns of species responses will vary with the nature of the environmental gradient.

Normand *et al.* (2009) tested the asymmetric abiotic stress limitation (AASL) hypothesis for three climatic gradients using 1577 European species. The hypothesis states that species have skewed responses to environmental gradients with a steep decline towards the extreme stressful conditions. The AASL hypothesis is in part based on Austin (1990) and Austin & Gaywood (1994); Fig. 3.3e is a graphical statement of the hypothesis. Three climatic gradients were modelled (minimum temperature, growing day degrees and water balance) using several methods (HOF curves, polynomial GLM and GAM) and two data sets. They concluded that the AASL hypothesis is supported for 'almost half of the studied species' (Normand *et al.* 2009).

The studies differ in the questions they ask, the methods they use, the gradients sampled and their length and the attention given to collective properties and discontinuities. However, Normand *et al.* (2009) shows that AASL hypothesis holds true for a large number of species in relation to climatic gradients. The continuum as conceived in Fig. 3.3e is supported. The analysis of patterns along a pH gradient by Peppler-Lisbach & Kleyer (2009) supports this conclusion but suggests that there may well be discontinuities in species responses at certain points along specific gradients. The results of Peper *et al.* (2011) for grazing and Heikkinen & Mäkipää (2010) for soil fertility gradients reinforce the possibility that species response patterns vary with the position on and the nature of the environmental gradient. These studies demonstrate that the potential exists to test models of vegetation composition using recently developed SDM regression methods.

3.7 Synthesis

Currently three research paradigms (Kuhn 1970) can be recognized in vegetation science as it concerns questions of whether vegetation is continuous or discontinuous and how it relates to environment. In Kuhnian paradigms, confirmatory studies, those providing supporting evidence, are more usual than tests of the basic assumptions of the paradigm whether the assumptions concern facts, theory or methodology. A willingness to recognize the strengths and weaknesses of each paradigm is needed to achieve a synthesis.

Traditional phytosociology constitutes one paradigm. The conceptual framework concerns the recognition and definition of the association and the hierarchical classification of associations; vegetation is assumed to be discontinuous. The two other recognizable paradigms tacitly accept the continuum concept; variation in vegetation composition is continuous and largely determined by environment. The two paradigms differ in the assumptions they make about species responses to environment and the methods selected to study vegetation variation. The multivariate analysis paradigm emphasizes the use of ordination and classification techniques (e.g. CCA). Three subordinate paradigms within this paradigm have been recognized here associated with different methods – CA, PCoA and NMDS (Section 3.5). Resolution of these methodological differences will require more explicit statements on vegetation theory. The emerging paradigm has been the use of statistical regression methods (SDM) to study the continuum (Normand *et al.* 2009; Peppler-Lisbach & Kleyer 2009; Heikkinen & Mäkipää 2010; Peper *et al.* 2011). This approach will allow the testing of assumptions about the species response to environmental variation.

The differences between these paradigms are less than many plant ecologists assume. The phytosociological association with its characteristic and differential species can be said to define a region in environmental space. This region of environmental space is more frequent in the landscape than others. The combinations of species characteristic of that environment are therefore frequent in the landscape and hence more recognizable (Figs 3.3, 3.4, 3.5). This hypothesis relates the association with a region in the multidimensional continuum and requires testing.

A synthesis of the three paradigms would help plant ecologists to focus on the unresolved issues of intrinsic causes of vegetation variation and the influence of environment on such variation. A possible framework is outlined here; the intention is not to suggest that this is the solution but to provide a topic for discussion.

Among the early pioneers of the direct gradient analysis of vegetation, Perring (1958, 1959) provided the most explicit conceptual framework. He proposed that vegetation properties were a function of various groups of environmental variables (factors), for example climate, or topography. On this basis vegetation studies should be undertaken with a stratified survey design based on such an explicit model of the possible processes involved. The variable groups were all indirect variables. These could be expressed as direct or resource variables (Fig. 3.9) using our increased knowledge of environmental processes. Accepting the interpretation presented in Fig. 3.2b, there is an interaction between regional climate and the microclimate as represented by aspect and slope. Each species shows an individualistic response shifting its topographic distribution depending on climate. The use of direct gradients might simplify the figure to a single gradient of moisture stress. Austin & van Niel (2011) discuss the use of a similar framework for use with SDMs applying our increased knowledge of biophysical processes and statistical modelling. GAM models as used by Leathwick provide rigorous quantitative descriptions of the relationships.

Questions of the existence of plant communities or of the relative importance of different direct gradients could be examined with appropriate stratified designs. There are problems of both theory and methodology that need to be addressed. The importance of history and geographical barriers in determining current vegetation composition needs to be examined. The methodology needed will have to incorporate spatial autocorrelation into statistical models or ordination techniques. Current methods often ignore interactions, while current theory tends to focus on a single dimension or continuum of variation. The pattern of variation in multidimensional space is a key issue; almost nothing is known about the shapes and orientation of species distributions in multidimensional environmental space. Species packing and distribution of species richness are other unknown patterns in this space.

Whether vegetation is discontinuous or continuous depends on the perspective of the viewer. Viewed from a landscape perspective it is often discontinuous. In environmental space it is usually thought to be continuous. Rigorous testing of vegetation patterns in this space has yet to be achieved. The descriptive patterns resulting from multivariate pattern analysis and statistical modelling take us only so far. Understanding these patterns is an essential ingredient in sustainable vegetation management. At the present time there are many unanswered questions in vegetation science but there are also too many unquestioned answers.

Acknowledgements

I thank P. Gibbons, C.J. Krebs, R.P. McIntosh, J. Reid, B. Wellington and the editors for comments on an earlier draft chapter.

References

Austin, M.P. (1987) Models for the analysis of species response to environmental gradients. *Vegetatio* 69, 35–45.

Austin, M.P. (1990) Community theory and competition in vegetation. In: *Perspectives in Plant Competition* (eds D. Tilman & J.B. Grace), pp. 215–237. Academic Press, San Diego, CA.

Austin, M.P. (1991) Vegetation theory in relation to cost-efficient surveys. In: *Nature Conservation: Cost-effective Biological Surveys and data analysis. Proceedings of a CONCOM Workshop* (eds C.R. Margules & M.P. Austin), pp.17–22. Canberra.

Austin, M.P. (1999a) The potential contribution of vegetation ecology to biodiversity research. *Ecography* 22, 465–484.

Austin, M.P. (1999b) A silent clash of paradigms: some inconsistencies in community ecology. *Oikos* 86, 170–178.

Austin, M.P. (2002) Spatial prediction of species distribution: an interface between ecological theory and statistical modelling. *Ecological Modelling* 157, 101–118.

Austin, M.P. & Gaywood, M.J. (1994) Current problems of environmental gradients and species response curves in relation to continuum theory. *Journal of Vegetation Science* 5, 473–482.

Austin, M.P. & Smith, T.M. (1989) A new model for the continuum concept. *Vegetatio* 83, 35–47.

Austin, M.P. & Van Niel, K.P. (2011) Improving species distribution models for climate change studies. *Journal of Biogeography* 38, 1–8.

Austin, M.P., Nicholls, A.O. & Margules, C.R. (1990) Measurement of the realized qualitative niche of plant species: examples of the environmental niches of five Eucalyptus species. *Ecological Monographs* 60, 161–177.

Austin, M.P., Belbin, L., Meyers, J.A., Doherty, M.D. & Luoto, M. (2006) Evaluation of statistical models used for predicting plant species distributions: role of artificial data and theory. *Ecological Modelling* **199**, 197–216.

Belbin, L. (1991) Semi-strong hybrid scaling, a new ordination algorithm. *Journal of Vegetation Science* **2**, 491–496.

Bray, J.R. & Curtis, J.T. (1957) An ordination of the upland forest communities of southern Wisconsin. *Ecological Monographs* **27**, 325–349.

Clarke, K.R. (1993). Non-parametric multivariate analyses of changes in community structure. *Australian Journal of Ecology* **18**, 117–143.

Clarke, K.R. & Gorley, R.N. (2006) *PRIMER v6: User Manual/Tutorial*. PRIMER-E, Plymouth.

Clarke, K.R., Somerfield, P.J. & Chapman, M.G. (2006) On resemblance measures for ecological studies, including taxonomic dissimilarities and a zero-adjusted Bray–Curtis coefficient for denuded assemblages. *Journal of Experimental Marine Biology and Ecology* **330**, 55–80.

Curtis, J.T. (1959) *The Vegetation of Wisconsin: An Ordination of Plant Communities*. University of Wisconsin Press, Madison, WI.

De'ath, G. (1999) Extended dissimilarity: a method of robust estimation of ecological distances from high beta diversity data. *Plant Ecology* **144**, 191–199.

Drake, J.A. (1991) Community-assembly mechanics and the structure of an experimental species ensemble. *American Naturalist* **137**, 1–25.

Dubayah, R. & Rich, P.M. (1995) Topographic solar radiation models for GIS. *International Journal of Geographical Information Systems* **9**, 405–419.

Elith, J. & Leathwick, J.R. (2009a) Species distribution models: ecological explanation and prediction across space and time. *Annual Review of Ecology, Evolution and Systematics* **4**, 677–697

Elith, J. & Leathwick, J.R. (2009b) The contribution of species modelling to conservation prioritization. In *Spatial Conservation Prioritization: Quantitative Methods & Computational Tools* (eds A. Moilanen, K.A. Wilson & H.P. Possingham), pp 70–93. Oxford University Press, Oxford.

Elith, J., Graham, C.H., Anderson, R.P. *et al.* (2006) Novel methods improve prediction of species distributions from occurrence data. *Ecography* **29**, 129–151.

Elith, J., Phillips. S.J., Hastle, T. *et al.* (2011) A statistical explanation of MaxEnt for ecologists. *Diversity and Distributions* **17**, 43–57.

Ellenberg, H. (1988) *Vegetation Ecology of Central Europe*. 4th ed. Cambridge University Press, Cambridge.

Faith, D.P., Minchin P.R. & Belbin L. (1987) Compositional dissimilarity as a robust measure of ecological distance. *Vegetatio* **69**, 57–68.

Franklin, J. (2009) *Mapping Species Distributions: Spatial Inference and Prediction*. Cambridge University Press, Cambridge.

Gauch, H.G. & Whittaker, R.H. (1972) Coenocline simulation. *Ecology* **53**, 446–451.

Gittins, R. (1965) Multivariate approaches to a limestone grassland community. 1. A stand ordination. *Journal of Ecology* **53**, 385–401.

Gleason, H.A. (1926) The individualistic concept of the plant association. *Bulletin of the Torrey Botanical Club* **53**, 1–20.

Gower, J.C. (1966) Some distance properties of latent root and vector methods used in multivariate analysis. *Biometrika* **53**, 325–338.

Greig-Smith, P. (1983) *Quantitative Plant Ecology*, 3rd edn. Blackwell Scientific Publications, Oxford.

Grime, J.P. (2001) *Plant Strategies, Vegetation Processes, and Ecosystem Properties*, 2nd edn. John Wiley & Sons, Chichester.

Guisan, A. & Zimmerman, N.E. (2000) Predictive habitat distribution models in ecology. *Ecological Modelling* **135**, 147–186.

Hastie, T. & Tibshirani, R. (1990) *Generalised Additive Models*. Chapman and Hall, London.

Heikkinen, J., & Mäkipää,R. (2010) Testing hypotheses on shape and distribution of ecological response curves. *Ecological Modelling* **221**, 388–399.

Huisman, J., Olff, H. & Fresco, L.F.M. (1993). A hierarchial set of models for species response analysis. *Journal of Vegetation Science* **4**, 37–46.

Jenny, H. (1941) *Factors of Soil Formation*. McGraw-Hill, New York, NY.

Jongman, R.G.H., ter Braak, C.J.F. & van Tongeren, O.F. (1995). *Data Analysis in Community and Landscape Ecology*. Cambridge University Press, Cambridge.

Kenkel, N.C. (2006) On selecting an appropriate multivariate analysis. *Canadian Journal of Plant Science* **86**, 663–676.

Kent, M. & Coker, P. (1992) *Vegetation Description and Analysis: A Practical Approach*. Belhaven Press, London.

Kuhn, T. S. (1970) *The Structure of Scientific Revolutions*, 2nd edn. The University of Chicago Press, Chicago.

Kumar, L., Skidmore, A.K. & Knowles, E. (1997) Modelling topographic variation in solar radiation in a GIS environment. *International Journal of Geographical Information Science* **11**, 475–497

Leathwick, J.R. (1995) Climatic relationships of some New Zealand forest tree species. *Journal of Vegetation Science* **6**, 237–248.

Leathwick, J.R. (1998) Are New Zealand's *Nothofagus* species in equilibrium with their environment? *Journal of Vegetation Science* **9**, 719–732.

Leathwick, J.R. (2001) New Zealand's potential forest pattern as predicted from current species-environment relationships. *New Zealand Journal of Botany* **39**, 447–464.

Leathwick, J.R. & Austin, M.P. (2001) Competitive interactions between tree species in New Zealand old-growth indigenous forests. *Ecology* **82**, 2560–2573.

Leathwick, J.R. & Mitchell, N.D. (1992) Forest pattern, climate and vulcanism in central North Island, New Zealand. *Journal of Vegetation Science* **3**, 603–616.

Leathwick, J.R., Whitehead, D. & McLeod, M. (1996) Predicting changes in the composition of New Zealand's indigenous forests in response to global warming: a modelling approach. *Environmental Software* **11**, 81–90.

Legendre, P. & Gallagher, E.D. (2001) Ecologically meaningful transformations for ordination of species data. *Oecologia* **129**, 271–280.

Legendre, P. & Legendre, L. (1998) *Numerical Ecology*, 2nd English edn. Elsevier Science, Amsterdam.

Lepš, J. & Šmilauer, P. (2003) *Multivariate Analysis of Ecological Data using CANOCO*. Cambridge University Press, Cambridge.

Margules, C.R., Nicholls A.O. &. Austin, M.P. (1987) Diversity of *Eucalyptus* species predicted by a multi variables environmental gradient. *Oecologia (Berlin)* **71**, 229–232.

McCullagh, P. & Nelder, J.A. (1989) *Generalized Linear Models*, 2nd edn. Chapman and Hall, London.

McCune, B. & Grace, J.B. (2002) *Analysis of Ecological Communities*. MjM Software Design, Gleneden Beach, Oregon.

McIntosh, R.P. (1985) *The Background of Ecology*. Cambridge University Press, Cambridge.

Minchin, P.R. (1987) An evaluation of the relative robustness of techniques for ecological ordination. *Vegetatio* **69**, 89–107.

Minchin, P.R. (1989) Montane vegetation of the Mt. Field Massif, Tasmania: a test of some hypotheses about properties of community patterns. *Vegetatio* **83**, 97–110.

Mueller-Dombois, D. & Ellenberg, H. (1974) *Aims and Methods of Vegetation Ecology*. John Wiley & Sons, Ltd, New York, NY.

Normand, S., Treier, U.A., Randin, C. *et al.* (2009) Importance of abiotic stress as a range-limit determinant for European plants: insights from species responses to climatic gradients. *Global Ecology and Biogeography* **18**, 437–449.

Økland, R.H. (1999) On the variation explained by ordination and constrained ordination axes. *Journal of Vegetation Science* **10**, 131–136.

Palmer, M.W. (1993) Putting things in even better order: The advantages of canonical correspondence analysis. *Ecology* **74**, 2215–2230.

Peper, J., Jansen, F., Pietzsch, D. & Manthey, M. (2011) Patterns of plant species turnover along grazing gradients. *Journal of Vegetation Science* **22**, 457–466.

Peppler-Lisbach, C. & Kleyer, M. (2009) Patterns of species richness and turnover along the pH gradient in deciduous forests: testing the continuum hypothesis. *Journal of Vegetation Science* **20**, 984–995.

Perring, F. (1958) A theoretical approach to a study of chalk grassland. *Journal of Ecology* **46**, 665–679.

Perring, F. (1959) Topographical gradients of chalk grassland. *Journal of Ecology* **47**, 447–481.

Perring, F. (1960) Climatic gradients of chalk grassland. *Journal of Ecology* **48**, 415–442.

Phillips, S.J., Anderson, R.P. & Schapire, R.E. (2006) Maximum entropy modeling of species geographic distributions. *Ecological Modelling* **190**, 231–259.

Podani, J. & Miklós, I. (2002) Resemblance coefficients and the horseshoe effect in principal coordinates analysis. *Ecology* **83**, 3331–3343.

Shipley, B. (2000) *Cause and Correlation in Biology: AUser's Guide to Path Analysis, Structural Equations and Causal Inference*. Cambridge University Press, Cambridge.

Shipley, B. & Keddy, P.A. (1987) The individualistic and community-unit concepts as falsifiable hypotheses. *Vegetatio* **69**, 47–55.

Swan, J.M.A. (1970) An examination of some ordination problems by use of simulated vegetational data. *Ecology* **51**, 89–102.

ter Braak, C.J.F. (1986) Canonical correspondence analysis: a new eigenvector technique for multivariate direct gradient analysis. *Ecology* **67**, 1167–1179.

ter Braak, C.J.F., Juggins, S., Birks, H.J.B. & van der Voet, H. (1993) Weighted averaging partial least squares regression (WA-PLS): definition and comparison with other methods for species-environment calibration. In: *Multivariate Environmental Statistics* (eds G.P. Patil & C.R. Rao), Vol. 6, pp. 525–560. North-Holland Publishing Company, Amsterdam.

Webb, D.A. (1954) Is the classification of plant communities either possible or desirable? *Botanisk Tidsskrift* **51**, 362–370.

Westhoff, V. & van der Maarel, E. (1978) The Braun-Blanquet approach. In: *Classification of Plant Communities*, 2nd. edn (ed. R.H. Whittaker), pp. 287–399. Junk, The Hague.

Whittaker, R.H. (1956) Vegetation of the Great Smoky Mountains. *Ecological Monographs* **26**, 1–80.

Whittaker, R.H. (1962) Classification of natural communities. *Botanical Review* **28**, 1–239.

Whittaker, R.H. (1967) Gradient analysis of vegetation. *Biological Review* **42**, 207–264.

Whittaker, R.H. (ed) (1978a) *Ordination of Plant Communities*. Junk, The Hague.

Whittaker, R.H. (ed) (1978b) *Classification of Plant Communities*. Junk, The Hague.

Wildi, O. (2010) *Data Analysis in Vegetation Ecology*. Wiley-Blackwell, Chichester.

Yee, T.W. & Mitchell, N.D. (1991) Generalised additive models in plant ecology. *Journal of Vegetation Science* **2**, 587–602.

Zuur, A.F., Ieno, E.N. & Smith, G.M. (2007). *Analysing Ecological Data*. Springer Science + Business Media, New York, NY.

4

Vegetation Dynamics

Steward T.A. Pickett[1], Mary L. Cadenasso[2] and Scott J. Meiners[3]
[1]Cary Institute of Ecosystem Studies, New York, USA
[2]University of California Davis, USA
[3]Eastern Illinois University, USA

4.1 Introduction

Succession is a fundamental concept in ecology. Simply, it is the change in species composition or in the three-dimensional architecture of the plant cover of a specified place through time. Such changes can occur on substrates that are newly created, or on those which are newly cleared or reduced in vegetation cover. The first case is labelled as primary succession, while the second, on which there is a legacy from prior vegetation, is labelled secondary succession. When vegetation dynamics was first codified by ecologists, they focused on three key features: (i) a discrete starting point; (ii) a clear directional trajectory; and (iii) an unambiguous end (Clements 1916). These three assumptions have been associated with the term 'succession'. With these limiting assumptions, succession becomes a special case of vegetation dynamics.

This chapter takes a broad view of vegetation dynamics that does not always accept the narrower assumptions of succession (Pickett *et al*. 2011). The larger focus helps solve many of the arguments and controversies about succession that have emerged from the failure of the narrower concept to portray the variety of patterns and causes of vegetation change in the field (Botkin & Sobel 1975). Controversies have focused on (i) a single, stable endpoint of vegetation change, (ii) the balance of internal community organization compared to the role of external events and constraints, and (iii) the determinism of transitions between subsequent communities over time. Using the larger concept of vegetation dynamics, ecologists can appreciate and understand the complexity found in real ecosystems (Cramer & Hobbs 2007), and can apply it in vegetation management (Davis *et al*. 2005).

Vegetation Ecology, Second Edition. Eddy van der Maarel and Janet Franklin.
© 2013 John Wiley & Sons, Ltd. Published 2013 by John Wiley & Sons, Ltd.

4.2   The causes of vegetation dynamics

4.2.1   Vegetation dynamics and natural selection

Vegetation dynamics has many causes (Glenn-Lewin *et al.* 1992). Causes in biology can often be cast in terms of conditional, or 'if-then' statements (Pickett *et al.* 2007). The general form of a conditional statement is this: if a certain condition holds, then a certain result will follow. Natural selection, for example, is a series of conditionals that leads to a consequence. If: (i) offspring vary; (ii) at least some of that variation is heritable; and (iii) more offspring are produced than can survive; then variation that matches the environmental conditions will tend to accumulate in a population (Mayr 1991). This conditional law requires multiple processes, is probabilistic and sets the bounds of change. The theory of evolution is a contingent, nested and probabilistic theory of a form that can be adopted for understanding the processes of vegetation change (Fig. 4.1).

As a law, vegetation change is based on the fundamental idea that the different capacities of plants to match the prevailing environment determines the nature of the plant assemblage that will exist in a place (Clark & McLachlan 2003).

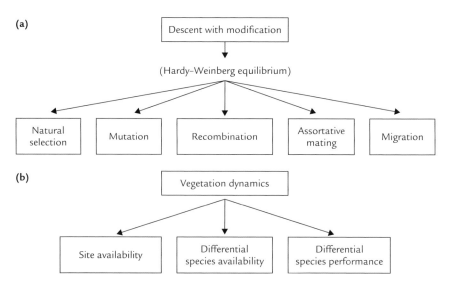

Fig. 4.1 Comparison of the hierarchical structures of the theory of evolution and the theory of succession or vegetation dynamics. Evolution (a) is summarized most generally as the phenomenon of descent with modification. The Hardy–Weinberg Law of Equilibrium embodies the processes that can affect evolutionary change between generations. Those factors – natural selection, mutation, recombination, assortative mating, and migration, among others – can result in heritable changes between generations. The process of succession (b) is represented most generally as vegetation dynamics. Changes in any one or any combination of site availability, differential availability of species, or differential performance of species can cause the structure or composition of vegetation to change through time.

The environment includes both abiotic factors and other organisms. At its core, vegetation dynamics depends on the behaviour of individual organisms as conditioned by the physical and biological environments (Brand & Parker 1995; Parker 2004; Eliot 2007). The law of vegetation dynamics has the form of a conditional statement. It states that if: (i) a site becomes available; (ii) species are differentially available at that site; or (iii) species perform differentially at that site; then the composition or structure of vegetation will change through time (Pickett & McDonnell 1989). In the following sections the causes of vegetation dynamics will be synthesized into a single organizing framework (Fig. 4.2), and related to various vegetation dynamics that ecologists have observed.

4.2.2 Site availability

Sites become available because disturbances disrupt established vegetation, or create new surfaces. The creation of new substrates can sometimes be relatively

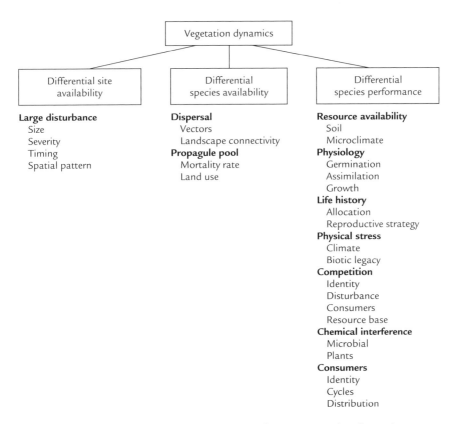

Fig. 4.2 Detailed nested hierarchy of successional causes, ranging from the most general phenomenon of community change, through the aggregated processes of site availability, and differentials in species availability and performance, to the detailed interactions, constraints, and resource conditions that govern the outcome of interactions at particular sites. (Based on Pickett *et al.* 1989.)

gradual, as when a salt marsh forms behind a new dune. Such an event can be labelled a disturbance because a previous environmental state is disrupted, leading to new conditions that generate a new vegetation structure or composition (Peters *et al.* 2011). In general terms, a disturbance is an event that alters the structure of vegetation or the substrate which vegetation is growing on (White & Pickett 1985; White & Jentsch 2001). Examples include certain kinds or intensities of fire, windstorms, stress-induced mortality of plants or herbivory. The dune example alerts ecologists to events that may be gradual on the human scale of decades, while perhaps being relatively abrupt on the scale of centuries to millennia.

The nature of the open site is governed by how intense the disturbance is, how different the created environmental conditions are relative to the prior state, how susceptible the biota are to the event (Johnson & Miyanishi 2007), how many layers of the prior vegetation are removed, or how deeply the substrate is stirred or buried (Walker 1999; Cramer & Hobbs 2007; Myster 2008). Structures that are disturbed by events capable of starting succession include, for example, forest canopies, grassland root mats and soil profiles. Some disturbance events can be very localized, such as the fall of a single tree in a forest, while others can be quite extensive, such as the opening of the forest canopy by hurricanes or typhoons.

The characteristics of a site following disturbance influence plant establishment, growth and interactions (Baeten *et al.* 2010; Shipley 2010). Disturbances affect the kinds and amounts of available resources that remain after the event, the degree to which biomass is removed or rearranged at the site, and the water and nutrient holding capacity of exposed substrates. Different disturbances may have contrasting effects on the resources available for colonizing plants. For example, fire may burn much of the organic matter at the soil surface, which will make a poorer resource base for recolonization than a windstorm that blows trees down but leaves the organic matter intact. Increased attention is being paid to above-ground and below-ground interactions, especially as conditioned by soil microbes and soil fauna (Reynolds *et al.* 2003; van der Putten *et al.* 2009).

4.2.3 Differential species availability

The way vegetation composition and structure changes after disturbance, or the emergence of new conditions in existing sites, depends on the ability of species to survive the disturbance or their ability to reach the site after the disturbance (Leck *et al.* 1989; Willson & Traveset 2000; Stearns & Likens 2002). Species may become available at the site in two ways. First, species may persist through the disturbance as seedlings, adults, seeds, tubers or the like. Second, they may arrive from elsewhere. Therefore, differential species availability depends on the characteristics of species to either survive or disperse to sites. Differential availability is important in both primary successions, including those created by shifting environmental conditions as well as physical disturbance (Glenn-Lewin & van der Maarel 1992) and in secondary successional sites.

Differential survival of individual plants after disturbances is determined by characteristics of both the species and the disturbance. Adults above some critical size may survive fires of low to moderate intensities due to thick, insulating bark. An example appears in *Sequoia sempervirens* during moderate ground fires. Adult *Sequoia* may not survive intense fires that spread into the tree crowns, however. Survival is also possible in some species possessing lignotubers that are sheltered in the relatively cooler soil environment during an intense fire that kills the above-ground parts of plants. Examples include the shrubs of chaparral or pine barrens (Forman & Boerner 1981). Survival of a population, although not of physiologically active individuals, can be accomplished by a dormant pool of seeds that is triggered to germinate by high temperatures or other post-fire conditions. Examples of this mechanism include seeds of annuals in chaparral, grasses in prairie, and trees of the pine family having serotinous cones. Thus, differential survival can be achieved by several mechanisms.

Differential species availability can depend on the ability of seedlings to tolerate unfavourable conditions for a time. The seedlings of some tree species are capable of persisting by growing slowly in deep shade, but can take advantage of the altered conditions and resource levels after the canopy and intervening layers of a forest are disturbed by wind. For example, a pool of *Prunus serotina* seedlings on the floor of undisturbed northern hardwood forests is limited by the low light availability beneath the canopy. After a blowdown of the canopy, the *Prunus* seedlings are released from suppression because of the increased light near the ground (Peterson & Pickett 1991).

Differential dispersal to open sites is determined by characteristics of species, the distances from seed sources to the available sites, or the activities of biotic and abiotic vectors that transport seeds. Some seeds disperse readily to open sites due to their small size or their wings or plumes (e.g. *Epilobium angustifolium*). Seeds also move with the help of animals. Dispersal by birds or bats that seek out forest gaps are examples of differential availability that depends on animals (see Chapter 6). While vagility of seeds and dispersal agents determine the potential of a species to enter a site, the spatial arrangements of potential colonizers in the surrounding landscape and the timing of their reproduction relative to the disturbance further constrains local colonization. However, because the spread of disturbance, or sizes of disturbed sites depend on landscape features, the probability of an open site being recolonized from dispersal of plants surviving in undisturbed habitats is also affected by landscape features. Differential dispersal thus results from a combination of landscape, plant, and vector characteristics, and is key to understanding and managing species invasions (Hobbs & Cramer 2007; Lockwood *et al.* 2005).

The neutral theory of plant communities (Hubbell 2001) emphasizes stochastic patterns of arrival to a site as the key driver of vegetation composition. This theory remains largely unsupported, as it is often difficult to statistically differentiate neutral and performance or niche based patterns of community structure (Chave 2004; Jabot *et al.* 2008). Attempts to relate the neutral and niche-based models have conceived them to represent two extremes along a continuum of neutrality (Gravel *et al.* 2006). If the neutral model is one end of a continuum that focuses on dispersal, then differential species performance is the niche-based complement. The nature of differential species performance is outlined in the next section. The nested causal hierarchy of vegetation dynamics (Fig. 4.2) also provides a way to achieve compromise between neutral and niche-based dynamics. Stochastic dispersal-mediated phenomena may operate in conjunction with more deterministic mechanisms of differential performance.

4.2.4 Differential species performance

Differential performance refers to the suite of activities that species use to acquire resources, grow, persist and reproduce (Bazzaz 1996; Loreau 1998; Luken 1990). Life history traits, relative growth rates, age to maturity, competitive ability, stress tolerance and herbivore and predator defence are some of the characteristics that will determine differential performance among species. Ecologists have examined some of the components of differential species performance for more than a century. Examples include the ability to tolerate shade compared to the demand for high levels of photosynthetically active radiation or the possession of thorns or secondary compounds that deter herbivores compared to species that are vulnerable to herbivores.

 The first example depends on the high light availability early in many successions compared to the low availability of light in older communities with closed canopies (Bazzaz 1996). Species that dominate early after disturbance in otherwise resource rich, forest environments often require high levels of light for maximum growth (Fig. 4.3). Such species have high light saturation levels of photosynthesis. In contrast, high photosynthetic efficiencies characterize the seedlings and juveniles of closed forest dominants. The forest dominants, in contrast to the early field dominants, often cannot tolerate high light levels or the rapid transpirational water loss associated with high photosynthetic rates. Though the approach was championed early by Grime (1979), there is now a growing literature addressing the traits of plants as they relate to community

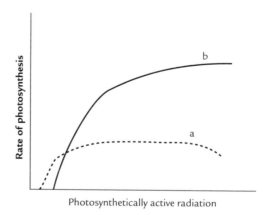

Fig. 4.3 Diagrammatic representation of the contrast between photosynthetic responses to varying availability of light of early and late-successional species. a. Shade tolerant species. b. Light-demanding species. (Following principles in Bazzaz 1996.)

assembly and vegetation dynamics (e.g. Westoby & Wright 2006; Aubin *et al.* 2009; Szabo & Prach 2009; Lebrija-Trejos *et al.* 2010; Shipley 2010; see also Chapter 12). Although succession is not the primary focus of this literature, there are clear parallels and shared utility of the approach.

A trait-based approach to succession allows partial separation of differential species performance from availability as it is less important which species make it to an area and more important which types of species are available to colonize. For example, there are five *Solidago* species within the Buell–Small Succession Study (BSS) fields. Further details are presented in Section 4.3.4. All of these species are relatively similar in life history, height, seed mass, etc., and all occupy roughly the same temporal range within succession (Pisula & Meiners 2010). Although their abundances differ among the fields, this is unlikely to reflect dramatic differences in successional trajectory, but rather vagaries of dispersal.

Trait-based, functional studies of succession may improve understanding of the diversity of mechanisms that drive vegetation change. In addition, functional approaches to community dynamics will allow direct comparison of systems which do not share species in order to determine whether the same individual drivers of community change operate in different situations. Ultimately, a trait-based approach can allow the development of broader hypotheses about the generality of individual drivers of vegetation change such as competition, herbivory or vegetative reproduction. Similarly, trait-based studies provide an opportunity evaluate vegetation dynamics across scales.

Nutrient contrasts also can drive differential species performance (Tilman 1991; Harpole & Tilman 2006). On substrates that initially lack a large nutrient pool, successful colonists often have the capacity to fix nitrogen (Kumler 1997), while species that dominate later exploit the higher levels of available nitrogen that have built up with the accumulation of humus and the increasing soil stratification resulting from initial colonizers (Vitousek *et al.* 1998). Differentials in nutrient use and adaptation are also found in very long, primary successions on new substrates, such as volcanoes. For example, on a chronosequence in Hawaii, phosphorus, which has a mineral rather than atmospheric cycle, declines through succession and shifts to a less biologically available form. The nutrient balances in the ecosystem and the plant populations shift accordingly (Vitousek 2004).

Differential performance can also be illustrated by contrasting susceptibility damage from animals (Krueger *et al.* 2009; Van Uytvanck *et al.* 2010). In sites that are exposed to large populations of browsers, species that are chemically or mechanically defended tend to dominate plant communities sooner than those woody species that are more palatable (see also Chapter 8). The effects of animals, whether invertebrates or vertebrates, have been relatively neglected over the history of succession studies. Experiments have increasingly showed the importance of herbivores in succession, however (Brown & Gange 1992; Bowers 1993; Facelli 1994; Meiners *et al.* 2000; Cadenasso *et al.* 2002). Interactions between plant toxicity and the role of herbivores in succession (Feng *et al.* 2009) or between physical stress and herbivores (Gedan *et al.* 2009) are emerging refinements to understanding the roles of consumers in succession.

4.2.5   A hierarchical framework of successional causes

Succession results from (i) the interaction of a site, either newly created or exposed by disturbance, (ii) a collection of species that can occupy that site and (iii) the interactions of the species that actually occupy the site. However, each disturbance event, the resultant characteristics of a site, and the specific characteristics and histories of the mixture of plants on that site can result in a unique trajectory of succession (Walker & del Moral 2003; Cramer & Hobbs 2007; Myster 2008; Koniak & Noy-Meir 2009; Baeten *et al.* 2010; Peters *et al.* 2011). Because there are so many factors influencing succession at a site it is important to have some way to organize the causes. Ecologists use a nested hierarchical framework to organize complex areas of study such as succession or evolution (Pickett & Kolasa 1989; Luken 1990; Pickett *et al.* 2007). Organizing factors into a hierarchical framework means that the general causes of succession must be broken down into more specific events and interactions. In such a causal hierarchy, the more specific processes are nested within the more general causes. The hierarchical framework of succession is similar to the hierarchy of evolutionary mechanisms (Fig. 4.1).

Using the hierarchical approach for succession presents the three general successional processes – site availability, differential species availability and differential species performance – as being composed of more specific causes or mechanisms (Fig. 4.2). The specific mechanisms within each factor are aspects of the abiotic environment, plants, animals and microbes, and their interactions. For succession to occur, at least one of the three general causes must operate. However, not all of the specific mechanisms that can contribute to the general causes will act in every succession. Nor will the detailed factors always act with the same intensity or relative importance. Exactly what factors dominate in a local succession depends on the history of the site and the species that reach the site. Yet the fact that we can organize the factors into a hierarchy and generalize them to three broad categories, suggests that there are broad expectations that can be drawn from succession (Glenn-Lewin & van der Maarel 1992). The causal hierarchy is a framework for explaining possible trajectories, processes, patterns and rates of succession in the field (Fig. 4.2). It also informs experiments that document the role of various factors for individual systems.

4.3   Succession in action: interaction of causes in different places

4.3.1   Complexity of successional patterns

The variety of actual successional patterns is immense. Complexity emerges from the combination of different mechanisms that can act in succession and the breadth of conditions that can affect those mechanisms. A second source of complexity in successional patterns is the breadth of conditions that can affect each of the successional causes. The different causes of succession do not operate under fixed conditions, but may each respond to important ecological gradients. For example, gradients may contrast high with low intensities of disturbance, or low to high levels of local species availability. Examples from contrasting environments show the richness of successional causes and trajectories, starting with successions shaped by large natural events continuing with processes that occur in more restricted sites, and ending with sites that experience shifts in management by people.

4.3.2   Vegetation dynamics in large, intensely disturbed sites

Floods and succession. Under the influence of large rivers, vegetation dynamics are driven by the timing, intensity and location of floods. Large floods move or deposit substantial amounts of sediment and organic debris. Such floods tend to occur infrequently. Moderate floods occur more frequently, and at least some flooding will probably occur every year at the beginning of the rainy season.

There are many effects of floods. Some effects are direct, resulting from the presence of water or the energy of moving water and the load of sediment and debris it carries. For example, flooding can kill plants that are intolerant of waterlogging, and the force of water and debris moving downstream can uproot woody plants (Sparks 1996). Other effects of floods can be indirect, such as the alteration of substrates (Moon *et al.* 1997). Substrates in which plants are rooted can be eroded away and sediment can be deposited in other areas. Both the direct and indirect effects of floods provide a heterogeneous template that can start, end, or change the course of vegetation dynamics.

The large rivers that flow through the Kruger National Park, South Africa, provide good examples of the diverse effects of flooding. From south to north in the park, there is a gradient of decreasing rainfall that determines whether the rivers flow continuously or only in the rainy season. In addition, the rivers flow through different substrates, so that in some stretches, the shape of the river channel is determined by bedrock, while in other sections, the flow interacts with deposits of sediment and vegetation (Moon *et al.* 1997; Rogers & Bestbier 1997). The new substrates laid down by flooding include sand and gravel bars, initially devoid of vegetation. Such sites support primary succession. Strictly, bedrock surfaces are not new, but if they do not support plants that survive the floods, then they too would reflect primary succession.

The vegetation sequences on bedrock and sediments differ. On bedrock, certain trees can establish in cracks and they can subsequently trap sediment. These trees tend to survive moderate floods and form a biological legacy. Clusters of stems and foliage further modify the habitat by trapping additional sediment in subsequent mild floods. With increasing sediment deposition, other trees and associated plants can establish. The earliest dominants have flexible stems and their branches and leaves can adopt a streamlined form in the current if they are submerged in moderate floods (van Coller *et al.* 1997). Such behaviour reduces the likelihood that the pioneering trees will be killed or severely damaged by later floods. These early dominants can thus survive modest floods, and continue to influence the site in ways that other species can exploit.

Succession in river channels can also occur on sediment deposited by floods. *Phragmites mauritianus* is the common colonist on newly deposited sediment in

South Africa. In many cases, plants are established from surviving rhizomes and buried stems. Once established, they trap additional sediment and further modify conditions. Some early woody dominants can resprout from stems of large trees that remain rooted although they are toppled by floodwaters and buried by sediment. The build-up of sediment has two effects. First, higher surfaces will be affected only by larger floods. Second, higher surfaces have a deeper water table. Tree species respond differently to both these effects of sediment accumulation and new species typically dominate as the sediment collects. Eventually spiny shrubs characteristic of upland vegetation and trees that cannot tolerate waterlogging emerge. In all these cases, conditions are modified by the early colonists and other species are better matched to the new environments. This net effect is called facilitation (see Chapter 7).

Floods of different intensities have different effects on succession. The typical sequences outlined above are those that are associated with floods of modest to high intensity but which occur relatively frequently. However, in 2000, extreme floods removed vegetation from both bedrock and sediment controlled sections of some of Kruger National Park's rivers. Following these more severe floods, the 'typical' sequence of vegetation dynamics was shown to be associated with only a particular part of the flooding regime. The 2000 floods were large, infrequent disturbances that made new sorts of sites available by increasing the amount of bedrock available, and set up new templates of woody debris that had not been observed earlier.

The intense floods of 2000 also set up unusual patterns of species availability because they removed some established 'upland' trees from the upper terraces near the rivers. Species availability associated with recent floods may also support novel successions because of the increase in exotic species as a result of human activities upstream of the park. All kinds of floods, representing the entire range of the temporal and spatial extent and volume, affect vegetation dynamics. The actual sequence of vegetation, expected patterns of species availability, and outcome of differential species performance, depends entirely on what part of the long-term flood regime has been studied. In all cases, the disturbances set up a spatially heterogeneous distribution of vacant substrates, surviving plants and living or dead biological legacies.

One of the key insights from the Kruger floods is that the pattern of vegetation dynamics observed depends on when one starts looking, and how long the observations last. The heterogeneity of substrate types available for the vegetation dynamics varies by flood intensity. Notably, the study of succession in these rivers continues to discover new patterns and interactions as observations encompass rarer events. However, the insight that what succession looks like depends on when the observations start, and where observations are framed in a complex disturbance regime can also guide our exploration of other cases of vegetation dynamics. There is no unambiguous point zero.

Tornado blowdown. Large areas of forest canopy can be blown down by hurricanes, large tornadoes and by downdrafts associated with large thunderstorms (Dale *et al.* 1999). Hurricanes or typhoons tend to be associated with coastal regions. Tornadoes may be spawned by hurricanes in some cases, but tornadoes are more commonly associated with convective storms located in interior regions of continents. Tornadoes tend to be temperate phenomena, while downdrafts also affect tropical habitats. Extremely severe windstorms of all types can occur on the order of once in a century or several centuries in mesic, closed canopy forest sites.

An example of a large forest blowdown is the result of the class 4 tornado in the Tionesta Scenic Area and adjacent Tionesta Research Natural Area in western Pennsylvania in spring of 1989 (Peterson & Pickett 1991). This forest had been free of large blowdowns for several hundred years, as indicated by the ages and architectures of many of the canopy trees. The canopy at the time of the storm was dominated by *Acer saccharum*, *Fagus grandifolia*, *Betula allegheniensis* and *Tsuga canadensis*.

The blowdown created a heterogeneous template for vegetation dynamics. Bare soil was exposed in the pits and on tip-up mounds created by the uprooting of canopy trees. The stacked boles of uprooted and snapped trees created a jumble of debris. Crowns of fallen trees created a cover of fine- and medium-sized woody debris and litter composed of broad leaves and needles. Dense patches of the fern *Dennstaedtia punctilobula* survived the storm in many places. The different habitats differentially favoured various species (Peterson *et al.* 1990). *Tsuga canadensis* seedlings died from desiccation on exposed soil but were protected from the depredations of deer beneath crown debris. In contrast, *B. allegheniensis* seedlings sprang up on bare soil patches and survived where soil was not waterlogged or unstable. Unfavourable sites for *B. allegheniensis* seedlings included pits, which accumulated standing water, and large soil plates eroding from tipped up root mats. Conspicuously absent were the pioneer species expected in northern hardwood forest in the eastern USA after large forest clearing events, such as *Prunus pensylvanica*. This species produces hard seeds that can survive in soil for a long time so that when a disturbance opens the canopy, the seeds are ready to germinate. However, the age of the pre-disturbance forest at Tionesta exceeded the life span of the dormant seed pool of *P. pensylvanica* and there were no viable seeds left in the soil to germinate after the disturbance. In younger forest areas near Tionesta affected by the same storm system, *P. pensylvanica* was important in the regenerating vegetation (Peterson & Carson 1996).

The Tionesta example highlights specific cases of site availability, differential species availability and differential species performance. In particular, the role of spatial heterogeneity created by the interaction of the tornado with the species composition and size of the pre-disturbance canopy was important. Differential species availability was expressed in the appearance of some species in a seedling pool on the forest floor following the disturbance, such as *T. canadensis* and *A. saccharum*. Seed banks were not important in Tionesta, although they were important for pioneers in a nearby, younger stand of similar forest. Seed rain was important for *B. allegheniensis* and some few individuals of *P. serotina* that colonized tip-up mounds. Differential species performance was expressed in drought tolerance, growth rate, interactions with herbivores and interactions between plant species. For example, sites that were protected from deer browsing by branch debris tended to support more *T. canadensis* and *F. grandifolia* seedlings than other sites, and dense patches of hay-scented fern inhibited growth of *P. serotina* and *B. allegheniensis* seedlings.

Volcanic eruption. Another example of a large, infrequent disturbance which initiated primary succession is the eruption of Mount St Helens in 1980 (Anderson & MacMahon 1985). New surfaces and substrates were among the important successional opportunities created by this event. This cone-shaped volcano had been dormant for centuries. The 1980 eruption blew off a large volume of one side of the mountain, created new surfaces in the form of mud and ash flows, displaced volumes of water from Spirit Lake in tidal wave proportions and deposited ash and coarser airborne debris widely. This single event thus produced a great variety of new substrates on which subsequent vegetation dynamics would play out. Although there were many sites in which all adult plants and seeds were killed, there were patches in some sites, such as the pumice plains, in which fast growing, nitrogen fixers emerged from a surviving, buried seed source. Other, wind-borne invaders, such as *Epilobium angustifolium*, colonized newly created sites. In a few places, animals, such as gophers (*Thomruya talpoides*), survived in their burrows and were available to interact with plants early in the succession.

The succession at Mount St Helens shows great heterogeneity of initial conditions created by the eruption (del Moral 1993). Some sites were completely new substrates, while others were highly modified, existing surfaces, and hence more akin to secondary rather than primary succession. Differential species availability played a role in both sorts of sites (Walker & del Moral 2003), although ecologists were not expecting there to be a pool of surviving seeds. Differential species performance appeared in the role of nitrogen fixers and the different life histories available immediately after the disturbance. Some patterns in the dynamics were expressions of different degrees of clonal growth and tolerance of relatively low versus high nutrient availabilities. In all cases, interaction of the plants with the heterogeneous template created by the disturbance was key. Different vegetation trajectories appeared on different patches, as was the case in the South African rivers and the Tionesta blowdown. Thus, primary and secondary successional behaviours can actually appear close to one another in space depending upon the way the substrates are disturbed, or how new substrates are laid down, and whether vegetative or sexual propagules survive the disturbance event.

4.3.3   Fine-scale vegetation dynamics

Vegetation dynamics can also respond to finer-scale and less intense events. Such events are often referred to as producing gaps in vegetation or substrate, or dependent on differential migration across a spatial matrix. Perhaps the earliest pioneer of this kind of model was Rutger Sernander (1936, reviewed by Hytteborn & Verwijst 2011). Sernander identified gaps created by windthrow in the primary boreal forest at Fiby, near Uppsala, Sweden, and the release of old but small 'dwarf trees' beneath the canopy as important parts of the forest regeneration process (Hytteborn & Verwijst 2011). The phenomenon of fine-scale vegetation dynamics is also acknowledged in the concept of 'pattern and process' (Watt 1947). Watt focused on vegetation dynamics resulting from the loss of individual plants from a closed canopy and the subsequent invasion of new individuals or release of seedlings that had been stagnant beneath the closed canopy, the phenomenon identified by Sernander (cf. Hytteborn & Verwijst 2011) as dwarf trees. A synonym is advanced regeneration.

The idea of gap phase replacement has been enlarged in several ways. First, disturbances can affect both the above-ground architecture of vegetation as well as the below-ground organization of a system (Pickett & White 1985). Second, gap phase dynamics can include openings of any scale (Prentice & Leemans 1990), as indicated by the concept of patch dynamics (Pickett & Thompson 1978). Another enlargement of the spatially dynamic pattern and process approach includes the movement of plants through a community in the concept of the 'carousel model' (van der Maarel & Sykes 1993). Like patch dynamics, the carousel emphasizes that vegetation change is not necessarily directional, and that species can occupy a patch or spatial cell at various times (Palmer & Rusch 2001). The carousel approach assumes that species can all occupy any site, but that they have differential mobilities in space (van der Maarel & Sykes 1993). It has been a major stimulus for improving the understanding of vegetation dynamics at any scale as a spatial phenomenon, not just as an interaction of neighbouring plants in small areas (van der Maarel 1996). A final general idea that has emerged from the study of gap dynamics is the regeneration niche (Grubb 1977). This concept recognizes that the niche, or physical and biological relationships, of young plants may be quite different from those which have ascended to the canopy. In other words, the regeneration niche and the adult niche may be unlike each other. Sernander's dwarf trees occupy a different multidimensional space in the realm of possible environmental relationships than adult plants of the same species. Below, we present several examples of the fine-scale vegetation dynamics that are closely aligned with Watt's (1947) conception of pattern and process and Sernander's (1936, in Hytteborn & Verwijst 2011) gap dynamics model.

Forest canopy gaps. One or a few trees can be removed from a forest canopy by several kinds of events. Wind may uproot or snap trees, lightning may kill trees, old trees may die, or parasites may kill one species in a mixed species stand, leaving a gap in the canopy. The resource availability in such gaps may be altered and environmental signals may change. For instance, in treefall gaps, substrate may be turned over by uprooting. Furthermore, water may be either more or less available, depending on whether the rainfall can better reach the forest floor compared to the rate of soil moisture removal by roots of neighbouring canopy trees or by understorey plants that remain in the gap. Nutrients, such as nitrogen, can become more available in the gap due to altered conditions for soil microbes or reduced root demand. Soil temperature extremes may increase, altering soil moisture availability or acting as a signal for germination of dormant seeds.

In an experiment conducted by cutting trees to create canopy gaps in the Kane Experimental Forest in western Pennsylvania, differential species performance was the general cause of successional dynamics following the experimental treatment (Collins & Pickett 1987, 1988). The experimental treatment mimicked a windstorm that snapped off the trees rather than uprooted them. Interactions between the causal factors of vegetation dynamics appeared in the experiment. Site availability was governed by the size and type of the experimental disturbance. Gaps were created without disturbing the forest floor and no new substrate was exposed, although the resources and regulators of the sites were altered. Greater alteration occurred in the larger, 10-m diameter experimental gaps than in the 5-m diameter gaps.

Differential species availability was based primarily on the existence of a pool of suppressed woody seedlings in the forest understorey (Collins & Pickett 1982). The altered conditions in the gaps changed the performance of understorey species and altered the rate of growth of tree seedlings that had been present before gap creation. The experimental treatment did not increase species richness of the understorey layer, indicating that species availability was not influenced by the experimental manipulation. Growth of some understorey broad-leaved species increased but, in general, the spread of the ferns and increase in height and cover of *Prunus serotina* seedlings far outstripped the enhanced performance of broad-leaved angiosperm herbs. The *P. serotina* seedlings had been 'idling' in the forest floor layer as advanced regeneration before the gaps were created. The change in resources, primarily light, as a result of opening the canopy allowed these seedlings to grow more rapidly. Therefore, the degree of differential species performance observed was modified by more specific mechanisms in the causal hierarchy. These more specific causes were competition with ferns, browsing by deer, the head start enjoyed by certain woody seedlings and the greater range of resources that were released in the larger gaps.

Desert soil disturbance. An example of fine-scale gap dynamics in which the substrate is disturbed comes from the Negev Desert of Israel (Boeken *et al.* 1995; Boeken & Shachak 1998; Shachak *et al.* 1999). In areas where soil lies downslope of rocky outcrops that supply runoff water, perennial geophytes – bulb bearing plants like tulips – can establish. Porcupines (*Hystrix indica*) exhume the bulbs of geophytes for food. In the process, they create a pit measuring 10–15 cm across and 15–20 cm deep. Such pits concentrate runoff water that flows from the rocks and intact soil upslope. Runoff water generally does not penetrate the surface of intact soil because the surface is cemented into a microphytic crust by the secretions of cyanobacteria, mosses and lichens. Pits collect the water and are, therefore, hot spots for water availability in this arid system. In addition, seeds and organic matter accumulate in the pits. As a result, the diversity and productivity of desert annual plants is greatly enhanced in the pits. The structure of the community in the pit changes through time as the pit fills with sediment carried by runoff water and wind. After it fills in, the pit again supports a microphytic crust and becomes a less effective trap for water and organic matter. Therefore the small site undergoes succession until it is indistinguishable from adjacent intact soil.

Sometimes the porcupines do not consume the entire bulb and the geophyte can resprout once the pit fills in somewhat with soil. At a later time the same site may be dug up by another porcupine. Geophytes also establish in new spots in the desert. The interaction of the porcupines, microphytic crust and filling of the pits creates a shifting mosaic of pits with their associated altered resource levels and enhanced availability of seeds of annual plants. This kind of dynamic is a pattern and process cycle like Sernander (1936, in Hytteborn & Verwijst 2011) and Watt (1947) envisioned.

The dynamics of porcupine diggings are also reminiscent of the carousel model (van der Maarel & Sykes 1993) mentioned in the introduction to this section. Again, non-directional vegetation change is emphasized, and migration and extinction in local patches or cells are crucial processes. Palmer & Rusch (2001) have defined the important quantitative indices relevant to the cycling of plant species through a spatial matrix. Unique to this model is 'carousel time' – the amount of time it takes a species to occupy all patches in the community. However, other variables assessing the residence time per patch, extinction rates and mobility of species are also required for complete understanding of fine-scale vegetation dynamics in terms of a carousel model.

4.3.4　Vegetation dynamics under changing management regimes

Vegetation dynamics are increasingly affected by human activities. As human societies modify or construct more and more systems, it becomes important to understand how human activities affect succession. Human domination of systems spans a range of management or control. One extreme is in national parks where managers control the nature and frequency of fire or the population densities and movements of herbivores. An example of a more intensively managed system is agriculture in which fields left fallow permanently or for various lengths of time exhibit change in vegetation. The planted, managed and volunteer vegetation in urban areas shows perhaps the strongest influence of humans on succession. In all these cases, some aspects of vegetation dynamics may be purposefully managed while other aspects are only indirectly influenced by human actions or the built environment. However, in all cases, managing vegetation for any purpose, whether aesthetic, productive or for ecosystem services, is essentially managing succession (Luken 1990). The entire scope of vegetation management is now being embedded in the context of global scale change (Vitousek *et al.* 1997). Vegetation dynamics in both wild and managed lands now is subject to the altered temperatures, nutrient levels, natural disturbance regimes, seasonal patterns and amounts of precipitation (van Andel & Aronson 2006). The same traits that affect vegetation dynamics are sensitive to the effects of global change (Chapin 2003).

Succession and management in riparian vegetation. In arid environments such as Kruger National Park, the band of structurally or compositionally distinct forest vegetation adjacent to the stream – the riparian – is an important component (Rogers 1995; Pickett & Rogers 1997).The relationship of management and succession in riparian zones involves changes in the amount and timing of flow in rivers and the role of introduced species. The sources of the major rivers in Kruger National Park lie well outside the park. They arise in the uplands of Mpumalanga and Limpopo Provinces. The flow of water in the rivers is influenced not only by the seasonal patterns and amounts of rainfall, but also by removal of water for various purposes upstream of the park. On the escarpment, water is removed by transpiration from forest plantations. Once the rivers reach the lowlands at the base of the escarpment, they are subject to use by orchard and row crop agriculture and by an increasing number and density of settlements.

A successional trend that is viewed as problematical in the riparian zones of the park is an encroachment of upland savanna plants into riparian habitats due to accumulation of sediment and lowering the water table, the process of terrestrialization. This process is driven by a reduction of flow in the rivers, due to the upstream removal of water and the attempts to control high flow events, and causes altered sediment dynamics. A retreat of the water table and an absence of flood-related mortality of upland-adapted species can alter the successional trajectories in riparian zones. This trend, evidenced by the invasion of small-leaved, spinescent trees and shrubs in the upper ranges of riparian zones is the result of terrestrialization. The practical management concern is that fires may spread into the now drier riparian zone resulting in a decline in primary productivity that supports certain key herbivore species dependent on these riparian zones. Such a shift in vegetation would also shift the pattern of movement and diversity of animal species in the park.

In addition to influencing water flow, human influence has also increased the availability of exotic species in South Africa. Such exotic species have broad tolerances and may alter the successions and contribute to terrestrialization regardless of the flood regime. In particular, changes in the availability or performance of species that typically do best on different riparian and in-channel geomorphologic features may be altered by exotic species, such that successions in the future, even those starting after severe floods, may have different trajectories than those of the past. Hence, human influence in this system has multiple layers and effects.

Post-agricultural succession. Post-agricultural successions have served as a model system for understanding succession, especially in the USA (Bazzaz 1986). This is because abandoned fields have been common, are easily manipulated and change relatively rapidly. In the eastern USA, land abandonment was especially common in the late 19th and early 20th centuries as farms were exhausted, or as more fertile or more easily tillable land became available farther west. However, such post-agricultural successions also appear in the tropics (Myster 2008), and have become more common in Europe with the alterations of agricultural policy in the European Union (Flinn & Vellend 2005).

Old-field succession has been most often studied by observing vegetation on fields having similar soils and that were abandoned after the same kind of crop and cultivation, but which differ in age. This strategy is used because it is difficult, costly and slow to observe vegetation change in one place over a long period of time. By substituting differences in age across fields separated in space for changes through time in one field, ecologists can study succession more quickly and conveniently. This research strategy is called space-for-time substitution, or more technically, chronosequence (Pickett 1989). The patterns derived from such space-for-time substitutions have been a staple of community ecology. In addition, studying replicate fields of the same age served to reduce the variation in the patterns. However, such studies also excluded the understanding of spatial heterogeneities in succession because the variations from place to place were assumed to be noise and only the mean was considered. Of course, space-for-time substitution assumes that fields abandoned at different times experience the same conditions through time. In spite of the limitations of space-for-time substitution, expectations of increasing species richness, a decline of exotic, often weedy, species and orderly transitions from herb to shrub and forest tree cover were often concluded from such studies. An analysis by Johnson and Miyanishi (2008) of classic successional patterns that were based on chronosequences confirmed that those cases were incorrect when compared to long-term trends on the sites.

The direct studies of successional change through time in specific fields are now yielding detailed information not available using space-for-time substitution (Meiners *et al.* 2001; Pickett *et al.* 2001). Information is now emerging from long-term studies of post-agricultural succession that use the same, permanently marked plots studied through time. One of these studies, the Buell–Small Succession Study (BSS) named after its founders, Drs Helen Buell, Murray Buell and John Small of Rutgers University in New Jersey, USA, was begun in 1958. The same 48 plots in each of 10 fields have been studied continuously since then. Because the study is continuous in time and extensive over space, different spatial scales of the process can be assessed and processes of plant species turnover be observed directly rather than inferred (Bartha *et al.* 2000).

Because of direct observation over 50 years, the BSS can discriminate among hypotheses that have been persistently controversial. An example is the controversy between the initial floristic composition hypothesis compared to the relay floristics hypothesis (Pickett *et al.* 2001). These two hypotheses vary in the role of differential species availability in determining succession. The initial floristics hypothesis predicts that species that will later dominate the community will be present from the start of the succession, while the relay floristics hypothesis posits that pioneer species dominate early but disappear, to be replaced by a flora of mid-successional species, which are in turn replaced by late successional species. Indeed, this ability to discriminate between these two competing hypotheses was one of the principal motivations for starting the study (H. Buell pers. comm.). In fact, aspects of both hypotheses have been supported over the 50-year study. In short, some species are present either long before or long after their period of dominance and some expected turnovers do not occur in specific plots.

Some species characteristic of later successional communities, at least in the context of the 50 years of record, do invade early. Herbaceous species that dominate in mid or late portions of the record are often present early (Fig. 4.4). In some cases, the rise to dominance is an expression of life history traits. For

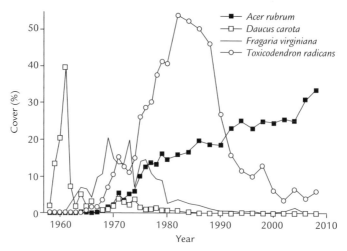

Fig. 4.4 Distribution of mean percentage cover through time of *Acer rubrum, Daucus carota, Fragaria virginiana* and *Toxicodendron radicans* in the 48 sample plots of field C3 of the Buell–Small Succession Study, illustrating the early arrival and long persistence of species common in the succession. (See further Myster & Pickett 1990).

example, short-lived perennials that dominate in years 5–10, are in fact present in low abundance earlier in the succession. Woody species – for example, the wind dispersed *Acer rubrum* – are often early invaders. However, not all individuals that invade early persist through succession. There is considerable turnover in individuals, inferred from periods of presence versus absence in particular plots through time. Though individuals are replaced, the species as a whole is present from the first or second years. In a somewhat more mesic field than those included in the permanent plot study, the ages of all woody stems present during year 14 of the succession were determined. The vast majority of *A. rubrum* individuals were themselves 14 years old. In other words, most surviving *A. rubrum* individuals had invaded early in the succession (Rankin & Pickett 1989). Other species, such as the wind-dispersed *Fraxinus americana*, showed increasing densities with age of the field, such that most individuals present in year 14 of the succession were younger than 14 years old.

Another instance of differential species availability is shown by the legacies of different abandonment treatments. Fields in the BSS varied by the last crop before abandonment and by treatment at time of release. Fields abandoned as hay fields maintain grass dominance for a long time, compared to fields that supported row crops at the time of abandonment. The legacy of the last crop can be detected in plant assemblages for *c.* 10 years after abandonment of the hay fields. In the abandoned hay fields, the grass species remain available to contribute to the succession, while in the ploughed fields species availability depends more heavily on dispersal to the site from external sources (Fig. 4.5).

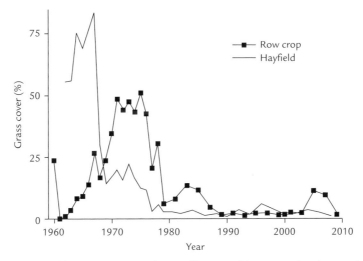

Fig. 4.5 The role of legacy in succession as illustrated by grass dominance in an abandoned hayfield (BSS field E1) compared to a plowed field abandoned from a row crop (field D3). An analysis of all hayfields and fields abandoned from row crops indicated a significant legacy effect of the hayfield grasses persisting for 10 years after abandonment (see Myster & Pickett 1990 for details).

In the permanent plot study, the colonization of other woody species tends to be delayed. For example, the bird-dispersed *Rosa multiflora* and *Rhus glabra* first appear in intermediate years. Another case of delayed dominance is seen where tree species colonize in an order that reflects dispersal mode, with species dispersed by birds and wind establishing first, followed by species that may be scatter hoarded by mammals, for example. The order of dominance may also reflect differential sensitivity to browsing by mammals. Among *Juniperus virginiana*, *Acer rubrum* and *Cornus florida*, the order of dominance is inversely related to the sensitivity of the species to browsing by mammals, so that browsing-resistant species dominate earlier in the fields (Cadenasso *et al.* 2002). In spite of such orderly patterns in differential sensitivity to browsing, forest canopy species can be present relatively early in the succession. These observations address processes that are features of both differential species availability and differential species performance.

One surprising feature in the long-term data is how commonly species remain present in the fields long after they decline in dominance. *Ambrosia artemisiifolia*, the most dominant herbaceous species in the early record in fields ploughed at abandonment, recurs in low abundance throughout the record. This is true also of some perennial species. Ground layer species such as *Poa compressa* or *Hieracium caespitosum* can be encountered late into the sequence. Some herbaceous species experience a second period of dominance when shrub canopies decline without being overtopped by trees. *Solidago* spp. usually dominates from 5 to 10 years after abandonment; however, in cases where *Rhus glabra* shrubs declined precipitously *c.* 20 years after they became dominant, *Solidago* assumed

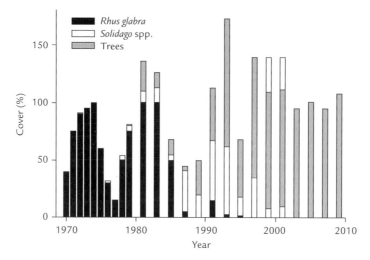

Fig. 4.6 Distribution of *Rhus glabra*, species of *Solidago*, and all species of trees in plot 10 of field D3 of the Buell-Small Succession Study from 1970 on. The field was abandoned as ploughed, bare ground after a row crop in 1960. The expected replacement of a dominant shrub, *R. glabra*, by overtopping trees does not appear. Instead, *R. glabra* declines without being overtopped, and the plot is subsequently dominated by patches of *Solidago*, followed by trees.

dominance again (Fig. 4.6). *Rosa multiflora* also declined either with or without an overtopping tree canopy. Because of its architectural complexity and the persistence of its dead stems, *Rosa multiflora* seems to have a more substantial legacy than *Rhus glabra*, and herbaceous species are slow to regain dominance in plots vacated by *R. multiflora*. Often the introduced invasive vine, *Lonicera japonica*, replaces the declining *R. multiflora*. A new invader, a disease of the *R. multiflora*, may play an increasing role in the shift of *R. multiflora* from dominance. The complex patterns of entry, persistence and demise of species in old-field succession combine aspects of differential availability and differential performance. In the realm of differential availability, mode of dispersal (whether wind, bird, or mammal), possession of seed dormancy and landscape position all feature prominently. In differential performance, life cycle, competition and interaction with consumers stand out in the examples above.

The differential performance of exotic and native species further characterizes post-agricultural succession. The proportion and cover of exotic species is expected to decline with succession. This is because many exotics that appear in early succession are agricultural weeds, adapted to the disturbance regimes of row crop agriculture. These species exploit relatively open sites and, as the layering and cover of successional communities increase with time, these species are restricted in abundance and frequency (Fig. 4.7). During the middle portion of the 50-year record, native perennials become more dominant than the agricultural weeds and ruderal plants (Meiners *et al.* 2002). However, as the fields begin to support a closed canopy of trees, exotic herbs are appearing in the

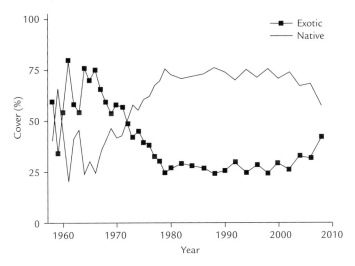

Fig. 4.7 Shift of dominance between exotic and native species over 50 years of the Buell–Small Succession Study in field C3. The proportion of each species group, as defined by Gleason & Cronquist (1991) is shown for each year. The upturn in exotic species later in succession is the result of colonization by the shade-tolerant species, *Alliaria petiolata* and *Microstegium vimineum*.

understorey. *Alliaria petiolata*, characteristic of forest edges in Europe where it is native, and increasingly common in deciduous forest of the eastern USA, is beginning to increase in those plots having a tree canopy. *Microstegium vimineum*, a shade-tolerant understory grass from Asia is also increasing rapidly across the site. Late in succession there is an increase in the relative abundance of exotic species because of these invaders (Fig. 4.7). It may be that forest-dwelling exotics, such as *Acer platanoides* or *Berberis thunbergii*, as well as *Alliaria* and *Microstegium*, will continue to increase in the community in the future, leading again to dominance by exotic species (Meiners *et al.* 2003).

The spatial pattern of vegetation in fields is an important aspect of succession (Gross *et al.* 1995), just as it was in the other cases of succession examined, such as the rivers or the tornado blowdowns at Tionesta. Spatial heterogeneity can exist within fields and also between fields based on the landscape context in which they exist. We will exemplify within-field heterogeneity first and then indicate a role for the larger landscape context.

In plots at the edge of fields nearest the remnant forest, woody species tend to invade earlier than in plots farther away from the forest (Myster & Pickett 1992). The forest may have both a direct and an indirect effect on species availability and species performance. Direct effects of forest edges probably result from the altering species availabilities of species dispersed by both wind and birds because edges provide a seed source for both. Other influences of the forest are indirect. For example, leaf litter from the forest reduces light availability at the surface in old fields, affecting establishment of light-sensitive species (Myster

& Pickett 1993). In addition, the presence of tree leaf litter in the fields affects competition between herbaceous dominants, and also the sensitivity of different species to predation. Meiners & Pickett (1999) examined both field and forest 'sides' of an old-field edge. They discovered that the boundary affected all major characteristics of the ground, shrub and seedling layers of both communities. Species richness and diversity increased from the forest to the edge and decreased slightly with distance into the field. Exotic species were most abundant in the forest within 20 m of the edge. Between-plot heterogeneity was greatest at the field edge (Meiners & Pickett 1999). Establishment probabilities of *Acer saccharum* and *Quercus palustris* increased with distance into the old field (McCormick & Meiners 2000). Spatial heterogeneity in the case of old-field–forest edges affects both differential species availability and differential performance.

The effect of herbivores and predators on differential performance of plants is proving to be important in succession. However, few studies have addressed the role of herbivores and predators in succession. In fields near the permanent plots, experimental fences that exclude large to medium herbivores have altered several features of successional communities. Soon after abandonment, exclusion of mammals by fine-meshed fences, with metal skirts sunk into the ground, affected plant species richness and evenness and substantially increased the success and survival of tree seedlings (Cadenasso *et al.* 2002). In addition, the architecture of the community was affected, with the maximum height of woody and herbaceous elements increased by the exclusion of mammals. The structure of the ground layer was also reduced in exclosures (Cadenasso *et al.* 2002).

Spatial heterogeneity also affects predation and herbivory. Predation upon *Quercus rubra* seedlings was concentrated at the forest–field edge (Meiners & Martinkovic 2002). Insect herbivory (Meiners *et al.* 2000), as well as mammalian seed predation (Meiners & LoGiudice 2003), was important for various species.

Differential species performance is an especially complex kind of successional cause since so many different processes can interact to affect it. In addition to herbivores and predators, the kind and interaction of resource types may be important. Resources such as space, light, water and nutrients influence the performance of species differentially. Experiments on the role of resources in population performance show that a mixture of resources governs the organization of old-field communities (Carson & Pickett 1990). The cover of different species in the understorey is significantly affected by different resources. For instance, *Fragaria virginiana* is affected by light, *Rumex acetosella* by nutrients and *Hieracium caespitosum* by water. Water availability has a major impact on community richness and composition late in the season in years with normal rainfall, while in drought years, water is a key controller of old-field community structure in general.

The trait-based approach described earlier in the chapter is relevant to post-agricultural succession. Concordant with successional shifts in composition and structure in a plant community are major shifts in the functional traits of the species in the community. These shifts in traits are strongly tied to mechanisms of species co-existence, colonization and replacement in plants (Weiher *et al.*

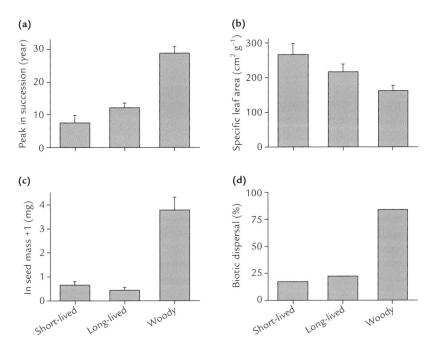

Fig. 4.8 Change in species characteristics among life-forms for the 85 most abundant species in the Buell–Small Succession Study. Species are separated into short-lived herbaceous species (annuals and biennials), long-lived herbaceous perennials and woody species following Gleason & Cronquist (1991). Four types of trait are illustrated: (a) peak in succession, defined as the year since abandonment when each species reached its peak cover during succession; (b) specific leaf area; (c) seed mass; (d) proportion dispersed by biotic vectors.

1999; Westoby & Wright 2006; Shipley 2007; Violle *et al.* 2007). Therefore, they reveal underlying mechanisms of differential species availability and performance that result in community dynamics. Data from the BSS provide an example of direct observation of trait transitions over time. As is usual for most successions, there is a shift from short-lived to long-lived species (Fig. 4.8a). As the dynamics of this site also includes succession to forest, there is also a shift from herbaceous to woody species, with woody species dominating late successional communities.

One major suite of traits that is often used to describe plant communities is the leaf–seed–height scheme of plant strategies (Westoby 1998; see also Chapter 12). While this is a small subset of the pool of traits that could be examined, it exposes the main determinants of differential performance. Leaf traits are typically expressed as specific leaf area (SLA), or the amount of leaf area generated by a given mass. SLA is strongly associated with growth, with high SLA plants exhibiting high relative growth rates and rapid rates of leaf turnover (Wright *et al.* 2004). In the BSS, specific leaf area decreases with successional transitions from short-lived herbaceous plants to woody plants, suggesting a decrease in growth rate (Fig. 4.8b). This functional shift reflects the delayed reproduction and increased maintenance costs for perennial and woody species and also helps to explain why they peak later in succession. Seed mass is directly related to the ability of a species to establish under competition or under environmental stress (Leishman *et al.* 2000). While large-seeded species may benefit from greater establishment success across a wide range of habitats, they incur a cost of reduced dispersal. Within the BSS there is a dramatic increase in seed mass between herbaceous and woody species (Fig. 4.8c). As most woody species become established once the herbaceous community has become closed, the increased seed mass should facilitate establishment, assuming the seeds escape predation. Associated with this dynamic is a shift from abiotic to biotic dispersal (Fig. 4.8d). While the minority of herbaceous species are biotically dispersed, the vast majority of woody species are dispersed by either birds or small mammals. The shift in dispersal vector allows larger-seeded plant species to partially avoid the costs of reduced dispersal. The third component of the suite of traits is height. As discussed previously, competition for light is a major driving force in succession. Potential plant height is associated with the ability of a plant to compete for light. Height increases from short-lived to long-lived herbs and from herbs to woody species in the BSS. While this trait change may not be particularly surprising as it is easily visible in the structure of the community, it reflects the underlying competitive regime of the plant community.

The richness of the record from the spatially extensive permanent plots of the BSS reveals a great deal about successional pattern. Experiments with different ages of successional communities in the same environment have exposed important interactions and mechanisms of succession. In addition, the insights from permanent plot and associated experimental studies relate to the general causes of succession. Site availability is clearly controlled by the agricultural history of the site and the season and action of disturbance at the time of abandonment. Although abandonment is often taken as a zero point in successional studies, clear legacies of the prior management and composition of the communities on the fields persist into succession (Myster & Pickett 1990). The presence of crop residue and survival of perennial species or propagules are important aspects of legacy. Differential species availability reflects the abandonment treatment, the distance to forest and adjacent field edges, and the season of abandonment. Differential species performance is based on life history attributes and longevity; different sensitivities to disturbance, light, water and nutrients; sensitivity to browsing and herbivory; and competitive ability among others. In other words, the same kinds of causes of succession seen after large, infrequent disturbances over which people have little control and which are also found at the fine-scale natural disturbance events, also act in post-agricultural old fields. Exotic species, management decisions on and off site and landscape features are important elements in all the successions discussed.

Lest it seem that the generalizations above are unique to the Buell–Small Succession Study, we briefly mention a study from a very different system, a large dune area near Oostvoorne, the Netherlands. This compelling example used aerial photographs for five dates between 1934 and 1980 (van Dorp *et al.* 1985). Vegetation maps were constructed from which successional trajectories could be evaluated quantitatively. The maps were sampled to derive successional pathways between different vegetation associations. The dynamics in all sites reflected release from heavy grazing in 1910. Although there were general trends from pioneer communities to woodland, multiple pathways of succession existed, and the spatial context strongly affected local transitions. As in the earlier examples in this section, initial floristic composition played a strong role in governing species availability. Together these examples show that vegetation dynamics has crucial spatial as well as temporal components. Patch dynamics, gap dynamics, the shifting mosaic and the carousel model are related conceptions intended to call attention to and sometimes operationalize the linkage between spatial and temporal interactions in plant assemblages.

4.4   Common characteristics across successions

The preceding sections have exemplified a wide variety of successional pathways and causes of succession. Within this variety, there are themes and insights that are common to all the examples and scales.

1   Exactly what the succession pathway looks like depends on when observations start. Starting a successional series after a large, infrequent disturbance yields a pathway that is affected by highly altered substrate availability and perhaps resources and propagules that remain in the site. Beginning with a smaller or less intense disturbance often leaves greater biological legacies. Succession studies frequently invoke an arbitrary zero point.

2   Few successions begin with a completely clean slate. There are two ways in which sites affect the subsequent successions on them. First are legacies that persist through the initiating disturbance. There are legacies that remain from the prior state of the community. Far from being the empty site suggested by the term that the earliest generation of theoreticians used – nudation – newly opened successional sites reflect structures, resources and reproductive potential left by some previously dominant community. Second are heterogeneities created by characteristics of the site or the disturbance, or the interaction of the two.

3   Sites can range from those that are relatively depauperate to those that are relatively rich in legacies. Classical terms identify the endpoints of this continuum (Fig. 4.9). The term primary succession is assigned to sites having new or newly exposed substrates (Walker *et al.* 2007), while secondary succession is assigned to sites that had previously supported a community. Although it is true that a completely new site can exist, such as a volcanic island that emerges from the open ocean, it may be useful to think of most sites as having characteristics that are some mix of primary and secondary (Walker & del Moral 2003).

4   How a successional trajectory is described depends on how long observations last. Classically, ecologists have called a succession complete when certain

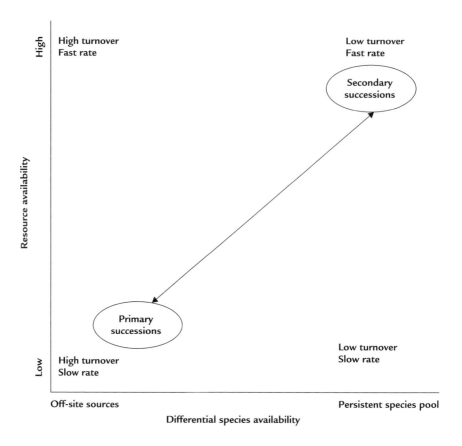

Fig. 4.9 Primary and secondary succession as extremes in a two-dimensional space representing continua of differential site resources and differential species availabilities. Breaking down the process into controls by propagule sources and control by resources exposes complexities not apparent from the simple, one-dimensional contrast between primary and secondary succession. Low resources are assumed to result in low rates of competitive interaction and low impact by herbivores and browsers. Local sources of propagules are assumed to make differential interactions more apparent while off-site propagule sources emphasize any time lags in the arrival of different species. This simple classification of successional patterns may be confounded by other specific mechanisms in the successional hierarchy (Fig. 4.2).

features, such as diversity or productivity, were maximized. However, important community dynamics continue in almost all communities beyond some idealized maximum state. For instance, continuing observations of forests well beyond the period of canopy closure typically exposes successions that take place in smaller patches, such as gaps opened by disturbance within the community.

4.4.1   Refining the concept

By combining the themes that have emerged from the examples in this chapter, a refined view of succession becomes clear. When ecologists first began to cement their growing knowledge of succession, they emphasized the directional and irreversible nature of the process. In part they did this because it matched the linear and seemingly goal-oriented patterns that were then being articulated in evolution and in geomorphology (Kingsland 2005). The first theories of succession emphasized the development of a community to be analogous to the growth and maturity of an individual organism. In contrast, the contemporary literature emphasizes a different conception of succession (Parker 2004). First, the pathways of succession do not necessarily follow a prescribed order. There is a high probability that short-lived, fecund species with extensive dispersal capacities will dominate sites soon after a disturbance. Likewise, species that grow slowly and allocate much of their assimilated resources to growth and structure as opposed to species that allocate resources to producing many, widely dispersable seeds, are likely to dominate in communities not recently disturbed. In between disturbances, biomass tends to accumulate, spatial structure of the community tends to become more heterogeneous and richness of species tends to increase as early successional and late successional species overlap in time. However, not all specific locations experience these probabilistically described trends. In some cases, pioneer species are not present in the seed bank to capture recently disturbed sites. In other cases, expected linear increases in species richness do not appear. Herbivorous animals can alter the patterns of succession that would appear if the only interactions were those among plants (Meiners & LoGiudice 2003). The fact that many important multitrophic level interactions involve soil animals and microbes is a challenge for future research (van der Putten *et al.* 2009; Chapter 9).

   In many cases, the soil resources are more limiting at the beginning of succession, while later, as plant biomass and structure accumulate, light becomes the limiting resource. This shift in limiting resources may be mirrored by a shift in species from those that can deal with low nitrogen levels to those that can deal with low light levels during their establishment phases (Tilman 1991). Events that reorganize the environment – its resources and the signals that govern species growth and reproduction – can appear with differing intensities through time. Thus observing a community through time shows the relationship of successional processes to both episodic events that originate from outside the community, as well as interactions within the community.

4.4.2   Net effects in succession

The multiplicity of successional trajectories have been summarized in 'models' of succession. Connell & Slatyer (1977) proposed that succession pathways could express three distinct kinds of turnover between species: facilitation, tolerance and inhibition. These alternative models helped expose the richness of successional processes, rather than a monolithic series of events. Early invaders have the capacity not only to facilitate the arrival or performance of later dominants, but pioneering species sometimes inhibit the invasion or performance of species that generally dominate later in succession. Their third model of succession, tolerance, is not just a straightforward neutral case. It can either reflect the meshing of life histories or the playing out of different environmental tolerances of the species without substantial interaction. Tolerance of the non-interactive kind is seen in those plots in which shrubs die with no apparent effect on the dominance of trees. Another complexity in models of succession is the mediation of effects of plant–plant interaction by herbivores or seed predators. In all cases, the models of succession are in fact the complex, net effects of many interactions. The complexity of successional models prefigures the contemporary focus on linking neutral and niche models (Chave 2004), mentioned earlier.

4.4.3 The differential processes of vegetation dynamics

The picture of succession that seems clear now is much more subtle than the classical view of succession (Odum 1969). The first subtlety applies to site availability. In order to understand why succession differs across the variety of kinds of conditions ecologists want to understand, it is clear that site characteristics differ considerably. Therefore, it seems wise to add 'differential' to the process of site availability as one of the fundamental successional processes. The recognition of differentials in site availability reminds ecologists that even if the availability and kinds of interactions among species are held constant, successions can differ in rate and composition simply because of differences in resource availability, landscape context and biological legacies present in different sites.

Succession has often been narrowly defined as the change in species composition of a community through time. More broadly, it is the change in both composition and structure. Therefore, the more inclusive definition emphasizes that communities can differ vastly because the architecture of the plants making them up differs.

Another generalization that emerges from the examples presented is that the communities involved in succession are spatially heterogeneous. This is not merely an inconvenience to investigators and managers but is part of the fundamental nature of communities. The ability of different species or different species groups to contribute to different trajectories in succession adds to the richness of those communities. It allows species to specialize on different resources or reflects their dependence on different kinds of sites and interactions. Spatial heterogeneity is also the result of the rich variety of kinds and intensities of disturbance that affect successions. Finally, spatial heterogeneities result from the landscape context in which successional sites are located. All these kinds of heterogeneity – internal and externally generated – are part and parcel of succession. Succession is as much a spatial phenomenon of extensive landscapes as it is a temporal process in local communities.

4.5 Summary

In summary, vegetation dynamics is governed by three general processes – differential site availability, differential species availability and differential species performance. These three processes interact in a spatially heterogeneous array that reflects the nature of the disturbance that punctuates community dynamics and the spatial neighbourhood of the landscape in which succession occurs. The general processes themselves are composed of more specific mechanisms that describe the detailed characteristics of sites, species dispersal and interactions. The pathways of succession that ecologists actually observe result from the specifics of each of these kinds of processes and how those processes interact. Wild and managed sites all support successions that combine these different processes in specific ways. This chapter has provided the conceptual tools that can be used to understand successional pathways observed in any specific situation. The study of succession is an example of how ecologists have to link generality of process with the specific constraints and opportunities that different sites provide. After more than 100 years of study, new combinations of factors and events are still being discovered that change our view of how succession occurs.

Acknowledgements

We are grateful to Sarah Picard for careful and efficient assistance with the long-term data set and analyses from the Buell–Small Succession Study. We thank Kirsten Schwarz for assistance in constructing the data figures. The Buell–Small Succession Study has been supported in part by student help provided by the Hutcheson Memorial Forest Center of Rutgers University and by a grant for Long-Term Research in Environmental Biology from the National Science Foundation, DEB-0424605. Our understanding of insights from the Kruger National Park were made possible by support of the Andrew W. Mellon Foundation of the River/Savanna Boundaries Programme. The original version of this chapter was dedicated to Professor Fakhri A. Bazzaz (1933–2008) on the occasion of his retirement from Harvard University, but now it sadly serves as a token of gratitude and memory.

References

Anderson, D.C. & MacMahon, J.A. (1985) Plant succession following the Mount St. Helens volcanic eruption: facilitation by a burrowing rodent, *Thomruya talpoides*. *American Midland Naturalist* **114**, 62–69.

Aubin, I., Ouellette, M.H., Legendre, P., Messier, C. & Bouchard, A. (2009) Comparison of two plant functional approaches to evaluate natural restoration along an old-field – deciduous forest chronosequence. *Journal of Vegetation Science* **20**, 185–198.

Baeten, L., Velghe, D., Vanhellemont, M. *et al.* (2010) Early trajectories of spontaneous vegetation recovery after intensive agricultural land use. *Restoration Ecology* **18**, 379–386.

Bartha, S., Pickett, S.T.A. & Cadenasso, M.L. (2000) Limitations to species coexistence in secondary succession. In: *Vegetation Science in Retrospect and Perspective* (eds P.S. White, L. Mucina & J. Lepš), pp. 55–58. Opulus Press, Uppsala.

Bazzaz, F.A. (1986) Life history of colonizing plants: some demographic, genetic, and physiological features. In: *Ecology of Biological Invasions of North America and Hawaii* (eds H.A. Mooney & J.A. Drake), pp. 96–110. Springer-Verlag, New York, NY.

Bazzaz, F.A. (1996) *Plants in Changing Environments: Linking Physiological, Population, and Community Ecology*. Cambridge University Press, New York, NY.

Boeken, B. & Shachak, M. (1998) Colonization by annual plants of an experimentally altered desert landscape: source-sink relationships. *Journal of Ecology* **86**, 804–814.

Boeken, B., Shachak, M., Gutterman, Y. & Brand, S. (1995) Patchiness and disturbance: plant community responses to porcupine diggings in the Central Negev. *Ecography* **18**, 410–422.

Botkin, D.B. & Sobel, M.J. (1975) Stability in time-varying ecosystems. *The American Naturalist* **109**, 625–646.

Bowers, M.A. (1993) Influence of herbivorous mammals on an old-field plant community: years 1–4 after disturbance. *Oikos* **67**, 129–141.

Brand, T. & Parker, V.T. (1995) Scale and general laws of vegetation dynamics. *Oikos* **73**, 375–380.

Brown, V.K. & Gange, A.C. (1992) Secondary plant succession: how is it modified by insect herbivory. *Vegetatio* **101**, 3–13.

Cadenasso, M.L., Pickett, S.T.A. & Morin, P.J. (2002) Experimental test of the role of mammalian herbivores on old field succession: community structure and seedling survival. *Journal of the Torrey Botanical Society* **129**, 228–237.

Carson, W.P. & Pickett, S.T.A. (1990) Role of resources and disturbance in the organization of an old-field plant community. *Ecology* **71**, 226–238.

Chapin, F.S. (2003) Effects of plant traits on ecosystem and regional processes: a conceptual framework for predicting the consequences of global change. *Annals of Botany* **91**, 455–463.

Chave, J. (2004) Neutral theory and community ecology. *Ecology Letters* **7**, 241–253.

Clark, J.S. & McLachlan, J.S. (2003) Stability of forest biodiversity. *Nature* **423**, 635–638.

Clements, F.E. (1916) *Plant Succession: An Analysis of the Development of Vegetation*. Carnegie Institution of Washington, Washington, DC.

Collins, B.S. & Pickett, S.T.A. (1982) Vegetation composition and relation to environment in an Allegheny hardwood forest. *American Midland Naturalist* **108**, 117–123.

Collins, B.S. & Pickett, S.T.A. (1987) Influence of canopy opening on the environment and herb layer in a northern hardwoods forest. *Vegetatio* **70**, 3–10.

Collins, B.S. & Pickett, S.T.A. (1988) Response of herb layer cover to experimental canopy gaps. *American Midland Naturalist* **119**, 282–290.

Connell, J.H. & Slatyer, R.O. (1977) Mechanisms of succession in natural communities and their role in community stability and organization. *The American Naturalist* **111**, 1119–1144.

Cramer, V.A. & Hobbs, R.J. (eds) (2007) *Old Fields: Dynamics and Restoration of Abandoned Farmland*. Island Press, Washington, DC.

Dale, V.H., Lugo, A.E., MacMahon, J.A. & Pickett, S.T.A. (1999) Ecosystem management in the context of large, infrequent disturbances. *Ecosystems* **1**, 546–557.

Davis, M.A., Pergle, J., Truscott, A. *et al.* (2005) Vegetation change: a reunifying concept in plant ecology. *Perspectives in Plant Ecology, Evolution and Systematics* **7**, 69–76.

del Moral, R. (1993) Mechanisms of primary succession on volcanoes: a view from Mount St. Helens. In: *Primary Succession on Land* (eds J. Miles & D.W.H. Walton), pp. 79–100. Blackwell Scientific Publications, Boston, MA.

Eliot, C. (2007) Method and metaphysics in Clements's and Gleason's ecological explanations. *Studies in History and Philosophy of Biological and Biomedical Sciences* **38**, 85–109.

Facelli, J.M. (1994) Multiple indirect effects of plant litter affect the establishment of woody seedlings in old fields. *Ecology* **75**, 1727–1735.

Feng, Z.L., Liu, R., DeAngelis, D.L. *et al.* (2009) Plant toxicity, adaptive herbivory, and plant community dynamics. *Ecosystems* **12**, 534–547.

Flinn, K.M. & Vellend, M. (2005) Recovery of forest plant communities in post–agricultural landscapes. *Frontiers in Ecology and the Environment* **3**, 243–250.

Forman, R.T.T. & Boerner, R.E.J. (1981) Fire frequency and the pine barrens of New Jersey. *Bulletin of the Torrey Botanical Club* **108**, 34–50.

Gedan, K.B., Crain, C.M. & Bertness, M.D. (2009) Small-mammal herbivore control of secondary succession in New England tidal marshes. *Ecology* **90**, 430–440.

Gleason, H.A. & Cronquist, A. (1991) *Manual of Vascular Plants of Northeastern United States and Adjacent Canada*, 2nd edn. The New York Botanical Garden, Bronx, NY.

Glenn-Lewin, D.C., Peet, R.K. & Veblen, T.T. (eds) (1992) *Plant Succession: Theory and Prediction*. Chapman and Hall, New York, NY.

Glenn-Lewin, D.C. & van der Maarel, E. (1992) Patterns and processes of vegetation dynamics. In: *Plant Succession: Theory and Prediction* (eds D.C. Glenn-Lewin, R.K. Peet & T.T. Veblen), pp. 11–59. Chapman and Hall, New York, NY.

Gravel, D, Canham, C.D., Beaudet, M. & Messier, C. (2006) Reconciling niche and neutrality: the continuum hypothesis. *Ecology Letters* 9, 399–409.

Grime, J.P. (1979) *Plant Strategies and Vegetation Processes*. John Wiley & Sons, Ltd, New York, NY.

Gross, K.L., Pregitzer, K.S. & Burton, A.J. (1995) Spatial variation in nitrogen availability in three successional plant communities. *Journal of Ecology* 83, 357–367.

Grubb, P.J. (1977) The maintenance of species-richness in plant communities: the importance of the regeneration niche. *Biological Reviews* 52, 107–145.

Harpole, W.S. & Tilman, D. (2006) Non-neutral patterns of species abundance in grassland communities. *Ecology Letters* 9, 15–23.

Hobbs, R.J. & Cramer, V.A. (2007) Old field dynamics: regional and local differences, and lessons for ecology and restoration. In: *Old Fields: Dynamics and Restoration of Abandoned Farmland* (eds V.A. Cramer & R.J. Hobbs), pp. 309–318. Island Press, Washington, DC.

Hubbell, S.P. (2001) *The Unified Neutral Theory of Biodiversity and Biogeography*. Princeton University Press, Princeton, NJ.

Hytteborn, H. & Verwijst, T. (2011) The importance of gaps and dwarf trees in the regeneration of Swedish spruce forests: the origin and content of Sernander's (1936) gap dynamics theory. *Scandinavian Journal of Forest Research* 26, 3–16.

Jabot, F., Etienne, R.S. & Chave, J. (2008) Reconciling neutral community models and environmental filtering: theory and an empirical test. *Oikos* 117, 1308–1320.

Johnson, E.A. & Miyanishi, K. (eds) (2007) *Plant Disturbance Ecology: The Process and the Response*. Academic Press, Burlington, MA.

Johnson, E.A. & Miyanishi, K. (2008) Testing the assumptions of chronosequences in succession. *Ecology Letters* 11, 419–431.

Kingsland, S.E. (2005) *The Evolution of American Ecology, 1890–2000*. Johns Hopkins University Press, Baltimore, MD.

Koniak, G. & Noy-Meir, I. (2009) A hierarchical, multi-scale, management-responsive model of Mediterranean vegetation dynamics. *Ecological Modelling* 220, 1148–1158.

Krueger, L.M., Peterson, C.J., Royo, A. & Carson, W.P. (2009) Evaluating relationships among tree growth rate, shade tolerance, and browse tolerance following disturbance in an eastern deciduous forest. *Canadian Journal of Forest Research–Revue Canadienne De Recherche Forestiere* 39, 2460–2469

Kumler, M.L. (1997) Nitrogen fixation in dry coastal ecosystems. In: *Dry Coastal Ecosystems: General Aspects* (ed. E. van der Maarel), pp. 421–436. Elsevier, Amsterdam.

Lebrija-Trejos, E., Perez-Garcia, E.A., Meave, J.A., Bongers, F. & Poorter, L. (2010) Functional traits and environmental filtering drive community assembly in a species-rich tropical system. *Ecology* 91, 386–398.

Leck, M.A., Parker, V.T. & Simpson, R.L. (eds) (1989) *Ecology of Soil Seed Banks*. Academic Press, San Diego, CA.

Leishman, M.R., Wright, I., Moles, A.T. & Westoby, M. 2000. The evolutionary ecology of seed size. In: *Seeds: The Ecology of Regeneration in Plant Communities* (ed. M. Fenner), pp 31–57. CAB International, Wallingford.

Lockwood, J.L., Cassey, P. & Blackburn, T. (2005) The role of propagule pressure in explaining species invasions. *Trends in Ecology and Evolution* 20, 223–228.

Loreau, M. (1998) Ecosystem development explained by competition within and between material cycles. *Proceedings of the Royal Society of London B* 265, 33–38.

Luken, J.O. (1990) *Directing Ecological Succession*. Chapman and Hall, New York, NY.

Mayr, E. (1991) *One Long Argument: Charles Darwin and the Genesis of Modern Evolutionary Thought*. Harvard University Press, Cambridge, MA.

McCormick, J.T. & Meiners, S.J. (2000) Season and distance from forest – old field edge effect and seed predation by white footed mice. *Northeastern Naturalist* 7, 7–16.

Meiners, S.J. & LoGiudice, K. (2003) Temporal consistency in the spatial pattern of seed predation across a forest–old field edge. *Plant Ecology* 168, 45–55.

Meiners, S.J. & Martinkovic, M.J. (2002) Survival of and herbivore damage to a cohort of Quercus rubra planted across a forest–old field edge. *American Midland Naturalist* 147, 247–256.

Meiners, S.J. & Pickett, S.T.A. (1999) Changes in community and population responses across a forest–field gradient. *Ecography* **22**, 261–267.

Meiners, S.J., Handel, S.N. & Pickett, S.T.A. (2000) Tree seedling establishment under insect herbivory: edge effects and inter-annual variation. *Plant Ecology* **151**, 161–170.

Meiners, S.J., Pickett, S.T.A. & Cadenasso, M.L. (2001) Effects of plant invasions on the species richness of abandoned agricultural land. *Ecography* **24**, 633–644.

Meiners, S.J., Pickett, S.T.A. & Cadenasso, M.L. (2002) Exotic plant invasion over 40 years of old field succession: community patterns and associations. *Ecography* **25**, 215–223.

Meiners, S.J., Cadenasso, M.L. & Pickett, S.T.A. (2003) Exotic plant invasions in successional systems: the utility of a long-term approach. In: *Proceedings US Department of Agriculture Interagency Research Forum on Gypsy Moth and Other Invasive Species 2002* (eds S.L.C. Fosbroke & K.W. Gottschalk). US Department of Agriculture, Forest Service, Northeastern Research Station 70–72, Newtown Square, PA.

Moon, B.P., van Niekerk, A.W., Heritage, G.L., Rogers, R.H. & James, C.S. (1997) A geomorphological approach to the ecological management of the rivers in the Kruger National Park: the case of the Sabie River. *Transactions of the Institute of British Geographers* **22**, 31–48.

Myster, R.W. (ed.) (2008) *Post-agricultural Succession in the Neotropics.* Springer, New York, NY.

Myster, R.W. & Pickett, S.T.A. (1990) Initial conditions, history and successional pathways in ten contrasting old fields. *American Midland Naturalist* **124**, 231–238.

Myster, R.W. & Pickett, S.T.A. (1992) Dynamics of associations between plants in ten old fields during 31 years of succession. *Journal of Ecology* **80**, 291–302.

Myster, R.W. & Pickett, S.T.A. (1993) Effects of litter, distance, density, and vegetation patch type on postdispersal tree seed predation in old fields. *Oikos* **66**, 381–388.

Odum, E.P. (1969) The strategy of ecosystem development. *Science* **164**, 262–270.

Palmer, M.W. & Rusch, G.M. (2001) How fast is the carousel? Direct indices of species mobility with examples from an Oklahoma grassland. *Journal of Vegetation Science* **12**, 305–318.

Parker, V.T. (2004) Community of the individual: implications for the community concept. *Oikos* **104**, 27–34.

Peters, D.P.C., Lugo, A.E., Chapin, F.S. *et al.* (2011) Cross-system comparisons elucidate disturbance complexities and generalities. *Ecosphere* **2**: art 81. doi:10.1890/ES11–00115.1.

Peterson, C.J. & Carson, W.P. (1996) Generalizing forest regeneration models: the dependence of propagule availability on disturbance history and stand size. *Canadian Journal of Forest Research* **26**, 45–52.

Peterson, C.J., Carson, W.P., McCarthy, B.C. & Pickett, S.T.A. (1990) Microsite variation and soil dynamics within newly created treefall pits and mounds. *Oikos* **58**, 39–46,

Peterson, C.J. & Pickett, S.T.A. (1991) Treefall and resprouting following catastrophic windthrow in an old-growth hemlock-hardwoods forest. *Forest Ecology and Management* **42**, 205–217.

Pickett, S.T.A. (1989) Space-for-time substitution as an alternative to long-term studies. In: *Long-term Studies in Ecology: Approaches and Alternatives* (ed. G.E. Likens), pp. 110–135. Springer-Verlag, New York, NY.

Pickett, S.T.A. & Kolasa, J. (1989) Structure of theory in vegetation science. *Vegetatio* **83**, 7–15.

Pickett, S.T.A. & McDonnell, M.J. (1989) Changing perspectives in community dynamics: a theory of successional forces. *Trends in Ecology & Evolution* **4**, 241–245.

Pickett, S.T.A. & Rogers, K.H. (1997) Patch dynamics: the transformation of landscape structure and function. In: *Wildlife and Landscape Ecology: Effects of Pattern and Scale* (ed. J.A. Bissonette), pp. 101–127. Springer-Verlag, New York, NY.

Pickett, S.T.A. & Thompson, J.N. (1978) Patch dynamics and the design of nature reserves. *Biological Conservation* **13**, 27–37.

Pickett, S.T.A. & White, P.S. (eds) (1985) *The Ecology of Natural Disturbance and Patch Dynamics.* Academic Press, Orlando, CA.

Pickett, S.T.A., Kolasa, J., Armesto, J.J. & Collins, S.L. (1989) The ecological concept of disturbance and its expression at various hierarchical levels. *Oikos* **54**, 129–136.

Pickett, S.T.A, Cadenasso, M.L. & Bartha, S. (2001) Implications from the Buell–Small Succession Study for vegetation restoration. *Applied Vegetation Science* **4**, 41–52.

Pickett, S.T.A., Kolasa, J. & Jones, C.G. (2007) *Ecological Understanding: The Nature of Theory and the Theory of Nature*, 2nd edn. Springer, New York, NY.

Pickett, S.T.A., Meiners, S.J. & Cadenasso, M.L. (2011) Domain and propositions of succession theory. In: *Theory of Ecology* (eds S.M. Scheiner & M.R. Willig), pp. 185–216. University of Chicago Press, Chicago, IL.

Pisula, N.L. & Meiners, S.J. (2010) Allelopathic effects of goldenrod species on turnover in successional communities. *American Midland Naturalist* **163**, 161–172.

Prentice, I.C. & Leemans, R. (1990) Pattern and process and the dynamics of forest structure: a simulation approach. *Journal of Ecology* **78**, 340–355.

Rankin, W.T. & Pickett, S.T.A. (1989) Time of establishment of red maple (*Acer rubrum*) in early oldfield succession. *Bulletin of the Torrey Botanical Club* **116**, 182–186.

Reynolds, H.L., Packer, A., Bever, J.D. & Clay, K. (2003) Grassroots ecology: plant–microbe–soil interactions as drivers of plant community structure and dynamics. *Ecology* **84**, 2281–2291.

Rogers, K.H. (1995) Riparian wetlands. In: *Wetlands of South Africa: Their Conservation and Ecology* (ed. G.I. Cowan), pp. 41–52. Department of Environmental Affairs, Pretoria.

Rogers, K.H. & Bestbier, R. (1997) *Development of a Protocol for the Definition of the Desired State of Riverine Systems in South Africa*. Department of Environmental Affairs and Tourism, Pretoria.

Shachak, M., Pickett, S.T.A., Boeken, B. & Zaady, E. (1999) Managing patchiness, ecological flows, productivity, and diversity in drylands. In: *Arid Lands Management: Toward Ecological Sustainability* (ed. T.W. Hoekstra), pp. 254–263. University of Illinois Press, Urbana, IL.

Shipley, B. (2007) Comparative ecology as a tool for integrating across scales. *Annals of Botany* **99**, 965–966.

Shipley, B. (2010) *From Plant Traits to Vegetation Structure: Chance and Selection in the Assembly of Ecological Communities*. Cambridge University Press, New York, NY.

Sparks, R.E. (1996) Ecosystem effects: positive and negative outcomes. In: *The Great Flood of 1993: Causes, Impacts, and Responses* (ed. S.A. Changnon), pp. 132–162. Westview Press, Boulder, CO.

Stearns, F. & Likens, G.E. (2002) One hundred years of recovery of a pine forest in northern Wisconsin. *American Midland Naturalist* **148**, 2–19.

Szabo, R. & Prach, K. (2009) Old-field succession related to soil nitrogen and moisture, and the importance of plant species traits. *Community Ecology* **10**, 65–73.

Tilman, D. (1991) Constraints and tradeoffs: toward a predictive theory of competition and succession. *Oikos* **58**, 3–15.

van Andel, J. & Aronson, J. (2006) *Restoration Ecology: The New Frontier*. Blackwell, Malden, MA.

van Coller, A.L., Rogers, K.H. & Heritage, G.L. (1997) Linking riparian vegetation types and fluvial geomorphology along the Sabie River within the Kruger National Park, South Africa. *African Journal of Ecology* **35**, 194–212.

van Dorp, D., Boot, R. & van der Maarel, E. (1985) Vegetation succession on the dunes near Oostvoorne, The Netherlands, since 1934, interpreted from air photographs and vegetation maps. *Vegetatio* **58**, 123–136.

van der Maarel, E. (1996) Pattern and process in the plant community: fifty years after A. S. Watt. *Journal of Vegetation Science* **7**, 19–28.

van der Maarel, E. & Sykes, M.T. (1993) Small-scale plant species turnover in a limestone grassland: the carousel model and some comments on the niche concept. *Journal of Vegetation Science* **4**, 179–188.

van der Putten, W.H., Bardgett, R.D. & de Ruiter, P.C. *et al.* (2009) Empirical and theoretical challenges in aboveground–belowground ecology. *Oecologia* **161**, 1–14.

Van Uytvanck, J., Van Noyen, A., Milotic, T., Decleer, K. & Hoffmann, M. (2010) Woodland regeneration on grazed former arable land: a question of tolerance, defence or protection? *Journal for Nature Conservation* **18**, 206–214.

Violle, C., Navas, M. & Vile, D. *et al.* (2007) Let the concept of trait be functional! *Oikos* **116**, 882–892.

Vitousek, P.M. (2004) *Nutrient Cycling and Limitation: Hawai'i as a Model System*. Princeton University Press, Princeton, NJ.

Vitousek, P.M., Aber, J. , Howarth, R.W. *et al.* (1997) Human alteration of the global nitrogen cycle: sources and consequences. *Ecological Applications* **7**, 737–750

Vitousek, P.M., Hedin, L.O., Matson, P.A., Fownes, J.H. & Neff, J. (1998) Within-system element cycles, input–output budgets, and nutrient limitation. In: *Successes, Limitations, and Frontiers in Ecosystem Science* (eds M.L. Pace & P.M. Groffman), pp. 432–451. Springer, New York, NY.

Walker, L.R. (ed) (1999) *Ecosystems of Disturbed Ground*. Elsevier, New York, NY.

Walker, L.R. & del Moral, R. (2003) *Primary Succession and Ecosystem Rehabilitation*. Cambridge University Press, New York, NY.

Walker, L.R., Walker, J. & Hobbs, R.J. (2007) Preface. In: *Linking Restoration and Ecological Succession* (eds L.R. Walker, J. Walker & R.J. Hobbs). Springer, New York, NY.

Watt, A.S. (1947) Pattern and process in the plant community. *Journal of Ecology* 35, 1–22.

Weiher, E., van der Werf, A., Thompson, K. *et al.* (1999) Challenging Theophrastus: a common core list of plant traits for functional ecology. *Journal of Vegetation Science* 10, 609–620.

Westoby, M. (1998) A leaf–height–seed (LHS) plant ecology strategy scheme. *Plant and Soil* 199, 213–227

Westoby, M. & Wright, I.J. (2006) Land-plant ecology on the basis of functional traits. *Trends in Ecology and Evolution* 21, 261–268.

White, P.S. & Jentsch, A. (2001) The search for generality in studies of disturbance and ecosystem dynamics. *Progress in Botany* 62, 399–449.

White, P.S. & Pickett, S.T.A. (1985) Natural disturbance and patch dynamics: an introduction. In: *The Ecology of Natural Disturbance and Patch Dynamics* (eds S.T.A. Pickett & P.S. White), pp. 3–13. Academic Press, Orlando, CA.

Willson, M.F. & Traveset, A. (2000) The ecology of seed dispersal. In: *Seeds: The Ecology of Regeneration in Plant Communities*, 2nd edn (ed. M. Fenner), pp. 85–110. CABI Publishing, New York, NY.

Wright, I.J., Reich, P.B., Westoby, M. *et al.* (2004) The worldwide leaf economics spectrum. *Nature* 428, 821–827.

5

Clonality in the Plant Community

Brita M. Svensson, Håkan Rydin and Bengt Å. Carlsson

Uppsala University, Sweden

5.1 Modularity and clonality

Plants are **modular** – they have a basic structure, the module, which reiterates and repeats itself throughout the plant body. This often results in an immense size variation between individuals of the same species. A sapling and a large tree, for example, may differ in size by a factor of 300. This is basically because they have different numbers of modules. The module can be seen as the fundamental functional unit of construction in most higher plants, and modularity is the basis for most forms of **clonality**, or vegetative reproduction.

A vascular land plant needs three structures to function – roots, leaves and stems connecting the first two. When there is only one root, as in a typical tree, we get a mainly vertically oriented plant, where all modules are connected to the same root via the main stem. Such a plant is modular but not clonal. In contrast, many horizontally oriented plants, such as wild strawberry, *Fragaria vesca*, have several main stems which have the ability to root at the nodes. Should the plant break apart, by natural processes or by injury, the fragments can therefore survive by themselves. All fragments are, thus, derived from the same zygote and form a **clone**. The smallest – actually or potentially – independent units of such **clonal fragments** are called **ramets**. This process of fragmentation can be called vegetative reproduction, and the species that exhibit this mode of reproduction are clonal. The longevity of the connection between ramets differs between species, and clonal fragments can, therefore, consist of one or several connected ramets. All ramets from a single zygote collectively make up one clone, also referred to as the genetic individual or **genet** (Fig. 5.1).

Since new ramets belonging to an old genet are in most respects identical to new ramets belonging to a young genet, clonal plants escape the size and age constraints of non-clonal plants. This means that general models and concepts

Vegetation Ecology, Second Edition. Eddy van der Maarel and Janet Franklin.
© 2013 John Wiley & Sons, Ltd. Published 2013 by John Wiley & Sons, Ltd.

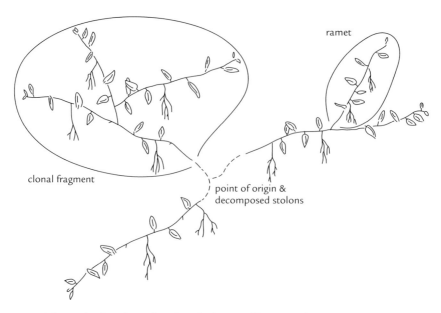

Fig. 5.1 Schematic drawing of a clonal plant to illustrate the terms genet, ramet and clonal fragment. This genet of a stoloniferous plant consists of three clonal fragments and a number of ramets.

of demographic and evolutionary processes, developed for non-clonal organisms, are often not directly applicable to clonal plants. For example, the concepts of fitness and generation time are ambiguous in clonal plants. In a population, gene frequencies can change over time not only by sexual reproduction, but also by the addition of asexually produced ramets. Among genets, the one that produces most ramets will spread its genes most efficiently by asexual means, and, when the ramets flower, also by sexual reproduction into the next generation. Another twist with clonality is that while ramets are basically genetically identical, somatic mutations may occur in meristems and propagate through clonal growth. The somatic mutations are inherited when a mutated meristem forms sexual organs and produces new zygotes.

Does evolution in clonal plants take place at the genet or ramet level? Different answers are obtained if we measure the fitness of a genotype as the number of new genets produced or as the number of ramets it contributes to future generations. One may argue that the potentially unlimited distribution of a genet in space and time, as well as the potential for within-genet genetic variation and selection (Pineda-Krch & Fagerström 1999), makes the ramet the more suitable unit for evolutionary studies in clonal plants. Chesson & Peterson (2002) suggested that fitness should be measured as the relative growth rate of the genet by adding up biomass portions of all ramets of a fragmented genet. This is theoretically appealing but practically difficult. Nevertheless, the hierarchical organization of all modular organisms must be recognized (Tuomi & Vuorisalo 1989) and clonal plants clearly also possess genet characteristics upon which selection can act.

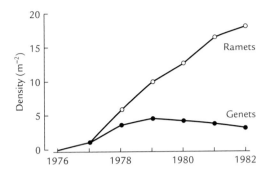

Fig. 5.2 *Solidago canadensis* is a natural invader in abandoned agricultural fields in Illinois, USA. Three years after abandonment the density of ramets (open symbols) and genets (closed symbols) diverge, and the number of ramets continues to increase whereas the number of genets decreases slowly. (Modified from Hartnett & Bazzaz 1985.)

A clone is often more or less fragmented, and we rarely know the genetic identity of the ramets. Therefore, the number (and fate) of genetic individuals in the population is not well known. Instead, the number of ramets and their fates usually form the basis for demographic studies. The fate of an individual ramet is in most cases determined by its size (e.g. number and length of leaves, number of internodes, height), or developmental stage (e.g. seedling, juvenile, adult ramet, flowering versus vegetative ramet). Models exploring the future behaviour of a ramet population are correspondingly based on size or stage, or a combination of the two. Population growth is of course affected by both sexual reproduction (increasing the number of genets) and asexual reproduction (increasing the number of ramets), and the two processes need not go in the same direction. A population characterized by an increasing ramet density may well lose genets in the absence of sexual reproduction; even an expanding population can show genetic impoverishment (Fig. 5.2).

Clonal plants appear in a huge variety of forms (Table 5.1), and since clonality might serve several functions – growth, vegetative reproduction, dispersal, and a means for increasing genet life-span – no clear-cut generalizations can be made as to how the role of clonal plants differs from that of non-clonal plants in the plant community. In this chapter we will focus on responses and behaviours that only clonal plant species are capable of, and put these in a community context. These behaviours concern the benefits of physiological integration among ramets, the ability to track resources in the habitat, the ability for short-range dispersal using competitively superior ramets and the longevity of genets enhancing population persistence. Many such routes to success have been demonstrated in several species, but for most types of behaviour it is also possible to find species that do not show the expected response, even though they seem to have a suitable clonal architecture.

A very different group of clonal plants are those that produce seeds asexually by various means, but that otherwise are quite similar to 'ordinary' non-clonal

Table 5.1 Examples of means of vegetative reproduction and dispersal in vascular plants, bryophytes and lichens. Tentative estimates of typical annual horizontal dispersal distances are given.

Clonal type	Description	Order of magnitude of annual dispersal (m)
Vascular plants		
Rhizome	Below-ground, horizontally extending stem with adventitious roots, stout, often acts as a storage organ. Similar structure appears above-ground on some epiphytes and woody monocotyledons ('aerial rhizome').[a]	$10^{-2}-10^{-1}$
Corm	Squat swollen stem, grows vertically in the soil, bears daughter corms (cormels).[a]	10^{-2}
Stem tuber	Swollen shoot with scale leaves each subtending one or more buds. Leaves present. Above- or below-ground.[a]	10^{-2}
Root tuber	As stem tuber, but leaves absent. Below-ground.[a]	10^{-2}
Bulb	Short, usually vertical stem axis bearing fleshy scale leaves.[a]	$10^{-3}-10^{-2}$
Stolon	Stem growing along the substrate surface or through surface debris. Long, thin internodes, bears foliage and adventitious roots.[a]	$10^{-2}-10^{0}$
Runner	Thin horizontal stem above-ground, one or more internodes, does not root between mother and daughter plant.[a]	$10^{-2}-10^{-1}$
Bulbil	A small bulb, e.g. on an aerial stem or developing in the axils of the leaves of a fully sized bulb. Often inaccurately applied to any small organs of vegetative multiplication such as axillary stem tubers.[a]	$10^{-2}-10^{-1}$
Turion	Detachable bud in water plants for survival during dry or cold periods ('winter bud').[a]	$10^{-1}-10^{2}$
Dropper	Detachable buds that are transported away from the mother plant at the end of a slender root-like structure.[a]	$10^{-2}-10^{-1}$
Prolification	A production of vegetative buds instead of flowers. Tiller production in sterile spikelets. 'False vivipary'.[a]	$10^{-2}-10^{-1}$
Root buds	Buds on roots capable of developing into a new shoot.[a]	$10^{-2}-10^{1}$
Apomixis	Seed produced without sexual fertilization.	$10^{-1}-10^{3}$
Fragment	Plant part breaks off and establishes. Particularly long dispersal distances on ice, in water and on sand.	$10^{-2}-10^{3}$
Layering	Aerial shoot bends down, touches the ground and produces adventitious roots.[a]	$10^{-2}-10^{-1}$

Table 5.1 (*Continued*)

Clonal type	Description	Order of magnitude of annual dispersal (m)
Bryophytes[b]		
Vegetative growth	Ramets formed after bifurcation or branching of stem or after expansion and separation of thallus parts.	10^{-3}–10^{-2}
Fragment	Detached shoot, leaf, or part or stem or thallus.	10^{-2}–10^{-1}
Bulbil	Axillary, detachable multicellular body, with leaf (or leaf-like) primordia.	10^{-2}–10^{-1}
Gemma	Multicellular body on stems or leaves, in splash-cups with dispersal assisted by rain-drops, or carried on specialized stalks.	10^{-2}–10^{-1}
Tuber	Swollen protuberance on rhizoid. Detaches after soil disturbance.	10^{-3}–10^{-2}
Lichens		
Vegetative growth	Ramets formed after expansion and separation of thallus parts.	10^{-4}–10^{-2}
Fragment	Any portion of the thallus.	10^{-2}–10^{-1}
Soredia	Bodies of algal cells surrounded by fungal hyphae (25–100 μm). Wind dispersed.	10^{-1}–10^{1}
Isidia	Small thallus outgrowths that break off.	10^{-2}–10^{-1}

[a]Bell (1991).
[b]Shaw & Goffinet (2000).

plants. Examples include many species of *Poa*, *Potentilla* and *Taraxacum*. They will be discussed only briefly in this chapter.

A number of workshops have provided a range of papers on clonal topics (Jackson *et al.* 1985; van Groenendael & de Kroon 1990; Callaghan *et al.* 1992; Soukupová *et al.* 1994; Oborny & Podani 1996; Price & Marshall 1999; Stuefer *et al.* 2001; Tolvanen *et al.* 2004; Sammul *et al.* 2008; Honnay & Jacquemyn 2010). The recent contributions are particularly useful in that they launch new ideas and hypotheses.

5.2 Where do we find clonal plants?

Clonality is widespread, both in terms of regional frequency and in local abundance. As an example, the ten most abundant species in Britain are clonal and cover 19% of the ground (Bunce & Barr 1988). In the temperate zone, 65–70% of the vascular plant species are clonal, and in some vegetation types the figure is as high as 80% (Klimeš *et al.* 1997). Exceptions to the widespread dominance of clonality are some forest types, steppes and, especially, artificial habitats (with

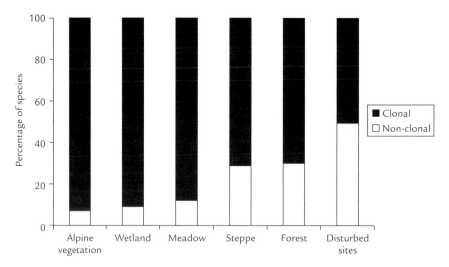

Fig. 5.3 Proportion of species that are clonal or non-clonal in different Central European vegetation types. (Based on data in Klimeš *et al.* 1997.)

a strong dominance of non-clonal annuals) (Fig. 5.3). In disturbed sites, opportunities for regeneration by seeds are ample and annuals often dominate because they reproduce quickly after establishment, are predominantly selfing (which makes them independent of pollinators) and often regenerate from the seed bank. Even though most clonal plant species are not favoured by disturbance, many disturbance-tolerant species exist that recover quickly after perturbations. One example is the common grass *Elytrigia repens*, a rhizomatous species that is a serious arable weed in large parts of its distribution area. On shorelines heavily affected by waves or ice-push, disturbances are often too severe for seedlings. Here clonals come to dominate since establishing ramets are initially firmly fixed to subterraneous organs of the mother plant and thereby withstand this kind of disturbance. Perturbations can cause clonal fragmentation, and thereby counteract some of the positive effects of clonality. However, the originally South American stoloniferous and invasive herb *Alternanthera philoxeroides* overcomes the negative effects of fragmentation by using stored carbohydrates in older parts of the fragmented pieces. These resources are translocated from older leaves and rhizomes into younger parts of the fragment (Dong *et al.* 2010). This significantly promotes the invasibility in frequently disturbed habitats, both aquatic and terrestrial, in subtropical China where this study was carried out.

Especially in primary succession, dispersal capacities of the pioneer species determine the initial species composition. Non-clonal annuals dominate as they have the advantages of a short life-cycle, often combined with selfing. However, the period with annual dominance is generally short, often only a few years (Rydin & Borgegård 1991) and clonals quickly come to dominate, even if the degree of dominance differs among habitat types (Prach & Pyšek 1994).

In secondary succession, many species can emerge from seed banks, and this may favour short-lived, often non-clonal species. Where the vegetation closes quickly, as in old-field succession, it may be easier to expand clonally than to establish from small seeds. If there is a stage with annual or non-clonal dominance this will generally be even shorter than in primary succession. Among grasses, rhizomatous species such as *Elytrigia repens* are successful in old fields. Among trees, clonal species with root suckers such as *Populus tremula* in northwest Europe or *P. tremuloides* in North America are often the first to arrive. Interestingly, these species are also superior colonizers in primary successions through their small wind-dispersed seeds (Rydin & Borgegård 1991).

It is generally held that seeds are needed for dispersal over long distances. Plants of the genus *Taraxacum* have small, plumed seeds which can travel long distances. Since *Taraxacum* is apomictic, this long-distance seed dispersal in fact represents clonal dispersal. One particular case where clonal long-range dispersal is common occurs among aquatic and shoreline species which may disperse successfully, also over long distances, with plant fragments. Two dioecious species may serve as examples. The North American aquatic *Elodea canadensis* is a widespread alien. In Europe it appears almost exclusively as female plants, whereas in Australia there are large regions with either only male or female plants (Spicer & Catling 1988). In a similar fashion, *Salix fragilis* occurs almost exclusively as male plants in some regions in Sweden. From medieval times this taxon has escaped (with floating twigs and branches that easily establish downstream) from trees planted at manors where male plants were preferentially used to avoid the heavy seed litter (Malmgren 1982). Another successful invader is the gynodioecious *Fallopia japonica*, which in the British Isles probably exists as a single male-sterile clone (Hollingsworth & Bailey 2000). See also Chapter 13.

The successful invasion of different alien species can give a clue to the advantage of clonality in different habitats. In the native vascular plant flora in Central Europe, 69% of the species are clonal, but among the aliens only 36% are clonal (Pyšek 1997). Aliens in natural communities are often clonal, but in artificial habitats they are more often non-clonal, in congruence with the large proportion of non-clonal species in disturbed habitats. We cannot say that successful invaders in general are clonal or non-clonal, but the example with *Alternanthera philoxeroides* above shows that clonal plants are not always at a disadvantage in artificial habitats. As an expected effect of the ease by which clonal plants spread in water, the percentage of clonal invaders is high in wetlands, and they actually dominate among aquatic aliens (Fig. 5.4). An extreme case is *Typha domingensis*, a rhizomatous macrophyte which under favourable conditions can expand its territory at a speed of $7\,\text{m}\cdot\text{yr}^{-1}$, occupying nearly 1 ha after 4 years solely via clonal spread. This expansion was measured in almost permanently flooded wetlands in northern Belize, after planting one single *T. domingensis* individual (Macek *et al.* 2010).

Both dispersal ability and the ability to achieve dominance in the community differ among species, and this affects their regional frequency and local abundance. We can see this among invading plant species: clonal invaders often reach relatively high abundance locally, whereas non-clonal invaders have a higher rate of regional spread.

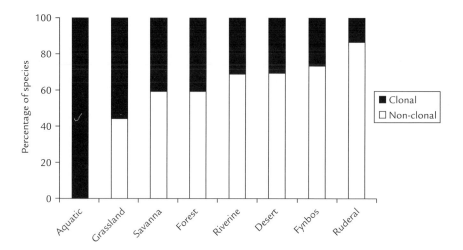

Fig. 5.4 Proportion of species that are clonal or non-clonal among aliens established in different habitat types in South African natural vegetation. (Modified from Pyšek 1997.)

However, there are also latitudinal differences: in the tropics aggressive invaders are often non-clonal, but in temperate zones they are more commonly clonal (Pyšek 1997).

5.3 Habitat exploitation by clonal growth

When diaspores (sexual or asexual) are detached, they are beyond the control of the mother plant. In contrast, clonality potentially allows the plant to exploit the environment via directional growth of, for example, rhizomes or stolons. The plant can move to a more favourable part of the habitat under two conditions: (i) if growth can be directed along gradients; and (ii) if the plant can minimize its elongation growth once it has reached a favourable patch. The latter means that as the plant moves towards favourable patches, it should gradually reduce carbon allocation for elongation.

Habitat exploitation in plants has been compared to foraging in animals, but the differences are obvious. According to the marginal value theorem (Begon *et al.* 1996), an animal will leave a patch at a certain profitability level. In clonal plants the response can be gradual: the lower the resource level, the higher should the tendency be to grow away from the patch. A second difference is that the plant does not leave the patch even if it produces ramets that do so. There is no consensus on the definition of foraging in plants, and Oborny & Cain (1997) suggested that the term should be restricted to cover morphological plasticity that is also selectively advantageous for resource acquisition. Since the focus in this chapter is on the role of clonality in plant communities rather than its role in determining plant fitness we will use the term forage in a wide sense.

A mechanistic problem with habitat exploitation is that in patches where biomass accumulation is lowest, the plant should have its maximum capacity for elongation growth to be able to leave for a better spot. Therefore, search behaviour cannot be achieved by increased growth, only by increased allocation to directional elongation. If the plant's growth is too small, it simply cannot produce stolons or rhizomes that are long enough to reach more productive patches. This constrains the ability of plants to explore their surroundings.

An example of a plant that seems to forage is *Glechoma hederacea*. It branches more sparsely and forms longer internodes under low-light or low-nutrient conditions than in more favourable habitats (Slade & Hutchings 1987a, b). While this leads to habitat exploitation, it is a plastic response that need not be selective. Quite a few clonal species have been tested for their ability to exploit the habitat, and it is clear that far from all species behave like *Glechoma*. Oborny & Cain (1997) found that only two out of 16 species fitted the *Glechoma* model and they offered three reasons:

1   Growth must be 'financed'. The outcome of experiments therefore depends on the range of resource levels tested. At very low resource levels there is very little growth, and long internodes do not occur. As resources increase there may first be a positive relationship – both biomass and internode length increase. At even higher levels biomass will continue to increase, but here internode length may decrease (Fig. 5.5). The *Glechoma* behaviour should therefore only be expected in habitats with a mosaic of patches with high and intermediate resource levels.
2   Evolutionary constraints. The morphology may be evolutionarily conservative, which means that not all strategies and responses to environmental heterogeneity can be realized.
3   Physiological integration alters the resources available to the ramet. If a ramet in a poor patch receives nutrients from ramets in richer patches it is not likely to respond to the local resource level.

The mechanisms of habitat exploitation are perhaps easiest to understand when light is the limiting resource. Elongation is promoted by shade and reduced by light as in the stoloniferous *Potentilla palustris*, growing in a range of habitats from wet nutrient-poor sites with base-rich water, including wet meadows and swamps. For this species, internode lengths were found to increase with increased vegetation height (Fig. 5.6; Macek & Lepš 2008). This enables the plant to leave the poorly lit patch but also to stay in lighter patches. Another plastic response to shade is increased leaf area, which means that carbon fixation can be maintained at a higher level than expected from the light flux alone. This certainly helps the plant to allocate more assimilates to clonal escape, and makes this mechanism more probable.

The mechanisms that could enable a plant to leave a patch with low nutrient and water supplies are not so intuitively obvious as the shade response mechanisms. In experimental units with low fertility, *Elytrigia repens* preferentially grew into vegetation-free patches, probably because transport of nutrients through the rhizomes resulted in directional growth into open areas (Kleijn &

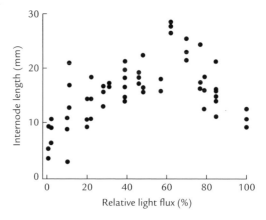

Fig. 5.5 Mean stolon internode length in *Trifolium repens* growing in the field under different levels of photosynthetically active radiation transmitted by natural canopies. (Redrawn from Thompson 1993.)

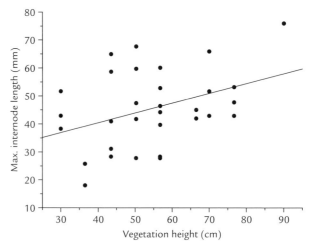

Fig. 5.6 Maximum internode lengths in *Potentilla palustris* when growing in vegetation of different heights in the Šumava Mts, Czech Republic. The line shows the best fit (linear regression; $P < 0.01$). (Redesigned from Macek & Lepš 2008.)

van Groenendael 1999). Field evidence for directional growth towards favourable patches is hard to find but two examples are available. Ramets of *Aechmea nudicaulis*, a perennial bromeliad inhabiting the spatially heterogeneous sandy coastal plains in Brazil, show directional growth towards bare sand environments when growing inside vegetation islands, where light is scarce (Sampaio *et al.* 2004). The stoloniferous *Potentilla palustris* instead adopts an escape strategy of linear growth, with longer stolon internodes in tall and dense vegetation (Macek & Lepš 2008). The concentration of ramets in favourable patches *could* happen even without a searching behaviour. The plants may move around in a random fashion, and mechanisms for active growth into good patches are not required, but growth is promoted here (Stoll *et al.* 1998; Piqueras *et al.* 1999). In the case of *A. nudicaulis* mentioned above, directional growth towards an open habitat was not found for ramets already growing in a lighter environment, indicating true directional growth (Sampaio *et al.* 2004).

The next step is to understand how the plants can stay in favourable patches. Some studies have demonstrated that clonal plants concentrate their ramets to fertile patches where they increase productivity without increasing growth in length of tubers or rhizomes. The growth morphology of the alpine sedge *Carex bigelowii* may give a clue to how this is possible. In this species, changed growth orientation is more important than internode length growth: higher nutrient levels lead to the production of more short-rhizome tillers and less emphasis on horizontal movement (Carlsson & Callaghan 1990). In the perennial clubmoss *Lycopodium annotinum*, longer horizontal segments are produced under favourable light and temperature conditions, and the increased growth makes the plant move away from the favourable patch. However, along the horizontal stem there are vertical segments, which are the main structures for carbon capture. The vertical segments are attached at constant intervals, which means that even if the horizontal apex grows out from the favourable patch the vertical segment stays put (Svensson *et al.* 1994).

5.4   Transfer of resources and division of labour

Transfer of resources requires that the connected ramets are physiologically integrated, and this varies widely among clonal plants, and can partly be explained by differences in vascular architecture (Marshall & Price 1997). Jónsdóttir & Watson (1997) distinguished between integrators (of four grades) and disintegrators. The relationships between longevity of ramet connectivity, longevity of ramets and generation time of ramets determine the degree of integration among species. Jónsdóttir & Watson (1997) suggested that there is a tendency that full integration is more common in stable and low-productive environments but we have too few data for reliable generalizations. In the community the benefits of integration are (i) support to new ramets (which may affect establishment and competition); (ii) recycling of nutrients and assimilates; and (iii) buffering of environmental heterogeneity.

A concept often discussed in the clonal literature is 'division of labour'. A non-clonal plant will generally allocate biomass to enable increased uptake of the limiting factor, for instance by increasing the root : shoot allocation ratio in nutrient-poor sites. In the 'division of labour' strategy the ramets allocate biomass to increase uptake of resources that are abundant, but only when these resources are scarce where the other ramets grow. This means that the clonal fragment (i.e. the integrated ramets collectively) follows the same allocation rule as a non-clonal plant, but the individual ramets do not. Division of labour may be programmed (as in *Carex bigelowii*; Carlsson & Callaghan 1990) or induced by a pronounced resource patchiness. We will focus on the induced response, since it is particularly important in a community context. Clonal plants with a capacity for division of labour have a high degree of morphological or physiological specialization to acquire locally abundant resources (Alpert & Stuefer 1997). Resources are then shared reciprocally between interconnected ramets, enhancing the performance of the whole clonal fragment. In other words, ramets will have different tasks, and resources are exchanged between ramets located in patches of different quality. Total clone yield is greater the larger the contrast is between patches of low and high quality (Hutchings & Wijesinghe 2008). This requires that the plant can assess and respond morphologically or physiologically to local variation in habitat quality. It also requires a high degree of physiological integration in the clonal fragment and results in a type of habitat exploitation that in many ways is opposite to foraging, and probably often more important than foraging.

In the stoloniferous *Fragaria chiloensis*, ramets in a connected system experiencing ample light and low levels of nitrogen specialize in developing leaves, whereas ramets in the opposite situation specialize in root growth. When the ramets are no longer connected, each ramet specializes in capturing the scarce resource (Alpert & Stuefer 1997). This was most pronounced in clones from patchy coastal dune sites, where light and nitrogen were strongly negatively associated. Clones from coastal grasslands, with a more uniform distribution of light and nitrogen, showed less division of labour (Fig. 5.7; Roiloa *et al.* 2007).

In addition to transfer of resources, attenuation of mechanical stress may be an equally important aspect of division of labour, for example in running water. *Potamogeton coloratus* and *Mentha aquatica* (when submerged) both showed a plastic response to increased water velocity by allocating growth to creeping stems, thereby avoiding being swept away (Puijalon *et al.* 2008).

A mature plant has a relation between root and shoot mass adjusted to the environment where the plant developed. When shoot parts are removed, the plant allocates resources to restore the relation. The same is true for clonal plants but we have to add a new dimension, the lateral growth of daughter ramets that also must be balanced with the vertical growth of the parent root/shoot system (Pitelka & Ashmun 1985). If the daughter ramet is situated further from the parent, the influence of the parent becomes reduced and the daughter ramet will develop a root/shoot balance for itself. In other words, selection on whole-plant plastic responses seems unlikely if not impossible due to the lack of central control and given the potential independence of ramets (de Kroon *et al.* 2005).

Another factor influencing the degree of division of labour is the architecture of the clonal system. So far, most studies have been concerned with clonal fragments consisting of two ramets (or two ramet clusters). Janaček *et al.* (2008) added a third – a parent ramet – and studied clonal integration experimentally in the rhizomatous peatland sedge *Eriophorum angustifolium*. In this system the parent ramet was the exclusive recipient of support from the daughter ramets, i.e., the strongest sink in the system received the strongest support, offsetting division of labour between the two daughter ramets. How and when clonal plant species divide labour, and how this affects community structure and function is an exciting and expanding research field.

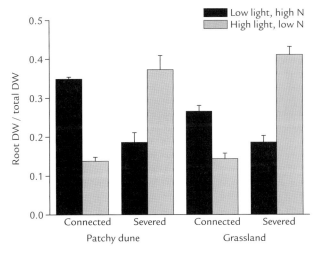

Fig. 5.7 Proportional dry mass of root (means) in clonal fragments of *Fragaria chiloensis* when grown in heterogeneous environments. The clonal fragments were either left intact ('connected') or were cut so that no transfer of resources was possible between the ramets ('severed'). Plant material was collected from two sites: one with large differences in structure between patches in patchy dune sites along the Californian coast (Patchy dune), and the other with less differences in coastal grassland (Grassland). Ramets were given high light (100%) vs. low light (10%) and high nitrogen (20 mg N·l^{-1} vs. low nitrogen (2 mg N·l^{-1}). Dark grey bars show ramets given low light and high N, light grey bars show ramets given high light and low N. When the ramets are connected, roots in the low light–high N patches allocate more resources to root production, and this was most pronounced in dune clones. These clones thus showed higher capacity for division of labour than the grassland clones. (Modified from Roiloa *et al.* 2007.)

5.5 Competition and co-existence in clonal plants

Since clonal growth is expressed in plant architecture, resource uptake, allocation and size, it is most likely that clonality should affect the competitive ability of the plant. But the relationship between clonality and competitive ability is not easy to generalize: it differs among clonal types and also depends on what aspects of competitive ability we are interested in.

Competition occurs when there is a negative effect on plants as they struggle to capture the same, limiting resource. In addition to resource competition where the struggle is for growth factors such as nutrients, water and light, we can also envisage competition for space (or ground area). In space competition, one plant covers the substrate and thereby prevents germination or rooting of other plants. It is also useful to realize that competitive ability has two components: competitive effect is the ability to take up resources and thereby reduce the amounts available for other plants, whereas competitive response is the ability to perform well even though resource levels are reduced by the competitors (Goldberg 1990). One such response involves the avoidance of competition, as in the stoloniferous herb *Glechoma hederacea* which invests in rapid expansion of unoccupied space when exposed to below-ground competition (Semchenko *et al.* 2007). The plant thus directs its growth away from the competitors.

Competing species are not likely to suffer equally. Competition for space is the most asymmetric form of interaction; the winner monopolizes all resources; pre-emptive, or interference competition are terms used to describe the situation. The first species to arrive holds its position and there will be no competitive replacement. If the order of arrival is decisive for species composition even after a long time we say that the community is founder controlled. Competition for light is also asymmetric and in many situations probably the only interaction that leads to competitive exclusion among plant species. In integrated clonal plants, young ramets can be supported from other parts of the clone, and will not suffer from being small. There is support for the notion that size-biased asymmetrical competition therefore is relatively unimportant in clonal plants (Suzuki 1994; Pennings & Callaway 2000), but there are also cases where this does not seem to hold.

The competition models of Grime (1979, 2001) and Tilman (1985) have bearings on the relationship between clonality and competitive ability. In Grime's CSR model (competitors, stress tolerators, ruderals), the competitors attain dominance in environments with little disturbance and low levels of stress. Stress in this model refers to abiotic conditions that reduce plant growth, for instance low nutrient levels. Since competition is for limiting resources, it is somewhat paradoxical that competition should be important where resources are abundant. The resolution must be that light competition is the dominant process: plants that can grow taller than their neighbours will win. It therefore seems that competitive ability in Grime's model mostly reflects competitive effect when light competition is the structuring force. The CSR classification of species is a synthesis of lateral spread and several other attributes (e.g. plant size, phenology, leaf area; cf. Hodgson *et al.* 1999). For this reason it is clear that there can be no simple relationship between clonality and competitive ability in the CSR space. Among the pure ruderals, species with strong lateral spread are lacking, whereas the pure competitors will, as a rule, be species with rapid clonal expansion (Fig. 5.8). Apart from these rather obvious extreme cases, the cluster of species moves from the ruderal part of the triangle via the stress tolerator part to the competitive part with increasing ability for clonal expansion. Klimeš *et al.* (1997) noted that the proportion of clonal species is higher than average in habitats with low nutrient levels and low temperature. In our diagram, this is reflected by rather many species with slow clonal expansion among the stress-tolerator species.

In Tilman's competition model (Tilman 1985), the best competitor is the one that can reduce resource levels to a lower level than other species, and maintain population growth at this lower level. Competition is not restricted to fertile patches, but can be important in all sorts of environments. According to this view there is not a single group of globally superior competitors, but, for

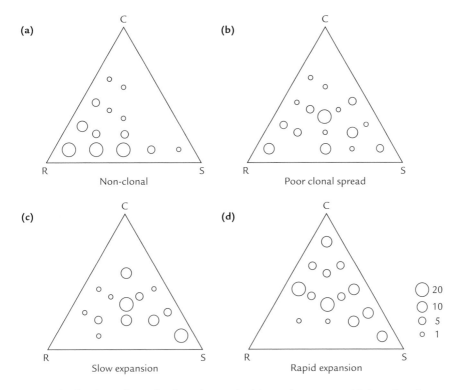

Fig. 5.8 Distribution of 255 herb and graminoid species among 19 functional types based on their position in the CSR triangle of Grime (1979, 2001) according to Hodgson *et al.* (1995). The area of each circle is proportional to the number of species. The position of each species in CSR space is taken from Hodgson *et al.* (1995), but we made the classification of species in clonal types independently, following the scheme in Klimeš (1999). **a.** Non-clonal plants. **b.** Plants with poor clonal spread; vegetative reproduction is occasional or does not result in clonal patches. **c.** Plants with a capacity for slow clonal expansion (<10 cm·yr^{-1}), or with a limited capacity to form local colonies. **d.** Plants with a capacity for rapid clonal expansion (>10 cm·yr^{-1}) or with a capacity to form large clonal patches.

instance, some species are strong light competitors, others will win when nitrogen is limiting. Root:shoot allocation and uptake efficiency will affect competitive ability, and therefore it is likely that clonality should affect competition. Growth of the plants will lead to resource depletion and ultimately to competitive exclusion. Through physiological integration and foraging, clonal plants may be very efficient competitors.

Various mechanisms may prevent the community from reaching the species composition predicted from equilibrium models based on competitive abilities. Disturbances such as wind, waves and trampling reset the system and open it up for re-colonization, and inferior species can persist if they are good dispersers. A competition–colonization trade-off is often assumed and the competitively weak species doomed to local extinction may disperse at random to occupy patches made available. Such 'escaping dispersal' can of course be through clonal growth as well as by seeds, and we must make a distinction between clonal attributes that confer competitive ability (e.g. large ramets with strong support from their mother plant) and clonal attributes that confer dispersal and escape from competition (e.g. bulbils, apomictic seeds, rapidly disintegrating runners that form seedling-like ramets).

The large variation in clonal morphological types has led to the distinction between 'guerilla' and 'phalanx' strategies (Lovett-Doust 1981). Guerilla species have long explorative internodes and branches only infrequently and typically spread by above-ground runners. In contrast, phalanx species spread as a front with dense and highly branched clusters with short internodes, most often below-ground. One can view the guerilla as colonizing and the phalanx as consolidating (de Kroon & Schieving 1990). The guerilla-phalanx distinction should be seen as a continuum with many intermediate types.

The presence of guerilla and phalanx species could affect competition in the community in many ways:

- The guerilla strategy could be a way to evade local interspecific competition. Since competition among plants only occurs between immediate neighbours, this will prevent the exclusion from the community of species that are competitively weak but which produce long runners. At the community scale this should slow down or even prevent competitive exclusion.
- The spreading behaviour of the guerilla species leads to increased interspecific contacts and mixing of species. The ramets quickly become independent and are subject to interspecific competition. This should speed up competitive exclusion at the local scale.
- It is generally held that phalanx species are strong in resource competition (they grow bigger to catch light and the new ramets are well fed with resources). Even when they encounter a superior competitor, their aggregation diminishes the degree of interspecific contacts, and it will take a long time for any other species to oust them. Phalanx growth that leads to reduced encounter probabilities among species will eventually result in spatial isolation of local dynamics and render global interactions in the community less important (Oborny & Bartha 1995).
- The longer-dispersing guerilla ramets carry fewer resources from the mother plant, and are most likely not particularly strong in resource competition. Instead they capture new space effectively, and may be good at pre-emptive competition. This could potentially lead to founder control in the community. If the phalanx species would gradually take over, the result would be dominance control. Founder control is possible if the guerilla ramet has established itself so well that it can withstand competition from arriving phalanx species.
- The shorter-dispersing phalanx ramets may suffer from intraspecific competition (including that from siblings), whereas the guerilla growth-form decreases intraspecific competition, but increases interspecific competition.

These mechanisms should be testable for pair-wise species interactions, but their role for community composition is difficult to assess or generalize. Whether a community is founder or dominance controlled is largely dependent on the rate of creation of open patches, the rate at which the species can reach these patches and the rate at which the phalanx species can outcompete the guerilla species. Several factors that counteract dominance control in competition between clonal plants have also been suggested (Herben & Hara 1997). First, there may be architectural constraints that prevent the plant to spread to dominance. Second, in low-nutrient sites there will be more root competition (which generally is symmetric; Blair 2001) and more spatial expansion. Mosaics of species will result. Third, if the competitive abilities of species are ranked as A > B > C, but C > A, there is intransitivity in competitive ability. Experiments have been performed to test the ability several grassland species to invade each others' turf (Silvertown *et al.* 1994, and other studies quoted therein), and it seems that a competitive hierarchy is more common than intransitivities. However, the rank order could change depending on grazing regime, and there need not be a simple relationship between the ability to invade another species territory and the ability to withstand invasion. In *Sphagnum* mosses, the ability of species to take over occupied area from each other by clonal expansion also varied between years (Rydin 1993).

The effect of aggregation has been modelled by cellular automata (Silvertown *et al.* 1992). A competitive hierarchy rapidly led to loss of species from the community, but only when the starting arrangement was random. Different arrangements with species clumping slowed down the processes and led to different outcomes, indicating that the spatial pattern may be more important than the competitive ranking or the abundance of the species.

For many species, clonality enables them to cope with different environments. Species with long-lived rhizomes, root systems or a dormant bud bank generally have broader niches than other species, in the sense that they occur in a wider range of habitats. It also seems that species with several modes of clonal growth have wider niches than those with only one mode (Klimeš & Klimešová 1999). The reasons for this are probably that clonal growth is a way to achieve high phenotypic plasticity and that physiological integration allows individual ramets to survive in suboptimal patches where they would have performed poorly on their own. An example of the plastic response among ramets in a clonal system is *Scirpus maritimus*. In this species, the ramets were shown to be plastically modified to specialize in sexual reproduction, vegetative growth or storage, depending on their position (Charpentier & Stuefer 1999). Such a species would then cover a large portion of the CSR strategy plane, and be able to cope with a large range of circumstances.

A plastic response together with a long life-span and the presence of a bud bank enable clonal species to survive as 'remnant populations' which can be sources for expansion when conditions become more favourable (Eriksson 1996). In addition, the extended longevity of genets is believed to enhance persistence of populations and thus increase community resilience (de Witte & Stöcklin 2010).

5.6 Clonality and herbivory

Clonal plants are susceptible to grazing, just as non-clonal plants are, and strategies have evolved to reduce its negative effects. Below we will describe some of these strategies and also some effects that grazing (including insect herbivory) has upon the clonal plant. First, clonality is a kind of risk-spreading (Eriksson & Jerling 1990); if there are many, say at least 10, ramets in the clonal fragment it is unlikely that all will be eaten.

In clonal as well as non-clonal plant species, the development and function of the different plant parts are restrained by interactions with other parts of the plant. The classical example is apical dominance – the inhibition exerted by the terminal bud on axillary buds. When the terminal bud is removed or damaged, apical dominance is released, and the axillary buds sprout. In *Lycopodium annotinum* (as in many other species) it is the bud closest to the no-longer-existing apex that starts to grow (Svensson & Callaghan 1988). This newly developed apex in turn becomes dominant and exerts apical dominance over buds situated proximally to it. This results in decreased competition between ramets within the clonal fragment (Callaghan *et al.* 1990).

Grazing has a large impact on the architecture of clonal plants by removing dominant apices together with green tissue. When the apex is grazed, trampled or otherwise damaged, some of the buds sprout which results in a proliferation of ramets. In clonal plants with lateral spread herbivory may thus be partly positive and will not kill the genet as in many non-clonal plants. The ability to recover can be amazing: Morrow & Olfelt (2003) suggested that *Solidago missouriensis* reappeared from rhizomes up to 10 years after they disappeared (apparently killed by a specialist herbivorous insect). Another consequence of herbivory may be that grazing speeds up the life-cycle, as in the example with *Carex stans* (Fig. 5.9; Tolvanen *et al.* 2001). Here, grazing not only induced formation of new ramets, but also increased the proportion of buds that

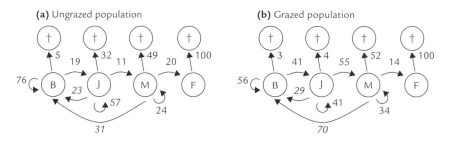

Fig. 5.9 Life-cycle graph of *Carex stans* in Canada. Population (a) from a sheltered, non-grazed habitat; (b) grazed by musk ox. Values are transition probabilities (%) between the life-stages juvenile tillers (J) and mature tillers (M), and vegetative reproduction, i.e. number of buds (B) produced, in italics. Also included are flowering tillers (F) even though they do not contribute to population growth, and the probability of dying (†). (Based on data in Tolvanen *et al.* 2001.)

developed to juveniles and subsequently to mature ramets. In ungrazed populations most buds remained dormant.

Apart from buds being released from their dormancy, depending on the level of physiological integration, sister ramets within the clonal fragment may or may not compensate for lost tissue by enhanced growth. When such a clone is grazed, resources are transported from the undamaged to the damaged part and new tissue for photosynthesis is produced. Such compensatory growth increases the chances of survival of the clonal fragment due to re-allocation of resources. Depending on clonal architecture and the degree of plasticity, species respond differently to clipping or other forms of experimental defoliation – the ecologist's way of simulating herbivory. The temperate sedge *Carex divisa* and the circumboreal spikesedge *Eleocharis palustris* illustrate this quite clearly. Both are perennial and rhizomatous, but while *C. divisa* fragments tend to spread directionally (the guerilla, or colonizing, strategy), *E. palustris* disperses in all directions, occupying space as it grows (intermediate to colonizing strategy). Experimental defoliation reduced branching in *C. divisa* but increased ramet density and this modification in architecture saves energy and results in reduced colonization but increased space occupation, i.e. a switch towards the consolidating strategy (Benot *et al.* 2010). Since *E. palustris* was not affected by defoliation it could rapidly colonize open gaps. Both strategies probably provide ecological advantages allowing their co-existence in grasslands.

Clonal woody plants have a mixed size and age structure of their above-ground parts, which is helpful in the defence against herbivory (Peterson & Jones 1997). First, this is because clonal woody plants are generally long-lived and have large reserves stored in below-ground tissue such as roots and rhizomes. Second, after disturbance, such as fire or a herbivore attack, ramets sprout from the underground bud reserve. This may happen infrequently. The result is a collection of ramets of different sizes and ages which differ in their attractiveness to herbivores.

Physiological integration also enables grazing-induced secondary metabolites to be transported to ungrazed parts of the clone. A chemical defence can thereby be built up at a lower cost than if the metabolite should always be present (Seldal *et al.* 1994). If heavily grazed, however, the whole clonal fragment may suffer – particularly if there is a high degree of physiological integration between ramets (Pitelka & Ashmun 1985). On the other hand, if the clonal fragment is poorly integrated, a ramet may be damaged without any effects on the other ramets, and the genet will not be at risk. Physical connections may also have negative consequences by providing pathways for the distribution of harmful systemic pathogens. This is another expanding research field (Stuefer *et al.* 2004), and Koubek & Herben (2008) discussed its significance for the evolution of clonality.

Acknowledgements

We thank our friend, the late Leoš Klimeš for providing data for Fig. 5.3, and Camilla Wessberg for compiling data for Fig. 5.8. Eddy van der Maarel and Petr Pyšek gave valuable comments on the manuscript.

References

Alpert, P. & Stuefer, J.F. (1997) Division of labour in clonal plants. In: *The Ecology and Evolution of Clonal Plants* (eds H. de Kroon & J. van Groenendael), pp. 137–154. Backhuys Publishers, Leiden.

Begon, M., Harper, J.L. & Townsend, C.R. (1996) *Ecology. Individuals, Populations and Communities*, 3rd edn. Blackwell Science, Oxford.

Bell, A.D. (1991) *Plant Form. An Illustrated Guide to Flowering Plant Morphology*. Oxford University Press, Oxford.

Benot, M.-L., Bonis, A. & Mony, C. (2010) Do spatial patterns of clonal fragments and architectural responses to defoliation depend on the structural blue-print? An experimental test with two rhizomatous Cyperaceae. *Evolutionary Ecology* 24, 1475–1487.

Blair, B. (2001) Effect of soil nutrient heterogeneity on the symmetry of belowground competition. *Plant Ecology* 156, 199–203.

Bunce, R.G.H. & Barr, C.J. (1988) The extent of land under different management regimes in the uplands and the potential for change. In: *Ecological Change in the Uplands* (eds M.B. Usher & D.B.A. Thompson), pp. 415–426. Blackwell Science, Oxford.

Callaghan, T.V., Svensson, B.M., Bowman, H., Lindley, D.K. & Carlsson, B.Å. (1990) Models of clonal plant growth based on population dynamics and architecture. *Oikos* 57, 257–269.

Callaghan, T.V., Svensson, B.M., Jónsdóttir, I.S. & Carlsson, B.Å. (eds) (1992) Clonal plants and environmental change. *Oikos* 63, 339–453.

Carlsson, B.Å. & Callaghan, T.V. (1990) Programmed tiller differentiation, intraclonal density regulation, and nutrient dynamics in *Carex bigelowii*. *Oikos* 58, 219–230.

Charpentier, A. & Stuefer, J.F. (1999) Functional specialization of ramets in *Scirpus maritimus*. *Plant Ecology* 141, 129–136.

Chesson, P. & Peterson, A.G. (2002) The quantitative assessment of benefits of physiological integration in clonal plants. *Evolutionary Ecology Research* 4, 1153–1176.

de Kroon, H., Huber, H., Stuefer, J.F. & van Groenendael, J.M. (2005) A modular concept of phenotypic plasticity in plants. *New Phytologist* 166, 73–82.

de Kroon, H. & Schieving, F. (1990) Resource partitioning in relation to clonal growth strategy. In: *Clonal Growth in Plants: Regulation and Function* (eds J.M. van Groenendael & H. de Kroon), pp. 113–130. SPB Academic Publishing, The Hague.

de Witte, L.C. & Stöcklin, J. (2010) Longevity of clonal plants: why it matters and how to measure it. *Annals of Botany* 106, 859–870.

Dong, B.-C., Yu, G.-L., Guo, W. *et al.* (2010) How internode length, position and presence of leaves affect survival and growth of *Alternanthera philoxeroides* after fragmentation? *Evolutionary Ecology* 24, 1447–1461.

Eriksson, O. (1996) Regional dynamics of plants: a review of evidence for remnant, source-sink and metapopulations. *Oikos* 77, 248–258.

Eriksson, O. & Jerling, L. (1990) Hierarchical selection and risk spreading in clonal plants. In: *Clonal Growth in Plants: Regulation and Function* (eds J.M. van Groenendael & H. de Kroon), pp. 79–94. SPB Academic Publishing, The Hague.

Goldberg, D.E. (1990) Components of resource competition in plant communities. In: *Perspectives on Plant Competition* (eds J.B. Grace & D. Tilman), pp. 27–49. Academic Press, San Diego, CA.

Grime, J.P. (1979) *Plant Strategies and Vegetation Processes*. John Wiley & Sons, Ltd, Chichester.

Grime, J.P. (2001) *Plant Strategies and Vegetation Processes*, 2nd edn. John Wiley & Sons, Ltd, Chichester.

Hartnett, D.C. & Bazzaz, F.A. (1985) The genet and ramet population dynamics of *Solidago canadensis* in an abandoned field. *Journal of Ecology* 73, 407–413.

Herben, H. & Hara, T. (1997) Competition and spatial dynamics of clonal plants. In: *The Ecology and Evolution of Clonal Plants* (eds H. de Kroon & J.M. van Groenendael), pp. 311–357. Backhuys Publishers, Leiden.

Hodgson, J.G., Grime, J.P., Hunt, R. & Thompson, K. (1995) *The Electronic Comparative Plant Ecology*. Chapman & Hall, London.

Hodgson, J.G., Wilson, P.J., Hunt, R., Grime, J.P. & Thompson, K. (1999) Allocating C-S-R plant functional types: a soft approach to a hard problem. *Oikos* 85, 282–294.

Hollingsworth, M.L. & Bailey, J.P. (2000) Evidence for massive clonal growth in the invasive *Fallopia japonica* (Japanese knotweed). *Botanical Journal of the Linnean Society* **133**, 463–472.

Honnay, O. & Jacquemyn, H. (eds) (2010) Clonal plants: beyond the patterns – ecological and evolutionary dynamics of asexual reproduction. *Evolutionary Ecology* (Special Issue) **24**, 1393–1397.

Hutchings, M.J. & Wijesinghe, D.K. (2008) Performance of a clonal species in patchy environments: effects of environmental context on yield at local and whole-plant scales. *Evolutionary Ecology* **22**, 313–324.

Janaček, Š., Kantorová, J., Bartoš, M. & Klimešová, J. (2008) Integration in the clonal plant *Eriophorum angustifolium*: an experiment with a three-member-clonal system in a patchy environment. *Evolutionary Ecology* **22**, 325–336.

Jackson, J.B.C., Buss, L.W. & Cook, R.E. (eds) (1985) *Population Biology and Evolution of Clonal Plants*. Yale University Press, New Haven, CT.

Jónsdóttir, I.S. & Watson, M.A. (1997) Extensive physiological integration: an adaptive trait in resource-poor environments? In: *The Ecology and Evolution of Clonal Plants* (eds H. de Kroon & J.M. van Groenendael), pp. 109–136. Backhuys Publishers, Leiden.

Kleijn, D. & van Groenendael, J.M. (1999) The exploitation of heterogeneity by a clonal plant in habitats with contrasting productivity levels. *Journal of Ecology* **87**, 873–884.

Klimeš, L. (1999) Small-scale plant mobility in a species-rich grassland. *Journal of Vegetation Science* **10**, 209–218.

Klimeš, L. & Klimešová, J. (1999) CLO-PLA2 – a database of clonal plants in central Europe. *Plant Ecology* **141**, 9–19.

Klimeš, L., Klimešová, J., Hendriks, R. & van Groenendael, J.M. (1997) Clonal plant architecture: a comparative analysis of form and function. In: *The Ecology and Evolution of Clonal Plants* (eds H. de Kroon & J.M. van Groenendael), pp. 1–29. Backhuys Publishers, Leiden.

Koubek, T. & Herben, T. (2008) Effect of systemic diseases on clonal integration: modelling approach. *Evolutionary Ecology* **22**, 449–460.

Lovett-Doust, L. (1981) Population dynamics and local specialization in a clonal perennial (*Ranunculus repens*). I. The dynamics of ramets in contrasting habitats. *Journal of Ecology* **69**, 743–755.

Macek, P., & Lepš, J. (2008) Environmental correlates of growth traits of the stoloniferous plant *Potentilla palustris*. *Evolutionary Ecology* **22**, 419–435.

Macek, P., Rejmánková, E. & Lepš, J. (2010) Dynamics of *Typha domingensis* spread in *Eleocharis* dominated oligotrophic tropical wetlands following nutrient enrichment. *Evolutionary Ecology* **24**, 1505–1519.

Malmgren, U. (1982) *Västmanlands flora*. SBT-förlaget, Lund. (In Swedish.)

Marshall, C. & Price, E.A.C. (1997) Sectoriality and its implications for physiological integration. In: *The Ecology and Evolution of Clonal Plants* (eds H. de Kroon & J.M. van Groenendael), pp. 79–107. Backhuys Publishers, Leiden.

Morrow, P.A. & Olfelt, J.P. (2003) Phoenix clones: recovery after long-term defoliation-induced dormancy. *Ecology Letters* **6**, 119–125.

Oborny, B. & Bartha, S. (1995) Clonality in plant communities – an overview. *Abstracta Botanica* **19**, 115–127.

Oborny, B. & Cain, M.L. (1997) Models of spatial spread and foraging in clonal plants. In: *The Ecology and Evolution of Clonal Plants* (eds H. de Kroon & J.M. van Groenendael), pp. 155–183. Backhuys Publishers, Leiden.

Oborny, B. & Podani, J. (eds) (1996) *Clonality in Plant Communities*. Opulus Press, Uppsala.

Pennings, S.C. & Callaway, R.M. (2000) The advantages of clonal integration under different ecological conditions: a community-wide test. *Ecology* **81**, 709–716.

Peterson, C.J. & Jones, R.H. (1997) Clonality in woody plants: a review and comparison with clonal herbs. In: *The Ecology and Evolution of Clonal Plants* (eds H. de Kroon & J.M. van Groenendael), pp. 263–289. Backhuys Publishers, Leiden.

Pineda-Krch, M. & Fagerström, T. (1999) On the potential for evolutionary change in meristematic cell lineages through intraorganismal selection. *Journal of Evolutionary Biology* **12**, 681–688.

Piqueras, J., Klimeš, L. & Redbo-Torstensson, P. (1999) Modelling the morphological response to nutrient availability in the clonal plant *Trientalis europaea*. *Plant Ecology* **141**, 117–127.

Pitelka, L.F. & Ashmun, J.W. (1985) Physiology and integration of ramets in clonal plants. In: *Population Biology and Evolution of Clonal Organisms* (eds J.B.C. Jackson, L.W. Buss & R.E. Cook), pp. 399–435. Yale University Press, New Haven, CT.

Prach, K. & Pyšek, P. (1994) Clonal plants – what is their role in succession? *Folia Geobotanica et Phytotaxonomica* 29, 307–320.

Price, E.A.C. & Marshall, C. (eds) (1999) Clonal plants and environmental heterogeneity – space, time and scale. *Plant Ecology* (Special Issue) 141: 1–206.

Puijalon, S., Bouma, T.J., van Groenendael, J.M. & Bornette, G. (2008) Clonal plasticity of aquatic plant species submitted to mechanical stress: escape versus resistance strategy. *Annals of Botany* 102, 989–996.

Pyšek, P. (1997) Clonality and plant invasions: can a trait make a difference? In: *The Ecology and Evolution of Clonal Plants* (eds H. de Kroon & J. van Groenendael), pp. 405–427. Backhuys Publishers, Leiden.

Roiloa, S.R., Alpert, P., Tharayil, N., Hancock, G. & Bhowmik, P.C. (2007) Greater capacity for division of labour in clones of *Fragaria chiloensis* from patchier habitats. *Journal of Ecology* 95, 397–405.

Rydin, H. (1993) Interspecific competition among *Sphagnum* mosses on a raised bog. *Oikos* 66, 413–423.

Rydin, H. & Borgegård, S.-O. (1991) Plant characteristics over a century of primary succession on islands: Lake Hjälmaren. *Ecology* 72, 1089–1101.

Sammul, M., Kull, T., Kull, K. & Novoplansky, A. (eds) (2008) Generality, specificity and diversity of clonal plant research. *Evolutionary Ecology* (Special Issue) 22, 273–492.

Sampaio, M.C., Araújo, T.F., Scarano, F.R. & Stuefer, J.F. (2004) Directional growth of a clonal bromeliad species in response to spatial habitat heterogeneity. *Evolutionary Ecology* 18, 429–442.

Seldal, T., Andersen, K.J. & Högstedt, G. (1994) Grazing-induced proteinase inhibitors: a possible cause for lemming population cycles. *Oikos* 70, 3–11.

Semchenko, M., John, E.A. & Hutchings, M.J. (2007) Effects of physical and genetic identity of neighbouring ramets on root-placement patterns in two clonal species. *New Phytologist* 176, 644–654.

Shaw, A.J. & Goffinet, B. (eds) (2000) *Bryophyte Biology*. Cambridge University Press, Cambridge.

Silvertown, J., Holtier, S., Johnson, J. & Dale, P. (1992) Cellular automaton models of inter-specific competition for space – the effect of pattern on process. *Journal of Ecology* 80, 527–534.

Silvertown, J., Lines, C.E.M. & Dale, M.P. (1994) Spatial competition between grasses – rates of mutual invasion between four species and the interaction with grazing. *Journal of Ecology* 82, 31–38.

Slade, A.J. & Hutchings, M.J. (1987a) The effects of nutrient availability on foraging in the clonal herb *Glechoma hederacea*. *Journal of Ecology* 75, 95–112.

Slade, A.J. & Hutchings, M.J. (1987b) The effects of light intensity on foraging in the clonal herb *Glechoma hederacea*. *Journal of Ecology* 75, 639–650.

Soukupová, L., Marshall, C., Hara, T. & Herben, T. (eds) (1994) *Plant Clonality: Biology and Diversity*. Opulus Press, Uppsala.

Spicer, K.W. & Catling, P.M. (1988) The biology of Canadian weeds. *Elodea canadensis* – Michx. *Canadian Journal of Plant Science* 68, 1035–1051.

Stoll, P., Egli, P. & Schmid, B. (1998) Plant foraging and rhizome growth patterns of *Solidago altissima* in response to mowing and fertilizer application. *Journal of Ecology* 86, 341–354.

Stuefer, J., Erschbamer, B., Huber, H. & Suzuki, J.-I. (eds) (2001) Ecology and evolutionary biology of clonal plants. *Evolutionary Ecology* (Special Issue) 15, 223–600.

Stuefer, J.F., Gómez, S. & van Mölken, T. (2004) Clonal integration beyond resource sharing: implications for defence signalling and disease transmission in clonal plant networks. *Evolutionary Ecology* 18, 647–667.

Suzuki, J. (1994) Shoot growth dynamics and the mode of competition of two rhizomatous *Polygonum* species in the alpine meadow of Mt Fuji. *Folia Geobotanica et Phytotaxonomica* 29, 203–216.

Svensson, B.M. & Callaghan, T.V. (1988) Apical dominance and the simulation of metapopulation dynamics in *Lycopodium annotinum*. *Oikos* 51, 331–342.

Svensson, B.M., Floderus, B. & Callaghan, T.V. (1994) *Lycopodium annotinum* and light quality: growth responses under canopies of two *Vaccinium* species. *Folia Geobotanica et Phytotaxonomica* 29, 159–166.

Thompson, L. (1993) The influence of natural canopy density on the growth of white clover, *Trifolium repens*. *Oikos* 67, 321–324.

Tilman, D. (1985) The resource-ratio hypothesis of plant succession. *The American Naturalist* **125**, 827–852.

Tolvanen, A., Schroderus, J. & Henry, G.H.R. (2001) Demography of three dominant sedges under contrasting grazing regimes in the High Arctic. *Journal of Vegetation Science* **12**, 659–670.

Tolvanen, A., Siikamäki, P. & Mutikainen, P. (eds) (2004) Population biology of clonal plants. *Evolutionary Ecology* (Special Issue) **18**, 403–694.

Tuomi, J. & Vuorisalo, T. (1989) Hierarchical selection in modular organisms. *Trends in Ecology and Evolution* **4**, 209–213.

van Groenendael, J.M. & de Kroon, H. (eds) (1990) *Clonal Growth in Plants: Regulation and Function*. SPB Academic Publishing, The Hague.

6

Seed Ecology and Assembly Rules in Plant Communities

Peter Poschlod[1], Mehdi Abedi[1], Maik Bartelheimer[1], Juliane Drobnik[1], Sergey Rosbakh[1] and Arne Saatkamp[2]

[1]University of Regensburg, Germany
[2]Aix-Marseille Université, IMBE, France

6.1 Ecological aspects of diaspore regeneration

6.1.1 Diaspore ecology and diaspore ecological traits

Diaspores (from the Greek 'diaspeiro' = I sow) or propagules are the reproduction units of plants. Diaspore regeneration comprises all ecological aspects of reproduction, dispersal and persistence of a diaspore bank and finally germination and establishment.

Dispersal is the movement of dispersal units away from their parent plants. Dispersal units are usually generative: seeds, fruits or spores but may also be vegetative: rhizomes, turions and bulbils (see Chapter 5). A special case is whole plant dispersal, where the entire plant is dispersed by wind, with seeds attached to the plant, such as in *Eryngium* spp. or *Boophane* spp. (in open habitats like deserts or steppes often called tumbleweed) or by water, as in *Lemna* (Bonn & Poschlod 1998).

Diaspores are also able to persist in the soil or above-ground (serotiny). These reservoirs are often termed 'soil diaspore bank' or 'above-ground diaspore bank'. In serotinous plants persistence lasts at least until diaspores are released (e.g. through fire). Persistence of the soil diaspore bank can last from a few weeks to at least several hundred years before germination (Leck *et al.* 1989).

Germination is the penetration of cells of the protonema or radicle through the spore, or the radicle through the seed coat, after imbibition and water uptake. In some cases, for example *Calla palustris* or *Scheuchzeria palustris*, cotyledons may appear first. Establishment is the stage when the gametophyte or seedling emergence is complete (Black *et al.* 2006) or rather is able to reproduce by forming antheridia and archegonia and to flower and reproduce, respectively.

Vegetation Ecology, Second Edition. Eddy van der Maarel and Janet Franklin.
© 2013 John Wiley & Sons, Ltd. Published 2013 by John Wiley & Sons, Ltd.

Plant and diaspore characters may affect dispersal in space (Tackenberg *et al.* 2003a; Römermann *et al.* 2005; Bruun & Poschlod 2006), diaspore bank persistence (Grime 1989; Bekker *et al.* 1998a; Gardarin *et al.* 2010; Saatkamp *et al.* 2011b), as well as germination and establishment (Baskin & Baskin 1998; Pearson *et al.* 2003; Moles & Westoby 2004). Understanding their functional role in dispersal, persistence, germination, and establishment may provide explanations for the occurrence of plants as well as for plant assemblages and mechanisms of vegetation dynamics (Weiher *et al.* 1999; Violle *et al.* 2007). Here we concentrate on seed plants and the respective seed ecological traits to explain these mechanisms.

6.1.2   The need for an integration of research on seed characteristics with community theory

Until recently, the global occurrence or the habitat niche of a plant species used to be explained by the life and growth-form of the mature or adult plant, or by its resistance against extreme environmental conditions, such as frost and drought. Although bioclimatic envelopes for a whole flora are still missing (Thompson *et al.* 1999; Morin *et al.* 2008), indicator values exist at least for the larger part of the Central European flora (Ellenberg *et al.* 2001; Landolt 2010) and also for Russia (Ramenskyi *et al.* 1956). Aspects of seed ecology are still sparsely considered for the explanation of distribution patterns of plant species and assemblies. However, recent theoretical concepts, for example the species pool-concept (Zobel 1997), the neutral theory of biodiversity (Hubbell 2001) and the metacommunity concept (Leibold *et al.* 2004), demonstrate the importance of dispersal as an important limiting factor for the local to continental distribution of plants. Morin *et al.* (2008) demonstrated how dispersal traits can help in understanding tree distribution ranges and diversity gradients on a continental scale and the formation of regional species pools. Also dormancy and germination requirements may shape the distribution of plants on global scales (Baskin & Baskin 1998; Tweddle *et al.* 2003; Walck *et al.* 2011) and locally (Fenner & Thompson 2005).

Dispersal is important, because seeds can escape from the immediate influence of the parent plant, and in this way avoid (1) intraspecific competition with the parent plant and with other seedlings from the same individual, (2) inbreeding and (3) predation by animals, which may be strongest near the parent plant, because seed density is usually highest there (Hildebrand 1873 in Bonn & Poschlod 1998; Janzen 1970; Howe & Smallwood 1982; Dirzo & Dominguez 1986; Hyatt *et al.* 2003; Petermann *et al.* 2008).

Seed persistence is important in temporally unpredictable environments, because after unfavourable years (environmental changes, catastrophes) a population may become extinct, whereas a persistent seed bank may buffer such years (Kalisz & McPeek 1993; Thompson 2000). Soil seed bank persistence can be correlated with the predictability of seedling mortality (Venable & Brown 1988), especially in annual communities (Venable 2007).

Dispersal enables species to recolonize unoccupied sites, as well as to colonize new suitable sites. Therefore, both dispersal potential and soil seed bank persistence are limiting factors in the dynamics of metapopulations (Husband & Barrett 1996; Poschlod 1996; Bonn & Poschlod 1998; Cain *et al.* 2000). Dispersal also affects the level of gene flow (Young *et al.* 1996) and therefore influences local adaptation and speciation (Harrison & Hastings 1996). A persistent seed bank may be a reservoir of genetic variability (Levin 1990; Vavrek *et al.* 1991; but see Honnay *et al.* 2007) to cope with future environmental changes. Both dispersal and seed bank persistence are especially important features in ephemeral habitats such as gaps in forests and irregularly drained areas in floodplains (Hanski 1987).

Specific germination requirements or dormancy patterns are important, since they allow a species (1) to find a suitable site to avoid competition and/or stress through limited resources ('gap detection mechanisms') such as diurnally fluctuating temperatures, light (far red : red ratio) and nitrates (Fenner & Thompson 2005) or (2) to spread the risk of being killed by environmental changes or catastrophes ('bet-hedging mechanisms' such as prolonged seed dormancy and 'seed-banking'; Cohen 1966; Philippi 1993; Evans *et al.* 2007; Venable *et al.* 2008).

However, simple seed characteristics such as size and mass also limit plant species occurrence and assembly (Turnbull *et al.* 1999; Leishman 2001). Seed size is consistently negatively correlated across species to seed production (number of seeds per unit of canopy; Moles *et al.* 2004). This evolutionary trade-off is an important background to understand the function of traits related to seed dispersal and persistence for diversity in plant communities. It may be also related to dormancy (Rees 1996). Seed size or mass may be a competitive advantage (Westoby *et al.* 1992; Westoby 1998) and be correlated to establishment success (Leishman 1999; Jensen & Gutekunst 2003; Moles & Westoby 2004, 2006) but also to species abundance (Murray & Leishman 2003). However, it has been shown that the higher mortality of seedlings from small-seeded species is compensated by a higher number of seeds produced. Whether or not differences in life-span between species with large and small seeds compensate for initial differences in mortality is still a matter of debate (Rees & Venable 2007; Westoby *et al.* 2009; Venable & Rees 2009).

Finally, further testing of hypotheses on community structure and diversity will be dependent on integrated data on dispersal, seed persistence, germination and subsequent establishment.

6.2   Brief historical review

Theophrast was the first to report on plant dispersal and germination, for example by stating that the germination of mistletoe seeds is enhanced by birds, and that germination is affected by climatic parameters, seed coat and seed age (Evenari 1980–1981). During the 18th century, different ways of dispersal were described by Rumphius and Linnaeus. Students of Linnaeus were probably the first to carry out dispersal experiments by feeding propagules of more than 800 species to various herbivores (Bonn & Poschlod 1998). In the second half of the

19th century, plant distribution patterns were analysed in more detail, and discussions on long-distance dispersal started. Darwin found that out of 87 species, 64 germinated after an immersion of 28 days, and some even of 137 days in sea water. He concluded that plants might be floated over large distances through oceans and seas. Darwin (1859) was also one of the first to realize that soil seed bank persistence was important as well, i.e. for the recolonization of sites. The first soil seed bank persistence studies, however, were strongly related to weedy species (Poschlod 1991). During the 20th century studies of seed dispersal, seed bank persistence and germination were developed (e.g. Bakker *et al.* 1996; Thompson *et al.* 1997; Bonn & Poschlod 1998).

6.3 Dispersal

6.3.1 Dispersal vectors, dispersal types, dispersal potential and distances

Classification systems of dispersal types are based on the morphology of the dispersal unit, which is interpreted as an adaptation to a specific dispersal mode (Table 6.1). However, the allocation to a certain dispersal mode lacks validation in most cases. Furthermore, the assignment to only one particular dispersal type is of limited value, since most propagules may be dispersed by several vectors. Hence, propagules of a species may cover a wide range of dispersal distances. Although Vittoz & Engler (2007) mentioned distance ranges for specific dispersal modes, a gliding scale for the dispersal potential, in terms of the proportion of seeds that reaches a certain distance (e.g. 100 m), is more realistic (Table 6.2). This was developed for several dispersal modes, e.g. wind (Fig. 6.1; Tackenberg *et al.* 2003a), animals (e.g. Bonn 2005, Bullock *et al.* 2011) and water (Römermann 2006).

As for dispersal distance, the terms 'short-distance' and 'long-distance dispersal' are often used, but without a consistent definition. Long-distance should be used if isolated populations are thereby connected (Hansson *et al.* 1992) or new habitats are colonized. Dispersal curves are leptocurtic, meaning that the majority of seeds are deposited within short distances from the source (Bullock & Clarke 2000, Bullock *et al.* 2011). Very long-distance dispersal events have rarely been reported (Fischer *et al.* 1996, Manzano & Malo 2006) and successful establishments have been derived mostly from species pool comparisons (Kirmer *et al.* 2008), which is speculative. Silvertown & Lovett-Doust (1993) stated that it is very improbable that we will ever know the longest distance covered by a successful seed, unless a new population of the species is discovered in an alien site. This was confirmed by field studies and modelling (Higgins & Richardson 1999) and for different dispersal syndromes such as wind (Nathan *et al.* 2002, Tackenberg 2003) and animals (Will & Tackenberg 2008) as well as humans (Wichmann *et al.* 2009). Nevertheless, these long-distance dispersal events may be of critical importance for the occurrence of natural populations and assembly of communities (Nathan 2006).

Table 6.1   Classification of dispersal types based on the dispersal vector, after Jackel
et al. (2006), see Bonn & Poschlod (1998) for a literature review. Data for the Central
European flora available through BioPop (Poschlod *et al.* 2003). Dispersal types and
vectors in bold and italics indicate a high potential for long distance dispersal;
barochory (dispersal by gravitation) is excluded from this system, because the
distinction between anemochory and barochory is gradual.

Dispersal type			Dispersal by . . .
Autochory	a.	Ballochory	ejection by the parent plant
	b.	Blastochory	dispersal by vegetative means such as stolons
	c.	Herpochory	creeping hygroscopic hairs of the diaspore
Anemochory	a.	Boleochory	swaying motion of the parent plant caused by external forces (wind, . . .)
	b.	*Meteorochory*	wind
	c.	Chamaechory	wind (dispersal unit – often whole plant – is blown over the surface)
Hydrochory	a.	Ombrochory	ejection caused by falling rain drops
	b.	*Nautochory*	water (dispersal unit – often whole plant or vegetative parts of it – is swimming on water-surface)
	c.	*Bythisochory*	flowing water (on the ground)
Zoochory	a.	Myrmekochory	ants
	b.	Ornithochory (Epizoo-, Endozoo-, Dysochory)	birds (on the body surface, via ingestion, via transport for nutrition)
	c.	*Mammaliochory* (Epizoo-, Endozoo-, Dysochory)	mammals (on the body surface, via ingestion, transport for nutrition)
	d.	*Others* (Epizoo-, Endozoo-, Dysochory)	man, other animals (snails, earthworms, . . .)
Hemerochory	a.	*Agochory*	human action(work, trade . . .)
	b.	*Speirochory*	impure seedcorn
	c.	*Ethelochory*	trading as seedcorn

6.3.2   Measurement of dispersal types and potential

Dispersal and dispersal potential of a certain vector may be measured directly,
indirectly (e.g. by genetic analysis) or through simulation models. Furthermore,
sowing experiments may show if dispersal is a limiting factor in certain cases
(Turnbull *et al.* 2000; Poschlod & Biewer 2005).

Direct measurement in the field includes the documentation of seed rain by
seed traps; these may be funnels in terrestrial or drift nets in aquatic habitats
(Bakker *et al.* 1996; Kollmann & Goetze 1998). A sampling design with transects
away from the seed source allows the assessment of dispersal distances (Bullock

Table 6.2   Classification of the dispersal potential (adapted from Tackenberg *et al.* 2003a).

Proportion of seeds exceeding reference distance	Indicator value for dispersal potential	
	DP value	Definition
<0.002	0	extremely low
0.002–0.004	1	very low
0.004–0.008	2	fairly low
0.008–0.016	3	moderately low
0.016–0.032	4	intermediately (low)
0.032–0.064	5	intermediately (high)
0.064–0.128	6	moderately high
0.128–0.256	7	fairly high
0.256–0.512	8	very high
>0.512	9	extremely high

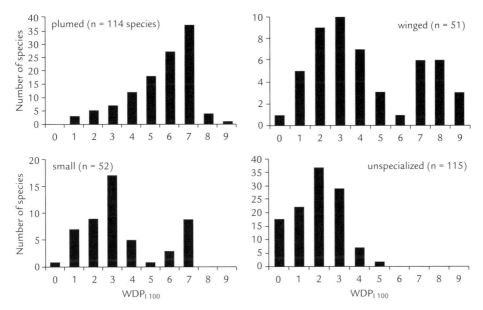

Fig. 6.1   Distribution of wind dispersal potential indicator values for a reference distance of 100 m (WDP$_{I\,100}$) for 335 plant species of different diaspore morphology after Tackenberg *et al.* (2003a). For a classification of WDP$_{I\,100}$ see Table 6.2.

& Clarke 2000) as well as the release of seeds or 'mimicries' at a certain point (Johansson & Nilsson 1993; Bill *et al.* 1999). However, seeds dispersed at long distances are difficult to trace due to the large sampling areas needed (Bullock *et al.* 2006).

Seed dispersal by wind has fascinated ecologists, but from a theoretical viewpoint; detailed field studies hardly exist. Measurements of seed dispersal in wind tunnels (Strykstra *et al.* 1998) can be informative but can hardly be extrapolated to natural conditions. Recent development of mechanistic simulation models of wind dispersal can only be validated by small data sets from single species (Bullock & Clarke 2000; Tackenberg 2003; Soons & Bullock 2008; Nathan *et al.* 2011). Only Tackenberg *et al.* (2003a) used a model to develop an assessment of wind dispersal potential of a larger flora.

Seed acquisition and transport by animals or humans is passive, either by attachment (e.g. to fur or feet), or through gut passage. Seeds and fruits can also be actively collected by animals for storage or as food (Jordano 2000; Türke *et al.* 2010). Passive transport of seeds on fur was studied on both dead and living animals. Alternatively, 'dummies' (Fischer *et al.* 1996) or 'machines' were used (Römermann *et al.* 2005; Tackenberg *et al.* 2006; see Plates 6.1 and 6.2). Dispersal distances can be determined by attachment experiments (Fischer *et al.* 1996; Manzano & Malo 2006). Dispersal distances were derived by recording the loss of marked seeds after distinct time periods or distances.

The study of passive transport of seeds by herbivores has a long tradition. As early as 1906, E. Kempski (Bonn & Poschlod 1998) fed cattle with seed-containing material and tested the viability of seeds in the dung after excretion. Dispersal by dung can also be studied by collecting dung in the field (Welch 1985). Subsequently, dispersal distances can be determined by recording movements during digestion and time period of digestion (Bakker *et al.* 1996). However, collecting quantitative data in the field is not realistic for entire plant communities, simply because it is extremely time consuming. Therefore, experimental and modelling approaches were developed (Will & Tackenberg 2008).

Finally, a comparative assessment of the dispersal potential for each vector needs standardized methods and measurements. Therefore, it is necessary to find out which plant characteristics, such as releasing height and seed production (Table 6.3), and which seed characteristics may be correlated to dispersal potentials.

Dispersal may not be successful because of secondary dispersal, predation, death before germination or unsuccessful germination. This problem can be approached by the genetical analysis of populations at different spatial scales (e.g. Willerding & Poschlod 2002). Maternal markers such as chloroplast or mitochondrial DNA allow the identification of mother plants and successfully established offspring (Ouborg *et al.* 1999).

6.3.3 Species and seed traits affecting dispersal potential

There are several parameters affecting the dispersal potential connected to different vectors (Table 6.3). Seed releasing height (Fischer *et al.* 1996; Tackenberg *et al.* 2003a; Thomson *et al.* 2011), seed production (Bruun & Poschlod 2006), and time and duration of seed release (Wright *et al.* 2008) are parameters that influence the dispersal potential of every vector. Specific vectors such as animals (e.g. migrating birds, grazing livestock), water (e.g. flooding events) or humans (e.g. seeding, mowing) are only available during distinct periods. Wind dispersal

Table 6.3 Potential long-distance dispersal vectors and parameters affecting (+) the dispersal potential.

| | | | Animals | | |
| | | | Ectozoo- | Endozoo- | |
Dispersal vector	*Wind*	*Water*	*chorous*	*chorous*	*Man*
Parameters affecting the dispersal potential					
Height of infructescence/releasing height	+	–	+	–	–
Seed production	+	+	+	+	+
Time and duration of seed release	+	+	+	+	+
Falling velocity	+	–	–	–	–
Buoyancy	–	+	–	–	–
Attachment capacity	–	–	+	–	–
Digestion tolerance	–	–	–	+	–

potential varies with the seasons and surrounding vegetation. A higher seed production increases the probability of an exceptional long-distance dispersal event.

Seed traits related to dispersal potential are seed mass or size, seed shape, seed surface and seed coat structure (Pakeman *et al.* 2002; Bonn 2005; Römermann *et al.* 2005; D'hondt & Hoffmann 2011). Although light or small seeds are said to be better dispersed over large distances (Westoby *et al.* 1996, Weiher *et al.* 1999), this correlation is only true for specific dispersal vectors (see Sections 6.3.4, 6.3.5, 6.3.6 and 6.3.7). The same is true for the surface of seeds and the seed coat structure (Table 6.4.).

6.3.4 Wind

Wind is probably the most common dispersal vector, since almost every propagule may be dispersed by wind (Tackenberg *et al.* 2003a). However, few seeds have high wind dispersal potential and are suited for long-distance dispersal. Surprisingly, many species commonly classified as wind-dispersed show low wind dispersal. Propagule traits correlated with a high wind dispersal potential are (1) low falling velocity, (2) shape and (3) surface structure combined with releasing height (Tables 6.3 and 6.4).

Wind speed as such is not an essential parameter. Wind movements related to local weather conditions are more important (Tackenberg 2003; Kuparinen *et al.* 2009), particularly updrafts (Nathan *et al.* 2002; Tackenberg *et al.* 2003b). Wind can also act as a secondary dispersal vector, moving propagules over the surface (Schurr *et al.* 2005).

Plant communities with a high proportion of species with high wind dispersal potential occur in open landscapes (tundras, alpine belts, grasslands, deserts) or

Table 6.4 Seed characteristics correlated to parameters affecting the dispersal potential in space and time.

	Seed mass/ seed size	Seed shape	Seed surface		Seed coat	
Seed traits			*Structure*	*'Hydro-phoby'*	*Thickness*	*Cells air-filled*
Parameters affecting the dispersal potential						
Seed production	+	−	−	−	−	−
Falling velocity	+	+	+	−	−	−
Buoyancy	+	−	+	+	−	+
Attachment capacity	−	−	+	−	−	−
Digestion tolerance	+	+	+	?	?	−
Parameters affecting seed bank persistence						
Persistence	+	+	−	−	+	−

ephemeral habitats (river banks). Many species from these habitats have a high seed production but a transient seed bank (e.g. *Myricaria* spp., *Salix* spp. and *Taraxacum* spp.).

6.3.5 Water

Water, particularly running water, may transport many propagules (Bill *et al.* 1999). Propagules may be vegetative diaspores rather than seeds (Boedeltje *et al.* 2003). This is especially true for free floating and submerged aquatic species. Propagules are transported either on the water surface or along the sediment surface (Table 6.1). Distances covered can reach over several kilometres and depend not only on the floating capacity or buoyancy of the seeds (Boedeltje *et al.* 2004) but also on their survival if transport along the ground is hampered by sand or gravel (Bill *et al.* 1999). The floating capacity of seeds depends not only on their specific weight but also on their surface structure and the hydrophoby of the fruit or seed coat (Poschlod 1990). Water dispersal is naturally limited to particular habitats, such as (in addition to aquatic habitats) banks of rivers and lakes, floodplains and marshes. Here the occurrence of species is often dependent on suitable conditions for germination and establishment (e.g. *Ranunculus lingua*; Johansson & Nilsson 1993).

6.3.6 Animals

Animals are probably the most important vector for extreme long-distance dispersal (Manzano & Malo 2006) in terrestrial and aquatic ecosystems. Many species transport seeds, for example mammals, birds, fishes, ants, beetles, earthworms, slugs. Vertebrates, notably sheep, may transport seeds up to several hundred kilometres (Manzano & Malo 2006) and fish may achieve several kilometres (Pollux *et al.* 2007). Invertebrates are not able to transport seeds over large distances (ants maximally 80 m, according to E. Ulbrich (Bonn & Poschlod 1998)). Diaspores of most trees and shrubs are dispersed by birds. Birds were thought to be responsible for the rapid migration and spread of woody species during the postglacial period. Both birds and mammals may have contributed to the rapid migration of fleshy fruited species. However, seed droppings by birds are mostly related to suitable resting places (McClanahan & Wolfe 1987). Therefore, seed rains depend on the vegetation structure built up by woody species and occur mostly locally (Kollmann & Pirl 1995), while only exceptionally dispersal of several kilometres may occur (Willson & Traveset 2000). Contrary to long-distance migration, the retention of low-density founder populations has been proposed. This refers to 'Reid's paradox of rapid plant migration' (Clark *et al.* 1998).

Various birds act as effective dispersal vectors by transporting and burying fruits of various woody species as food storage. Since only part of these fruits are later recovered, seeds may germinate at distant sites, sometimes several kilometres away from the mother tree (Johnson & Webb 1989). For fruit trees in Amazonian flood plains, fishes are one of the most important vectors (Goulding 1983).

Large herbivores are regarded as more effective with respect to the number of species (Janzen 1982, Malo & Suárez 1995; Pakeman 2001), especially non-woody plants. Herrera (1989) supposed that carnivorous species may also have acted as important dispersal vectors during the postglacial time.

In aquatic habitats, especially in the tropics, fish disperse fruits and seeds (Goulding 1983, Pollux 2011) but in contrast to the situation with the animals mentioned earlier, gut passage does not seem to enhance germinability (Pollux *et al.* 2007).

Seed size, seed shape, a hard seed coat and the palatability of the species itself affect the probability of endozoochorous dispersal (Janzen 1971, Pakeman *et al.* 2002). Dispersal distances are a function of migration speed and retention time, which is between one hour and 4–5 days for birds (with a maximum of 12–13 days) and between 6 hours and 10 days for mammals (with a maximum of 70 days; see Bonn & Poschlod 1998). For endozoochorous dispersal, the dung heap may act as a safe site for some species (Malo & Suárez 1995) or increase the probability of the establishment of nutrient-demanding species in acid and/or nutrient-poor sites (Cosyns *et al.* 2006).

6.3.7 Humans

Since the Neolithic period, landscape and vegetation have changed, the forests became exploited and agriculture was developed. The use of the natural resources by humans has resulted in the distribution of plants to new sites, both regionally and worldwide. In Europe its intensity has fluctuated because of major changes in climate and the migration of people. Most species have spread, particularly during the climatic optima of the Neolithic and the period of the Roman Empire (Poschlod unpublished data). After the European discovery and takeover of settlements in new continents, European land use became worldwide. In the industrializing countries of the Northern hemisphere, especially in Europe, the industrial revolution, followed by increasingly intensive agricultural and forestry practices, lead to a tremendous reduction in the diversity of natural habitats and dispersal processes. This may have resulted in the first human-induced extinctions of plant populations and species (Fig. 6.2).

Agricultural practices include processes leading to environmental changes with a high dispersal potential within and between habitats on a local and regional level. These include sowing of uncleaned seed, fertilizing, irrigating, harvesting and mowing (Fig. 6.2). From archaeological findings it became obvious that sowing of uncleaned seed contributed considerably to the development of species-rich arable weed communities and that the recent sowing of cleaned seed has reduced this diversity (Poschlod & Bonn 1998). A classical example is the history of *Agrostemma githago*. It was introduced in Neolithic Europe and was spread widely by uncleaned seed. Today, it is extinct in many places and is now re-established by sowing seeds in fields in open-air museums.

Natural fertilizers used in traditional agriculture were very diverse – for example manure, composts, sods of heathland and/or forest, and freshwater mud – and contained many seeds. This source was reduced considerably by the use of modern artificial mineral fertilizers. Slurry and sewage waste are still used but contain seeds of very few species (e.g. *Chenopodium* spp.) which are able to survive extreme conditions (Poschlod & Bonn 1998).

Different practices of harvesting cereals since the Neolithic contributed to the diversity of the arable weed flora due to dispersal adaptations of the respective species (Poschlod & Bonn 1998). Traditional mowing with the scythe was not suitable for the dispersal of seeds. In contrast, mowing machineries may transport larger quantities of seeds (Strykstra *et al.* 1996). Grasslands were traditionally developed with the help of introduced hayseed. This has been documented for species-rich calcareous grasslands in abandoned vineyards and litter meadows (Poschlod & WallisDeVries 2002; Poschlod *et al.* 2009). Furthermore, meadows were often irrigated or flooded.

Whether seeds can be dispersed by agricultural practices is dependent on certain seed characteristics. For instance, only seeds of similar size to the uncleaned cereal seeds will be dispersed.

Domestic livestock, which has been found since the Neolithic, has probably the biggest impact on propagule dispersal in artificial semi-natural and agricultural landscapes (Poschlod & Bonn 1998). Flocks of domestic livestock, particularly sheep, were guided over large distances from summer to winter pastures and back; this in known as transhumance. Distances of 100–300 km one-way covered in south-west Germany and up to 800 km from southern to northern Spain have been reported. Also transport to market places (Poschlod & WallisDeVries 2002) may imply seed dispersal.

From two studies on cattle and sheep, we know that more than 50% of the local flora in an area with calcareous grassland was transported either through ecto- or endozoochory (Fischer *et al.* 1996). Ectozoochory includes not only seeds with a sticky surface but also seeds with a coarse or even a smooth surface, even if the first category predominates. Dispersal distances covered do not only

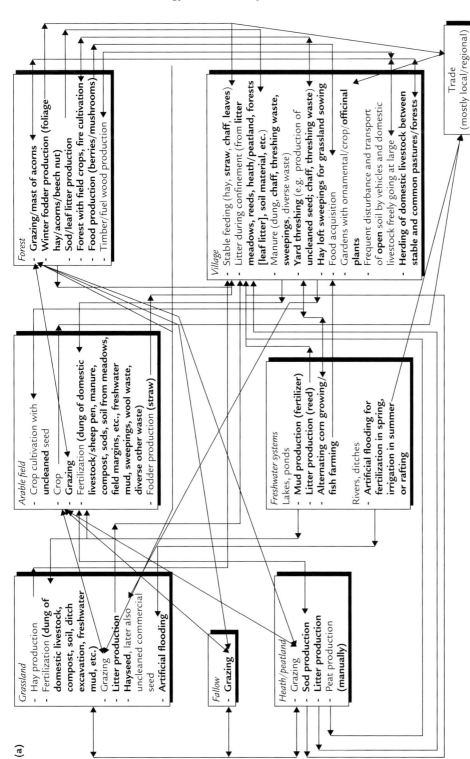

Fig. 6.2 (a) Processes in the ancient, traditional artificial landscape more or less relevant for dispersal. Bold: forms of traditional management relevant for dispersal which got lost today; arrows, direction of dispersal. (b) Processes in the present modern artificial landscape relevant for dispersal. Arrows, direction of dispersal; dotted line, reduced dispersal relevance compared to the ancient, traditional artificial landscape. (From Poschlod & Bonn 1998.)

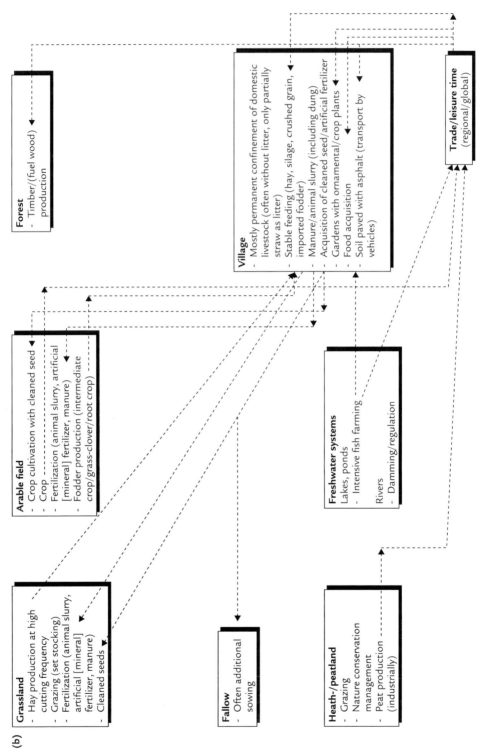

Fig. 6.2 *(Continued)*

depend on attachment capacity (Couvreur *et al.* 2005; Römermann *et al.* 2005) but also on the process of attachment and release (Will & Tackenberg 2008). Most propagules drop off shortly after their attachment, and comparatively few after a longer period (Fischer *et al.* 1996; Bullock *et al.* 2011). Domesticated dogs are also known to be effective dispersal vectors (Heinken 2000).

Nowadays, traffic and trade may be the most effective dispersal vectors over very large distances, which may lead to invasions by non-indigenous plants (von der Lippe & Kowarik 2007).

Through trade (especially of garden plants) many neophytes could spread and establish elsewhere (Di Castri *et al.* 1990). Recent distribution maps show rapid migrations along roads (e.g. Ernst 1998). Seeds transported this way mostly germinate along road verges (Hodkinson & Thompson 1997) and populations of established species are usually not connected with populations in other habitats. On the other hand, dispersal by traffic may contribute effectively to urban biodiversity (von der Lippe & Kowarik 2008).

6.4 Soil seed bank persistence

6.4.1 Classification and importance of seed bank persistence

Soil seed banks may be divided into three groups (Thompson *et al.* 1997, Walck *et al.* 2005):

Transient: seeds that persist in the soil no longer than before the second germination season starts, while often all seeds germinate during the first season or die. *Myricaria* and *Salix* seeds may even survive only for a few days or weeks (Densmore & Zasada 1983; van Splunder *et al.* 1995; Chen & Xie 2007).
Short-term persistent: seeds that persist in the soil at least until the second germination season, but no longer than before the sixth season.
Long-term persistent: seeds that persist in the soil to at least the sixth season.

Some long-term persistent seed banks may persist much longer, up to hundreds of years (Priestley 1986) or exceptionally probably more than 1000 years (Sallon *et al.* 2008). The cut-off level of 5 years between the second and third category was chosen, because many burial experiments do not last longer (Bakker *et al.* 1996). Poschlod & Jackel (1993) and Poschlod *et al.* (1998) elaborated the seed bank classification on the basis of the dynamics of the seed banks and the seed rain, combining seasonal behaviour and depth distribution. However, this only works for dry calcareous grassland communities in Central Europe.

Plant communities with a high proportion of species with a transient seed bank are usually in a climax or stable successional state, for instance forests and semi-natural grasslands (Hopfensberger 2007). Communities with a high proportion of long-term persistent seed banks are found in frequently and/or regularly disturbed habitats, such as arable fields and river banks (Leck *et al.* 1989).

In arid zones, short-lived species rely on a long-term persistent seed bank. The importance of seed-banking relates to the species-specific risk of reproductive failure (Venable 2007).

6.4.2   Measurement of soil seed bank persistence

An exact measurement of persistence is only possible through long-term burial experiments. These started in the late 19th century (Telewski & Zeevaart 2002), but were mostly not performed under natural conditions (Priestley 1986). Another direct, but not equally exact, method is the radiocarbon-dating of viable seeds, ideally of dead parts such as the pericarp or testa (McGraw *et al.* 1991). This method recently allowed the proof that the claim of the oldest record of seed persistence, *Lupinus arcticus* (Porsild *et al.* 1967), was erroneous (Zazula *et al.* 2009).

Since burial experiments are time consuming and radiocarbon dating is expensive, methods of direct seed bank sampling had to be developed to estimate the number of viable seeds (Bakker *et al.* 1996; ter Heerdt *et al.* 1996, 1999; Bernhardt *et al.* 2008), and persistence (Bakker *et al.* 1996; Thompson *et al.* 1997). One can also compare the composition of the seed bank with that of the actual vegetation, but this may be not very reliable. Germination characteristics of species with a persistent soil seed bank may also be included (see Section 6.4.4). A comprehensive key to classify species according to the persistent/ transient categories was given by Grime (1989).

The degree of persistence can be estimated indirectly by analysing seed distribution along the soil profile: the higher the proportion of viable seeds in deeper layers, the more persistent the seed bank. Since seed bank sampling on a single date may miss transient species, seasonal seed bank dynamics should be followed (Thompson & Grime 1979; Poschlod & Jackel 1993). Persistence can also be followed along a successional series (Poschlod 1993). However, persistence is not entirely specific for a species, population or individual.

Thompson *et al.* (1998) described a longevity index *LI*. This index is basically estimated along a continuous scale from transient to persistent. It reads:

$$LI = \text{(short-term + long-term persistent records)/(transient + short-term +}$$
$$\text{long-term persistent records)}.$$

The index varies between 0 and 1, where 0 means no records of persistence and 1 only persistence records.

A general problem is that, according to the methods used, data on persistence of seeds in soil vary in quality. Data based on seedling emergence from soil samples are positively correlated with seed production and hence negatively with seed size (Bekker *et al.* 1998a; Saatkamp *et al.* 2009). Data from burial experiments are not or positively related to seed size (Moles & Westoby 2006). They will probably give larger persistence rates than data from seedling emergence studies while these depend on soil conditions and seed material (Schafer & Kotanen 2003).

6.4.3   Environmental filters affecting soil seed bank persistence

Environmental factors may affect the persistence of soil seed banks. Light becoming available after soil disturbance is probably the most important factor reducing seed longevity (Saatkamp *et al.* 2011a). Seeds (of weeds and crops) will deteriorate more rapidly in organic, acidic peat soil than in mineral soil of neutral pH (Lewis 1973). Seeds can survive longer under either wet (Bekker *et al.* 1998c) or dry conditions (Murdoch & Ellis 2000). The highest amounts of viable seeds were found in sediments of bogs, lakes and ponds (Skoglund & Hytteborn 1990; Poschlod 1995, Poschlod *et al.* 1996). Nutrients (nitrates) reduce persistence as well by affecting dormancy and releasing germination (Bekker *et al.* 1998b). The impact of pathogens may be habitat-specific (O'Hanlon-Manners & Kotanen 2006).

For most species records of transient, short-term or long-term persistent soil seed banks vary considerably according to the method of recording (Thompson *et al.* 1997). Therefore, calculation of a longevity index using data from databases may be of limited value (Saatkamp *et al.* 2009).

6.4.4   Seed traits correlated with seed bank persistence

Several seed and plant traits have evolved that enable a plant to build up a persistent soil seed bank as a reservoir for population buffering or for re-colonization after disturbance. Notably, seed banks with small (and rounded) seeds are often persistent (Thompson *et al.* 1993; Bekker *et al.* 1998a). Small seeds also enter the soil faster (Benvenuti 2007). They can, however, only emerge successfully if they stay located close to the surface (Bond *et al.* 1999) because small-seeded species are often dependent on light for germination, which prevents them from germinating in deeper soil (Milberg *et al.* 2000). Finally, small seeds are produced in greater numbers (Jakobsson & Eriksson 2000; Moles *et al.* 2004), which increases the chance of survival of at least some seeds until the next disturbance and germination season.

There is a high diversity of regeneration strategies in plants. Small seeds like those of *Salix* may be extremely short-lived and large seeds like those of *Nelumbo* and Fabaceae – having a thick and impermeable seed coat effectively protecting them from predation – (Thompson 2000), are very long-lived. Seed size tends to be either not related (Leishman & Westoby 1998; Saatkamp *et al.* 2009) or positively related to longevity (Moles & Westoby 2006) because higher mortality rates in smaller seeds may be compensated by higher seed production. Seed number and size may be confounded when persistence is studied in soil seed bank samples, but not when mortality is determined from defined buried seed samples (Saatkamp *et al.* 2009). It is thus helpful to distinguish between *persistence*, which is mainly used for soil seed banks with an undefined seed input, and *longevity*, which is used for individual seeds or defined seed populations.

Dormancy and dormancy cycling enhance the formation of a seed bank: the inclusion of seeds in the soil (Saatkamp *et al.* 2011b), reduces the percentage of germination during the germination season (Venable 2007) and prevents seeds from germinating under unfavourable conditions (Walck *et al.* 2005; Baskin

& Baskin 2006). Seeds of some plants detect daily temperature fluctuations (Thompson & Grime 1983), and this capability effectively prevents them from germination once buried (Benech-Arnold *et al.* 2000). However, this cannot be extrapolated to larger sets of species and burial depths (Saatkamp *et al.* 2011a). Also, seed coat thickness is a good predictor of the maximum longevity of seeds, especially for longer burial periods (Davis *et al.* 2008; Gardarin *et al.* 2010).

6.5   Germination and establishment

6.5.1   Dormancy types and germination filters

Most seeds experience a period of dormancy. For successful germination this dormancy has to be broken by environmental stimuli such as cold or warm periods and moist or dry conditions (Table 6.5). Dormancy is a mechanism for avoiding germination during unsuitable environmental conditions such as cold or dry periods. The proportion of species with dormant seeds is much higher in temperate and arctic than in subtropical and tropical zones (Baskin & Baskin 2003). Often, dormancy is broken by environmental cues related to the periods of unsuitable conditions, for example arctic and temperate species have physiological dormancy, which is broken by longer periods of low temperatures (cold stratification), resulting in seed germination in spring. Similarly, physiologically dormant seeds in arid regions after-ripen under high temperatures and dry conditions. The interplay of dormancy and seasonal changes in temperature and moisture conditions results in specific germination niches for a local flora (e.g. Merritt *et al.* 2007).

There are many parameters related to climate and habitat that affect germination, germination rate and establishment: temperature, precipitation, light, soil physics and soil chemistry (Fenner & Thompson 2005). Furthermore, certain environmental conditions during maturation may affect germination characteristics which persist as maternal effects in ripe seed (Baskin & Baskin 1998).

Germination and germination rate may vary along a temperature gradient with species having a very wide or a narrow temperature niche. In the latter case, germination will occur only at low or only at high temperatures, respectively (Grime *et al.* 1981). Fire may affect germination through heat and smoke (Keeley & Fotheringham 2000). Diurnal temperature fluctuations have evolved as a gap detection mechanism (Thompson & Grime 1983) and the detection of diurnally constant temperatures is a mechanism to find a safe, protected site such as that surrounding a nurse plant (Kos & Poschlod 2007). Light parameters include mainly light quality (e.g. the ratio red:far-red, R : FR) and the physical and chemical parameters of soil affecting germination include moisture, surface texture, reaction, nitrate content and salinity (Fenner 2000).

6.5.2   Light

Light influences germination in a complex manner (Pons 1992, Milberg *et al.* 2000). Light may be a prerequisite *per se* for germination but there are also

Table 6.5 Dormancy types and their occurence in biomes (after Baskin & Baskin 1998, 2003). + = positive effect; – = negative effect; no = no effect; CS = cold stratification; WS = warm stratification; ABA = abscisic acid, GA = giberellic acid; DS = dry storage (shortening cold stratification period); SC = scarification (promoting germination); D = dry conditions; MW = moist and warm conditions.

Type	Levels of dormancy	Mechanisms to break dormancy	Biomes
Physiological (PD)	Deep	3 to 4 months CS; GA no	In tropical and subtropical zones 20–50% except hot deserts (nearly 60%), in temperate and arctic zones
	Intermediate	2 to 3 months CS; ds + or no; GA + or –	
	Non-deep	4 to 6 weeks CS (0–10°C) or WS (>15°C); SC + or no; GA +	50–90% except broad-leaved evergreen forest and matorral (around 40%)
Morphological (MD)	–	No (waiting until embryo is fully developed); D + or no; MW +, no or –	Low importance through all vegetation zones (0–<5%)
Morphophysiological (MPD)	Deep – simple, simple epicotyl, simple double, complex	WS and CS, WS and CS, CS and WS and CS, CS	Low importance through all vegetation zones (<10%) except in tropical montane, temperate deciduous forest and boreal vegetation zone (10–20%)
	Intermediate – simple, complex	WS and CS, CS	
	Non-deep – simple, complex	WS or CS, CS	
Physical (PY)	–	Mechanical SC +; acid SC + or no; heat +, no or –;	Through all vegetation zones but dominant in Matorral as well as in tropical deciduous forest, savannas and hot deserts
Combinational (PY + PD)	Non-deep		Low importance through all vegetation zones (0–<5%)

many species whose seeds germinate in darkness (Grime *et al.* 1981); this often refers to a transient seed bank. Often a few seconds' exposure to light is sufficient to trigger germination (Hartmann & Mollwo 2000). Day length may also play a role in the detection of the suitable germination season (Densmore 1997). As for light quality, the R:FR ratio controls the emergence of seeds of species (Benech-Arnold *et al.* 2000, Kyereh *et al.* 1999). Seed germination may be inhibited by a low R:FR ratio which is tightly linked to the red-light absorption and far red emission by leaf canopies or dense vegetation structures. Therefore, seeds from species in open habitats such as grasslands or heathlands, often do not germinate (or have significantly lower germination rates) at a low R:FR ratio (van Tooren & Pons 1988), whereas species from forests do germinate, although this may depend on seed size. Small-seeded forest species require a higher R:FR ratio than large-seeded species (Jankowska-Blaszczuk & Daws 2007).

6.5.3 Temperature, temperature fluctuations and the seasonal germination niche

Differences in germination responses to temperature are found along broad environmental gradients such as from south to north, from oceanic to continental, or altitudinal. Species of tundra or alpine vegetation germinate exclusively or at least better at relatively high temperatures (Baskin & Baskin 1998; Fig. 6.3) whereas species from boreal and temperate vegetation have a wider germination range. They may germinate just above freezing temperatures and emerge in (early) spring under the snow or soon after snowmelt. Other species germinate at higher temperatures and thus emerge in summer (Baskin *et al.* 2000). The

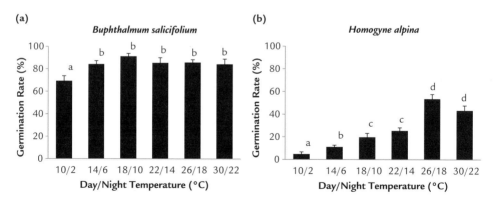

Fig. 6.3 Seed germination rates of two grassland species (Asteraceae) along a temperature gradient: a lowland (*Buphthalmum salicifolium*; altitudinal distribution 100–1700 m a.s.l.) and a subalpine-alpine (*Homogyne alpina*; 1700–2400 m a.s.l.) species. Different minor letters indicate statistically significant differences in Kruskall–Wallis comparisons followed by pairwise *U*-tests and Bonferroni correction (mean ± SE, *n* = 8). Unpublished data.

largest proportions of species with physiological dormancy are found in tundra, steppe and cold deserts (Baskin & Baskin 1998).

The variation in temperature-related germination niches can also be observed along local environmental gradients and within plant communities (e.g. arable weed communities). Weeds of fields with winter cereals germinate mainly at temperatures from 3 to 20 °C (e.g. *Buglossoides arvensis, Consolida regalis*) or 3–30 °C (e.g. *Aphanes arvensis* and *Legousia speculum-veneris*), whereas weeds from fields with summer cereals or root crops do not germinate below 7 °C (e.g. *Galinsoga ciliata* and *Matricaria discoidea*) or 15 °C (e.g. *Chenopodium polysper-mum* and *Setaria viridis*) and may still germinate at 35 °C (Otte 1994). In pro-ductive grasslands, species tend to germinate at lower temperatures than in less productive grasslands, probably to cope with competition for light later on in the season (Olff *et al.* 1994).

Dormancy and germination temperature requirements may result in a specific seasonal germination niche (Schütz 2000; Merritt *et al.* 2007) that may have evolved in response to the occurrence of regeneration niches. Species may also establish in seasonally available gaps. Kahmen & Poschlod (2008) have shown that the number of seedlings from grassland species germinating in autumn decreased when the the litter layer in autumn was not reduced, while species germinating in spring were not affected since the litter layer was decomposed during the winter period. Early spring fires in grasslands increased the abundance of species with physiological (Drobnik *et al.* 2011) or physical dormancy (Poschlod *et al.* 2011).

Constant or diurnally fluctuating temperatures may also affect the occur-rence of species. In the savanna of the Kalahari, many species occur only under *Acacia* or other trees, where diurnal temperature fluctuations are much lower than in the open space beneath the trees. Many species occurring under a canopy do not germinate at high temperature fluctuations, which is seen as a mechanism to 'detect suitable habitat conditions via comparatively constant temperatures' (Kos & Poschlod 2007). In temperate regions, many wetland species germinate only when temperatures are fluctuating, which does not occur during flooding but does occur under low water level conditions (Thompson & Grime 1983).

Also, diurnal temperatures in gaps fluctuate more strongly. Most low-competitive plants (e.g. annuals and biennials and disturbance indicators), are sensitive to such fluctuations, which is seen as a gap detection mechanism (Silvertown & Smith 1988). Trampling or management measures such as grazing may produce regeneration niches for these plants (Harper 1977; Bullock *et al.* 1994).

6.5.4 Precipitation, hydrology and soil moisture

Precipitation and soil moisture clearly affect species richness and composition of plant communities. Species may have desiccation-sensitive seeds, which occur in much larger numbers in tropical and subtropical zones as well as in humid zones than in temperate or arctic or arid zones (Fig. 6.4; Tweddle *et al.* 2003). Species with such seeds cannot form a persistent seed bank (Roberts 1973). The

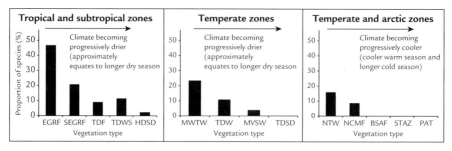

Fig. 6.4 Proportion of species with desiccation-sensitive seeds in vegetation types of different climatic zones. Vegetation types: EGRF, evergreen rain forest; SEGRF, semi-evergreen rain forest; TDF, tropical deciduous forest; TDWS, tropical dry woodland and savanna; HDSD, hot desert and semi-desert; MWTW, moist, warm temperature woodland; TDW, temperate deciduous woodland; MV, matorral vegetation; TDSD, temperate desert and semi-desert; NTW, northern temperate woodland; NCMF, northern conifer dominated montane forest; BSAF, boreal and northern temperate subalpine forest; STAZ, southern temperate subalpine zone; PAT, polar and alpine tundra. (After Tweddle *et al.* 2003.)

decrease in desiccation sensitive seeds along the gradient presented in Fig. 6.4 is related to an increase in desiccation-tolerant seeds.

Finch-Savage & Leubner-Metzger (2006) developed hydrothermal time models to estimate seed germination in the field on the basis of hydrological and temperature factors. Soil seed bank cultivation studies have shown that the frequency of irrigation (ter Heerdt *et al.* 1999) and different hydrological regimes may result in different species compositions (Weiher & Keddy 1995). Functional traits related to specific hydrological regimes lead to sensitivity to flooded or hypoxic conditions and fluctuating temperatures. Seeds from aquatic species such as *Nuphar lutea* and *Nymphaea alba* germinate better or even exclusively under hypoxic conditions (Smits *et al.* 1990), whereas seeds of typical reed species such as *Phragmites australis* will not germinate at all when submerged, even at very low water levels (Spence 1964). Other traits are germination speed and root elongation rate as an adaptation to water availability, which was shown for hot deserts and savannas (Kos & Poschlod 2010). Species associated with the sub-canopy habitat need a longer time to germinate and their seedlings have lower root elongation rates than species from the open space between trees (Fig. 6.5).

6.5.5 Soil chemistry (soil reaction, nutrients and salinity)

Soil reaction is the filtering of the species composition of plant communities by soil chemicals which become toxic at low pH values, including aluminium (Grime & Hodgson 1969, Rorison 1973), the predominant nitrogen compounds (NH_4^+, NO_3^-; Bogner 1968; Gigon 1971), or elements such as iron and manganese that are only available to the plant in low amounts at high pH values (Lambers *et al.* 2008). These soil parameters usually affect the establishment of plants very early on in the seedling stage. Toxic aluminium concentrations, which

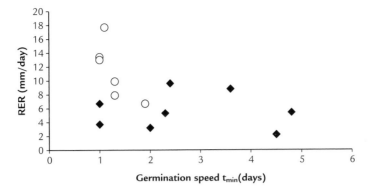

Fig. 6.5 Germination speed (t_{min}, minimum number of days) and root growth rates (RGR, mm/day) in subcanopy (◆) and matrix species (○) of the Kalahari savanna (South Africa). (After Kos & Poschlod 2010.)

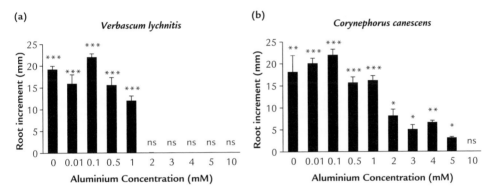

Fig. 6.6 Seedling root lengths (increment from the 3rd to 12th day after onset of experiment; mean ± SE, $n = 5$) of two dry sandy grassland species at increasing aluminium concentrations. The species are contrasting in their soil pH demand: *Verbascum lychnitis* – R = 7; *Corynephorus canescens* – R = 3. Asterisks indicate statistically significant differences from zero in t-tests: * $p < 0.05$; ** $p < 0.01$; *** $p < 0.001$. Unpublished data.

are common in acidic soils, can either prevent germination or damage the root as soon as it protrudes (Fig. 6.6).

Germination does not depend on nutrients, except for nitrate. Nitrate, which may become available in gaps, may initiate germination (Hilhorst & Karssen 2000) and its availability may serve as a gap detection mechanism (Pons 1989). In nutrient-poor ecosystems, the release of nitrogen (e.g. by dung deposition or by fire) may influence species composition by stimulating germination (Luna & Moreno 2009) and thereby depleting the seed bank of such species (see earlier).

Salinity is another factor affecting germination and establishment. Germination of halophyte seeds is promoted by saline conditions, particularly of succulent halophytes (Khan & Gus 2006).

6.5.6 Fire

There are many ecological adaptations of seeds in species found in fire-prone habitats. Examples are seed release through fire in serotinous plants, breaking of physical and physiological dormancy by heat and stimulation of germination via smoke (Brown & van Staden 1997; Keeley & Fotheringham 2000). The decisive chemical compound of smoke is butenolide 3-methyl-2Hfuro[2,3-c] pyran-2-one (carricinolide), which is related to gibberellin (Flematti *et al.* 2004). Reactions to either heat or smoke are not only species specific but also habitat specific. Species of the Mediterranean vegetation are sensitive to heat (Moreira *et al.* 2010), whereas smoke is more important in the stimulation of germination in similar communities in the South African fynbos (Brown 1993) and Australian heathland (Dixon *et al.* 1995).

6.6 Ecological databases on seed ecological traits

Several databases are available on seed traits of many species. Data on seed traits and germination traits of species of the flora of North and Central Europe are contained in the databases Electronic Comparative Plant Ecology (ECPE; Hodgson *et al.* 1995) and BioPop (Poschlod *et al.* 2003; Jackel *et al.* 2006). They are available through the TRY initiative (Kattge *et al.* 2011).

In BioPop, traits on dispersal in space refer to a vector-based dispersal classification for the Central European flora developed within DIASPORUS (Bonn *et al.* 2000). Within the database LEDA (Kleyer *et al.* 2008), dispersal potential data were derived from many Central European species, which are also available through BioPop. Soil seed bank persistence was classified according to the database of the north-west European Flora (see Section 6.5.1; Thompson *et al.* 1997), and were complemented for a large number of species in LEDA. The BioPop data contain information on dormancy type, treatments to break dormancy, germination temperature, sensitivity to diurnal temperature fluctuations, light requirement and seasonal germination niche of species from the temperate decidous forests of Central Europe. Two other large databases on germination data concern the UK germination toolbox within the seed information data base SID (Liu *et al.* 2008) for species from Central Europe and the database on seed germination of the Russian flora (Nikolaeva *et al.* 1985).

ECPE, BioPop and LEDA also contain data on seed production, seed size, mass, shape and surface, and SID also contains data on seed size (Liu *et al.* 2008).

6.7 Seed ecological spectra of plant communities

Plant communities may have very specific dispersal spectra (Molinier & Müller 1938; Dansereau & Lems 1957; Willson *et al.* 1990) and different habitat types have different dispersal and soil seed bank persistence spectra (Bekker *et al.* 1998d; Hodgson & Grime 1990). Studies on dispersal have two shortcomings: classifications according to seed morphology are often incorrect, and the dispersal of individual species is often polychorous. Poschlod *et al.* (2005) compared three different vegetation types in a more detailed way using the database DIASPORUS (Bonn *et al.* 2000). They showed that differences between plant communities are not as pronounced as previously claimed by other authors. Some differences were still obvious: hemerochory was more dominant in arable field vegetation than in grasslands and forests, and the predominant modes of dispersal in grasslands and forests are anemochory and zoochory. In fact, the predominant mode of dispersal depends on the actual availability of dispersal vectors in the respective communities and landscapes. Since landscape and land use are changing more rapidly nowadays, and many dispersal processes related to traditional land-use types are now lost (Poschlod & Bonn 1998), a realistic assessment of actual dispersal processes in many plant communities is not possible or can only be carried out on a local scale.

Only one study (Bekker *et al.* 1998d) compared seed bank persistence spectra in different plant communities, and showed clear differences in Dutch plant communities. In Central Europe, habitats with frequent disturbances – such as arable fields and ephemerally dry wetlands – contain communities of plants with a higher longevity index as compared to those of grasslands and forests (Fig. 6.7). This implies a higher similarity between seed bank and above-ground vegetation (Hopfensberger 2007).

There is also only one study which compares the germination ecology spectra of different habitats (Grime *et al.* 1981). Species of disturbed fertile and skeletal habitats (rocks, walls, roofs) germinated faster than woodland species. Species from skeletal and grassland habitats had wider germination niches related to temperature variation, whereas woodland species were restricted to a narrow range of intermediate temperatures.

Plant communities may also show specific spectra related to other seed traits. Seed production is higher in arable weed communities than in grasslands or woodlands (Fig. 6.7), whereas species from woodlands have larger seeds than species from grasslands and arable weed communities (Fig. 6.7).

6.8   Seed ecological traits as limiting factors for plant species occurrence and assembly

Theories explaining species occurrences are mostly concerned with environmental filters acting on the mature or adult plant. Seed ecological aspects have been widely ignored, although seed ecological traits are strongly related to environmental filters on global, regional and local scales.

The geographical distribution of plant species and floras of biomes may be affected by germination traits related to climatic factors (e.g. demands of cold stratification or certain temperature intervals for germination and precipitation) acting differently on dessication-sensitive or insensitive seeds, or germination speed and/or root growth rate (Fig. 6.8).

On a local scale, species niches as well as species assemblies are affected by light quality, soil physics and chemistry through their impact on specific

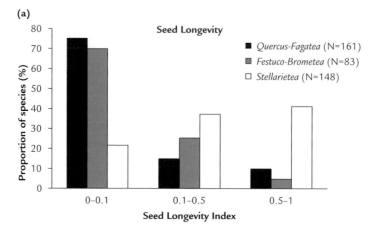

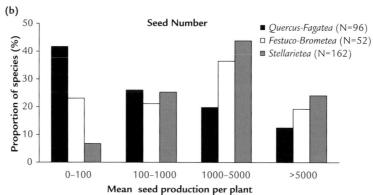

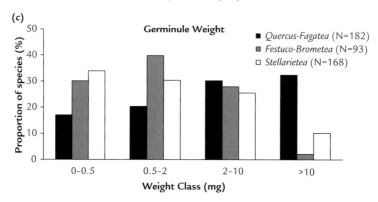

Fig. 6.7 Seed ecological spectra for agricultural weed communities (*Stellarietea*), semi-natural grasslands (*Festuco-Brometea*) and forests (*Querco-Fagetea*) of Central Europe (*n*, number of species analysed in each case). (Calculation of seed bank longevity index (0–1) following Thompson *et al* 1998; phytosociological classification of the species according to Ellenberg *et al*. 2001; seed ecological data from BioPop (Poschlod *et al*. 2003; Jackel *et al*. 2006) and LEDA (Kleyer *et al*. 2008).)

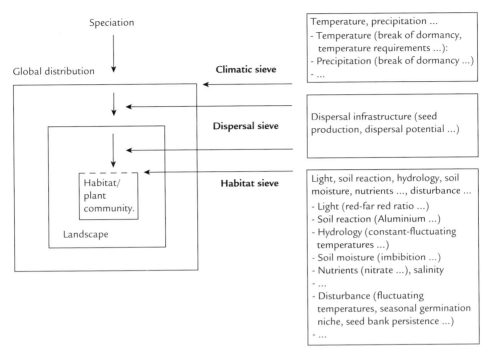

Fig. 6.8 Seed ecological traits affecting the distribution of a plant species, plant community composition and plant species coexistence.

germination demands. As mentioned before, a long-term persistent seed bank allows species survival in seasonally or occasionally disturbed habitats (Fig. 6.8).

Furthermore, the availability of seeds through dispersal or persistence in the soil is responsible for the local occurrence of species and species composition and richness in plant communities. Tackenberg (2001) showed that the frequency of species occurrences in potentially suitable habitats – in this case 70 isolated dry grasslands on porphyry outcrops in an agricultural landscape matrix – was strongly correlated with their dispersal potential (Fig. 6.9). The distribution of plant species and the species richness of plant communities will be influenced by dispersal limitations, as was recently confirmed by Schurr *et al.* (2007), Römermann *et al.* (2008) and Ozinga *et al.* (2009) and experimentally validated by Bugla & Poschlod (2006). An exclosure experiment showed that the dispersal infrastructure, i.e. the set of locally available dispersal processes and vectors, is limiting species composition and richness of plant communities. The community under study was richer in species when more dispersal vectors were operational in the exclosures, with goats and sheep as major agents. This confirms the statement by Poschlod & Bonn (1998) that the maintenance or re-establishment of plant communities, which have developed under historical, traditional land use, is not possible when the former dispersal processes are no longer operational.

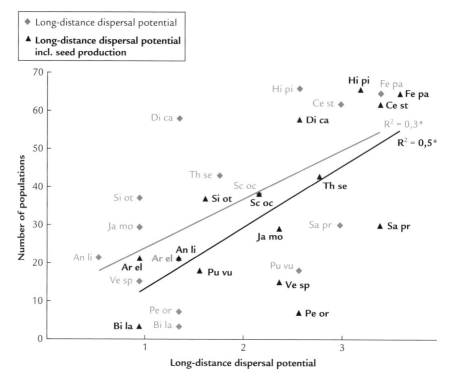

Fig. 6.9 Frequency of plant species growing in the *Thymo-Festucetum* in the porphyry hill landscape near Halle (Germany) is correlated with their dispersal potential in space. Selection of 15 species which (a) grow only in the *Thymo-Festucetum*, (b) do not differ in their essential functional traits except the dispersal potential (perennial hemicrypto- and chamaephytes without persistent seed bank). Classification of the potential for long distance by mechanistic simulation models (wind) or rule-based (zoochory, hemerochory). (From Tackenberg 2001.) For species codes (bold) see Poschlod et al. (2005).

Finally, many restoration experiments have shown that their success depends on dispersal potential and soil seed bank persistence of the characteristic species from the target communities (Bakker *et al.* 1996, Poschlod *et al.* 1998), but also on suitable germination niches (Poschlod & Biewer 2005). The latter study was concerned with the restoration of *Molinion* litter meadows, which are amongst the top species-rich communities in Central Europe. The meadows were drained and fertilized during the 1960s and 1970s; later on fertilization changed from litter to liquid manure. In the 1990s, an effort was made to restore the habitat conditions by nutrient impoverishment and rewetting. However, although all target species were still found in the area, this approach was not successful, because the dispersal potential of the respective species was very low, their soil seed bank was transient or short-term persistent, and suitable gaps were lacking. Return to the former species composition and richness of the target communities was only successful after artificial introduction of species by sowing and hay spreading. The number of characteristic species established after 4 years was significantly higher when a larger number of germination niches was created by harrowing (Poschlod & Biewer 2005).

6.9   Seed ecological traits and species co-existence in plant communities

Among the hypotheses on small-scale species' co-existence in plant communities, the most relevant one is the resource-ratio hypothesis by Tilman (e.g. 1988, see also Chapter 8), which states that species co-exist through niche differentiation and different demands on resource availability. As an alternative to this hypothesis, van der Maarel & Sykes (1993) indicated on the basis of long-term field studies that existing theoretical models do not fully explain species co-existence, at least not in the case of open, dry, species-rich calcareous grassland. They concluded that 'all species of this plant community have the same habitat niche . . . ; the essential variation amongst the species is their individual ability to establish or re-establish by making use of favourable conditions appearing in microsites in an unknown, complex spatio-temporal pattern.' They suggested a carousel model to describe the fine-scale mobility of species. This model, which is yet phenomenological, includes a turnover rate, the speed with which a species moves around in the community. Clearly, this rate will depend on dispersal on a fine scale, i.e. the short-distance dispersal capacity, as well as soil seed bank persistence and on the availability of gaps or suitable germination niches. See also Chapter 4.

The turnover hypothesis was experimentally tested in different grassland communities. In a grassland management experiment in south-west Germany, species turnover was strongly related to seed production (dispersal in space), but not to seed bank persistence (Fig. 6.10). Amongst the species that became extinct, those with a low seed production were over-represented and species with a high seed production were under-represented. Amongst the species which immigrated during the experiment, species with a high seed production were strongly over-represented. The importance of dispersal for species co-existence was also claimed in a simulation model for fire-prone Mediterranean-type shrublands in Western Australia which included seven traits – regeneration mode, seed production, seed size, maximum crown diameter, drought tolerance, dispersal mode and seed bank type (Esther *et al.* 2008).

On the other hand, Stöcklin & Fischer (1999) showed that plants with longer-lived seeds have lower local extinction rates in grassland remnants over a 35-year period. Angert *et al.* (2011) stress the importance of persistent seed banks for co-existence in desert annual communities calling it storage effect. Finally, Zobel *et al.* (2000) demonstrated not only seed availability, but also microsite availability to be responsible for species turnover and co-existence, respectively.

In conclusion, the co-existence of plant species is determined by many seed ecological traits.

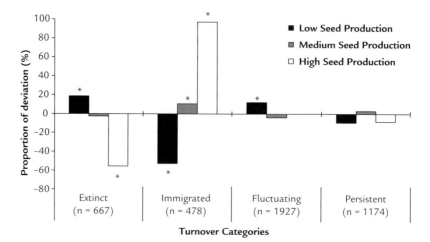

Fig. 6.10 Difference of the observed frequency of species with different seed production in four turnover categories of permanent plots (5 × 5 m²) in grasslands running from 1975 to 2000 (fallow experiments Baden-Wuerttemberg, Germany; six sites; 11 treatments; 65 permanent plots [not every treatment on each site]) compared to the expected frequency. Classes of seed production per ramet: 1, low seed production (1–1000), $n = 972$; 2, medium seed production (1000–10000), $n = 3089$; 3, high seed production (>10000), $n = 185$. (From Poschlod *et al.* 2005.)

References

Angert, A.L., Huxman, T.E., Chesson, P. & Venable, D.L. (2011) Functional tradeoffs determine species coexistence via the storage effect. *Proceedings of the National Academy of Sciences of the United States of America* 106, 11641–11645.

Bakker, J.P., Poschlod, P., Strykstra, R.J., Bekker, R.M. & Thompson, K. (1996) Seed banks and seed dispersal: important topics in restoration ecology. *Acta Botanica Neerlandica* 45, 461–490.

Baskin, C.C. & Baskin, J.M. (1998) *Seeds: Ecology, Biogeography, and Evolution of Dormancy and Germination.* Academic Press, San Diego, CA.

Baskin, C.C. & Baskin, J.M. (2006) The natural history of soil seed banks of arable land. *Weed Science* 54, 549–557.

Baskin, C.C., Milberg, P., Andersson, L. & Baskin, J.M. (2000) Germination studies of three dwarf shrubs (*Vaccinium*, Ericaceae) of Northern Hemisphere coniferous forests. *Canadian Journal of Botany* 78, 1552–1560.

Baskin, J.M & Baskin, C.C. (2003) Classification, biogeography, and phylogenetic relationships of seed dormancy. In: *Seed Conservation: Turning Science into Practice* (eds R.D. Smith, J.B. Dickie, S.H. Linnington, H.W. Pritchard & R.J. Probert), pp. 517–544. Kew, Royal Botanic Gardens.

Bekker, R.M., Bakker, J.P., Grandin, U. *et al.* (1998a) Seed size, shape and vertical distribution in the soil: indicators of seed longevity. *Functional Ecology* 12, 834–842.

Bekker, R.M., Knevel, I.C., Tallowin, J.B.R., Troost, E.M.L. & Bakker, J.P. (1998b) Soil nutrient input effects on seed longevity: a burial experiment with fen meadow species. *Functional Ecology* 12, 673–682.

Bekker, R.M., Oomes, M.J.M. & Bakker, J.P. (1998c) The impact of groundwater level on soil seed bank survival. *Seed Science Research* 8, 399–404.

Bekker, R.M., Schaminée, J.H.J., Bakker, J.P. & Thompson, K. (1998d) Seed bank characteristics of Dutch plant communities. *Acta Botanica Neerlandica* 47, 15–26.

Benech-Arnold, R.L., Sánchez, R.A., Forcella, F., Kruk, B.C. & Ghersa, C.M. (2000) Environmental control of dormancy in weed seed banks in soil. *Field Crops Research* **67**, 105–122.

Benvenuti, S. (2007) Natural weed seed burial: effect of soil texture, rain and seed characteristics. *Seed Science Research* **17**, 211–219.

Bernhardt, K.-G., Koch, M., Kropf, M., Ulbel, E. & Webhofer, J. (2008) Comparison of two methods characterising the seed bank of amphibious plants in submerged sediments. *Aquatic Botany* **88**, 171–177.

Bill, H.-C., Poschlod, P., Reich, M. & Plachter, H. (1999) Experiments and observations on seed dispersal by running water in an Alpine floodplain. *Bulletin of the Geobotanical Institute ETH* **65**, 13–28.

Black, M., Bewley, D. & Halmer, P. (eds.) (2006) *The Encyclopedia of Seeds. Science, Technology and Uses.* CABI, Wallingford.

Boedeltje, G., Bakker, J.P., Bekker, R.M., van Groenendael, J. & Soesbergen, M. (2003) Plant dispersal in a lowland stream in relation to occurrence and three specific life-history traits of the species in the species pool. *Journal of Ecology* **91**, 855–866.

Boedeltje, G., Bakker, J.P., ten Brinke, A., van Groenendael, J. & Soesbergen, M. (2004) Dispersal phenology of hydrochorous plants in relation to discharge, seed release time and buoyancy of seeds: the flood pulse concept supported. *Journal of Ecology* **92**, 786–796.

Bogner, W. (1968) Experimentelle Prüfung von Waldbodenpflanzen auf ihre Ansprüche an die Form der Stickstoffernährung. *Mitteilungen des Vereins für Forstliche Standortskunde und Forstpflanzenzüchtung* **18**, 3–45.

Bond, W.J., Honig, M. & Maze, K.E. (1999) Seed size and seedling emergence: an allometric relationship and some ecological implications. *Oecologia* **120**, 132–136.

Bonn, S. (2005) Dispersal of plants in the Central European landscape – dispersal processes and assessment of dispersal potential examplified for endozoochory. PhD Thesis, University of Regensburg.

Bonn, S. & Poschlod, P. (1998) *Ausbreitungsbiologie der Blütenpflanzen Mitteleuropas.* Quelle & Meyer, Wiesbaden.

Bonn, S., Poschlod, P. & Tackenberg, O. (2000) 'Diasporus' – a database for diaspore dispersal – concept and applications in case studies for risk assessment. *Zeitschrift für Ökologie und Naturschutz* **9**, 85–97.

Brown, N.A.C. (1993) Promotion of germination of fynbos seeds by plant-derived smoke. *New Phytologist* **123**, 575–583.

Brown, N.A.C. & van Staden, J. (1997) Smoke as a germination cue: a review. *Plant Growth Regulation* **22**, 115–124.

Bruun, H.H. & Poschlod, P. (2006) Why are small seeds dispersed through animal guts: large numbers or seed size per se? *Oikos* **113**, 402–411

Bugla, B. & Poschlod, P. (2006) Biotopverbund für die Migration von Pflanzen – Förderung von Ausbreitungsprozessen statt 'statischen' Korridoren und Trittsteinen. Das Fallbeispiel 'Pflanzenarten der Sandmagerrasen' in Bamberg, Bayern. *Naturschutz und Biologische Vielfalt* **17**, 101–117.

Bullock, J.M. & Clarke, R.T. (2000) Long distance seed dispersal by wind: measuring and modelling the tail of the curve. *Oecologia* **124**, 506–521.

Bullock, J.M., Clear Hill, B., Dale, M.P. & Silvertown, J. (1994) An experimental study of the effects of sheep grazing on vegetation change in a species-poor grassland and the role of seedling recruitment. *Journal of Applied Ecology* **31**, 493–507.

Bullock, J.M., Galsworthy, S. J., Manzano, P. *et al.* (2011) Process-based functions for seed retention on animals: a test of improved descriptions of dispersal using multiple data sets. *Oikos* **120**, 1201–1208.

Bullock, J.M., Shea, K. & Skarpaas, O. (2006) Measuring plant dispersal: an introduction to field methods and experimental design. *Plant Ecology* **186**, 217–234.

Cain, M.L., Milligan, B.G. & Strand, A.E. (2000) Long-distance seed dispersal in plant populations. *American Journal of Botany* **87**, 1217–1227.

Chen, F.-Q. & Xie, Z.-Q. (2007) Reproductive allocation, seed dispersal and germination of *Myricaria laxiflora*, an endangered species in the Three Gorges Reservoir area. *Plant Ecology* **191**, 67–75.

Clark, J.S., Fastie, C., Hurtt, G. *et al.* (1998) Reid's paradox of rapid plant migration. *BioScience* **48**, 13–24.

Cohen, D. (1966) Optimizing reproduction in a randomly varying environment. *Journal of Theoretical Biology* **12**: 119–129.

Cosyns, E., Bossuyt, B., Hoffmann, M., Vervaet, H. & Lens, L. (2006) Seedling establishment after endozoochory in disturbed and undisturbed grasslands. *Basic and Applied Ecology* 7, 360–369.

Couvreur, M., Verheyen, K. & Hermy, M. (2005) Experimental assessment of plant seed retention times in fur of cattle and horse. *Flora* 200, 136–147.

Dansereau, P. & Lems, K. (1957) The grading of dispersal types in plant communities and their ecological significance. *Contributions de l'Institut Botanique de l'Université de Montréal* 71, 1–52.

Darwin, C. (1859) *On the Origin of Species by Means of Natural Selection, or the Preservation of Favoured Races in the Struggle for Life.* John Murray, London.

Davis, A.S., Schutte, B.J., Iannuzzi, J. & Renner, K.A. (2008) Chemical and physical defense of weed seeds in relation to soil seedbank persistence. *Weed Science* 56, 676–684.

Densmore, R.V. (1997) Effect of day length on germination of seeds collected in Alaska. *American Journal of Botany* 84, 274–278.

Densmore, R. & Zasada, J.C. (1983) Seed dispersal and dormancy patterns in northern willows: ecological and evolutionary significance. *Canadian Journal of Botany* 61, 3207–3216.

D'hondt, B. & Hoffmann, M. (2011) A reassessment of the role of simple seed traits in mortality following herbivore ingestion. *Plant Biology* 13, 118–124.

Di Castri, F., Hansen, A.J. & Debussche, M. (eds) (1990) *Biological Invasions in Europe and the Mediterranenan Basin.* Kluwer, Dordrecht.

Dirzo, R. & Dominguez, C.A. (1986) Seed shadows, seed predation and the advantages of disersal. In: *Frugivores and Seed Dispersal* (eds A. Estrada & T.H. Fleming), pp. 237–249. Dr W. Junk, Dordrecht.

Dixon, K.W., Roche, S. & Pate, J.S. (1995) The promotive effect of smoke derived from burnt native vegetation on seed germination of Western Australian plants. *Oecologia* 101, 185–192.

Drobnik, J., Römermann, C., Bernhardt-Römermann, M. & Poschlod, P. (2011) Adaptation of plant functional group composition to management changes in calcareous grassland. *Agriculture, Ecosytems and Environment* 145, 29–37.

Ellenberg, H.,Weber, H. E., Düll, R.,Wirth, V. &Werner, W. (2001) Zeigerwerte von Pflanzen in Mitteleuropa. 3rd edn. *Scripta Geobotanica* 18, 1–258.

Ernst, W.H.O. (1998) Invasion, dispersal and ecology of the South African neophyte *Senecio inaequidens* in The Netherlands: from wool alien to railway and road alien. *Acta Botanica Neerlandica* 47, 131–151.

Esther, A., Groeneveld, J., Enright, N.J. et al. (2008) Assessing the importance of seed immigration on coexistence of plant functional types in a species-rich ecosystem. *Ecological Modelling* 213, 402–416.

Evans, M.E.K., Ferrière, R., Kane, M.J. & Venable, D.L. (2007) Bet hedging via seed banking in desert evening primroses (*Oenothera*, Onagraceae): demographic evidence from natural populations. *The American Naturalist* 169, 184–194.

Evenari, M. (1980–1981) The history of germination research and the lesson it contains for today. *Israel Journal of Botany* 29, 4–21.

Fenner, M. (2000): *Seeds. The Ecology of Regeneration in Plant Communities*, 2nd edn. CABI, Wallingford.

Fenner, M. & Thompson, K. (2005): *The Ecology of Seeds.* Cambridge University Press, Cambridge.

Finch-Savage, W.E. & Leubner-Metzger, G. (2006) Seed dormancy and the control of germination. *New Phytologist* 171, 501–523.

Fischer, S.F., Poschlod, P. & Beinlich, B. (1996) Experimental studies on the dispersal of plants and animals on sheep in calcareous grasslands. *Journal of Applied Ecology* 63, 1206–1221.

Flematti, G.R., Ghisalberti, E.L., Dixon, K.W. & Trengove, R.D. (2004) A compound from smoke that promotes seed germination. *Science* 305, 977.

Gardarin, A., Dürr, C., Mannino, M.R., Busset, H. & Colbach, N. (2010) Seed mortality in the soil is related to seed coat thickness. *Seed Science Research* 20, 243–256.

Gigon, A. (1971) Vergleich alpiner Rasen auf Silikat- und auf Karbonatboden; Konkurrenz- und Stickstofformenversuche sowie standortkundliche Untersuchungen im Nardetum und im Seslerietum bei Davos. *Veröffentlichungen des Geobotanischen Institutes ETH, Stiftung Rübel, Zürich*, 48, 1–164.

Goulding, M. (1983) The role of fishes in seed dispersal and plant distribution in Amazonian floodplain ecosystems. *Sonderbände des Naturwissenschaftlichen Vereins in Hamburg* 7, 271–283.

Grime, J.P. (1989) Seed banks in ecological perspective. In: *Ecology of Soil Seed Banks* (eds. M.A. Leck, V.T. Parker & R.L. Simpson), pp xv–xxii. Academic Press, London.

Grime, J.P., Mason, G., Curtis, A.V. *et al.* (1981) A comparative study of germination characteristics in a local flora. *Journal of Ecology* **69**, 1017–1059.

Grime, J.P. &. Hodgson, J.G. (1969) An investigation of the ecological significance of lime chlorosis by means of large-scale comparative experiments. In: *Ecological Aspects of the Mineral Nutrition of Plants* (ed. I.H. Rorison), pp.67–99. Blackwell, Oxford.

Hanski, I. (1987) Colonization of ephemeral habitats. In: *Colonization, Succession and Stability* (eds. A.J. Gray, M.J. Crawley & P.J. Edwards), pp. 155–185. Blackwell, Oxford.

Hansson, L., Söderström, L. & Solbreck, C. (1992) The Ecology of Dispersal in Relation to Conservation. In: *Ecological Principles of Nature Conservation* (ed. L. Hansson), pp. 162–200. Elsevier, London.

Harper, J.L. (1977) *Population Biology of Plants*. Academic Press, London.

Harrison, S. & Hastings, A. (1996) Genetic and evolutionary consequences of metapopulation structure. *Trends in Ecology & Evolution* **11**, 180–183.

Hartmann, K. M. & Mollwo, A. (2000) The action spectrum for maximal photosensitivity of germination. *Naturwissenschaften* **87**, 398–403.

Heinken, T. (2000) Dispersal of plants by a dog in a deciduous forest. *Botanisches Jahrbuch Systematik* **122**, 449–467.

Herrera, C.M. (1989) Frugivory and seed dispersal by carnivorous mammals, and associated fruit characteristics, in undisturbed Mediterranean habitats. *Oikos* **55**, 250–262.

Higgins, S.I. & Richardson, D.M. (1999) Predicting plant migration rates in a changing world: the role of long-distance dispersal. *The American Naturalist* **153**, 464–475.

Hilhorst, H.W.M. & Karssen, C.M. (2000) Effect of chemical environment on seed germination. In: *Seeds. The Ecology of Regeneration in Plant Communities* (ed. M. Fenner), pp. 293–310. CABI, Wallingford.

Hodgson, J.G. & Grime, J.P. (1990) The role of dispersal mechanisms, regenerative strategies and seed banks in the vegetation dynamics of the British landscape. In: *Species Dispersal in Agricultural Habitats* (eds. R.G.H. Bunce & D.C. Howard), pp. 61–81. Belhaven, London.

Hodgson, J.G., Grime, J.P., Hunt, R. & Thompson, K. (1995) *Electronic Comparative Plant Ecology*. Chapman & Hall, London.

Hodkinson, D.J. & Thompson, K. (1997) Plant dispersal: the role of man. *Journal of Applied Ecology* **34**, 1484–1496.

Honnay, O., Bossuyt, B., Jacquemyn, H., Shimono, A. & Uchiyama, K. (2007) Can a seed bank maintain the genetic variation in the above ground vegetation? *Oikos* **117**: 1–5.

Hopfensberger, K. (2007) A review of similarity between seed bank and standing vegetation across ecosystems. *Oikos* **116**, 1438–1448.

Howe, H.F. & Smallwood, J. (1982) Ecology of seed dispersal. *Annual Review Ecology Systematics* **13**, 201–228.

Hubbell, S. (2001) The unified neutral theory of biodiversity and biogeography. *Monographs in Population Biology* **32**, 1–448.

Husband, B.C. & Barrett, S.C.H. (1996) A metapopulation perspective in plant population biology. *Journal of Ecology* **84**, 461–469.

Hyatt, L.A., Rosenberg, M.S., Howard, T.G. *et al.* (2003) The distance dependence prediction of the Janzen–Connell hypothesis: a meta-analysis. *Oikos* **103**, 590–602.

Jackel, A.-K., Dannemann, A., Tackenberg, O., Kleyer, M. & Poschlod, P. (2006) BIOPOP – Funktionelle Merkmale von Pflanzen und ihre Anwendungsmöglichkeiten im Arten-, Biotop- und Naturschutz. *Naturschutz und Biologische Vielfalt* **32**, 1–168.

Jakobsson, A. & Eriksson, O. (2000) A comparative study of seed number, seed size, seedling size and recruitment in grassland plants. *Oikos* **88**, 494–502.

Jankowska-Blaszczuk, M. & Daws, M.I. (2007) Impact of red : far red ratios on germination of temperate forest herbs in relation to shade tolerance, seed mass and persistence in the soil. *Functional Ecology* **21**, 1055–1062.

Janzen, D.H. (1970) Herbivores and the number of tree species in tropical forests. *The American Naturalist* **104**, 501–528.

Janzen, D.H. (1971) Seed predation by animals. *Annual Review of Ecology and Systematics* **2**, 465–492.

Janzen, D.H. (1982) Differential seed survival passage rates in cows and horses, surrogate Pleistocene dispersal agents. *Oikos* 38, 150–156.

Jensen, K. & Gutekunst, K. (2003) Effects of litter on establishment of grassland plant species: the role of seed size and successional status. *Basic and Applied Ecology* 4, 579–587.

Johansson, M.E. & Nilsson, C. (1993) Hydrochory, population dynamics and distribution of the clonal aquatic plant Ranunculus lingua. *Journal of Ecology* 81, 81–91.

Johnson, W.C. & Webb, T. (1989) The role of blue jays (*Cyanocitta cristata* L.) in the postglacial dispersal of fagaceous trees in eastern North America. *Journal of Biogeography* 16, 561–571.

Jordano, P. (2000) Fruits and frugivory. In: *Seeds. The Ecology of Regeneration in Plant Communities* (ed. M. Fenner), pp. 125–165. CABI Publishing, Oxon, New York.

Kahmen, S. & Poschlod, P. (2008) Does germination success differ with respect to seed mass and germination season? Experimental testing of plant functional trait responses to grassland management. *Annals of Botany* 101, 541–548.

Kalisz, S. & McPeek, M.A. (1993) Extinction dynamics, population growth and seed banks. An example using an age-structured annual. *Oecologia* 95, 314–320.

Kattge, J. and 112 others (2011): TRY – a global database of plant traits. *Global Change Biology* 17, 2905–2935.

Keeley, J.E. & Fotheringham, C.J. (2000) Role of fire in regeneration from seed. In: *Seeds. The Ecology of Regeneration in Plant Communities*, 2nd edn (ed. M. Fenner), pp. 311–330. CABI, Wallingford.

Khan, M.A. & Gus, B. (2006) Halophyte seed germination. In: *Ecophysiology of High Salinity Tolerant Plants* (eds. M.A. Khan & D.J. Weber), pp. 11–30. Springer, Dordrecht.

Kirmer, A., Tischew, S., Ozinga, W.A. *et al.* (2008) Importance of regional species pools and functional traits in colonisation processes: predicting re-colonisation after large-scale destruction of ecosystems. *Journal of Applied Ecology* 45, 1523–1530.

Kleyer, M., Bekker, R.M., Knevel, I.C. *et al.* (2008) The LEDA traitbase: a database of life-history traits of the Northwest European flora. *Journal of Ecology* 96, 1266–1274.

Kollmann, J. & Goetze, D. (1998) Notes on seed traps in terrestrial plant communities. *Flora* 193, 31–40.

Kollmann, J. & Pirl, M. (1995) Spatial pattern of seed rain of fleshy-fruited plants in a scrubland–grassland transition. *Acta Oecologia* 16, 313–329.

Kos, M. & Poschlod, P. (2007) Seeds use temperature cues to ensure germination under nurse-plant shade in xeric Kalahari savannah. *Annals of Botany* 99, 667–675.

Kos, M. & Poschlod, P. (2010) Why wait? Trait and habitat correlates of variation in germination speed among Kalahari annuals. *Oecologia* 162, 549–559.

Kuparinen, A., Katul, G., Nathan, R. & Schurr, F.M. (2009) Increases in air temperature can promote wind-driven dispersal and spread of plants. *Proceedings of the Royal Society B* 276, 3081–3087.

Kyereh, B., Swaine, M.D. & Thompson, J. (1999) Effect of light on the germination of forest trees in Ghana. *Journal of Ecology* 87, 772–783.

Lambers, H., Chapin III, F.S. & Pons, T.L. (2008) *Plant Physiological Ecology*, 2nd edn. Springer, New York, NY.

Landolt, E. (2010) *Flora Indicativa. Ökologische Zeigerwerte und Biologische Kennzeichen zur Flora der Schweiz und der Alpen*. Bern: Haupt.

Leck, M.A., Parker, V.T. & Simpson, R.L. (1989) *Ecology of Soil Seed Banks*. Academic Press, London.

Leibold, M.A., Holyoak, M., Mouquet, N. *et al.* (2004) The metacommunity concept: a framework for multi-scale community ecology. *Ecology Letters* 7, 601–613.

Leishman, M.R. (1999) How well do plant attributes correlate with establishment ability? Evidence from a study of 16 calcareous grassland species. *New Phytologist* 141, 487–496.

Leishman, M.R. (2001) Does the seed size/number trade-off model determine plant community structure? An assessment of the model mechanisms and their generality. *Oikos* 93, 294–302.

Leishman, M.R. & Westoby, M. (1998) Seed size and shape are not related to persistence in soil in Australia in the same way as in Britain. *Functional Ecology* 12, 480–485.

Levin, D.A. (1990) The seed bank as a source of genetic novelty in plants. *The American Naturalist* 135, 563–572.

Lewis, J. (1973) Longevity of crop and weed seeds: survival after 20 years in soil. *Weed Research* 13, 179–191.

Liu, K., Eastwood, R.J., Flynn, S., Turner, R.M. & Stuppy, W.H. (2008). *Seed Information Database* (release 7.1, October 2011). http://www.kew.org/data/sid (accessed 30 April 2012).

Luna, B. & Moreno, J.M. (2009) Light and nitrate effects on seed germination of Mediterranean plant species of several functional groups. *Plant Ecology* **203**, 123–135.

Malo, J.E. & Suárez, F. (1995) Herbivorous mammals as seed dispersers in a Mediterranean dehesa. *Oecologia* **104**, 246–255.

Manzano, P. & Malo, J.E. (2006) Extreme long-distance seed dispersal via sheep. *Frontiers in Ecology and the Environment* **4**, 244–248.

McClanahan, T.R. & R.W. Wolfe (1987) Dispersal of ornithochorous seeds from forest edges in central Florida. *Vegetatio* **71**, 107–112.

McGraw, J.B., Vavrek, M.C. & Bennington, C.C. (1991) Ecological genetic variation in seed banks. I. Establishment of a time-transect. *Journal of Ecology* **79**, 617–626.

Merritt, D.J., Turner, S.R., Clarke, S. & Dixon, K.W. (2007) Seed dormancy and germination stimulation syndromes for Australian temperate species. *Australian Journal of Botany* **55**, 336–344.

Milberg, P., Andersson, L. & Thompson, K. (2000) Large-seeded species are less dependent on light for germination than small-seeded ones. *Seed Science Research* **10**, 99–104.

Moles, A.T., Falster, D.S., Leishman, M.R. & Westoby, M. (2004) Small-seeded species produce more seeds per square metre of canopy per year, but not per individual per lifetime. *Journal of Ecology* **92**, 384–396.

Moles, A.T. & Westoby, M. (2004) Seedling survival and seed size: a synthesis of the literature. *Journal of Ecology* **92**, 372–383.

Moles, A.T. & Westoby, M. (2006) Seed size and plant strategy across the whole life cycle. *Oikos* **113**, 91–105.

Molinier, R. & Müller, P. (1938) La dissémination des espèces végétales. *Revue Générale de Botanique* **50**, 1–761.

Moreira, B., Tormo, J., Estrelles, E. & Pausas, J.G. (2010) Disentangling the role of heat and smoke as germination cues in Mediterranenan Basin flora. *Annals of Botany* **105**, 627–635.

Morin, X., Viner, D. & Chuine, I. (2008) Tree species range shifts at a continental scale: new predictive insights from a process based model. *Journal of Ecology* **96**, 784–794.

Murdoch, A.J. & Ellis, R.H. (2000) Dormancy, Viability and Longevity. In: *Seeds. The Ecology of Regeneration in Plant Communities* (ed. M. Fenner), pp. 183–214. CABI, Wallingford.

Murray, B.R. & Leishman, M.R. (2003) On the relationship between seed mass and species abundance in plant communities. *Oikos* **101**, 643–645.

Nathan, R. (2006) Long distance dispersal of plants. *Science* **313**, 786–788.

Nathan, R., Katul, G.G., Horn, H.S. *et al.* (2002) Mechanisms of long-distance dispersal of seeds by wind. *Nature* **418**, 409–413.

Nathan, R., Katul, G.G., Bohrer, G. *et al.* (2011) Mechanistic models of seed dispersal by wind. *Theoretical Ecology* **4**, 113–132.doi: 10.1007/s12080-011-0115-3.

Nikolaeva, M.G., Razumova, M.V. & Gladkova, V.N. (1985) *Spravochnik po prorashchivaniyo pokoyashchikhsya semyan* (reference book on dormant seed germination). Nauka, Leningrad.

O'Hanlon-Manners, D.L. & Kotanen, P.M. (2006) Losses of seeds of temperate trees to soil fungi: effects of habitat and host ecology. *Plant Ecology* **187**, 49–58.

Olff, H., Pegtel, D.M., van Groenendael, J.M. & Bakker, J.P. (1994) Germination strategies during grassland succession. *Journal of Ecology* **82**, 69–77.

Otte, A. (1994) Die Temperaturansprüche von Ackerwildkräutern bei der Keimung – auch eine Ursache für den Wandel im Artenspektrum auf Äckern (dargestellt am Beispiel der Landkreise Freising und München). *Aus Liebe zur Natur (Stiftung zum Schutz gefährdeter Pflanzen) Schriftenreihe* **5**, 103–122.

Ouborg, N.J., Piquot, Y. & van Groenendael, J.M. (1999) Population genetics, molecular markers and the study of dispersal in plants. *Journal of Ecology* **87**, 551–568.

Ozinga, W.A., Römermann, C., Bekker, R.M. *et al.* (2009) Dispersal failure contributes to plant losses in NW Europe. *Ecology Letters* **12**, 66–74.

Pakeman, R.J. (2001) Plant migration rates and seed dispersal mechanisms. *Journal of Biogeography* **28**, 795–800.

Pakeman, R.J., Digneffe, G. & Small, J.L. (2002) Ecological correlates of endozoochory by herbivores. *Functional Ecology* **16**, 296–304.

Pearson, T.R.H., Burslem, D.F.R.P., Mullins, C.E. & Dalling, J.W. (2003) Functional significance of photoblastic germination in neotropical pioneer trees: a seed's eye view. *Functional Ecology* **17**, 394–402.

Petermann, J. S., Fergus, A.J.F., Turnbull, L.A. & Schmid, B. (2008) Janzen–Connell effects are widespread and strong enough to maintain diversity in grasslands. *Ecology* **89**, 2399–2406.

Philippi, T. (1993) Bet-hedging germination of desert annuals beyond the first year. *The American Naturalist* **142**, 474–487.

Pollux, B.J.A. (2011) The experimental study of seed dispersal by fish (ichthyochory). *Freshwater Biology* **56**, 197–212.

Pollux, B.J.A., Ouborg, N.J., van Groenendael, J.M. & Klaassen, M. (2007) Consequences of intraspecific seed-size variation in *Sparganium emersum* for dispersal by fish. *Functional Ecology* **21**, 1084–1091.

Pons, T.L. (1989) Breaking of seed dormancy by nitrate as a gap detection mechanism. *Annals of Botany* **63**, 139–143.

Pons, T.L. (1992) Seed responses to light. In: *Seeds: The Ecology of Regeneration in Plant Communities* (ed. M. Fenner), pp. 259–284. CABI, Wallingford.

Porsild, A.E., Harington, C.R. & Mulligan, G.A. (1967) *Lupinus arcticus* Wats. grown from seeds of Pleistocene age. *Science* **158**, 113–114.

Poschlod, P. (1990) Vegetationsentwicklung in abgetorften Hochmooren des bayerischen Alpenvorlandes unter besonderer Berücksichtigung standortskundlicher und populationsbiologischer Faktoren. *Dissertationes Botanicae* **152**, 1–331.

Poschlod, P. (1991) Diasporenbanken in Böden – Grundlagen und Bedeutung. In: *Populationsbiologie der Pflanzen* (eds B. Schmid & J. Stöcklin), pp. 15–35. Birkhäuser, Basel, Boston, Berlin.

Poschlod, P. (1993) Die Dauerhaftigkeit von generativen Diasporenbanken in Böden von Kalkmagerrasenpflanzen und deren Bedeutung für den botanischen Arten- und Biotopschutz. *Verhandlungen der Gesellschaft für Ökologie* **22**, 229–240.

Poschlod, P. (1995) Diaspore rain and diaspore bank in raised bogs and its implication for the restoration of peat mined sites. In: *Restoration of Temperate Wetlands* (eds B.D. Wheeler, S.C. Shaw, W.J. Fojt. & R.A. Robertson), pp. 471–494. John Wiley & Sons Ltd, Chichester.

Poschlod, P. (1996) Das Metapopulationskonzept – eine Betrachtung aus pflanzenökologischer Sicht. *Zeitschrift für Ökologie und Naturschutz* **5**, 161–185.

Poschlod, P. & Biewer, H. (2005) Diaspore and gap availability limiting species richness in wet meadows. *Folia Geobotanica* **40**, 13–34.

Poschlod, P. & Bonn, S. (1998) Changing dispersal processes in the central European landscape since the last ice age – an explanation for the actual decrease of plant species richness in different habitats. *Acta Botanica Neerlandica* **47**, 27–44.

Poschlod, P. & Jackel, A.-K. (1993) Untersuchungen zur Dynamik von generativen Diasporenbanken von Samenpflanzen in Kalkmagerrasen. I. Jahreszeitliche Dynamik des Diasporenregens und der Diasporenbank auf zwei Kalkmagerrasenstandorten der Schwäbischen Alb. *Flora* **188**, 49–71.

Poschlod, P. & WallisDeVries, M. (2002) The historical and socioeconomic perspective of calcareous grasslands – lessons from the distant and recent past. *Biological Conservation* **104**, 361–376.

Poschlod, P., Bonn, S. & Bauer, U. (1996) Ökologie und Management periodisch abgelassener und trockenfallender kleinerer Stehgewässer im schwäbischen und oberschwäbischen Voralpengebiet. *Veröffentlichungen Projekt Angewandte Ökologie* **17**, 287–501.

Poschlod, P., Kiefer, S., Tränkle, U., Fischer, S. and Bonn, S. (1998) Plant species richness in calcareous grasslands as affected by dispersability in space and time. *Applied Vegetation Science* **1**, 75–90.

Poschlod, P., Kleyer, M., Jackel, A.-K., Dannemann, A. & Tackenberg, O. (2003) BIOPOP – a database of plant traits and internet application for nature conservation. *Folia Geobotanica et Phytotaxonomica* **38**, 263–271.

Poschlod, P., Tackenberg, O. & Bonn, S. (2005) Plant dispersal potential and its relation to species frequency and coexistence. In: *Vegetation Ecology* (ed. E. van der Maarel), pp. 147–171. Blackwell, Oxford.

Poschlod, P., Baumann, A. & Karlik, P. (2009) Origin and development of grasslands in central Europe. In: *Grasslands in Europe – of High Nature Value* (eds P. Veen, R. Jefferson, J. de Smidt & J. van der Straaten), pp. 15–25. KNNV Publishing, Zeist.

Poschlod, P., Hoffmann, J. & Bernhardt-Römermann, M. (2011) Population structures of *Helianthemum nummularium* and *Lotus corniculatus* as affected by grassland management. *Preslia* **83**, 421–435.

Priestley, D.A. (1986) *Seed Aging.* Cornell University Press, Ithaca, NY.

Ramenskyi, L.G., Zazenkin, I.A., Tschishikov, O.N. & Antipin, N.A. (1956) *Ekologhiceskaia Ocenka Kormovih Ugodii po Rstitelnom Pokrova* [*Ecological Evaluation of Grazed Lands by Their Vegetation*]. Selkhozgiz, Moscow.

Rees, M. (1996) Evolutionary ecology of seed dormancy and seed size. *Philosophical Transactions of the Royal Society of London, Series B* **351**, 1299–1308.

Rees, M. & Venable, D.L. (2007) Why do big plants make big seeds? *Journal of Ecology* **95**, 926–936.

Roberts, E.H. (1973) Predicting the storage life of seeds. *Seed Science and Technology* **1**, 499–514.

Römermann, C. (2006) Patterns and processes of plant species frequency and life-history traits. *Dissertationes Botanicae* **402**, 1–117.

Römermann, C., Tackenberg, O. & Poschlod, P. (2005) How to predict attachment potential of seeds to sheep and cattle coat from simple morphological seed traits. *Oikos* **110**, 219–230.

Römermann, C., Tackenberg, O., Jackel, A.-K. & Poschlod, P. (2008) Eutrophication and fragmentation are related to species' rate of decline but not to species rarity – results from a functional approach. *Biodiversity and Conservation* **17**, 591–604.

Rorison, I. H. (1973) The effects of extreme soil acidity on the nutrient uptake and physiology of plants. In: *Acid Sulphate Soils* (ed. H. Dost), pp. 223–253. Proceedings of the Second International Symposium on Acid Sulphate Soils, ILRI Publication 18.

Saatkamp, A., Affre, L., Dutoit, T. & Poschlod, P. (2009) The seed bank longevity index revisited: limited reliability evident from a burial experiment and database analyses. *Annals of Botany* **104**, 715–724.

Saatkamp, A., Affre, L., Baumberger, T., Gasmi, A., Dumas, P.-J., Gachet, S. & Arène, F. (2011a) Soil depth detection by seeds and diurnally fluctuating temperatures: different dynamics in 10 annual plants. *Plant and Soil* **349**, 331–340.

Saatkamp, A., Affre, L., Dutoit, T. & Poschlod, P. (2011b) Germination traits explain soil seed persistence across species – the case of Mediterranean annual plants in cereal fields. *Annals of Botany* **107**, 415–426.

Sallon, S., Solowey, E., Cohen, Y. *et al.* (2008) Germination, genetics, and growth of an ancient date seed. *Science* **320**, 1464.

Schafer, M. & Kotanen, P.M. (2003) The influence of soil moisture on losses of buried seeds to fungi. *Acta Oecologia* **24**, 255–263.

Schurr, F.M., Bond, W.J., Midgley, G.F. & Higgins, S.I. (2005) A mechanistic model for secondary seed dispersal by wind and its experimental validation. *Journal of Ecology* **93**, 1017–1028.

Schurr, F.M., Midgley, G.F., Rebelo, A.G. *et al.* (2007) Colonization and persistence ability explain the extent to which plant species fill their potential range. *Global Ecology and Biogeography* **16**, 449–459.

Schütz, W. (2000) The importance of seed regeneration strategies for the persistence of species in the changing landscape of Central Europe. *Zeitschrift für Ökologie und Naturschutz* **9**, 73–83.

Silvertown, J.W. & Lovett Doust, J. (1993) *Introduction to Plant Population Biology.* Blackwell, Oxford.

Silvertown, J. & Smith, B. (1988) Gaps in the canopy: the missing dimension in vegetation dynamics. *Vegetatio* **77**, 57–60.

Skoglund, J. & Hytteborn, H. (1990) Viable seeds in deposits of the former lakes Kvismaren and Hornborgasjon, Sweden. *Aquatic Botany* **37**, 271–290.

Smits, A.J.M., Avesaath, P.H. & Velde, G. (1990) Germination requirements and seed banks of some nymphaeid macrophytes: *Nymphaea alba* L., *Nuphar lutea* (L.) Sm. and *Nymphoides peltata* (Gmel.) O. Kuntze. *Freshwater Biology* **24**, 315–326.

Soons, M.B. & Bullock, J.M. (2008) Non-random seed abscission, long-distance wind dispersal and plant migration rates. *Journal of Ecology* **96**, 581–590.

Spence, D.H.N. (1964) The macrophytic vegetation of freshwater lochs, swamps and associated fens. In: *The Vegetation of Scotland* (ed. J.H. Burnett), pp. 306–425. Oliver and Boyd, Edinburgh, London.

Stöcklin, J. & Fischer, M. (1999) Plants with longer-lived seeds have lower local extinction rates in grassland remnants 1950–1985. *Oecologia* **120**, 539–543.

Strykstra, R.J., Bekker, R.M. & Verweij, G.L. (1996) Establishment of *Rhinanthus angustifolius* in a successional hayfield after seed dispersal by mowing machinery. *Acta Botanica Neerlandica* **45**, 557–562.

Strykstra, R., Pegtel, D.M. & Bergsma, A. (1998) Dispersal distance and achene quality of the rare anemochorous species *Arnica montana* L.: implications for conservation. *Acta Botanica Neerlandica* **47**, 45–56.

Tackenberg, O. (2001) Methoden zur Bewertung gradueller Unterschiede des Ausbreitungspotentials von Pflanzenarten – Modellierung des Windausbreitungspotentials und regelbasierte Ableitung des Fernausbreitungspotentials. *Dissertationes Botanicae* **347**, 1–138.

Tackenberg, O. (2003) A model for wind dispersal of plant diaspores under field conditions. *Ecological Monographs* **73**, 173–189.

Tackenberg, O., Poschlod, P. & Bonn, S. (2003a) Assessment of wind dispersal potential in plant species. *Ecological Monographs* **73**, 191–205.

Tackenberg, O., Poschlod, P. & Kahmen, S. (2003b) Dandelion seed dispersal – The horizontal wind speed doesn't matter for long distance dispersal – it are updrafts. *Plant Biology* **5**, 451–454.

Tackenberg, O., Römermann, C., Thompson, K. & Poschlod, P. (2006) What does seed morphology tell us about external animal dispersal? Results from an experimental approach measuring retention times. *Basic and Applied Ecology* **7**, 45–58.

Telewski, F.W. & Zeevaart, J.A.D. (2002) The 120-yr period for Dr Beal's seed viability experiment. *American Journal of Botany* **89**, 1285–1288.

ter Heerdt, G.N.J., Verweij, G.L., Bekker, R.M. & Bakker, J.P. (1996) An improved method for seed bank analysis: seedling emergence after removing the soil by sieving. *Functional Ecology* **10**, 144–151.

ter Heerdt, G.N.J., Schutter, A. & Bakker, J.P. (1999) The effect of water supply on seed-bank analysis using the seedling-emergence method. *Functional Ecology* **13**: 428–430.

Thompson, K. (2000) The functional ecology of soil seed banks. In: *Seeds. The Ecology of Regeneration in Plant Communities* (ed. M. Fenner), pp. 215–235. CABI Publishing, Wallingford.

Thompson, K. & Grime, J. P. (1979) Seasonal variation in the seed banks of herbaceous species in ten contrasting habitats. *Journal of Ecology* **67**, 893–921.

Thompson, K. & Grime, J. P. (1983) A comparative study of germination responses to diurnally fluctuating temperatures. *Journal of Applied Ecology* **20**, 141–156.

Thompson, K., Band, S.R. & Hodgson, J.G. (1993) Seed size and shape predict persistence in soil. *Functional Ecology* **7**, 236–241.

Thompson, K., Bakker, J.P. & Bekker, R.M. (1997) *The Soil Seed Banks of North West Europe: Methodology, Density and Longevity*. Cambridge University Press, Cambridge.

Thompson, K., Bakker, J.P., Bekker, R.M. & Hodgson, J.G. (1998) Ecological correlates of seed persistence in soil in the NW European flora. *Journal of Ecology* **86**, 163–169.

Thompson, R.S., Anderson, K.H. & Bartlein, P.J. (1999) *Atlas of Relations Between Climatic Parameters and Distributions of Important Trees and Shrubs in North America*. US Geological Survey, Denver, CO.

Thomson, F.J., Moles, A.T., Auld, T.D. & Kingsford, R. (2011) Seed dispersal distance is more strongly correlated with plant height than with seed mass. *Journal of Ecology* **99**, 1299–1307.

Tilman, D. (1988) *Dynamics and Structure of Plant Communities. Monographs in Population Biology*, 26. Princeton University Press, Princeton, NJ.

Türke, M., Heinze, E., Andreas, K., Svendsen, S.M., Gossner, M.M. & Weisser, W.W. (2010) Seed consumption and dispersal of ant-dispersed plants by slugs. *Oecologia* **163**, 681–693.

Turnbull, L.A., Rees, M. & Crawley, M.J. (1999) Seed mass and the competition/colonization trade-off: a sowing experiment. *Journal of Ecology* **87**, 899–912.

Turnbull, L.A., Crawley, M.J. & Rees, M. (2000) Are plant populations seed-limited? A review of seed sowing experiments. *Oikos* **88**, 225–238.

Tweddle, J.C., Dickie, J.B., Baskin, C.C. & Baskin, J.M. (2003) Ecological aspects of seed desiccation sensitivity. *Journal of Ecology* **91**, 294–304.

van der Maarel, E. & Sykes, M. T. (1993) Small-scale plant species turnover in a limestone grassland: the carousel model and some comments on the niche concept. *Journal of Vegetation Science* **4**, 179–188.

van Splunder, I., Coops, H., Voesenek, L.A.C.J. & Blom, C.W.P.M. (1995) Establishment of alluvial forest species in floodplains: the role of dispersal timing, germination characteristics and water level fluctuations. *Acta Botanica Neerlandica* **44**, 269–278.

van Tooren, B.F. & Pons, T.L. (1988) Effects of temperature and light on the germination in chalk grassland species. *Functional Ecology* **2**, 303–310.

Vavrek, M.C., McGraw, J.B. & Bennington, C.C. (1991) Ecological genetic variation in seed banks. III. Phenotypic and genetic differences between plants from young and old seed populations of *Carex bigelowii*. *Journal of Ecology* **79**, 645–662.

Venable, D.L. (2007) Bet hedging in a guild of desert annuals. *Ecology* **88**, 1086–1090.

Venable, D.L. & Brown, J.S. (1988) The selective interaction of dispersal, dormancy and seed size as adaptations for reducing risks in variable environments. *The American Naturalist* **131**, 360–384.

Venable, D.L. & Rees, M. (2009) The scaling of seed size. *Journal of Ecology* **97**, 27–31.

Venable, D.L., Flores-Martinez, A., Muller-Landau, H.C., Barron-Gafford, G. & Becerra, J.X. (2008) Seed dispersal of desert annuals. *Ecology* **89**, 2218–2227.

Violle, C., Navas, M.L., Vile, D. *et al.* (2007) Let the concept of trait be functional! *Oikos* **116**, 882–892.

Vittoz, P. & Engler, R. (2007) Seed dispersal distances: a typology based on dispersal modes and plant traits. *Botanica Helvetica* **117**, 109–124.

Von der Lippe, M. & Kowarik, I. (2007) Long-distance dispersal of plants by vehicles as driver of plant invasions. *Conservation Biology* **21**, 986–996.

Von der Lippe, M. & Kowarik, I. (2008) Do cities export biodiversity? Traffic as dispersal vector across urban-rural gradients. *Diversity and Distributions* **14**, 18–25.

Walck, J.L., Baskin, J.M., Baskin, C.C. & Hidayati, S.N. (2005) Defining transient and persistent seed banks in species with pronounced seasonal dormancy and germination patterns. *Seed Science Research* **15**, 189–196

Walck, J.L., Hidayati, S.N., Dixon, K.W., Thompson, K. & Poschlod, P. (2011) Climate change and plant regeneration from seed. *Global Change Biology* **17**, 2145–2161.

Weiher, E. & Keddy, P.A. (1995) The assembly of experimental wetland plant communities. *Oikos* **73**, 323–335.

Weiher, E., van der Werf, A., Thompson, K. *et al.* (1999) Challenging Theophrastus: a common core list of plant traits for functional ecology. *Journal of Vegetation Science* **10**, 609–620.

Welch, D. (1985) Studies in the grazing of heather moorland in north-east Scotland. IV. Seed dispersal and plant establishment in dung. *Journal of Applied Ecology* **22**, 46–72.

Westoby, M. (1998) A leaf–height–seed (LHS) plant ecology strategy scheme. *Plant and Soil* **199**, 213–227.

Westoby, M., Jurado, E. & Leishman, M. (1992) Comparative evolutionary ecology of seed size. *Trends in Ecology and Evolution* **7**, 368–372.

Westoby, M., Leishman, M.R. & Lord, J.R. (1996) Comparative ecology of seed size and dispersal. *Philosophical Transactions of the Royal Society of London, Series B* **351**, 1309–1318.

Westoby, M., Moles, A.T. & Falster, D.S. (2009) Evolutionary coordination between offspring size at independence and adult size. *Journal of Ecology* **97**, 23–26.

Wichmann, M.C., Alexander, M.J., Soons, M.B. *et al.* (2009) Human-mediated dispersal of seeds over long distances. *Proceedings of the Royal Society B* **276**, 523–532.

Will, H. & Tackenberg, O. (2008) A mechanistic simulation model of seed dispersal by animals. *Journal of Ecology* **96**, 1011–1022.

Willerding, C. & Poschlod, P. (2002) Does seed dispersal by sheep affect the population genetic structure of the calcareous grassland species *Bromus erectus*? *Biological Conservation* **104**, 329–337.

Willson, M.F. & Traveset, A. (2000) The ecology of seed dispersal. In: *Seeds. The Ecology of Regeneration in Plant Communities* (ed. M. Fenner), pp. 85–110. CABI Publishing, Wallingford.

Willson, M.F., Rice, B.L. & Westoby, M. (1990) Seed dispersal spectra, a comparison of temperate plant communities. *Journal of Vegetation Science* **1**, 547–562.

Wright, S.J., Trakhtenbrot, A., Bohrer, G. *et al.* (2008) Understanding strategies for seed dispersal by wind under contrasting atmospheric conditions. *Proceedings of the National Academy of Sciences of the United States of America* **105**, 19084–19089.

Young, A. G., Boyle, T. & Brown, T. (1996) The population genetic consequences of habitat fragmentation for plants. *Trends in Ecology & Evolution* **11**, 413–418.

Zazula, G.D., Harington, C.R., Telka, A.M. & Brock, F. (2009) Radiocarbon dates reveal that *Lupinus arcticus* plants were grown from modern not Pleistocene seeds. *New Phytologist* **182**, 788–792.

Zobel, M. (1997) The relative role of species pools in determining plant species richness: an alternative explanation of species coexistence. *Trends in Ecology & Evolution* **12**, 266–269.

Zobel, M, Otsus, M., Liira, J., Moora, M. & Möls, T. (2000) Is small-scale species richness limited by seed availability or microsite availability. *Ecology* **81**, 3274–3282.

7

Species Interactions Structuring Plant Communities

Jelte van Andel

University of Groningen, The Netherlands

7.1 Introduction

A plant community is composed of individuals of different species that have arrived and established at the site and persist there until they become locally extinct. The presence of species in a plant community depends, apart from the availability of propagules and safe sites, on environmental resources (nutrients, water, light) and conditions (climate, soil pH, human impact) for growth and reproduction, whereas the species' abundances in the community can be modified by a variety of interspecific interactions structuring the community, both in space and in time. Interactions between species do not only affect community structure, but also provide the community with emergent properties as compared to the sum of the individual plants (cf. Looijen & van Andel 1999). The absence of species in a plant community can be due to failure of dispersal or/and lack of appropriate resources and conditions. Ozinga *et al.* (2009) have shown that losses in plant diversity in north-western Europe are at least as much due to a degraded dispersal infrastructure as to effects of, for example, eutrophication.

Plants do not only interact with other plants in the community, but also with a large number of fungal and animal species (e.g. van Dam 2009). Different types of interaction and their importance in structuring plant communities, in space and time, will be presented and discussed. They will first be defined (Section 7.2) and then treated one after the other (Sections 7.3–7.7). Thereafter, the complexity of species interactions will be illustrated by referring to a number of both direct and indirect interactions in ecosystems that may act together or change in different phases of development (Section 7.8). Finally, in Section 7.9, the notion of 'assembly rules' will be discussed.

Vegetation Ecology, Second Edition. Eddy van der Maarel and Janet Franklin.
© 2013 John Wiley & Sons, Ltd. Published 2013 by John Wiley & Sons, Ltd.

7.2 Types of interaction

When individuals of two species meet, the interaction between the two results in a positive (advantageous), negative (disadvantageous) or indifferent effect on the fitness of either or both species, as compared to a control situation with no interaction (see Table 7.1). Several fitness components have been used to measure species interactions, such as survival, biomass and reproductive capacity.

Competition between organisms can be direct (interference for space) or indirect (via exploitation of limiting resources). Usually the detrimental effects are asymmetric, one species being affected more than the other. Different mechanisms of direct competition are known, for example: (i) through claiming a territory by clonal propagation, where one of the species may take 'priority' over others (cf. Yapp 1925); (ii) through allelopathy, that is the release of organic compounds from one plant species that are detrimental to other species. Competition for resources, called resource competition or exploitation competition, is an indirect interaction. Begon *et al.* (2005) presented a useful working definition of exploitation competition: 'An interaction between individuals, brought about by a shared requirement for a resource in limited supply, and leading to a reduction in the survivorship, growth and/or reproduction of the competing individuals concerned'. The latter part could be summarized by stating that the process leads to a reduction in one or more fitness components, either at the individual level or at the level of a population (Goldberg *et al.* 1999). Section 7.3 focuses on indirect competition, for resources.

Allelopathy can be considered to be a form of interference competition, brought about by chemical signals, i.e. organic compounds produced and released by one species of plants which reduce the germination, establishment, growth, survival or fecundity of other species (Calow 1998). Allelopathy will be addressed in Section 7.4 by referring to a number of examples in the understorey of forests.

Parasitism, like predation and herbivory (not dealt with in this chapter), is a direct and one-sided relationship where one of the species (the consumer) benefits, whereas the other (the resource species) suffers, similar to a consumer–resource relationship (Calow 1998). Section 7.5 presents a number of examples

Table 7.1 Simplified presentation of different interactions between two species (A and B), when they meet or do not meet: disadvantage (–), advantage (+), or indifference (0).

| | Meeting | | Not meeting | |
	Species A	Species B	Species A	Species B
Competition	–	–	0	0
Allelopathy	0	–	0	0
Parasitism	+	–	–	0
Facilitation	0	+	0	0
Mutualism	+	+	–	–

related to inter-plant parasitism, and the effects of fungal parasites and nematodes on plants.

Facilitation implies that plants of a species modify the abiotic environment in such a way that it becomes more suitable for the establishment, growth and/or survival of other species, either in space or in time. The effects are always indirect, via an impact on environmental factors, i.e. by providing shade or shelter for other plants, or by transforming physical or chemical soil conditions or by acting as a shelter against harsh above-ground conditions. In this sense, it is contrary to allelopathy. In Section 7.6 nursery phenomena and hydraulic lift in plant communities will be illustrated.

In mutualism, there are two-sided benefits for the interacting species. The species facilitate each other. Mutualism can be defined as an interaction between individuals of different species that lead to an increase of fitness of both parties, based on mutual assistance in resource supply. Section 7.7 presents examples of plant–mycorrhiza, plant–pollinator, and plant–ant interactions. Many examples are known of asymmetric mutualism, one species benefiting more than the other, and the relationship may also turn into parasitism. Johnson *et al.* (1997) and Neuhauser & Fargione (2004) speak of a mutualism–parasitism continuum. Mutualism can therefore be considered as a temporarily balanced antagonism, a phase in the co-evolution of two species which can develop in different ways, or even fluctuate over the years depending on the community contexts (Thompson & Fernandez 2006).

7.3 Competition

As mentioned, competition between organisms can be direct (for space or territory) or indirect (for resources). The present section will deal with resource competition. The effects of competition for light are in general more one-sided (one species shadowing the other) than those of competition for soil nutrients (both species sharing a limiting resource). If a large plant competes for nitrogen with a small plant, each of them has a negative impact on the other. In view of the absolute amount of nitrogen taken up, resource competition can be asymmetric: **contest competition**, but in terms of the relative loss of fitness of each of the participating plants it can be symmetric: **scramble competition**. Mostly, loss of fitness between two parties is measured in a relative way, as compared to the fitness of an organism without competitors, and not in terms of the total amount of resource captured. The interaction may result in competitive exclusion, but competing species may also remain co-existing.

7.3.1 Early experiments on resource competition

Gause (1934) performed his classic experiments with three *Paramecium* species, two by two feeding on either the same food resource (bacteria) or on different food resources (bacteria and yeasts). From these experiments Gause's principle of competitive exclusion was derived, implying that the number of species that can co-exist cannot exceed the number of limiting resources. Competition theory was at that time mathematically related to population growth, in terms of the Lotka-Volterra logistic growth curves, which suggest that r (per capita growth rate) and K (carrying capacity) are important parameters of success. MacArthur & Wilson (1967) elaborated this approach by formulating the concept of r- and K-selection, resulting in a colonizing strategy vs. a competitive or maintenance strategy. According to this concept, competitive ability is assumed to have been evolved at the expense of the colonizing ability of species; there would be a trade-off between r- and K-characteristics. Similarly, Grime (1974) proposed a trade-off between competitive ability and stress tolerance, which would have resulted in three strategies: competitors, stress tolerators and ruderals. Fitness of individuals is considered to be the outcome of these strategies, expressed in the relative importance of different fitness components (e.g. generative vs. vegetative reproduction of plants).

Another line of research originates from within-species density experiments in agronomy. The 'self-thinning law', proposed by Yoda *et al.* (1963), describes the relationship between plant density and plant biomass in monospecific stands of annual crops. This resulted in the notion of optimal sowing or planting densities in agriculture and forestry. As a follow-up, the question was asked to what extent mixed cropping could increase the production as compared to monocultures. De Wit (1960) developed an experimental technique, the so-called replacement series, and a mathematical analysis to investigate niche overlap and niche differences, by comparing yields under intraspecific and interspecific competition in annual crops. This implicitly recognized Gause's ecological concept of niche separation or differentiation between co-existing species or varieties; if they occupy different niches, they can together exploit the resources to a larger extent than can be done by each of the species or varieties alone. In this case, biomass of harvestable parts of the crop could easily be used as a fitness component to measure the interaction effects. The replacement series have been varied in several ways, both in agronomy and in ecology, for example by using different densities or by applying an additive approach instead of a replacement at a single density (Gibson *et al.* 1999), or by applying multispecies mixtures (Austin *et al.* 1985, McDonnell-Alexander 2006). In most experiments, the competing species are mixed homogeneously. However, when the mixtures are arranged in mosaic patches, they behave as monocultures (e.g. van Andel & Nelissen 1981). In general, the applicability of replacement series to ecological problems, as compared to agricultural problems, is limited (see Joliffe 2000 for a review). Several experimental methods for studying plant competition, as well as the interpretation of experimental results, are well explained in Gurevitch *et al.* (2002).

7.3.2   Mechanisms of resource competition

Grime (1979) related the competitive ability of plants to their maximum relative growth rate in the early phase of development and plant morphological characteristics in the adult phase, both determined by 'maximum resource capture'. Tilman (1982), who repeated experiments following Gause, now using unicellular algae, explained competitive hierarchies on the basis of resource depletion;

'minimum resource requirement' would determine the winner. For two limiting resources, it was actually the ratio between the supplies of the two resources that determined which species would be favoured after equilibrium conditions had set, and under which conditions the two species could co-exist.

Note that the two mechanisms, maximum resource capture vs. minimum resource requirement, are not *a priori* mutually exclusive. Indeed, when we search for mechanisms of competition between organisms of different species, both competing species should be taken into account. Goldberg (1990) proposed a distinction between effect and response; plants competing for a resource have both an effect on the abundance of the resource and a response to changes in abundance of the resource. Individual plants can be good competitors by rapidly depleting a resource, or by being able to continue growth at depleted resource levels. Both the effect and response components of competition must be significant and of appropriate sign for competition to occur, she proposed. Indeed, observations and experiments by Suding *et al.* (2004) revealed the long-term co-existence of a species with a rapid potential growth strategy and a slow-growing species with a nutrient retention strategy, due to a combination of effects of the species on the nitrogen supply and responses to modifications of the supply rates. Individuals of different species can therefore be ranked in competitive ability either by how strongly they suppress other individuals (net competitive effect) or by how little they respond to the presence of competitors (net competitive response). Species A can gain dominance in the early phase of an experiment, or in an early-successional stage, at non-equilibrium transient conditions, because it has the highest growth rate by capturing a greater amount of a limiting resource, whereas species B can ultimately be favoured at equilibrium conditions or in a climax phase, because it uses the resource more efficiently, i.e. it has a lower minimum requirement. Eventually, species B may competitively exclude species A, that is, if it survived as a subordinate species in the transient period, and if an equilibrium can be achieved anyway (Fig. 7.1). Such a transient period of plant succession may last several tens of years.

Which plant characteristics determine the competitive ability of plant species? Tilman (1985) proposed a trade-off between below-ground and above-ground competitive abilities as related to a resource ratio gradient (i.c. light versus nutrients), but Grime (2001) argued and proved that, while this may hold for within-species phenotypic plasticity in response to a nutrient gradient, it is not applicable to differences between species. Berendse & Elberse (1990), extending the classical competition theory of de Wit (1960), suggested a trade-off between nutrient acquisition efficiency and nutrient use efficiency; for nitrogen these efficiencies are defined as the efficiency with which the acquired nitrogen is used for carbon assimilation, and the efficiency with which the assimilated carbon is used for the acquisition of nitrogen. Gaudet & Keddy (1988) used 44 wetland plant species to test whether competitive ability, determining the winner, could be predicted from plant traits. Plant biomass explained 63% of the variation in competitive ability, and plant height, canopy diameter, canopy area and leaf shape explained most of the residual variation. However, Goldberg *et al.* (1999) did not find a significant relationship between competition intensity

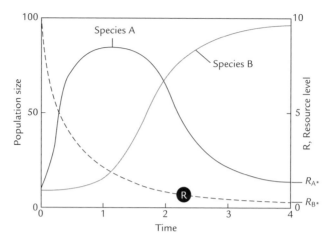

Fig. 7.1 Population responses of two species (A and B) competing for a single limiting resource (R), showing that species A can be dominant in an early phase of competition because it can use the resource rapidly, whereas species B can take over due to its lower minimum resource requirement (R^*). Population size, resource level and time are given in arbitrary units; these will vary depending on the organism. (After Tilman 1988.)

and standing crop. One may suggest that measures of plant architecture are as important as measures of plant biomass, when dealing with competition for light in the canopy. This may similarly hold for root architecture, as soil resources are seldom homogeneously distributed. For roots, the distinction between scale and precision in resource foraging (Campbell *et al.* 1991) seems to be associated with maximum resource capture and minimum resource requirement respectively.

7.3.3 Competition and succession

Several hypotheses have been formulated to indicate the relative importance of competition along productivity gradients, either in space or in time (Fig. 7.2). Tilman (1985, 1990) attempted to predict both competition and succession in productivity gradients from a similar set of assumptions. The resource-ratio hypothesis predicts a change from mainly nutrient competitors to mainly light competitors during succession from bare soil. Competition could be equally intense along productivity gradients, although the resource for which competition occurs may change. Indeed, van der Veen (2000) showed that root competitive intensity experienced by seedlings of seven different species in a primary succession in a coastal salt marsh was negatively related with standing crop, whereas shoot competitive intensity was positively related. Gleeson & Tilman (1990), however, showed an increase in proportional root biomass with successional age for secondary succession on poor soil. Similar results were obtained

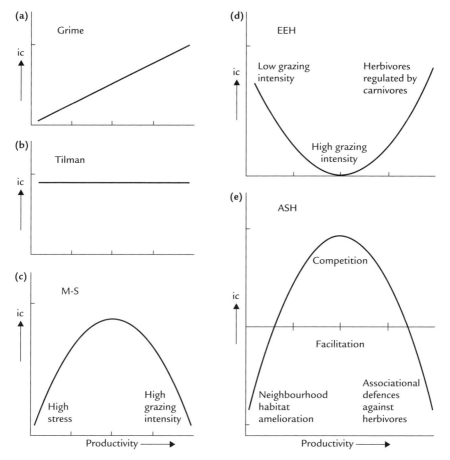

Fig. 7.2 The importance of competition (ic) along a productivity gradient according to (a) Grime (1979), (b) Tilman (1985), (c) Menge & Sutherland (1987), (d) the exploitation ecosystems hypothesis (EEH; Oksanen *et al.* 1981), and (e), the abiotic stress hypothesis (ASH; Callaway & Walker 1997). (As presented by van der Veen 2000; reproduced by permission of the author.)

from an additional study of 46 different successional species. Although absolute leaf biomass increased with successional age, proportion leaf biomass decreased almost twofold, because absolute root biomass increased almost twice as absolute leaf biomass. Their data suggested that the first 40–60 years of succession are a period of strong competition for soil nitrogen, which they considered a period of transient dynamics of competitive displacement, with a pattern that is, at least in part, caused by a trade-off between maximal growth rate and competitive ability for nitrogen and, in part, a trade-off between colonization ability (seed production) and competitive ability for nitrogen.

A plant community can become fixed at any stage of a successional sere, for example due to positive-feedback switches (Wilson & Agnew 1992; Aerts 1999).

This 'inhibition of succession' may result from several mechanisms. For example, small pioneer plants of *Littorella uniflora* are capable of inhibiting succession in nutrient-poor wet dune slacks due to radial oxygen loss from the roots, which prevents the accumulation of organic matter, thus keeping nutrient availability at a low level (Grootjans *et al.* 1998; Adema *et al.* 2002). As soon as this positive feedback system does no longer work properly, the amount of organic matter and nutrients increases which may soon result in an 'alternative stable state' with dominance of tall competitive monocots such as *Calamagrostis epigejos*. Several ecosystems are known where nutrient enrichment of the soil results in strong competitors for light taking over dominance and inhibit further succession, for example *Molinia caerulea* in wet heathlands (Berendse & Elberse 1990), *Brachypodium pinnatum* in calcareous grasslands (Bobbink *et al.* 1989), and *Elymus athericus* in coastal salt marshes (Bakker 1989). In the latter cases, maximum resource capture according to Grime (1979, 2001) can explain the inhibition of succession by dominant competitors.

7.3.4 Co-existence of competing species

Experiments on plant interactions generally focus on only a few species under more or less homogeneous environmental conditions. The increasing interest in biodiversity issues motivated researchers to wonder how subordinate species can remain co-existing in competitive plant communities. This resulted in a renewed interest in environmental heterogeneity and unpredictability. Environmental heterogeneity has gained attention as a cause of species co-existence, through increasing niche complementarity (e.g. Tilman 1994). Modelling approaches and some experimental evidence have explained the co-existence of competing species by plant-induced soil heterogeneity (so-called negative plant-soil feedbacks) that may trigger cyclic population dynamics in multispecies communities (Bonanomi *et al.* 2005, 2008). Hence the competitive ability of a species varies on different substrates. Huisman & Weissing (1999) offered a solution to the so-called 'plankton paradox' (implying that the number of species sometimes exceeds the number of limiting resources), which is based on the dynamics of the competition itself. They showed that (i) resource competition models can generate oscillations and chaos when species compete for three or more resources and (ii) these oscillations and chaotic fluctuations in species abundances allow co-existence of many species on a limited number of resources. In recent competition theory it has been shown that competitive hierarchies or co-existence among a number of species may very well depend on the initial species composition and abundance, which may imply chaos and unpredictability. Both equilibrium outcomes (competitive exclusion or regular oscillations) and non-equilibrium outcomes (irregular oscillations, co-existence of competing species) could be explained by using the same resource competition model. Benincà *et al.* (2008) have experimentally shown, by observations of a laboratory mesocosm with a complex plankton food web from the Baltic Sea over a period of over 7 years, that species abundances showed striking fluctuations over several orders of magnitude. All the species persisted in this chaos, but predictability remained limited to a period of 14 days only.

7.4 Allelopathy

There are many thousands of organic compounds in plants that have been termed secondary substances from a physiological point of view, but which seemed to have an ecological role in interaction with other species. Many such substances, such as tannins from *Pteridium aquilinum* and volatile oils from *Eucalyptus* species, are supposed to have an anti-herbivore function as well and inhibit fungal infections as long as the plant organ is alive. After leaf fall, they may retard litter decomposition rates, thus contributing to the accumulation of organic matter and affecting the germination and establishment of other plants, that is, causing allelopathy. Also, root exudates may be involved in allelopathy, in that they are detrimental to plants of other species. Rice (1974) mentioned several allelopathic compounds, including organic acids and alcohols, fatty acids, quinones, terpenoids and steroids, phenols, cinnamic acids and derivates, coumarins, flavonoids and tannins, amino-acids and polypeptides, alkaloids and cyanohydrins. Not all of them have been proved to play such a role, as it is always difficult to distinguish between allelopathy and other competitive effects between species.

Here phenolics as referred to as an example (see e.g. Kuiters 1990, Hättenschwiler & Vitousek 2000, and references therein). They include simple phenols, phenolic acids and polymeric phenols (condensed tannins, flavonoids). Once released in the soil environment, they influence plant growth directly by interfering with plant metabolic processes and by effects on root symbionts, and indirectly by affecting site quality through interference with decomposition, mineralization and humification. Effects of phenolics on plants include almost all metabolic processes, such as mitochondrial respiration, rate of photosynthesis, chlorophyll synthesis, water relations, protein synthesis and mineral nutrition. Phenolic substances affect plant performance especially under acidic, nutrient-poor soil conditions. In calcareous soils, most phenolic compounds are rapidly metabolized by microbial activity and adsorption is high.

In the Swedish boreal forest, the ground-layer vegetation in late post-fire successions is frequently dominated by dense clones of the dwarf shrub *Empetrum hermaphroditum*. This species is largely avoided by herbivores. It produces large quantities of phenolics, in particular batatasin-III, which is held responsible for the low plant litter decomposition rates, resulting in humus accumulation and reduced nitrogen availability (Nilsson *et al.* 1998). The litter exerts strong negative effects on tree seedling establishment and growth of, for example, *Pinus sylvestris*. Zackrisson *et al.* (1996) proposed that charcoal particles can act as foci both for microbial activity (biodegradation) and chemical deactivation of phenolic compounds through adsorption. Indeed, charcoal was shown to absorb phytotoxic active phenolic metabolites from an *E. hermaphroditum* solution, and wildfires were shown to play an important role in boreal forest dynamics. With prolonged absence of fire in mesic and nutrient-poor sites, the boreal forest can become dominated by *Picea abies* and *E. hermaphroditum*, while fire at intervals of 50 to 100 years may lead to the dominance of *Pinus sylvestris* and the ground-layer species *Vaccinium vitis-idaea* and *V. myrtillus*. This knowledge was used to experimentally disentangle the effects of allelopathy (via litter) and root competition (for resources) between living plants. Nilsson (1994) tried to determine the relative impacts of chemical inhibition and resource competition by *E. hermaphroditum* on seedling growth of *P. sylvestris* by adding fine powdered pro-analysis activated carbon as an adsorbent to the soil surface to remove the allelopathic effect, while exclusion tubes were used to subject pine seedlings to allelopathy in the absence of below-ground competition by *E. hermaphroditum*. Both allelopathy and root competition had a strong, negative influence on seedling growth of *P. sylvestris*.

As mentioned, not all potentially toxic compounds cause allelopathic effects. Recently, Ens *et al.* (2009) applied a comprehensive bioassay, using extracts from plant leaves and roots of dominant shrubs, and from soil underneath their canopies. They were able to distinguish between phytotoxicity, allelopathy and indirect soil effects. The allelopathic suppression of a number of indigenous plant species in Australia by the invasive exotic bitou bush *Chrysanthemoides molinifera* ssp. *rotundata*, planted for restoration purposes after mining, was stronger than the allelopathy caused by the indigenous dominant *Acacia longiflora* var. *sophorae*. The effects were mainly due to phenolic compounds, affecting both the plant community and the microbial soil community.

7.5   Parasitism

A parasite exploits resources from the host, to the latter's disadvantage. It depends on a host for its fitness, whereas the host can live without the relationship; it only suffers from the parasite if it is present. It is not in the interest of the parasite to kill its host, but it may occur, for example in the case of *Cuscuta* species. Plants can be parasitized by other plants, by fungi, or by animal species.

7.5.1   Inter-plant parasitism

Parasitism between plants is a widespread phenomenon (Kuijt 1969). Currently, over 4000 species of parasitic plants are known, occurring in only 19 families. Parasitism in the plant kingdom does occur among trees, shrubs, long-lived perennials and annuals, and all parasites are dicots in only a few lineages. Examples of families with parasitic plants are the Convolvulaceae (including the previous Cuscutaceae with *Cuscuta*), the Loranthaceae (mistletoes, with *Loranthus*), the Lauraceae (with *Cassytha*), the Orobanchaceae (with *Orobanche*, broom-rapes, and *Striga*, the latter formerly classified under Scrophulariaceae), and the Santalaceae (with *Santalum album*, a parasitic tree producing sandalwood, well-known from Indonesia and Malaysia). The latter family currently also includes the previous Viscaceae (with *Viscum*).

Holoparasites (such as *Cuscuta*, *Orobanche* and several orchid species) exploit both root and photosynthesis products from the host; they do not contain chlorophyll and are heterotrophic. Hemiparasites (such as *Rhinanthus* and *Striga* species) exploit the root products only and are capable of photosynthesis themselves as they contain chlorophyll. All plant parasites are connected with the roots or shoots of their host plants by means of a haustorium. Water, minerals, and a wide variety of organic substances are transported through this organ. It is always a one-way flow, but the degree of dependence varies; some species can be grown to flower and set seed without a host, whereas others do not even germinate without a host stimulus (after germination, *Striga* can only survive for 4–5 days without a host). There are many differences with regard to the host-dependence of the parasite. Strict host specificity does not seem to exist. The effect on the host is variable, too; it can be dramatic or hardly measurable and difficult to detect in other cases.

Pennings & Callaway (1996) investigated the impact of *Cuscuta salina*, a common and widespread obligate parasitic annual in saline locations on the west coast of North America. Their results suggest that the parasite is an important agent affecting the dynamics and diversity of vegetation. Because it prefers to parasitize the dominant salt-marsh species *Salicornia virginica*, *C. salina* indirectly facilitates the rare species *Limonium californicum* and *Frankenia salina*, thus increasing plant diversity, and possibly initiating plant vegetation cycles. For other hemiparasites such as annual species of *Rhinanthus*, *Odontites*, *Euphrasia* and *Melampyrum*, it is clear that the parasites depend on a host vegetation to some extent, but in which way do they affect the vegetation? Is the vegetation open because of the presence of the parasite, or is the parasite present because the vegetation is rather low and open (ter Borg 1985)? The effects of hemiparasitism on the plant community may be negative, neutral or positive (Pennings & Callaway 2002). Grasses and legumes are mostly strongly reduced by *Rhinanthus* species, whereas non-leguminous dicots mostly benefit from the presence of the hemiparasite (Ameloot *et al.* 2005). Damaging effects of annual plant parasites on crop plants are known for hemiparasitic species of the genus *Striga* (e.g. on tropical cereals) and holoparasitic species of the genus *Orobanche* (e.g. on tobacco).

7.5.2   Fungal parasites on plants

In the early 1930s, an epidemic wasting disease of eelgrass (*Zostera marina*) decimated populations of eelgrass along the Atlantic coast of North America and Europe. Several causes have been brought forward (den Hartog 1987; Mühlstein *et al.* 1991; van der Heide *et al.* 2007). There is evidence for the pathogenic slime mould *Labyrinthula macrocystis* or *L. zosterae* acting as the causative agent, but the fungal attack may have been facilitated locally by human impact on an increasing turbidity of the water. In the Dutch Wadden Sea, for example, the massive wane of some 15 000 ha of evergreen submarine eelgrass beds coincided with altered hydrodynamics and increased suspended sediment turbidity caused by the construction of the Enclosure Dam between the provinces North Holland and Fryslân. Natural recovery seemed to take a very long time, and the eelgrass beds could only partly become re-established; the system seemed to have changed into an 'alternative stable state'.

   In terrestrial ecosystems pathogenic fungi, selectively parasitizing plant species in a plant community, may accelerate vegetation succession. A classic example, given by Baxter & Wadsworth (1939), is the willow rust *Melampsora bigelowii* that killed many seedlings of *Salix pulchra* and *S. alexensis*, pioneer species which formed nearly pure stands on gravel banks in the river Yukon in Alaska, once the ice had receded. This might have accelerated a succession to *Betula* and *Picea*. Several other examples are referred to by Dobson & Crawley (1994), for example the phenomenon that fungal blights removed *Castanea dentata* from the eastern deciduous forests of the USA, *Tsuga mertensiana* from the Pacific north-west of Canada and the USA, *Ulmus* species from much of western Europe, and a whole range of species from *Eucalyptus* forests of western Australia. In each of these cases, the removal of a dominant species led to the development of forests dominated by less competitive earlier-successional species.

7.5.3   Nematodes feeding on plants

A comparatively small fraction of the highly diverse group of free-living nematodes are feeding on plant roots, sometimes also stems. In the ecological literature these herbivorous nematodes are predominantly called plant- or root-feeders (e.g. Vandegehuchte *et al.* 2010), whereas in agronomy and nematology these nematodes are called plant parasites (e.g. Baldwin *et al.* 2004). Root-feeding nematodes are one component of soil-borne diseases. Other components may be fungal or bacterial pathogens; even mycorrhizal fungi can sometimes reduce plant growth. Soil-borne diseases seemed to be involved in the degeneration of the grass *Ammophila arenaria* and the shrub *Hippophae rhamnoides*, two species that dominate the coastal foredunes of the Netherlands (see van der Putten & van der Stoel 1998, and references therein). *A. arenaria* is widely planted for sand stabilization. Soil-borne enemies seemed to be responsible for the reduced vitality of *A. arenaria*. There was a correlation with root-feeding nematode occurrence, however, inoculation studies showed that root-feeding nematodes alone could not explain the observed reduction of plant performance. Anyway, the reduced vitality of *A. arenaria* favours *Festuca rubra* ssp. *arenaria* and later-successional plant species. Ectoparasitic nematodes of the genus *Longidorus* are capable of damaging the root system of *H. rhamnoides*, including nitrogen-fixing nodules, and the related mycorrhizal system, thus reducing the uptake of phosphate and other nutrients. This damage might contribute to an acceleration of succession to, for example, the shrubs *Sambucus nigra*, *Ligustrum vulgare* and *Rosa rubiginosa* on calcareous soils, or *Empetrum nigrum* on acid soils, but those assumptions need further testing.

   In general, the spatial and temporal dynamics of above-ground and below-ground herbivores, plant pathogens and their antagonists are linked and can differ in space and time (Vandegehuchte *et al.* 2010). This affects the temporal interaction strengths and impacts of above-ground and below-ground higher-trophic-level organisms on plants (see van der Putten *et al.* 2009 for a review).

7.6 Facilitation

Several ecological communities, such as mangrove forests, seagrass beds, conifer forests and semi-arid plant communities, have been shown to be governed by facilitation (Bruno & Bertness 2001). Examples are the facilitation in primary successions through pioneering plants with N_2-fixing micro-organisms in their root systems, be it *Rhizobium* or *Frankia* species, described in many textbooks (e.g. Krebs 2008). Nevertheless, research on facilitation has only recently gained adequate attention. Bruno *et al.* (2003) predicted that the inclusion of facilitation in ecological theory 'will change many basic predictions and will challenge some of our cherished paradigms'. For example, facilitation enlarges the realized niche of the beneficiary species even beyond the boundary of its fundamental niche (see Chapter 2 for the niche concept), whereas competition reduces the realized niche (Fig. 7.3A). Michalet *et al.* (2006) suggested that, while competition is

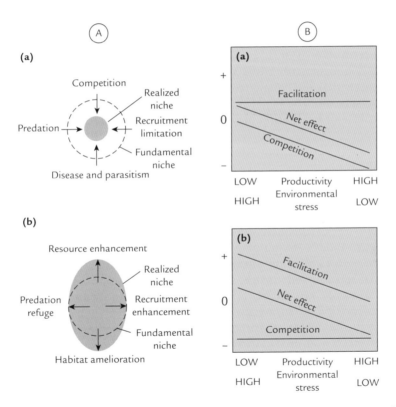

Fig. 7.3 A. Competition, predation and parasitism generally reduces the size of the fundamental niche of a species, whereas the realized niche can be larger than the fundamental niche if facilitation is involved. (Aa) without facilitation, (Ab) with facilitation. B. Facilitation may affect competitive abilities of species along a productivity gradient. (Ba) facilitation weak, constant, (Bb) facilitation strong, variable. (After Bruno *et al.* 2003; reproduced by permission of Elsevier.)

supposed to shape the right-hand side of Grime's (2001) unimodal diversity curve, facilitation may expand the range of stress-tolerant competitive species into harsh physical conditions, thus promoting diversity at the left-hand side; only at extremely severe conditions would species diversity be reduced. Research on facilitative processes has recently been reviewed by Brooker *et al.* (2008); see also the special feature of *Journal of Ecology*, introduced by Brooker & Callaway (2009). The effects may last much longer than the lifetime of the facilitating organisms. Here some examples of facilitation will be presented that are not or hardly mentioned in these reviews, notably nursery phenomena and hydraulic lift.

7.6.1 Nursery phenomena

Nursery is a phenomenological expression of facilitation. The mechanisms that may act have in some cases been discovered through field manipulations, showing that there are effects on nutrients, light, temperature, humidity, wind and other abiotic factors. Positive spatial associations between seedlings of one species and sheltering adults of another species are common in a wide range of environments, and have been referred to as the 'nurse plant syndrome' (reviewed by Callaway & Walker 1997). In many of these cases, seedlings of beneficiary species are found spatially associated with nurse plants, whereas adults are not, which suggests that the balance of competition and facilitation shifts among the various life stages of the beneficiary and the benefactor.

A major role of facilitation between higher plant species, particularly in semi-arid environments, was reported by Pugnaire *et al.* (1996). In south-eastern Spain, for example, the leguminous shrub *Retama sphaerocarpa* strongly improves its own environment, facilitates the growth of *Marrubium vulgare* and other understorey species, and at the same time obtains benefits from sheltering herbs underneath. The interaction between the two species is indirect, associated with differences in soil properties and with improved nutrient availability underneath shrubs compared with plants grown on their own. But it may also work the other way around, N_2-fixing plants being the beneficiary. In a study on facilitation between coastal dune shrubs in California, Rudgers & Maron (2003) showed that seedling emergence, survival and growth of a nitrogen-fixing shrub (*Lupinus arboreus*) was facilitated by a prostrate form of a non-nitrogen-fixing shrub (*Baccharis pilularis*), which implied the possibility of a cascading effect of facilitation in the coastal plant community. In an experimental field study in a sand dune succession in Ontario (Canada), Kellman & Kading (1992) showed that establishment of *Pinus strobus* and *P. recinosa* was facilitated by trees of *Quercus rubra* of at least 35 years old. This effect could be attributed to shading effects, which might imply an improved moisture and temperature regime for seed germination and early seedling survival. In a post-fire matorral shrubland in northern Patagonia (Argentina), Raffaele & Veblen (1998) showed experimentally that two shrub species facilitate the vegetative re-sprouting of herbaceous and woody plants. *Schinus patagonicus* proved to be the most favourable nurse, due to producing more shade and humidity, and the magnitude of facilitation may have been reduced as a result of cattle browsing. In the ecotone between the Rocky Mountain forests and Great Plain grasslands in Montana, USA, Bau-meister & Callaway (2006) proved using a factorial experiment that *Pinus flexilis* facilitated the two later-successional understorey species *Pseudotsuga menziesii* and *Ribes cereum* by a hierarchical set of factors; primarily by providing shade, and, once shade was provided, then also by protection from strong winds. In the humid tropics of Mexico, Guevara *et al.* (1986, 1992) demonstrated that large isolated trees – either left as remnants after forest clearance or in a frag-mented agricultural landscape – enabled the germination and establishment of woody species which otherwise would not succeed in open pasture conditions. Fruit-eating birds, using the trees as perching sites (another type of facilitation), disperse seeds from elsewhere and create 'regeneration nuclei' under the canopy. The tree-induced conditions of shade, soil moisture and soil fertility further facilitate the development of high species richness (Fig. 7.4).

7.6.2 Hydraulic lift

The phenomenon of hydraulic lift, reviewed by Horton & Hart (1998), implies that deep-rooted plants take water from deeper, moister soil layers at daytime and transport it through their roots to upper, drier soil layers, where the roots release it at night, after the stomata have closed. Hydraulic-lifted water can

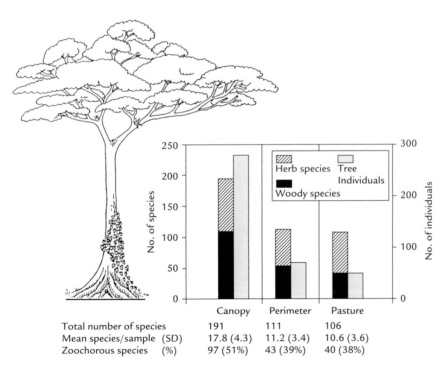

	Canopy	Perimeter	Pasture
Total number of species	191	111	106
Mean species/sample (SD)	17.8 (4.3)	11.2 (3.4)	10.6 (3.6)
Zoochorous species (%)	97 (51%)	43 (39%)	40 (38%)

Fig. 7.4 Large isolated trees facilitate species richness in a former rain forest site in the Sierra de Los Tuxtlas, Vera Cruz, Mexico. This site has later been used as pasture. (After Guevara *et al.* 1992.)

benefit the plant that lifts it, but may also benefit neighbouring more shallow-rooted plants, which was proven by using deuterated water (Caldwell & Richards 1989; Dawson 1993; Armas *et al.* 2010). The phenomenon has been demonstrated for several species, for example the desert shrub *Prosopis tamarugo*, the semi-arid sagebrush *Artemisia tridentata*, and the sugar maple *Acer saccharum* in a mesic forest (references in Horton & Hart 1998), more recently also in *Quercus suber* trees in a savanna-like Mediterranean ecosystem (Kurz-Besson *et al.* 2006). The nocturnal increase in soil water potential could be several orders of magnitude greater than that expected from simple capillary water movement from deep to shallow soil. In a field study in a mesic forest, water hydraulic-lifted by *A. saccharum* supplied up to 60% of the water used by neighbouring shallow-rooted species. Plants that used hydraulically lifted water were able to maintain higher transpiration rates and experienced less water stress than plants that did not and *A. saccharum* seedlings that performed hydraulic lift were able to achieve higher daily integrated carbon gain than plants in which hydraulic lift was experimentally suppressed (Dawson 1993). Yoder & Nowak (1999) documented hydraulic lift for the first time for a CAM species, *Yucca schidigera*, a native plant in the Mojave Desert. The pattern of diel flux in soil water potential for the CAM species was temporarily opposite to that of the C_3 species investigated. The authors suggested that, because CAM plants transport water to shallow soils during the day when surrounding C_3 and C_4 plants transpire, CAM species that lift water may influence water relations of surrounding species to a greater extent than hydraulic-lifting C_3 or C_4 species.

Several authors pointed to the phenomenon of 'reverse hydraulic lift'. Burgess *et al.* (1998) measured sap flow in the roots of *Grevillea robusta* and *Eucalyptus camaldulensis* that could be interpreted as hydraulic lift. After this, however, hydraulic redistribution of water occurred, facilitating root growth in dry soils and modifying resource availability. Water can move down the taproot of trees when the surface soil layers are wetter than the deeper soil layers. Similarly, Smith *et al.* (1999) used measurements of reverse flow in tree roots to demonstrate the opposite process to hydraulic lift: the siphoning of water downwards by root systems of trees spanning the gradient in water potential between a wet surface and dry subsoil. They suggested a competitive advantage for trees over their neighbours in dry environments where plants are reliant on seasonal rainfall for water. Reverse hydraulic lift has been suggested to facilitate root growth through the dry soil layers underlaying the upper profile where precipitation penetrates, thus allowing roots to reach deep sources of moisture in water-limited ecosystems.

The effects of hydraulic lift on plant community structure are only moderate. In Section 7.8.2, literature is discussed that demonstrates that the facilitating effects of hydraulic lift may be counteracted by competition.

7.7   Mutualism

For mutualism, the facilitation is bidirectional between two species, but the benefit may be asymmetric. Mutualistic relationships can be facultative (leguminous plants can live with or without *Rhizobium*), or obligate, i.e. a condition for survival, also known as symbiosis, as in many lichens which are based on a symbiosis between a fungal and an algal component. Plant–mycorrhiza interactions can be considered a mutualistic symbiosis, plant–pollinator and plant–ant interactions a non-symbiotic mutualism. As mentioned in Section 7.2, mutualism can also turn into parasitism.

7.7.1   Plant–mycorrhiza interactions

For vascular plants in general, mutualistic relationships with mycorrhizal fungi are of utmost importance; see Ozinga *et al.* (1997) for a review. Many experimental investigations have shown that both plant and fungal symbionts benefit from the reciprocal exchange of mineral and organic resources. In general, mycorrhizal fungi assist plants in the uptake of nutrients and water from the soil, and plants provide the associated fungi with carbohydrates (see Chapter 9). There are different types of mycorrhizal fungi. The majority, *c.* 80%, of species of temperate, subtropical and tropical plant communities are colonized by fungi with arbuscular mycorrhiza (AM; formerly known as vesicular-arbuscular mycorrhiza). AM fungi are presumably especially efficient in the uptake of inorganic P (and other relatively immobile ions such as Cu, Zn and ammonium) and are capable of increasing the P uptake more in nutrient-rich patches than in soils with a uniform P distribution (Cui & Caldwell 1996). There is ample evidence that in such communities a vigorous semi-permanent group of fungal symbionts with low 'host' specificity is involved in an infection process which effectively integrates compatible species into extensive mycelial networks (Francis & Read 1994). Ectomycorrhizal fungi (ECM) occur mainly on woody plants and only occasionally on herbaceous and graminoid plants. Ericoid mycorrhizas (EM) occur mainly in Ericales and are physiologically comparable with ECM. ECM and EM fungi are especially effective in N-limited ecosystems; enzymatic degradation by these fungi has been shown for proteins, cellulose, chitin and lignin. Non-mycorrhizal plants occur mainly in very wet or saline ecosystems and in ecosystems with a high nutrient availability and/or with recently disturbed soil.

Orchid–mycorrhiza relations are a special case (Dijk *et al.* 1997). After germination, the developing protocorm (the first heterotrophic and subterranean phase of orchid development) is entirely dependent on mycorrhizal fungi. Mycorrhizal infection is restricted to subterranean tissues only, i.e. to the subepidermal zone of the protocorm and root parenchyma. The primary function of mycorrhizal infection in the juvenile phase lies in the transport of C-compounds to the developing seedlings. Translocation of sugars towards protocorms has been demonstrated by radioactive labelling in classic studies. Apart from interfering with the carbon metabolism, mycorrhizal infection has a pronounced influence on the uptake of mineral macronutrients (P and N). As soon as the first leaf of the orchid seedling produces chlorophyll (not all orchid species do so), it becomes independent of the mycorrhizal fungus. In symbiotic protocorm–fungus cultures a range of interactions could be met, from a loss of mycorrhiza via normal mycorrhizal infections to pathogenic effects (the fungus parasitizing the protocorm), depending on the nutrient status of the medium. Rasmussen &

Rasmussen (2009) suggested that the orchid is the only party that profits; indeed, the fungus does not need the protocorm.

In Section 7.8.3 some examples will be given of more complex mycorrhizal fungal networks affecting plant community structure.

7.7.2 Plant–pollinator interactions

In their review on 'endangered mutualisms', Kearns *et al.* (1998) pointed out that over 90% of modern angiosperm species are pollinated by animals. Among the flower-visiting animal species are insects, lizards, birds, bats and small marsupials. Even if there are impressive specific relationships such as between yuccas and yucca moths or between figs and fig wasps, specialist relationships are relatively rare and plant–pollinator interactions are only seldom specific, i.e. to the species level. Relatively few plant–pollinator interactions are absolutely obligate in a strict sense (Johnson & Steiner 2000; Waser & Ollerton 2006). Many flowers that have developed a specialization in floral traits are still often visited by diverse assemblages of animals. The most basic evolutionary outcome that is common across both plants and pollinators is the efficiency of both in exploiting what is for each a valuable or critical resource. Both parties are opportunistic and flexible. As a result, mutualism is neither symmetrical nor cooperative; the exploitation may even be skewed towards a consumer–resource relationship between the two parties. Rather than a mutualistic interest between two-species relations, the plants and pollinators in an ecosystem often form together a mutualistic plant–pollinator network (Bascompte *et al.* 2006; Bosch *et al.* 2009).

Habitat fragmentation and other effects of land use, such as agriculture, grazing, herbicide and pesticide use, and the introduction of non-native species, have a crisis-like impact on plant–pollinator systems. For plants, the fitness consequences of habitat fragmentation depend on the amount of gene flow still possible between local populations, as well as within populations. Kwak *et al.* (1998) illustrated that the flow of pollen and genes in fragmented habitats depends not only on the investigated plant populations as such, but also on the neighbouring species of the plant communities and the flowering phenologies of the component species (cf. also Lázaro *et al.* 2009). In general, changes in the species composition of a plant community have a great impact on pollination and pollen flow due to the differences in pollination efficiency and flight distances. A reduction in local flower population size of all or several component plant species causes a decrease in the richness of the assemblage of insect pollinators as well, which affects pollination quantity and pollination quality. Pollinator communities may adapt more quickly to reduction in population sizes and habitat fragmentation than plant communities (Taki & Kevan 2007). If necessary, insects visit several plant species to meet their energy demands, at the same time increasing the chance of heterospecific pollen deposition on the stigmas. This often results in a reduction of seed set and greater inbreeding in the plant population which can only be counteracted through gene flow between local populations. In small habitat fragments, less-attractive plant species may receive fewer pollen visits and a higher proportion of heterospecific pollen grains, thereby reducing pollination success and gene flow. The introduction of alien plant species may have a further negative impact on insect visitation and seed set of co-flowering 'focal species', as compared to the effects of native plant species in the community (Morales & Traveset 2009).

7.7.3   Plant–ant interactions

Among the invertebrates, only ants have a major role in seed dispersal; it is called myrmecochory. In almost all biomes (see Chapter 15), thousands of plant species produce seeds with food bodies (elaiosomes), specialized for ant dispersal in 'diffuse' multispecies interactions. Well known are the conspicuous aggregates of epiphytes called 'ant-gardens' in Amazonian rainforests, where arboreal ants collect seeds of several epiphyte species and cultivate them in nutrient-rich nests. Workers of the ant *Camponotus femoratus* have been shown to be attracted to odorants emanating from seeds of *Peperomia macrostachya*, and chemical cues may also elicit seed-carrying behaviour (Youngsteadt *et al.* 2008).

Worldwide some hundred plant species, called 'myrmecophytes', mainly shrubs and trees in the tropics and subtropics (for example acacias), produce structures that accommodate the inhabitation of ant colonies (Beattie 1989). The ant species belong to scores of families, and one or more ant species (a guild) may be associated with a plant species. They live above-ground in different plant organs, in special chambers called 'domatia'. Plants may produce food rewards such as extrafloral nectaries to attract the associated ants, and in turn can absorb nutrient ions, especially of nitrogen and phosphorus, from the decaying waste of the ant colony. The presence of fungi and bacteria within domatia may facilitate nutrient breakdown and transport. The plants benefit also indirectly from the presence of ants. They may protect them against herbivores (leaf-feeding insects, stem-boring beetles, vertebrate browsers), seed predators and their eggs, fungal spores, vines, encroaching vegetation and epiphytes (Rosumek *et al.* 2009). Palmer & Brody (2007) showed that the defence of host plants may differ substantially among ant species, depending on their aggressiveness, and also between vegetative and reproductive structures.

The benefit of plant–ant relations is not always mutual. Depending on abiotic conditions (for example shade or non-shade), mutualism may turn into a one-sided benefit for the ants only (Kersch & Fonseca 2005). Competition between plant-inhabiting ant species, the dominant species for example pruning the tree branches to prevent invasion by other ant species, may even turn out to be at the expense of the host plant. In mutualistic communities, networks comprised of ant interactions with extrafloral nectar-bearing plants in the Sonoran Desert, Chamberlain & Holland (2009) have shown that the number of ant-species interactions per plant species (that is, their degree) follows particular power distributions, and that the degree–body size relationship for ants in such ant–plant networks is consistent with that of predators in predator–prey networks, possibly suggesting similar underlying processes at work.

7.8   Complex species interactions affecting community structure

The different types of interaction have been discussed one after the other, but in plant communities several interactive mechanisms may be at work simultaneously, the intensity of a particular interaction may change during succession or even within a season, and so-called 'third parties' may affect the interactions between two species. It is now generally recognized that many interactions in ecological communities are variable in strength and complex. Network relationships, including several feedback systems, are the rule rather than the exception. Further research on interactions will probably develop in this direction.

7.8.1 Simultaneous or intermittent facilitation and competition

Callaway & Walker (1997) provided many examples illustrating that species interactions may involve a complex balance of competition and facilitation. *Quercus douglasii* trees had the potential to facilitate understorey herbs by adding considerable amounts of nutrients to the soil beneath their canopies. However, experimental tree root exclusion increased understorey biomass under trees with high shallow-root biomass, but this had no effect on understorey biomass beneath trees with low shallow-root biomass. Thus, the overall effect of an overstorey tree on its herbaceous understorey was determined by the balance of both facilitation and competition. Another example is that shifts in facilitation and competition among aerenchymous wetland plants occur as temperatures change in anaerobic substrates. *Myosotis laxa*, a small herb common in wetlands of the northern Rocky Mountains, benefited from soil oxygenation when grown with *Typha latifolia* at low soil temperatures in greenhouse experiments. At higher soil temperatures, the significant effects of *Typha* on soil oxygen disappeared (presumably because of increased microbial and root respiration) and the interaction between *Myosotis* and *Typha* became competitive. In the field, the overall effect of *Typha* on *Myosotis* was positive, as *Myosotis* plants growing next to transplanted *Typha* were larger and produced more fruits than those isolated from *Typha*. In the subalpine environment of the central Caucasus Mountains, the relationship between two co-dominant species changed even within a season from facilitation to competition (Kikvidze *et al.* 2006).

In a Tanzanian semi-arid savanna ecosystem, Ludwig *et al.* (2004) showed that the beneficiary effect of hydraulic lift from *Acacia tortilis* trees to grasses can be overruled by competition for water that the grasses experience from the same trees. This would imply that, while the phenomenon of lift has been proven to exist, the net ecological effects may be of little importance. Something similar holds for the positive effects of nutrient enrichment of the soil by tree litter fall, versus the negative effects from competition for nutrients, but in this case the beneficiary effects may prevail. A conceptual model of these complex interactions is presented in Fig. 7.5. Similarly, Armas *et al.* (2010) have shown that two semi-arid evergreen shrub species in an arid coastal sand dune system in Spain may co-exist due to contrasting effects. Hydraulic-lifted water from the deep-rooting *Pistacia lentiscus* facilitates the shallow-rooting *Juniperus phoenicea*, but this positive effect is counterbalanced when *Pistacia* brings saline water to the soil surface in drought periods, which is harmful to *Juniperus*.

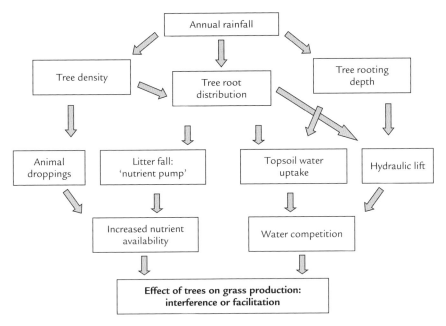

Fig. 7.5 Conceptual model showing the determinants of facilitation and competition in a semi-arid tree-grass savanna ecosystem in Tanzania. (After Ludwig 2001; reproduced by permission of the author.)

7.8.2   Interactions along environmental gradients

Connell & Slatyer (1977) proposed to distinguish between facilitation, competition and inhibition as models of succession, with different mechanisms involved (e.g. Glenn-Lewin & van der Maarel 1992; van Andel *et al.* 1993). These mechanisms are not mutually exclusive and may act together or one after the other. For example, facilitation may affect competitive abilities of species along an environmental gradient in such a way as to keep it at a low level all along the gradient (Bruno *et al.* 2003; see Fig. 7.3B). The stress-gradient hypothesis, brought forward by Bertness & Callaway (1994) and elaborated by Maestre *et al.* (2009) – assuming that facilitation is more common in conditions with high abiotic stress, whereas competition prevails in more benign conditions – predicts that the relative frequency of facilitation and competition varies along productivity gradients. From an experimental test in an alpine altitudinal gradient, Dullinger *et al.* (2007) draw the conclusion that the relationships between small-scale co-occurrence patterns of vascular plants and environmental severity are weak and variable, and may differ among indicators of severity, growth-forms and scales. In a primary succession on a coastal beach plain, van der Veen (2000) has experimentally shown that resource competition can be important from the beginning of a primary succession onwards, but changing from below-ground to above-ground. In conclusion, the stress-gradient hypothesis is interesting, but it requires further tests for generalization.

7.8.3   Mediators of species interactions: third parties

Miller (1994) argued that the success of species in a community is affected not only by direct interactions between species, but also by indirect interactions among groups of species, for example if a third species modifies the conditions for interaction between two other species. This phenomenon is called mediation (Price *et al.* 1986; Allen & Allen 1990; Pennings & Callaway 1996). Mediation by parasites is very common in nature and must be regarded as one of the major types of interaction in ecological systems, comparable in importance to direct competition, predation, parasitism, or mutualism (Price *et al.* 1986).

The presence of mycorrhizae has been shown to change the outcome of plant competition in many cases, both for AM plants and for ECM plants, and is thus a determinant of plant community structure (van der Heijden *et al.* 1998). In a microcosm experiment Grime *et al.* (1987) demonstrated that ^{14}C could be transported through a mycorrhizal network from dominant to subordinate species, which led to an increase in biomass of the inferior competitors. Mycorrhizal linkages were also shown to transport ^{15}N and ^{32}P within and between plant species (Chiarello *et al.* 1982; Finlay *et al.* 1988), for example from dying roots of one species, to developing roots of another species. The presence of AM fungi has been shown to make intraspecific competition more severe, and decrease the strength of interspecific competition (Moora & Zobel 1996). Tropical mycorrhizal AM-fungal communities have the potential to differentially influence seedling recruitment among host species and thereby affect community composition (Kiers *et al.* 2000). AM-symbiosis has also been shown to alleviate the unfavourable effects on plant growth of stresses such as heavy metals, soil compaction, salinity and drought (Miransari 2010).

Effects of herbivores on plant structure and succession are well known. Brown & Gange (1989) were among the first to pay attention to the effects of interacting above-ground and below-ground plant consumers (herbivores and pathogens) on plant succession. Three major life history groupings – annual herbs, perennial herbs and perennial grasses – responded differently, with a considerable effect on the pattern of early succession. Effects on the rate and direction of succession apparently differ between above-ground and below-ground herbivores. In summary, above-ground herbivores, ranging from insects to mammals, can feed on shoots and roots. As far as their above-ground effects are concerned, they are known to retard succession, while optimizing their food supply at a particular successional stage. Below-ground herbivores feed on roots, a process known to accelerate succession, at least in early stages. Plant pathogens, both above- and below-ground, may accelerate succession still further, if they kill dominant plants. For further reading see Wardle's (2002) book and the review by van der Putten *et al.* (2009).

If a population of dominant herbivores is strongly reduced, the effects on the vegetation may be dramatic. A well-known example was provided by the infection of rabbits (*Oryctolagus cuniculus*) by the *Myxoma* virus in southern England (see Dobson & Crawley 1994). Myxomatosis was introduced into Australia in 1950 and into France in 1952, from where it spread throughout western Europe, reaching Britain in 1953. The initial *Myxoma* virus in 1953 was a highly virulent strain and the 1950s rabbit population was reduced by about 99% in a few years. The rabbits remained extremely scarce for the following 15 years. Once rabbits had almost disappeared, acorns buried in grassland by jays had a vastly greater chance of producing seedlings and becoming established. The reduced rabbit grazing was responsible for the transformation of Silwood Park from an open grass parkland in 1955 into an oak woodland (*Quercus robur*) with occasional clearings within 15–20 years. This change was irreversible, even after the recovery of the rabbit population in the 1970s.

7.9 Assembly rules

The term 'assembly rules' was coined by Diamond (1975), who used it to deterministically explain the structure of stable communities, based on niche-related processes. Weiher & Keddy (1999) proposed to envisage two basic kinds of plant community pattern, with different causes:

1 Environmentally mediated patterns, i.e. correlations between species due to their shared or opposite responses to the physical environment.
2 Assembly rules, i.e. patterns due to interactions between species, such as competition, allelopathy, facilitation, mutualism, and all other biotic interactions that we know about in theory, and actually affect communities in the real world.

Currently, all these processes, including the arrival of propagules, their germination and establishment, and their interactions with co-occurring species, are included in the notion of assembly rules. Indeed, Belyea & Lancaster (1999) emphasized that there is no principle difference in assembly rules concerned with plant dispersal, plant responses to abiotic factors, and plant-plant responses in the community. A further step in the clarification was made by Cavender-Bares *et al.* (2009), who distinguished three perspectives on the dominant factors that influence community assembly, composition and diversity: (i) the classic perspective that communities are assembled mainly according to niche-related processes; (ii) the perspective that community assembly is largely a neutral process in which species are ecologically equivalent; and (iii) the perspective that emphasizes the role of historical factors in dictating how communities are assembled, with a focus on speciation and dispersal rather than on local processes. Note that these different points of view are not mutually exclusive (cf. Myers & Harms 2009; Vergnon *et al.* 2009), and that it is useful to investigate the relative importance of the different hypothetical processes (Bossuyt *et al.* 2005). Zobel (1997) formalized the process of assembly by proposing that local communities are assembled from a regional species pool, representing the total of species available for colonization and defined within a large biogeographic or climatic region. Assembly rules thus indicate constraints or environmental filters determining which species can occur in the community and which combinations are irrelevant (Fig. 7.6).

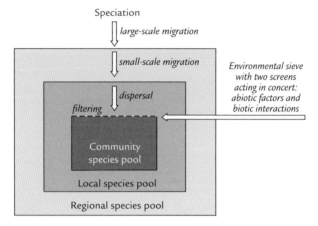

Fig. 7.6 The role of large-scale and small-scale processes in determining different species pools. (After Zobel 1997.)

Without referring to the term 'assembly rules', several models and hypotheses have been proposed to explain the co-existence of species assemblages in plant communities. The 'resource balance hypothesis of plant species diversity', presented by Braakhekke & Hooftman (1999), relates to competition and niche differences and suggests a static equilibrium. On the basis of a model of competition for multiple resources and related experimental tests, these authors have given evidence for the idea that opportunities for plant species diversity are favoured when the actual resource supply ratios of many resources are balanced according to the optimum supply ratios for the vegetation as a whole. Their theory predicts that diversity will be relatively low when biomass production of the whole vegetation is limited by a single nutrient, while it can be high when there is co-limitation by several nutrients. While non-spatial models predict that no more consumer species can co-exist at equilibrium than there are limiting resources, a similar model that includes neighbourhood competition and random dispersal among sites predicts stable co-existence of a potentially unlimited number of species on a single resource (Tilman 1994). Co-existence occurs because species with sufficiently high dispersal rates persist in sites not occupied by superior competitors. It requires limiting similarity and two-way and three-way interspecific trade-offs among competitive ability, colonization ability and longevity. Co-existence can, however, also be explained by non-equilibrium models, emphasizing dynamic dispersal phenomena rather than niche separation to explain species richness. Fine-scale repeated observations by van der Maarel & Sykes (1993) in species-rich alvar grassland vegetation, revealed that species co-existence at a coarse-grained scale may result from a relatively fast turnover of species at a finer scale, a process which they labelled with the term 'carousel model' (see also Chapter 3). Similarly, Gigon & Leutert (1996) explained co-existence of a large number of plant species by postulating the 'dynamic keyhole–key model', assuming that plant species diversity (the keys) in a plant community is matched by the diversity of microsites (the keyholes), which both change in the course of time.

It is still an open question whether we can really speak of 'rules' as a set of principles or laws that predict the development of specific biological communities, as compared to development that is attributable to random processes. The advantage of the search for assembly rules is that it makes ecological knowledge explicit in terms of predictions that can be tested. The rules can be considered a challenge to explicitly formalize our knowledge of decisions that are implicitly taken by plants in response to their environment during the process of plant community development (e.g. Chapter 12).

References

Adema, E.B., Grootjans, A.P., Petersen, J. & Grijpstra, J. (2002) Alternative stable states in a wet calcareous dune slack in The Netherlands. *Journal of Vegetation Science* 13, 107–114.

Aerts, R. (1999) Interspecific competition in natural plant communities: mechanisms, trade-offs and plant-soil feedbacks. *Journal of Experimental Botany* 50, 29–37.

Allen, E.B. & Allen, M.F. (1990) The mediation of competition by mycorrhizae in successional and patchy environments. In: *Perspectives on Plant Competition* (eds J.B. Grace & D. Tilman), pp. 367–389. Academic Press, London.

Ameloot, E., Verheyen, K. & Hermy, M. (2005) Meta-analysis of standing crop reduction by *Rhinanthus* spp. and its effect on vegetation structure. *Folia Geobotanica* 40, 289–310.

Armas, C., Padilla, F.M., Pugnaire, F.I. & Jackson, R.B. (2010) Hydraulic lift and tolerance to salinity of semiarid species: consequences for species interactions. *Oecologia* 162, 11–21.

Austin, M.P., Groves, R.H., Fresco, L.F.M. & Kave, P.E. (1985) Relative growth of six thistle species along a nutrient gradient with multispecies competition. *Journal of Ecology* 73, 676–684.

Bakker, J.P. (1989) *Nature Management by Grazing and Cutting*. Kluwer Academic Publishers, Dordrecht.

Baldwin, J.G., Nadler, S.A. & Adams, B.J. (2004) Evolution of plant parasitism among nematodes. *Annual Review of Phytopathology* 42, 83–106.

Bascompte, J., Jordano, P. & Olesen, J.M. (2006) Asymmetric coevolutionary networks facilitate biodiversity networks. *Science* 312, 431–433.

Baumeister, D. & Callaway, R.M. (2006) Facilitation by *Pinus flexilis* during succession: a hierarchy of mechanisms benefits other plant species. *Ecology* 87, 1816–1831.

Baxter, D.V. & Wadworth, F.H. (1939) Forest and fungus succession in the lower Yukon Valley. *Bulletin of the University of Michigan, School of Forestry and Conservation, Ann Arbor* 9.

Beattie, A. (1989) Myrmecotrophy: plants fed by ants. *Trends in Ecology & Evolution* 4, 172–176.

Begon, M., Townsend, C.R. & Harper, J.L. (2005) *Ecology: From Individuals to Ecosystems*, 4th edn. Wiley-Blackwell, Oxford.

Belyea, L.R. & Lancaster, J. (1999) Assembly rules within a contingent ecology. *Oikos* 86, 402–416.

Benincà, E., Huisman, J., Heerkloss, R. *et al.* (2008) Chaos in a long-term experiment with a plankton community. *Nature* 451, 822–826.

Berendse, F. & Elberse, W.Th. (1990) Competition and nutrient availability in heathland and grassland ecosystems. In: *Perspectives on Plant Competition* (eds J.B. Grace & D. Tilman), pp. 93–116. Academic Press, London.

Bertness, M. & Callaway, R.M. (1994) Positive interactions in communities. *Trends in Ecology & Evolution* 9, 191–193.

Bobbink, R., den Dubbelden, J. & Willems, J.H. (1989) Seasonal dynamics of phytomass and nutrients in chalk grassland. *Oikos* 55, 216–224.

Bonanomi, G., Giannino, F. & Mazzoleni, S. (2005) Negative plant-soil feedback and species co-existence. *Oikos* 111, 311–321.

Bonanomi, G., Rietkerk, M., Dekker, S.C. & Mazzoleni, S. (2008) Islands of fertility induce co-occurring negative and positive plant-soil feedbacks promoting co-existence. *Plant Ecology* 197, 207–218.

Bossuyt, B., Honnay, O. & Hermy, M. (2005) Evidence for community assembly constraints during succession in dune slack plant communities. *Plant Ecology* **178**, 201–209.

Bosch, J., González, A.M.M., Rodrigo, A. & Navarro, D. (2009) Plant–pollinator networks: adding the pollinator's perspective. *Ecology Letters* **12**, 409–419.

Braakhekke, W.G. & Hooftman, D.A.P. (1999) The resource balance hypothesis of plant species diversity in grassland. *Journal of Vegetation Science* **10**, 187–200.

Brooker, R.W. & Callaway, R.M. (2009) Facilitation in the conceptual melting pot. *Journal of Ecology* **97**, 1117–1120. [Intro to a Special Feature – see also other papers in this issue.]

Brooker, R.W., Maestre, F.T., Callaway, R.M. *et al.* (2008) Facilitation in plant communities; the past, the present, and the future. *Journal of Ecology* **96**, 18–34.

Brown, V.K. & Gange, A.C. (1989) Differential effects of above- and below-ground insect herbivory during early plant succession. *Oikos* **54**, 67–76.

Bruno, J.F. & Bertness, M.D. (2001) Habitat modification and facilitation in benthic marine communities. In: *Marine Community Ecology* (eds M.D. Bertness, S.D. Gaines & M.E. Hay), pp. 201–218. Sinauer, Sunderland, MA.

Bruno, J.F., Stachowicz, J.J. & Bertness, M.D. (2003) Inclusion of facilitation into ecological theory. *Trends in Ecology & Evolution* **18**, 119–125.

Burgess, S.S.O., Adams, M.A., Turner, N.C. & Onk, C.K. (1998) The redistribution of soil water by tree root systems. *Oecologia* **115**, 306–311.

Caldwell, M.M. & Richards, J.H. (1989) Hydraulic lift: water efflux from upper roots improves effectiveness of water uptake by deep roots. *Oecologia* **79**, 1–5.

Callaway, R.M. & Walker, L.R. (1997) Competition and facilitation: a synthetic approach to interactions in plant communities. *Ecology* **78**, 1958–1965.

Calow, P. (ed) (1998) *The Encyclopedia of Ecology & Environmental Management*. Blackwell Science, Oxford.

Campbell, B.D., Grime, J.P. & Mackey, J.M.L. (1991) A trade-off between scale and precision in resource foraging. *Oecologia* **87**, 532–538.

Cavender-Bares, J., Kozak, K.H., Fine, P.V.A. & Kembel, S.W. (2009) The merging of community ecology and phylogenetic biology. *Ecology Letters* **12**, 693–715.

Chamberlain, S.A. & Holland, J.N. (2009) Body size predicts degree in ant–plant mutualistic networks. *Functional Ecology* **23**, 196–202.

Chiarello, N., Hickman, J.C. & Mooney, H.A. (1982) Endomycorrhizal role for interspecific transfer of phosphorus in a community of annual plants. *Science* **217**, 941–943.

Connell, J.H. & Slatyer, R.O. (1977) Mechanisms of succession in natural communities and their role in community stability and organization. *The American Naturalist* **111**, 1119–1144.

Cui, M. & Caldwell, M.M. (1996) Facilitation of plant phosphate acquisition by arbuscular mycorrhizas from enriched soil patches. *New Phytologist* **133**, 453–460, 461–467.

Dawson, T.E. (1993) Hydraulic lift and water use by plants: implications for water balance, performance and plant–plant interactions. *Oecologia* **95**, 565–574.

den Hartog, C. (1987) Wasting disease and other dynamic phenomena in *Zostera* beds. *Aquatic Botany* **27**, 3–14.

de Wit, C.T. (1960) On Competition. *Agricultural Research Report, Wageningen University* **66.8**, 1–82.

Diamond, J.M. (1975) Assembly of species communities. In: *Ecology and Evolution of Communities* (eds M.L. Cody & J.M. Diamond), pp. 342–444. Harvard University Press, Cambridge.

Dijk, E., Willems, J.H. & van Andel, J. (1997) Nutrient responses as a key factor to the ecology of orchid species. *Acta Botanica Neerlandica* **46**, 339–363.

Dobson, A. & Crawley, M. (1994) Pathogens and the structure of plant communities. *Trends in Ecology & Evolution* **9**, 393–398.

Dullinger, S., Kleinbauer, I., Gottfried, M. *et al.* (2007) Weak and variable relationships between environmental severity and small-scale co-occurrence in alpine plant communities. *Journal of Ecology* **95**, 1284–1295.

Ens, E.J., French, K. & Bremner, J.B. (2009) Evidence for allelopathy as a mechanism of community composition change by an invasive exotic shrub, *Chrysanthemoides monilifera* ssp. *rotundata*. *Plant and Soil* **316**, 125–137.

Finlay, R.D., Ek, H., Odham, G. & Söderström, B. (1988) Mycelial uptake, translocation and assimilation of nitrogen from ¹⁵N-labelled ammonium by *Pinus sylvestris* plants infected with four different ecto-mycorrhizal fungi. *New Phytologist* **110**, 59–66.

Francis, R. & Read, D.J. (1994) The contribution of mycorrhizal fungi to the determination of plant community structure. *Plant and Soil* **159**, 11–25.

Gaudet, C.L. & Keddy, P.A. (1988) A comparative approach to predicting competitive ability from plant traits. *Nature* **334**, 242–243.

Gause, G.F. (1934) *The Struggle for Existence*. Waverly Press, Baltimore, MD.

Gibson, D.J., Connolly, J., Hartnett, D.C. & Weidenhamer, J.D. (1999) Designs for greenhouse studies of interactions between plants. *Journal of Ecology* **87**, 1–16.

Gigon, A. & Leutert, A. (1996) The dynamic keyhole-key model of co-existence to explain diversity of plants in limestone and other grasslands. *Journal of Vegetation Science* **7**, 29–40.

Gleeson, S.K. & Tilman, D. (1990) Allocation and the transient dynamics of succession on poor soils. *Ecology* **71**, 1144–1155.

Glenn-Lewin, D.C. & van der Maarel, E. (1992) Patterns and processes of vegetation dynamics. In: *Plant Succession* (eds D.C. Glenn-Lewin, R.K. Peet & Th.T. Veblen), pp. 11–59. Chapman & Hall, London.

Goldberg, D.E. (1990) Components of resource competition in plant communities. In: *Perspectives on Plant Competition* (eds J.B. Grace & D. Tilman), pp. 27–49. Academic Press, London.

Goldberg, D.E., Rajaniemi, T., Gurevitch, J. & Stewart, O.A. (1999) Empirical approaches to quantifying interaction intensity: competition and facilitation along productivity gradients. *Ecology* **80**, 1118–1131.

Grime, J.P. (1974) Vegetation classification by reference to strategies. *Nature* **250**, 26–31.

Grime, J.P. (1979) *Plant Strategies and Vegetation Processes*. John Wiley & Sons, Ltd, Chichester.

Grime, J.P. (2001) *Plant Strategies, Vegetation Processes, and Ecosystem Properties*, 2nd edn. John Wiley & Sons, Ltd, Chichester.

Grime, J.P., Mackey, J.M., Hillier, S.H. & Read, D.J. (1987) Floristic diversity in a model system using experimental microcosms. *Nature* **328**, 420–422.

Grootjans, A.P., Ernst, W.H.O. & Stuyfzand, P.J. (1998) European dune slacks: strong interactions between vegetation, pedogenesis and hydrology. *Trends in Ecology & Evolution* **13**, 96–100.

Guevara, S., Purata, S.E. & van der Maarel, E. (1986) The role of remnant forest trees in tropical secondary succession. *Vegetatio* **66**, 77–84.

Guevara, S., Meave, J., Moreno-Casasola, P. & Laborde, J. (1992) Floristic composition and structure of vegetation under isolated trees in neotropical pastures. *Journal of Vegetation Science* **3**, 655–664.

Gurevitch, J., Scheiner, S.M. & Fox, G.A. (2002) *The Ecology of Plants*. Sinauer Associates, Sunderland.

Hättenschwiler, S. and Vitousek, P.M. (2000) The role of polyphenols in terrestrial ecosystem nutrient cycling. *Trends in Ecology & Evolution* **15**, 238–243.

Horton, J.L. & Hart, S.C. (1998) Hydraulic lift: a potentially important ecosystem process. *Trends in Ecology & Evolution* **13**, 232–235.

Huisman, J. & Weissing, F.J. (1999) Biodiversity of plankton by species oscillations and chaos. *Nature* **402**, 407–410.

Johnson, N.C., Graham, J.H. & Smith, F.A. (1997) Functioning of mycorrhizal associations along the mutualism–parasitism continuum. *New Phytologist* **135**, 575–585.

Johnson, S.D. & Steiner, K.E. (2000) Generalization versus specialization in plant pollination systems. *Trends in Ecology & Evolution* **15**, 140–143.

Joliffe, P.A. (2000) The replacement series. *Journal of Ecology* **88**, 371–385.

Kearns, C.A., Inouye, D.W. & Waser, N.M. (1998) Endangered mutualisms: the conservation of plant–pollinator interactions. *Annual Review of Ecology and Systematics* **29**, 83–112.

Kellman, M. & Kading, M. (1992) Facilitation of tree seedling establishment in a sand dune succession. *Journal of Vegetation Science* **3**, 679–688.

Kersch, M.F. & Fonseca, C.R. (2005) Abiotic factors and the conditional outcome of an ant–plant mutualism. *Ecology* **86**, 2117–2126.

Kiers, E.T., Lovelock, C.E., Krueger, E.L. & Herre, E.A. (2000) Differential effects of tropical arbuscular mycorrhizal fungal inocula on root colonization and tree seedling growth: implications for tropical forest diversity. *Ecology Letters* **3**, 106–113.

Kikvidze, Z., Khetsuriani, L., Kikodze, D. & Callaway, R.M. (2006) Seasonal shifts in competition and facilitation in subalpine plant communities of the central Caucasus. *Journal of Vegetation Science* 17, 77–82.

Krebs, C.J. (2008) *Ecology: The Experimental Analysis of Distribution and Abundance*, 6th edn. Benjamin Cummings, San Francisco, CA.

Kuijt, J. (1969) *The Biology of Parasitic Flowering Plants*. University of California Press, Berkeley and Los Angeles, CA.

Kuiters, A.T. (1990) Role of phenolic substances from decomposing forest litter in plant–soil interactions. *Acta Botanica Neerlandica* 39, 329–348.

Kurz-Besson, C., Otieno, D., Lobo do Vale, R. *et al.* (2006) Hydraulic lift in cork oak trees in a savannah-type Mediterranean ecosystem and its contribution to the local water balance. *Plant and Soil* 282, 361–378.

Kwak, M.M., Velterop, O. & van Andel, J. (1998) Pollen and gene flow in fragmented habitats. *Applied Vegetation Science* 1, 37–54.

Lázaro, A., Lundgren, R. & Totland, Ø. (2009) Co-flowering neighbors influence the diversity and identity of pollinator groups visiting plant species. *Oikos* 118, 691–702.

Looijen, R.C. & van Andel, J. (1999) Ecological communities: conceptual problems and definitions. *Perspectives in Plant Ecology, Evolution and Systematics* 2, 210–222.

Ludwig, F. (2001) Tree Grass Interactions on an East African Savanna: The Effects of Competition, Facilitation and Hydraulic Lift. Tropical Resource Management Papers 39, PhD Thesis, Wageningen University, Wageningen.

Ludwig, F., Dawson, T.E., Prins, H.H.T., Berendse, F. & de Kroon, H. (2004) Below-ground competition between trees and grasses may overwhelm the facilitative effects of hydraulic lift. *Ecology Letters* 7, 623–631.

MacArthur, R. & Wilson, E.O. (1967) *The Theory of Island Biogeography*. Princeton University Press, Princeton, NJ.

Maestre, F.T., Callaway, R.M., Valladares, F. & Lortie, C.J. (2009) Refining the stress-gradient hypothesis for competition and facilitation in plant communities. *Journal of Ecology* 97, 199–205.

McDonnell-Alexander, M.P. (2006) Spatial Nutrient Heterogeneity and Plant Species Co-existence. PhD Thesis, University of Groningen.

Menge, B.A. & Sutherland, J.P. (1987) Community regulation: variation in disturbance, competition, and predation in relation to environmental stress and recruitment. *The American Naturalist* 130, 730–757.

Michalet, R., Brooker, R.W., Cavieres, L.A. *et al.* (2006) Do biotic interactions shape both sides of the humped-back model of species richness in plant communities? *Ecology Letters* 9, 767–773.

Miller, T.E. (1994) Direct and indirect species interactions in an early old-field plant community. *The American Naturalist* 143, 1007–1025.

Miransari, M. (2010) Contribution of arbuscular mycorrhizal symbiosis to plant growth under different types of soil stress. *Plant Biology* 12, 563–569.

Morales, C.L. & Traveset, A. (2009) A meta-analysis of impacts of alien vs. native plants on pollinator visitation and reproductive success of co-flowering native plants. *Ecology Letters* 12, 716–728.

Mühlstein, L.K., Porter, D. & Short, F.T. (1991) *Labyrinthula zosterae* sp. nov., the causative agent of wasting disease of eelgrass, *Zostera marina*. *Mycologia* 83, 180–191.

Myers, J.A. & Harms, K.E. (2009) Seed arrival, ecological filters, and plant species richness: a meta-analysis. *Ecology Letters* 12, 1250–1260.

Moora, M. & Zobel, M. (1996) Effect of arbuscular mycorrhiza on inter- and intraspecific competition of two grassland species. *Oecologia* 108, 79–84.

Neuhauser, C. & Fargione, J.E. (2004) A mutualism–parasitism continuum and its application to plant–mycorrhizae interactions. *Ecological Modelling* 177, 337–352.

Nilsson, M.-C. (1994) Separation of allelopathy and resource competition by the boreal dwarf shrub *Empetrum hermaphroditum* Hagerup. *Oecologia* 98, 1–7.

Nilsson, M.-C., Gallet, C. & Wallstedt, A. (1998) Temporal variability of phenolics and batatasin-III in *Empetrum hermaphroditum* leaves over an eight-year period: interpretations of ecological function. *Oikos* 81, 6–16.

Oksanen, L., Fretwell, S.D., Arruda, J. & Niemalä, P. (1981) Exploitation ecosystems in gradients of primary productivity. *The American Naturalist* 118, 240–261.

Ozinga, W.A., van Andel, J. & McDonnell-Alexander, M.P. (1997) Nutritional soil heterogeneity and mycorrhiza as determinants of plant species diversity. *Acta Botanica Neerlandica* **46**, 237–254.

Ozinga, W.A., Römermann, C., Bekker, R.M. *et al.* (2009) Dispersal failure contributes to plant losses in NW Europe. *Ecology Letters* **12**, 66–74.

Palmer, T.M. & Brody, A.K. (2007) Mutualism as reciprocal exploitation: African plant-ants defend foliar but not reproductive structures. *Ecology* **88**, 3004–3011.

Pennings, S.C. & Callaway, R.M. (1996) Impact of a parasitic plant on the structure and dynamics of salt marsh vegetation. *Ecology* **77**, 1410–1419.

Pennings, S.C. & Callaway, R.M. (2002) Parasitic plants: parallels and contrasts with herbivores. *Oecologia* **131**, 479–489.

Price, P.W., Westoby, M., Rice, B. *et al.* (1986) Parasite mediation in ecological interactions. *Annual Review of Ecology and Systematics* **17**, 487–505.

Pugnaire, F.I., Haase, P. & Puigdefábregas, J. (1996) Facilitation between higher plant species in a semiarid environment. *Ecology* **77**, 1420–1426.

Raffaele, E. & Veblen, T.T. (1998) Facilitation by nurse shrubs of resprouting behavior in a post-fire shrubland in northern Patagonia, Argentina. *Journal of Vegetation Science* **9**, 693–698.

Rasmussen, H.N. & Rasmussen, F.N. (2009) Orchid mycorrhiza: implications of a mycophagous life style. *Oikos* **118**, 334–345.

Rice, E.L. (1974) *Allelopathy*. Academic Press, New York, NY.

Rosumek, F.B., Silveira, F.A.O., Neves, F. de S. *et al.* (2009) Ants on plants: a meta-analysis of the role of ants as plant biotic defenses. *Oecologia* **160**, 537–549.

Rudgers, J.A. & Maron, J.L. (2003) Facilitation between coastal dune shrubs: a non-nitrogen fixing shrub facilitates establishment of a nitrogen-fixer. *Oikos* **102**, 75–84.

Smith, D.M., Jackson, N.A., Roberts, J.M. & Ong, C.K. (1999) Reverse flow of sap in tree roots and downward siphoning of water by *Grevillea robusta*. *Functional Ecology* **13**, 256–264.

Suding, K.N., Larson, J.R., Thorsos, E., Steltzer, H. & Bowman, W.D. (2004) Species effects on resource supply rates: do they influence competitive interactions? *Plant Ecology* **175**, 47–58.

Taki, H. & Kevan, P.G. (2007) Does habitat loss affect the communities of plants and insects equally in plant–pollinator interactions? *Biodiversity and Conservation* **16**, 3147–3161.

ter Borg, S.J. (1985) Population biology and habitat relations of some hemiparasitic Scrophulariaceae. In: *The Population Structure of Vegetation* (ed. J. White), pp. 463–487. Dr W. Junk Publishers, Dordrecht.

Thompson, J.N. & Fernandez, C.C. (2006) Temporal dynamics of antagonism and mutualism in a geographically variable plant–insect interaction. *Ecology* **87**, 103–112.

Tilman, D. (1982) *Resource Competition and Community Structure*. Princeton University Press, Princeton, NJ.

Tilman, D. (1985) The resource ratio hypothesis of plant succession. *The American Naturalist* **125**, 827–852.

Tilman, D. (1988) *Plant Strategies and the Dynamics and Structure of Plant Communities*. Princeton University Press, Princeton, NJ.

Tilman, D. (1990) Constraints and trade-offs: toward a predictive theory of competition and succession. *Oikos* **58**, 3–15.

Tilman, D. (1994) Competition and biodiversity in spatially structured habitats. *Ecology* **75**, 2–16.

van Andel, J. & Nelissen, H.J.M. (1981) An experimental approach to the study of species interference in a patchy vegetation. *Vegetatio* **45**, 155–163.

van Andel, J., Bakker, J.P. & Grootjans, A.P. (1993) Mechanisms of vegetation succession: a review of concepts and perspectives. *Acta Botanica Neerlandica* **42**, 413–433.

van Dam, N.M. (2009) How plants cope with biotic interactions. *Plant Biology* **11**, 1–5.

Vandegehuchte, M.L., de la Peña, E. & Bonte, D. (2010) Interactions between root and shoot herbivores of *Ammophila arenaria* in the laboratory do not translate into correlated abundances in the field. *Oikos* **119**, 1011–1019.

van der Heide, T., van Nes, E.H., Geerling, G.W. *et al.* (2007) Positive feedbacks in seagrass ecosystems – implications for success in conservation and restoration. *Ecosystems* **10**, 1311–1322.

van der Heijden, M.G.A., Boller, T., Wiemken, A. & Sanders, I.R. (1998) Different arbuscular mycorrhizal fungal species are potential determinants of plant community structure. *Ecology* **79**, 2082–2091.

van der Maarel, E. & Sykes, M.T. (1993) Small-scale plant species turnover in a limestone grassland: the carousel model and some comments on the niche concept. *Journal of Vegetation Science* **4**, 179–188.

van der Putten, W.H. & van der Stoel, C.D. (1998) Plant parasitic nematodes and spatio-temporal variation in natural vegetation. *Applied Soil Ecology* **10**, 253–262.

van der Putten, W.H., Bardgett, R.D., de Ruiter, P.C. *et al.* (2009) Empirical and theoretical challenges in aboveground-belowground ecology. *Oecologia* **161**, 1–14.

van der Veen, A. (2000) Competition in Coastal Sand Dune Succession. PhD Thesis, University of Groningen.

Vergnon, R., Dulvy, N.K. & Freckleton, R.P. (2009) Niches versus neutrality: uncovering the drivers of diversity in a species-rich community. *Ecology Letters* **12**, 1079–1090.

Wardle, D.A. (2002) *Communities and Ecosystems – Linking the Aboveground and Belowground Components*. Princeton University Press, Princeton, NJ.

Waser, N.M. & Ollerton, J. (eds) (2006) *Plant–Pollinator Interactions: From Specialization to Generalization*. The University of Chicago Press, Chicago, IL.

Weiher, E. & Keddy, P. (eds) (1999) *Ecological Assembly Rules*. Cambridge University Press, Cambridge.

Wilson, J.B. & Agnew, A.D.Q. (1992) Positive-feedback switches in plant communities. *Advances in Ecological Research* **23**, 263–337.

Yapp, R.H. (1925) The interrelations of plants in vegetation, and the concept of 'association'. *Veröffentlichungen Geobotanisches Institut Rübel Zürich* **3**, 684–706.

Yoda, K., Kira, T., Ogawa, H. & Hozumi, K. (1963) Self-thinning in overcrowded pure stands under cultivated and natural conditions. *Journal of Biology, Osaka City University* **14**, 107–129.

Yoder, C.K. & Nowak, R.S. (1999) Hydraulic lift among native plant species in the Mojave Desert. *Plant and Soil* **215**, 93–102.

Youngsteadt, E., Nojima, S., Häberlein, C., Schulz, S. & Schal, C. (2008) Seed odor mediates an obligate ant–plant mutualism in Amazonian rainforests. *Proceedings of the National Academy of Sciences of the United States of America* **105**, 4571–4575.

Zackrisson, O., Nilsson, M.-C. & Wardle, D.A. (1996) Key ecological function of charcoal from wildfire in boreal forest. *Oikos* **77**, 10–19.

Zobel, M. (1997) The relative role of species pools in determining plant species richness: an alternative explanation of co-existence? *Trends in Ecology & Evolution* **12**, 266–269.

8

Terrestrial Plant-Herbivore Interactions: Integrating Across Multiple Determinants and Trophic Levels

Mahesh Sankaran[1] and Samuel J. McNaughton[2]

[1]Tata Institute of Fundamental Research, Bangalore, India
[2]Syracuse University, New York, USA

8.1 Herbivory: pattern and process

Carbon fixed by the Earth's primary producers supports life at all other trophic levels. This carbon follows one of three trophic routes in ecological time: it may accumulate in plant tissue, be consumed by herbivores or be channelled into the decomposer pathway as litter. In most ecosystems, the bulk of primary production enters the decomposer pathway (Cebrian 1999). In others however, herbivory can be substantial, with herbivores consuming as much as 83% of the above-ground foliage production (McNaughton *et al*. 1989).

Which factors determine the fate of fixed carbon? Across ecosystems, herbivory levels have been linked to ecosystem productivity. More productive systems on average support greater herbivore biomass (Fig. 8.1a). Larger herbivore loads in these systems mean that greater absolute amounts of plant biomass are consumed (Fig. 8.1b), resulting in greater secondary productivity (production of herbivore tissue; Fig. 8.1c). A direct positive correlation between ecosystem productivity and herbivore biomass, as suggested by McNaughton *et al*. (1989), is consistent with theories of bottom-up control of trophic structure. Here, organisms at each trophic level are assumed to be food-limited and increases in resource availability to plants therefore translates to increased biomass of organisms at higher trophic levels. However, for a given level of primary production, herbivore biomass and consumption can vary almost 1000-fold between ecosystems (Fig. 8.1a, b), indicating that ecosystem production is only one of many factors regulating herbivory patterns (McNaughton *et al*. 1989; Cebrian 1999; Cebrian & Lartigue 2004).

While bottom-up forces, i.e. resource availability, ultimately constrains both the number and productivity of different trophic levels in an ecosystem,

Vegetation Ecology, Second Edition. Eddy van der Maarel and Janet Franklin.
© 2013 John Wiley & Sons, Ltd. Published 2013 by John Wiley & Sons, Ltd.

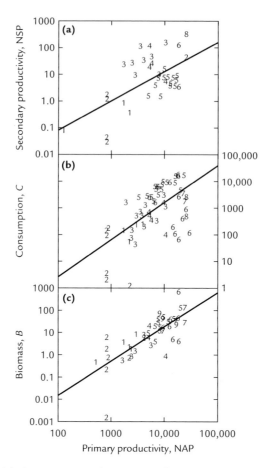

Fig. 8.1 Relationship between net above-ground primary productivity (*x*-axis) and (a) herbivore biomass, (b) consumption and (c) net secondary productivity. Units are in kJ·m^{-2}·yr^{-1} except for biomass which is kJ·m^{-2}. Ecosystems: 1, desert; 2, tundra; 3, temperate grassland; 4, temperate successional old field; 5, tropical grassland; 6, temperate forest; 7, tropical forest; 8, salt marsh; 9, agricultural tropical grassland. (From McNaughton *et al.* 1989.)

influences imposed by organisms at higher trophic levels (top-down forces) are also believed to be important in regulating herbivore biomass patterns in ecosystems. Formalized as the hypothesis of exploitation ecosystems (EEH; Oksanen *et al.* 1981), this viewpoint contends that when ecosystems are productive enough to support carnivores, predators, rather than plant production, control herbivore populations. Consequently, herbivore biomass should not be related to primary productivity in these systems (Moen & Oksanen 1991; Oksanen & Oksanen 2000). In unproductive areas incapable of supporting viable herbivore populations, plant biomass should increase with increasing productivity (Fig. 8.2, zone I). When productivity increases above the threshold required to support herbivores, herbivory should maintain plant biomass at a constant level with all

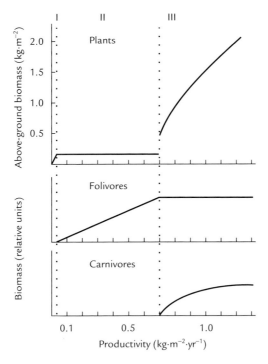

Fig. 8.2 Patterns in plant, herbivore and carnivore biomass across gradients of primary productivity as predicted by the Oksanen hypothesis. Roman numerals indicate the predicted number of trophic links for a given range of plant productivity. (From Oksanen & Oksanen 2000.)

increases in productivity going toward supporting greater herbivore loads (Fig. 8.2, zone II). Where production is sufficient to support carnivores, predation on herbivores should free plants from the constraints of herbivory such that plant and carnivore biomass increase with productivity, while herbivore biomass remains constant (Fig. 8.2, zone III).

Plant and herbivore biomass patterns along arctic–alpine productivity gradients in the tundra landscape of northern Norway corroborate the EEH (Aunapuu *et al.* 2008). In unproductive habitats characterized by two trophic levels, plant biomass was constant, but herbivore biomass increased, with increasing productivity. In contrast, in productive habitats with three trophic levels, plant and predator biomass varied spatially with productivity, but herbivore biomass did not (Aunapuu *et al.* 2008). Similarly, distribution patterns of deer biomass across North America also seem to endorse predictions of exploitation ecosystems (Crete 1999). From the high-arctic to the transition zone between the tundra and forest, where resource availability constrains the number of trophic levels to two, cervid biomass increases with productivity. Within the wolf range in the boreal zone, deer biomass remains relatively constant, whereas south of the wolf range and in wolf-free areas, cervid biomass increases with plant productivity. Indeed, some of the most well-documented examples of predator control in terrestrial ecosystems comes from 'natural experiments' involving the loss, and subsequent reintroductions, of apex predators in North America (Beschta & Ripple 2009). The loss of large predators such as gray wolf (*Canis lupus*) and cougar (*Puma concolor*), and subsequent increases in ungulate populations in the late 1800s and early 1900s resulted in dramatic reductions in the recruitment of deciduous trees in several national parks in the western USA including Olympic, Yosemite, Yellowstone, Zion and Wind Cave National Parks (Fig. 8.3; Beschta & Ripple 2009). Where previously extirpated predators have been reintroduced, such as gray wolves in Yellowstone National Park, woody browse

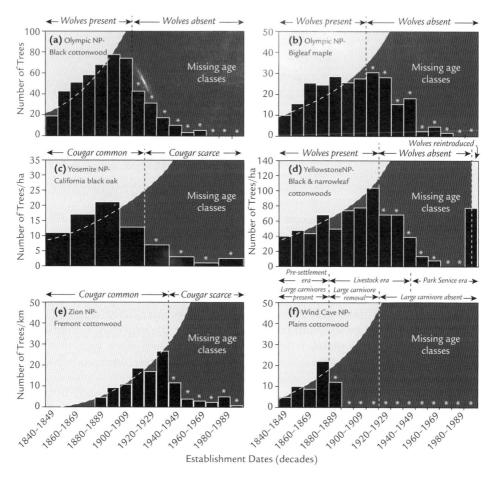

Fig. 8.3 Establishment dates of woody browse species in five National Parks in the USA from 1840 to 2000 for (a) black cottonwood and (b) bigleaf maple in Olympic National Park, (c) California black oak in Yosemite National Park, (d) black and narrowleaf cottonwood in Yellowstone National Park, (e) Fremont cottonwood in Zion National Park, and (f) Plains cottonwood in Wind Cave National Park. Decreases in establishment are apparent following the loss of large predators (significant decreases in observed tree frequencies are indicated with an asterisk (*), as also is the recovery in establishment following the reintroduction of wolves in Yellowstone NP (d). (From Beschta & Ripple 2009.)

species have begun to recover (Fig. 8.3d), suggesting strong top-down limitation in these systems (Beschta & Ripple 2009).

Paucity of data from systems free of human impact and difficulties with experimentally manipulating whole predator communities in terrestrial ecosystems has hampered more widespread corroboration of these ideas, and available data both support and contradict different predictions of the EEH, precluding a consensus. For example, in a two-link system in the Norwegian Arctic, Wegener & Odasz-Albrigtsen (1998) found no evidence to indicate that Reindeer (*Rangifer tarandus platyrhynchus*), in the absence of predators, regulated plant standing crop to a constant low level independent of productivity. Contrary to Oksanen & Oksanen (2000; Fig. 8.2, zone II), plant standing crop differed almost threefold between different grazed vegetation types and grazer exclusion had no discernible effect on plant biomass.

Classical theories such as the EEH emphasize consumption of prey as the primary mechanism by which predators influence ecosystem structure (Elmhagen *et al.* 2010). However, predators can also influence ecosystem processes via non-lethal effects, for example by inducing behavioural changes in herbivores (Ripple & Beschta 2004). Herbivores exist in a 'landscape of fear' within which they have to balance demands for food and safety, and changes in herbivore behaviour due to predation risk, either actual or perceived, can affect ecosystem processes in various ways (Ripple & Beschta 2004). For example, following wolf reintroductions to Yellowstone National Park, willow (*Salix* spp.) and cottonwoods (*Populus* spp.) were subject to less browsing pressure by elk in high-predation risk sites with limited visibility or with terrain features that impeded escape compared to low-risk areas (Ripple & Beschta 2004; but see Kauffman *et al.* 2007). Similarly, Riginos & Grace (2008) demonstrated that native herbivores in a semi-arid savanna in Kenya, with the exception of elephant, exhibited a strong preference for areas of low tree densities because of greater visibility in these areas rather than any vegetation characteristics associated with low tree densities. For all but the largest species, top-down behavioural effects of predation avoidance mediated habitat use with resulting cascading effects on herbaceous vegetation (Riginos & Grace 2008).

Broad generalizations of top-down effects in ecosystems are further complicated by the fact that predators are not a homogeneous guild, and interactions among top predators and mesopredators can have effects that cascade through lower trophic levels (Elmhagen *et al.* 2010). In such 'interference ecosystems', predators can be functionally divided into two groups, top predators and mesopredators (Fig. 8.4). Top predators can suppress large herbivores as well as mesopredators, and thus indirectly release smaller herbivores which are the primary prey of mesopredators, as well as some plants, from top-down control (Mesopredator Release Hypothesis (MRH); Fig. 8.4; Elmhagen *et al.* 2010). In Finland, recolonizing lynx (*Lynx lynx*) (a top predator) have been shown to suppress red fox (*Vulpes vulpes*) (a mesopredator) and thereby have an indirect positive impact on mountain hares (*Lepus timidus*) (Elmhagen *et al.* 2010). This is in accordance with the predictions of the MRH, suggesting that top-down interference as well as bottom-up productivity must be taken into account to understand the nature of trophic control in ecosystems (Elmhagen *et al.* 2010).

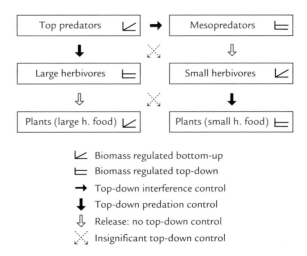

Fig. 8.4 Conceptual illustration of interference ecosystems where predators can be functionally divided into two groups: top predators and mesopredators. Top predators suppress both large herbivores and mesopredators and indirectly release small herbivores and some plants from top-down control. The heterogeneity within trophic levels in such systems implies that trophic levels do not respond uniformly to top-down or bottom-up control. (From Elmhagen *et al.* 2010.)

Ultimately, neither simple donor-controlled nor consumer-controlled models, by themselves, are likely to fully explain herbivory patterns across productivity gradients since they ignore the inherent heterogeneity among species that characterizes trophic levels in natural systems (Chase *et al.* 2000). Differences within trophic levels in plant (tissue chemistry, nutritional quality and compensatory ability), herbivore (body size, foraging behaviour, interference, anti-predator strategies) and predator characteristics (self-regulation, competition for resources other than food, intra-guild predation) all interact to influence herbivore consumption patterns (e.g. McNaughton *et al.* 1989; Cebrian 1999; Oksanen & Oksanen 2000). More complex models that explicitly incorporate such heterogeneity in their formulations better explain observed patterns of plant and herbivore biomass, as well as herbivore effects on vegetation dynamics and composition, across productivity gradients (Chase *et al.* 2000). Furthermore, most simple models of trophic-dynamics essentially treat herbivores as passive conduits of energy flow through ecosystems. However, herbivores are more than just inert components of ecosystems; herbivory constitutes an integral control of plant production.

In reality, natural communities are likely to be characterized by concurrent bottom-up and top-down control, the relative strengths of which depend on the interplay between characteristics of organisms in different trophic levels (see Power 1992). For a plant–herbivore system, the absolute flux of production consumed by herbivores is indicative of the strength of bottom-up control, i.e. the extent to which plant productivity limits herbivore abundance. In contrast, the fraction of primary productivity consumed reflects the importance of herbivores as controls of plant biomass in ecosystems (top-down control; Cebrian 1999). Plant nutritional quality has been implicated as an important determinant of the latter, acting to regulate the relative amounts of carbon that flow through the herbivore *vs.* decomposer pathway (Cebrian 1999). Nutritional quality is often positively correlated with plant relative growth rates or turnover rates. Communities composed of plants with high relative growth rates tend to lose a greater percentage of primary production to herbivores, and channel a lower percentage as detritus (Fig. 8.5). Presumably, the high tissue nutrient concentrations, specifically nitrogen and phosphorus, required to support fast growth

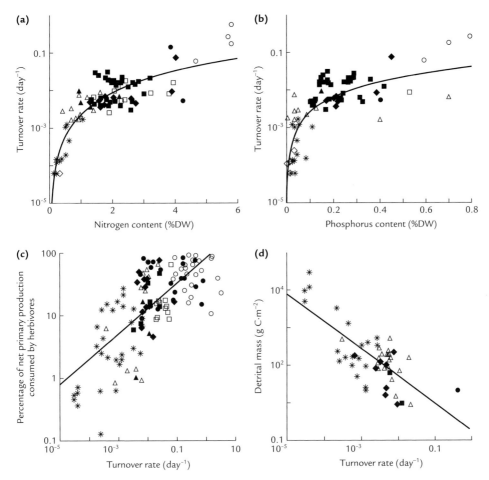

Fig. 8.5 Relationship between plant turnover rates or relative growth rates and (a) tissue nitrogen concentrations; (b) tissue phosphorus content; (c) fraction of production consumed by herbivores; (d) amounts of detritus produced across ecosystems. Open circles, phytoplankton; filled circles, benthic microalgae; open squares, macroalgal beds; filled diamonds, freshwater macrophyte meadows; filled squares, sea grass meadows; filled triangles, marshes; open triangles, grasslands; open diamonds, mangroves; asterisks, forests. (Adapted from Cebrian 1999.)

also renders such plants more attractive to herbivores, resulting in greater relative amounts of herbivory in such communities.

Such broad-scale considerations, although invaluable for inferring general trends, obscure within-system specifics of herbivory patterns. Within any community, all plants are not created alike, and herbivores typically face an autotrophic environment that is chemically heterogeneous, both in terms of nutrient quality and feeding deterrents in plant tissue. This heterogeneity is evident at all spatial scales: between tissues within a plant, between genotypes and populations of a species, between species and between communities of different plant species. Plant nutritional quality also varies temporally, both across seasons and over the life cycle of a plant. In addition, herbivores also confront a food base that is nutritionally inadequate. Plant tissues contain a preponderance of low-quality substances such as structural carbohydrates, cellulose and toxins, but a dearth of nutrients such as nitrogen and phosphorus. Nutrient concentrations, particularly nitrogen and phosphorus in herbivore tissue exceed those in plants, sometimes even 5–10× (Hartley & Jones 1997). The necessity to overcome such stoichiometric imbalances, coupled with the need to avoid plant toxins, has led to a proliferation of feeding strategies in herbivores aimed at maximal exploitation of their food sources.

At its simplest, herbivory is just heterotrophic consumption of plant tissue. Yet, this seemingly straightforward interaction induces suites of responses in eater and eaten alike, and has been a driving force behind the adaptive radiation of both plants and herbivores. From a long-term co-evolutionary perspective, two major groups of present-day terrestrial vascular plants and their affiliated herbivore fauna (McNaughton 1983a) may be recognized: the first, more ancient group, includes non-graminoid plants characterized by diverse and toxic secondary chemistry, and their relatively specialized insect herbivores. The other, more-recent group comprises graminoids, by comparison pharmacologically inert, and their allied general-purpose mammalian and orthopteran herbivores. Within these broad evolutionary lines, herbivores vary widely in how they exploit food sources. Most terrestrial herbivores display some measure of feeding selectivity for different plant species. Monophytophagous insects that feed exclusively on a single species occupy one end of the spectrum, and large bulk-feeding mammals that are more catholic in their diets, the other. Herbivores are also fastidious about the plant parts they consume, the degree of selectivity varying with herbivore body size, morphology of mouth parts and digestive system properties (McNaughton 1983a). Feeding mechanisms used and plant organs consumed not only provide a useful way to functionally classify herbivores (Table 8.1), but also govern plant responses to herbivory.

Over evolutionary time, plants have been selected to reduce the impacts herbivores exert upon them, while herbivores have been selected to maximally exploit their food sources without being overly destructive. These reciprocal effects have led to a proliferation of traits such as physical and chemical defences in plants that operate to reduce or tolerate bouts of herbivory. Herbivores, for their part, have evolved elaborate physiological and behavioural mechanisms to breach plant defences such that no plant is totally immune to herbivory at all stages of its life.

Table 8.1　A functional classification of herbivores based on feeding modes and feeding targets. Also included are a few representative taxa for each functional class.[a]

Plant organ used	Feeding mode	Representative taxa
Foliage	Bulk feeders with grinding and chewing mouth-parts	Mammalian herbivores, some birds
		Orthopterans
		Hymenoptera / Lepidoptera larvae
	Leaf miners that feed on the mesophyll without destroying the epidermis	Lepidoptera and Hymenoptera
		Diptera (family Agromyzidae)
	Strip miners that rasp through the epidermis and underlying mesophyll	Coleoptera
		Lepidoptera
Twig and branch feeders	Stem miners and borers	Coleoptera
		Lepidoptera larvae (family Cossidae)
		Hymenoptera larvae (family Cephidae)
Sap feeders	Xylem and phloem sap feeders	Homoptera
		Heteroptera
Root feeders	Bulk feeders	Fossorial vertebrates particularly rodents
		Vertebrates that feed on roots after disturbing the soil surface
	Young root and root hair feeders	Collembola
		Diplura
		Nematoda (order Tylenchida)
	Internal chewers that feed on roots & storage organs	Insects
	External feeders that consume roots or root epidermal tissue	Insects
	Cell content feeders that either fully or partially enter plants (endo- and semi-endoparasites) or feed from outside plants (ectoparasites)	Nematoda (orders Tylenchida, Dorylaimida & Aphelenchida)
	Sap feeders	Aphids / cicadas
Propagule feeders	Flower, fruit and seed feeders	Mammals / birds / insects (Bruchidae and Megastigmidae)

[a]See McNaughton (1983a) and Mortimer *et al.* (1999) for more details.

8.2　Coping with herbivory

8.2.1　Avoidance or tolerance

Plants deal with herbivory in two basic ways: they try to avoid it or alternatively, tolerate it. Avoidance of herbivore damage can be achieved through investment in mechanical defences, production of secondary compounds, or by escape in space and time. When herbivory is inevitable, plants may instead 'tolerate' herbivory through adaptations that maintain growth and reproduction following damage. Although these alternative strategies are not mutually exclusive, their relative importance varies depending on plant life history, frequency of herbivory and the prevalence of physiological or resource-constraints that impede simultaneous investment in both (Rosenthal & Kotanen 1994). From a herbivore's perspective, these alternative strategies have different selective influences as 'tolerance' does not reduce herbivore fitness and so they are under no evolutionary pressure to overcome it (Rosenthal & Kotanen 1994).

8.2.2   Use of secondary chemicals

Among the principal variables affecting a plant's susceptibility to herbivory is the presence of 'secondary' compounds. These are, by definition, not directly involved in the primary metabolism of the plant, i.e., not common to all plants but restricted to select plant groups (Pichersky & Gang 2000). Of the 20 000–60 000 odd genes estimated to exist in plant genomes, 15–25% may code for products involved in secondary metabolism (Pichersky & Gang 2000). They comprise an exceptionally diverse set of chemicals, many of which are known to have deleterious effects on herbivores. The roles of secondary metabolites are not solely restricted to anti-herbivore defence. Many serve other functions including UV absorption, attraction of pollinators and seed dispersers, and drought and salt tolerance (Hartley & Jones 1997; Pichersky & Gang 2000).

Plant secondary chemicals have varied and diverse effects (McNaughton 1983a). They repel herbivores, inhibit their feeding, mask a plant's nutritional suitability, reduce digestibility of plant tissue and are, in some cases, toxic. Some are effective in small doses, while others function in a dosage dependent manner. They may simultaneously deter several different herbivores, and concurrently serve as attractants for other herbivores, pollinators or seed dispersers. They can stimulate production of secondary compounds in neighbouring plants, or act as allelochemicals to inhibit the growth of neighbours. Their effects overstep trophic boundaries when adapted herbivores successfully appropriate them for their own defence purposes, or when predators and parasitoids use them as cues to locate herbivores. Their presence can also alter the decomposability of plant litter, thereby modifying nutrient recycling rates. As a consequence of these diverse roles, secondary chemicals are important mediators of both herbivore- and decomposer-based food webs (McNaughton 1983a).

8.2.3   Avoidance of herbivory

It is commonly assumed that plants incur a resource-cost of defence investment since defence diverts resources away from other potential uses. Selection should therefore favour plants that optimally allocate resources to defence, both in terms of quantity and quality, to maximize their fitness (see Hartley & Jones 1997). Plant investment in defence at any point in time can be simplistically envisioned as a series of 'decisions': whether or not to invest in defences at all, what proportion of resources to allocate to defence, and what kind of defences to invest in.

Several plants maintain background levels of defence compounds at all times (constitutive or passive defences). Others induce production of defence compounds following herbivory or some cue of impending herbivory. High probability of herbivore attack has been implicated as the driving force favouring investment in constitutive defences (Agrawal & Karban 1999). When probability of herbivore attack is low, plants would benefit by inducing defences only when needed, diverting resources to other functions in the mean time. Factors besides saving of allocation costs may also favour induction of defences (Agrawal & Karban 1999). For example, several specialist herbivores that have successfully breached plant defences employ the very same defence compounds to locate host plants. In such cases, constitutive defences make a plant more apparent to herbivores, and induction may be favoured as a means to reduce specialist herbivory. Induced defence may also be favoured over a constitutive strategy if it:

1 simultaneously confers resistance against several different enemies;
2 increases variability in food quality thereby reducing herbivore performance;
3 increases herbivore movement and subsequent predation or parasitism on the herbivore;
4 reduces autotoxicity;
5 is less deleterious to natural enemies of herbivores relative to constitutive defenses;

or

6 reduces pollinator deterrence (Agrawal & Karban 1999).

Hypotheses to explain the amounts and type of chemical defences deployed by plants invoke a variety of factors such as the probability of herbivore attack, resource availability, kinds of limiting resources, and internal physiological constraints and trade-offs between allocation to growth and defence (Hartley & Jones 1997). Suffice it to say, a consensus is still lacking because demonstration of appropriate fitness benefits has largely thwarted ecologists on account of manifold problems with identifying as well as measuring direct and indirect cost–benefit components.

8.2.4 From avoidance to tolerance

Despite the formidable arsenal of defences that plants have erected against herbivores, most plants are not totally immune from herbivory. It stands to reason that plants have evolved ways to deal with or tolerate these bouts of herbivory. The term 'compensation' has often been used synonymously with 'tolerance', particularly with reference to the re-growth capacity of plants following damage. Mechanisms of plant tolerance, albeit complex and interrelated, can be broadly classified as intrinsic and extrinsic mechanisms (McNaughton 1983a, b). Genetically determined responses, specific to species or related sets of species, resulting from physiological or development changes are considered intrinsic mechanisms. These include increased photosynthetic rates following damage, the ability to alter growth-form through tillering or branching, reallocation of assimilates from storage organs to meristems, changes in root:shoot ratios, modification of hormonal balance, reductions in rates of tissue senescence and increased nutrient uptake following damage.

In contrast, extrinsic mechanisms of 'tolerance' are not species specific and stem from modification of a plant's immediate environment by herbivory. For example, vertebrate grazing can often result in an increase in light-use efficiency of the remaining ungrazed tissue by reducing mutual leaf shading. Removal of older, less efficient plant tissue can increase overall photosynthetic rates of plants. Vertebrate grazing can also conserve soil water status by reducing transpiration surface area, which can influence the subsequent ability of plants to compensate for tissue loss. In addition, plant growth following herbivory may also be stimulated by nutrients recycled in more readily available forms such as dung and urine.

8.2.5   A continuum of compensatory responses

Is adequate compensation inevitable in the presence of tolerance mechanisms? The traditional literature distinguishes three contrasting views on the tolerance capabilities of plants (Fig. 8.6). The first assumes herbivory is always detrimental (Fig. 8.6, Line A), the second that herbivory is only detrimental above a critical threshold (Line B), and the third that moderate levels of herbivory can actually result in overcompensation by the plant (Line C). Obviously, intensity of herbivory is an important factor determining plant responses, and no plant is likely to tolerate herbivory above a critical threshold. Below this threshold, to assume plant responses are fixed, and plants respond in only one of these three ways, is to treat plants as divorced from all biotic and abiotic components of the ecosystem besides herbivores. Rather than treat a plant's response as deterministic,

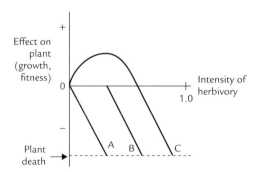

Fig. 8.6   Hypothesized effects of herbivory on plant growth and fitness. (From McNaughton 1983a.)

tolerance to herbivory must be viewed as a continuum of potential responses from under-compensation to over-compensation. Substantial evidence has accumulated in recent years which suggests, at least for vegetative growth, that the level of compensation achieved by plants in nature is contingent on several factors including plant species identity and prevalent environmental conditions (Whitham *et al.* 1991).

Among the factors influencing the ability of plants to tolerate herbivory, intrinsic growth rates are a key determinant (Whitham *et al.* 1991). Slow growth rates make it harder for a plant to replace damaged tissue in a timely manner. The ability to compensate for damage is also contingent on plant phenological status and timing of herbivory. Herbivory during the seedling stage, before root systems and photosynthetic machinery are established, is more likely to have a detrimental effect and result in mortality or under-compensation than herbivory following plant establishment. Similarly, plants in the seed setting stage are also likely to under-compensate following herbivory. The ability of a plant to compensate generally declines the later herbivory occurs during the growing season, primarily because plants have less time to recover before the end of the growing season. Plant responses are also contingent on stored reserves of carbon and nutrients present at the time of herbivory; the greater the reserves, the higher the probability of compensating for the damage. Similarly, a plant is more likely to compensate when nutrients, water and light are not limiting in the post-herbivory environment, and if it does not have to compete with other plants for these resources.

Besides these factors, type and frequency of herbivore damage, spatial distribution of herbivore damage within the plant, as well as the number of different herbivore species that feed in concert or successively on a plant, all go to determine whether a plant successfully compensates for herbivore damage (Whitham *et al.* 1991). Just as all plants are not created alike, neither are all herbivores. What effects different herbivores have on plants will depend on the type of resource the herbivore consumes, and how damaging the removal of that specific resource is for the plant (Meyer 1993). Furthermore, in certain instances, damage by one species of herbivore can render plants more susceptible to attack by other herbivore species, while in other cases, susceptibility to one pest is associated with resistance to others (Whitham *et al.* 1991). Compensatory responses of plants in such situations will depend on the damage inflicted by each species and whether different species have additive or opposing effects on plant properties. Also, compensatory ability is likely to be negatively correlated with frequency of herbivory. The greater the recovery period between herbivory bouts, the more likely a plant is to compensate.

Provided conditions are right, plants can overcompensate for tissue loss from herbivory. Overcompensation in response to mammalian herbivory has been demonstrated for plants in the Serengeti ecosystem, where seasonal migratory patterns of herbivores result in conditions that favour stimulation of plant productivity (Fig. 8.7). Ungulate herbivores in this system track pulses of primary productivity associated with rainfall. Herbivory occurs early in the growing season and the migratory nature of herbivores provides plants sufficient time to recover between herbivory bouts. High plant growth rates, coupled with increased

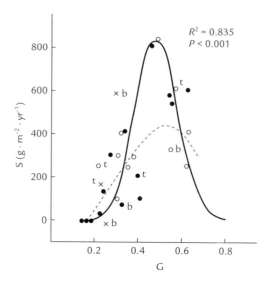

Fig. 8.7 Relationship between grazing intensity (G) and grazer stimulation of above-ground primary productivity (S). Open circles, short grasslands; filled circles, medium-height grasslands; crosses, tall grasslands. The dotted line fits all sites, while the solid line fits topographically similar sites. (From McNaughton 1985.)

nutrient availability from herbivore dung and urine, results in conditions conducive to compensation.

While several studies have shown that plants can equally or over-compensate for tissue loss to herbivores, enhancing vegetative components of fitness, fewer studies have demonstrated increases in terms of the sexual component. Indeed, the majority of studies have documented decreased seed set following herbivore damage. However, *Ipomopsis aggregata* (Paige & Whitham 1987) and *Gentianella campestris* (Lennartsson *et al.* 1998) are examples for over-compensation through seed output following herbivory. Higher seed output following grazing can result if herbivory overcomes a genetic and/or developmental constraint of plants (e.g. removal of apical dominance), or if plants withhold reproductive resources until a herbivory event in situations where there is a high probability of initial attack, but a low probability of secondary attack (Whitham *et al.* 1991). However, the idea that herbivores actually 'benefit' plants and increase their reproductive fitness by eating them has been strongly contested in the literature (Belsky *et al.* 1993). Part of the controversy stems from differential interpretations of the notion of mutualism (de Mazancourt *et al.* 2001). Mutualism in ecological time (when the performance of each partner is immediately negatively impacted following removal of the other) differs from evolutionary mutualism (over evolutionary time each partner reaches a level of performance not attainable in the absence of the partnership). Agrawal (2000) provided an effective parable to demonstrate the concept. Consider a plant that has the 'ideal' potential to produce 1000 seeds. In a environment damaged predictably by migratory herbivores, consider a genotype that employs herbivory as a cue and phenologically splits its reproductive output 20% pre-herbivory and 80% post-herbivory. In the absence of herbivory, the plant produces 200 seeds, while in the presence of herbivores seed output is 800. An evolutionary consideration (comparison with the 'ideal' plant) suggests a negative impact of herbivores on plant fitness. On the other hand, in ecological time, fitness of plants is higher in the presence of herbivores than in their absence.

Besides directly influencing amounts of resources available for reproductive allocation, herbivory can also indirectly influence plant fitness if either preference or efficiency of pollinators and dispersers is altered following floral or foliar herbivory. Experimental damage in *Oenothera macrocarpa* reduced fruit set by 18% and seed set by 33% (Mothershead & Marquis 2000). Rather than a direct effect through reduced resource availability, herbivory decreased female reproduction by altering floral traits and subsequently changing preference and efficiency of pollinators. Such indirect interactions, although relatively unstudied, are critical to understanding plant fitness consequences of herbivory.

8.3 The continuum from symbiotic to parasitic

8.3.1 Effects of three common symbionts

Terrestrial plants live intimately linked with several micro-organisms, the relationships between which range from mutualistic to parasitic (see Chapter 7). For plants, benefits of mutualistic associations range from an increased ability to acquire limiting nutrients to enhanced capabilities of withstanding abiotic stresses. Such alliances often alter the nutritional status of plants, and in doing so, modulate interactions between plants and herbivores.

8.3.2 Mycorrhizae

Symbiotic associations between plants and mycorrhizal fungi are ubiquitous in nature, such associations being especially important in nutrient-poor communities. Plants provide mycorrhizae with carbon, and duly obtain several benefits from mycorrhizal infection including increased nutrient uptake, improved water relations and greater tolerance to pathogens (see Chapter 9). By improving plant nutritional status, mycorrhizae can make plants more attractive to herbivores, increasing a plant's susceptibility to attack. Alternately, mycorrhizal colonization can also potentially reduce a plant's susceptibility to herbivory if enhanced nutrient uptake relative to carbon cost permits greater plant allocation to anti-herbivore defences. Induction of defence compounds that follow infection of plant roots by fungal hyphae, and secondary compounds synthesized by the mycorrhizae themselves, can also act to enhance plant resistance to herbivores. Besides altering plant resistance, mycorrhizal colonization can also improve plant tolerance to herbivory if it enhances a plant's ability to acquire limiting nutrients post-herbivory. Consistent with these potential alternate outcomes, experimental studies of mycorrhizal colonization have demonstrated both increases and decreases in host-plant resistance to herbivory (Gehring & Whitham 1994).

Just as plant–herbivore interactions are influenced by mycorrhizae, herbivores too can affect how a plant interacts with its mycorrhizal symbionts. As much as 10–60% of a plant's photosynthate might be required to support mycorrhizae (Gehring & Whitham 1994). Consequently, when tissue loss to herbivores is high, costs of supporting mycorrhizae can far outweigh benefits, shifting the relationship from mutualistic to parasitic. Many studies have documented reduced mycorrhizal colonization following herbivory, while others show no significant effects or positive effects of herbivory on mycorrhizal colonization (Gehring & Whitham 1994). No significant effects, and possibly increased myc-orrhizal colonization, can result if herbivory induces shifts in mycorrhizal com-munities favouring species or morphotypes with lower carbon requirements (Saikkonen *et al.* 1999). Ultimately, the specific outcome is dependent on how herbivory interacts with prevailing environmental conditions to alter the cost–benefit ratio of association for both involved parties.

8.3.3   The trade-off of N-fixation

Nitrogen limits plant growth in many terrestrial ecosystems. Plants have evolved several adaptations to cope with this limitation, including forming symbiotic associations with nitrogen-fixing bacteria. Where nitrogen is limiting, plants involved in symbiotic associations should have a competitive advantage over non-fixers. Yet, nitrogen-fixing species do not reach widespread dominance. Limita-tion by nutrients other than nitrogen, inability to quickly colonize early successional sites and the high energetic costs of fixing atmospheric nitrogen are potential reasons for the lack of widespread dominance by N-fixers. However, herbivores also play critical roles in the observed rarity of N-fixers in several ecosystems.

Bentley & Johnson (1991) compared alkaloid content and growth rates of *Lupinus succulentus* plants grown under low nitrogen concentrations against defoliated and undamaged plants provided with either inorganic nitrogen or with N_2-fixing bacteria (Fig. 8.8). Unlike plants provided with supplemental inorganic nitrogen, leaf damage in N_2-fixing plants reduced both alkaloid con-centrations and growth rates, suggesting herbivory costs to N_2-fixation. Although N-fixers may invest substantially in nitrogenous defence compounds while undamaged (Fig. 8.8), their ability to tolerate herbivory can be compromised once damaged. Presumably, leaf tissue loss reduces photosynthate available to support N_2-fixing bacteria resulting in N_2-fixing plants becoming nitrogen stressed under conditions of high herbivory.

Indirect evidence for a herbivory-cost of N-fixation comes from studies that report increased abundance of N-fixers following herbivore removal (Ritchie *et al.* 1998, Sirotnak & Huntly 2000). Experimental exclusion of voles (*Microtus* spp.) from a site in Yellowstone National Park resulted in increased legume abundance within exclosures (Sirotnak & Huntly 2000). However, responses were not consistent across all sites suggesting that herbivore effects on N-fixers can vary over space and time and may be contingent on specific site conditions. Since fixation represents a substantial source of nitrogen input into many systems, the interaction between herbivores and N-fixers can directly and indirectly affect several aspects of community and ecosystem function.

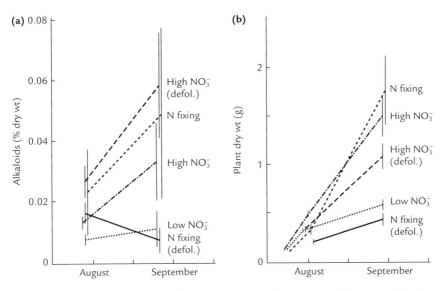

Fig. 8.8 Effects of defoliation on alkaloid concentrations (a) and biomass (b) of *Lupinus succulentus* plants grown under the indicated nitrogen nutrition treatments. (Adapted from Bentley & Johnson 1991.)

8.3.4 Fungal endophyte associations

Plants also form mutualistic associations with fungal endophytes that grow intercellularly in leaf and stem cells and infect plants asymptomatically. All plant species examined to date have been found to harbour fungal endophytes (Saikkonen *et al*. 1998, Faeth 2002, Arnold *et al*. 2003). Most endophytes form localized infections in leaves, stems and other plant parts and are transmitted horizontally between plants via spores. A smaller fraction of relatively species-poor endophytes, mostly found in pooid grasses, form systemic infections in above-ground plant tissues and are transmitted vertically via seeds.

Endophytes receive nutrients and protection from plants, and in turn are thought to confer plants with increased resistance to herbivores, pathogens and drought, enhanced competitive ability and increased germination success (Saikkonen *et al*. 1998).

Effects of such associations on plant–herbivore interactions have received relatively little attention in the ecological literature, but a recent meta-analysis provides support for the hypothesis of a defensive mutualism between grasses and their vertically transmitted fungal endophytes, potentially through the production of multiple alkaloid compounds by endophytes (Saikkonen *et al*. 2010). In contrast, the nature of the relationship between trees and their horizontally transmitted endophytes appears much more variable, ranging from negative to positive (Saikkonen *et al*. 2010). Thus, while endophytes can increase tree resistance to herbivores, they may also enhance foliage quality for herbivores (Saikkonen *et al*. 2010). Besides herbivores, endophytes can also influence the nature of interactions between plants and pathogens. For example, horizontally

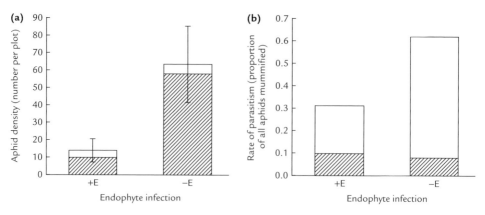

Fig. 8.9 (a) Differential density responses of two species of aphids, *Rhopalosiphum padi* (shaded bars) and *Metopolophium festucae* (empty bars) to presence (+E) and absence (−E) of fungal endophytes in *Lolium multiflorum* plants. Responses are significant only for *R. padi*. (b) Endophyte treatment effects on total aphid parasitism rates. Shaded bars represent proportion of emerged primary parasitoids and open bars, secondary parasitoids. (From Omacini *et al.* 2001.)

transmitted endophytes have been shown to significantly decrease leaf necrosis and leaf mortality in the neotropical cacao tree (*Theobroma cacao*) due to an important foliar pathogen *Phytophthora* spp., with the protection primarily localized to endophyte infected tissues (Arnold *et al.* 2003). Fungal endophytes can also influence the nature of interactions between herbivores and organisms at higher trophic levels. In an experiment involving *Lolium multiflorum* plants, Omacini *et al.* (2001) showed that fungal endophyte infection decreased aphid densities on plants threefold (Fig. 8.9a). However, responses differed between aphid species. Endophyte infection also influenced rates of aphid parasitism (Fig. 8.9b). While hatching rates of primary parasatoids (those that attack aphids directly) did not differ between endophyte infected and uninfected treatments, hatching rates of secondary parasatoids (those that attack primary parasatoids) was significantly higher in endophyte-free plants (Fig. 8.9b). Fungal endosymbionts may therefore be important modulators of plant–herbivore interactions and food web structure. However, there seems to be much specificity in the nature and outcome of interactions between endophytes and their host plants (Hartley & Gange 2009), and our understanding of the implications of this widespread mutualism is far from complete.

8.4 Community level effects of herbivory

8.4.1 Herbivores and plant species diversity

Herbivores have varied effects on the plant richness of communities, working to either increase, decrease or cause no significant changes in plant richness (see also Chapter 11). Nutrient and water availability, and evolutionary history of grazing are some of the variables hypothesized to have a regulatory effect on herbivore mediation of plant richness (Milchunas *et al.* 1988; Proulx & Mazumder 1998). Comparative studies of a broad array of herbivores and habitat types suggest that herbivore effects on plant species richness may be contingent on nutrient availability. Richness often declines with grazing in nutrient-poor ecosystems, while the outcome is reversed in nutrient-rich ecosystems (Proulx & Mazumder 1998). Presumably, ability to tolerate low nutrient conditions is the primary factor controlling plant species richness in nutrient-poor ecosystems. When grazed, species intolerant of herbivory are removed from the system. Since few species remain in the pool tolerant to both herbivory and low nutrient conditions, colonization is low and species diversity declines under grazing. In nutrient-rich systems on the other hand, competitive ability rather than stress tolerance is presumed to be the variable defining plant species richness. In this case, grazing on competitive dominants relaxes competitive interactions, permitting co-existence of inferior competitors. Diversity therefore increases with grazing in such systems. However, increases in grazing intensity beyond a critical threshold, even in nutrient-rich ecosystems, can cause diversity to decline.

Over and above such broad generalizations, herbivore effects are likely to be specific to spatial scales of inquiry (Olff & Ritchie 1998; Stohlgren *et al.* 1999). At small scales, herbivore mediation of competitive interactions may be the dominant process influencing species diversity, while colonization–extinction dynamics may be more important at larger scales. At small scales, herbivores can enhance plant diversity by (1) selectively consuming competitive dominants, permitting establishment of inferior species, (2) increasing heterogeneity through soil disturbances and permitting species coexistence, and (3) reducing individual plant size and allowing for greater species packing within a given area. In the absence of grazing, dominants may grow bigger, exclude sub-dominants, and lower plant diversity at small scales. However, if overall rates of colonization and extinction are not altered, such differences may not be evident at large scales (Stohlgren *et al.* 1999). Species excluded by grazers at small scales may still persist in 'grazing-safe sites' at larger scales. However, if grazing pressure is strong enough, intolerant species may be weeded out altogether from the regional species pool, lowering diversity at larger scales (Stohlgren *et al.* 1999).

8.4.2   Effects of herbivore diversity

Natural communities typically contain several herbivores that vary in body size and differ in their feeding strategies and selectivity. Such a diversity of herbivores can have additive or complementary effects on plant species diversity (Ritchie & Olff 1999). When multiple herbivores feed on the same species, their effects will be additive. Simultaneous feeding on competitive dominants can increase diversity, while that on competitive inferiors decreases it. When herbivores feed on different species, their effects may be complementary, serving to maintain plant diversity in a quasi-stable state.

The potential for different herbivore species to have additive or complementary effects on plant community diversity is also dependent on whether herbivore species are likely to facultatively diverge or converge in their diets in the presence of other herbivores. Compensatory effects arising from herbivore diet shifts in the presence of competitors has been experimentally demonstrated for grasshoppers feeding on Minnesota old-field plants (see Ritchie & Olff 1999). The proportion of grasses and forbs in the diets of three different grasshopper species changed in the presence of the other species as opposed to when alone. In contrast, additive effects of diverse herbivores can occur if feeding by a particular herbivore makes a plant more attractive to other herbivore species. How such individual effects translate to additive effects at the community level is unclear. Additive effects at the community level can arise if, for example, several different herbivores cue in on a particular plant species following specialist herbivore outbreaks on that species.

Theoretical syntheses suggest that herbivore effects should vary predictably across soil fertility and moisture gradients (Olff & Ritchie 1998; Ritchie & Olff 1999). The underlying premise is that tissue nutrient concentrations and palatability of dominant species differ depending on the particular limiting resource, which in turn, determines the characteristics of the herbivore community and their consumption patterns. Where dominant plant species tend to be palatable, i.e. have high tissue nutrient concentrations, multiple herbivores can consume the same species in an additive fashion. On the other hand, in communities characterized by abundant low-quality plants and rare high-quality plants, effects of multiple herbivores can lead to compensatory effects. In such cases, large herbivores potentially consume the dominant low-quality plants permitting the co-existence of both high quality plants as well as the smaller bodied herbivores that feed on them. Compensatory effects of herbivore diversity in these situations arise from herbivores of different body sizes consuming different plant species. Approaches such as these provide a fruitful avenue of pursuit since they integrate ecosystem level constraints on plant traits with herbivore feeding selectivity as a function of body size to predict plant community responses to grazing by a diverse herbivore assemblage.

Few studies have, however, experimentally tested the effects of herbivore richness or body-size diversity on plant communities. Excluding small herbivores while retaining large ones is rarely feasible in field studies, and most studies progressively exclude larger bodied herbivores. Bakker *et al.* (2006) used a long-term multi-site experiment across a 10-fold productivity gradient in North America and Europe to look at the combined effects of productivity and herbivore body size on plant diversity. In accordance with the results of Proulx & Mazumder (1998), they found that herbivore effects on plant diversity switched from negative at low productivity to positive at high productivity (see Section 8.4.1 above), but only when large herbivore species (>30 kg) were present in the assemblage (Fig. 8.10; Bakker *et al.* 2006). In sites with only small herbivores, there were no consistent effects of grazing on plant species diversity across productivity gradients (Fig 8.10), suggesting that large herbivores play key regulatory roles in grazing ecosystems.

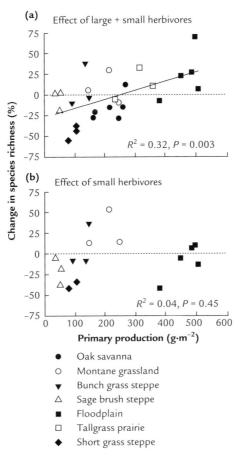

Fig. 8.10   Change in plant species richness across productivity gradients for systems where the herbivore assemblage comprises (a) both large (>30 kg) and small herbivores, and (b) only small herbivores. Where large herbivores are present (a), plant species richness declines under grazing when plant productivity is low and increases under grazing when productivity is high. When only small herbivores are present (b), there is no consistent effect of grazing on plant species richness. (From Bakker *et al.* 2006.)

Besides directly consuming plant tissue and altering the competitive balance between species, herbivores can also influence plant community diversity by impacting seed dispersal and colonization patterns, which in turn may depend on herbivore body size. For example, Bakker & Olff (2003) looked at the effects of a large and small herbivore, cattle and rabbits, on recruitment of subordinate herbs in a floodplain grassland in the Netherlands. They concluded that both cattle and rabbits had a major impact on the dispersal and colonization of subordinate species in this grassland, but for different reasons. Cattle were important for seed dispersal, dispersing more than 10 times the number of seeds when compared to the smaller bodied rabbits (Bakker & Olff 2003). In contrast, rabbits were critical for the establishment process, and played important roles as creators of soil disturbances (e.g. bare patches), which enhanced seedling establishment (Bakker & Olff 2003).

8.4.3 Herbivory and plant succession

Herbivores also influence successional rates and successional trajectories of communities (Ritchie & Olff 1999). Herbivores that feed preferentially on late successional species tend to retard succession. By the same token, selective herbivory on early successional species can hasten establishment of late successional species, thereby accelerating succession. As with species diversity patterns, the presence of a diverse herbivore assemblage can have additive or complementary effects on successional trends. Herbivore assemblages that feed on species characteristic of the same successional state have additive effects on plant species replacement patterns. In such cases, effects of diverse assemblages may be similar to those of individual herbivores, either accelerating or retarding successional rates. On the other hand, when different members of the herbivore assemblage feed on species characteristic of different successional stages, they can 'arrest' plant communities at intermediate stages of succession.

Besides influencing successional rates, herbivores also regulate successional trajectories, thereby defining the qualitative nature of mature plant communities. Seedling herbivory, in particular, is an important pathway through which such herbivore effects are manifested (Crawley 1997; Hanley 1998). Plants are particularly vulnerable to tissue loss at this stage in their life cycle and even if herbivory does not result in mortality, it can reduce seedling vigour thereby influencing its competitive ability and chances of long-term survival. The magnitude of such effects can be substantial. For example, in Panamanian forests, mammalian herbivory can cause as much as a sixfold reduction in tree seedling survivorship for certain species (Asquith *et al.* 1997). Several herbivore guilds, from nematodes to large mammals, have been shown to have deleterious effects on seedling survival and establishment; the case of mollusc herbivores on seedling dynamics in temperate systems being particularly well documented (see Crawley 1997; Hanley 1998). Differences between species in herbivore-induced seedling mortality is the primary mechanism through which herbivores influence plant community development and species composition patterns. Differential herbivore-induced mortality can arise from interspecific variation in seedling palatability, size and morphology, as also from differences in abundance, spatial distribution and timing of seedling emergence (Hanley 1998). However, herbivore effects on seedling establishment need not always be detrimental (Crawley 1997). Besides direct negative effects on vulnerable species that arise from increased mortality or reduced competitive vigour following tissue consumption, herbivores can also have indirect positive effects on seedling establishment of other species. Seedling establishment, particularly for species avoided by herbivores, may be favoured when herbivores enhance microsite suitability through physical disturbances to the environment or when consumption of plant tissue opens up canopies, reduces competition, reduces litter loads and increases light availability at the soil surface, thereby creating opportunities for recruitment (see also Chapter 6).

Although the role of herbivory at the seedling stage in influencing successional rates and trajectories is well recognized, it is still poorly understood how its importance changes relative to other factors such as plant competition as succession proceeds. One study that experimentally manipulated herbivory and competition across a successional gradient in a salt marsh found no consistent differences in the relative intensities of either competition or herbivory across different successional stages (Dormann *et al.* 2000). Further, the combined impacts of herbivory and plant competition varied across species, increasing over succession for some species but not for others (Dormann *et al.* 2000). Nevertheless, in many instances, particularly in productive sites where herbivores are not able to maintain the vegetation in a suitable grazing condition, the importance of herbivory typically decreases as succession proceeds (van der Wal *et al.* 2000). For example, Brent geese (*Branta bernicla*) have been shown to be progressively excluded from older salt marshes as succession proceeds (van der Wal *et al.* 2000). Geese in older successional sites were confronted with a high proportion of non-preferred species, and displayed a significant reduction in the time spent foraging when compared to early successional sites, eventually abandoning older successional sites (van der Wal *et al.* 2000).

8.5   Integrating herbivory with ecosystem ecology

The interaction between plants and herbivores has important repercussions for patterns of energy and nutrient flow through the ecosystem because herbivore consumption of plant tissue, plant nutrient uptake and litter decomposition rates are intimately linked. Ecosystem level studies of energy and nutrient cycling have reported diametrically opposite effects of herbivory on ecosystem processes. Herbivores enhance nutrient cycling in certain systems, and retard it in others. Augustine & McNaughton (1998) identified four mechanisms by which herbivores influence energy and nutrient flow in ecosystems: (1) by altering species composition of communities and hence, the quality of litter inputs from uneaten plants, (2) by consuming plant nutrients and returning them to the soil in more readily available forms such as dung and urine, (3) by altering inputs from eaten plants to the soil through changes in the root system, litter quality and other non-detrital inputs such as root exudates, and (4) by altering plant and soil micro-environment. While the latter three enhance nutrient cycling rates, species compositional changes can either enhance or retard nutrient cycling rates. The eventual outcome depends on whether effects of altered species composition offset the effects of the latter three processes.

Species compositional changes influence ecosystem nutrient cycling by modifying the quality of litter inputs to the soil (Fig. 8.11). Herbivory favouring unpalatable, slow-growing species with well-defended or nutrient-poor tissues, results in litter that is of poor quality. Such litter, containing high amounts of structural tissue or secondary chemicals, is broken down more slowly by

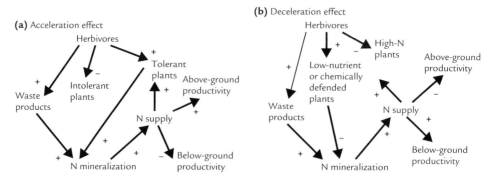

Fig. 8.11 Hypothesized mechanisms by which herbivore feeding preferences can accelerate (a) or decelerate (b) rates of nutrient cycling through an ecosystem. Arrows indicate the net indirect effect of herbivores on the abundance of plants or the rate of the process. (From Ritchie *et al.* 1998.)

micro-organisms, reducing rates at which limiting nutrients are recycled between different ecosystem components. When herbivory favours fast growing, palatable species with high tissue nutrient concentrations, the litter produced is easily broken down by micro-organisms and nutrient cycling rates are amplified.

In systems comprising both palatable and unpalatable plant species, herbivores are likely to consume proportionately more tissue from palatable species. Considerations of herbivore intake alone suggests that palatable species must be at a competitive disadvantage in these situations, leading to eventual domination by unpalatable species in these communities. Why then, does herbivory not cause all plant communities to be dominated by unpalatable species? Obviously, greater tissue loss to herbivory is insufficient to tilt the competitive balance in favour of unpalatable species in all systems. Intrinsic and extrinsic mechanisms by which plants tolerate herbivory, intensity and frequency of herbivory, as well as prevalent environmental conditions, all interact to determine the nature of herbivore-induced community change. Conditions favouring the persistence of palatable species in a community include high nutrient levels in the system, intermittent herbivory rates such as those resulting from migratory habits of herbivores, early-season and post-fire herbivory, asynchronous phenology of palatable and unpalatable species, herbivore body size dichotomy and herding behaviour of herbivores (Whitham *et al.* 1991; Augustine & McNaughton 1998; Ritchie & Olff 1999). Compositional shifts favouring unpalatable species are more likely in systems that are nutrient poor, contain sedentary herbivores that feed selectively on high-quality plants, forage singly or in small groups, and subject plants to chronic levels of herbivory (Augustine & McNaughton 1998).

Herbivore body size dichotomy can also be important in regulating the balance between palatable and unpalatable species in a community (Olff & Ritchie 1998; Ritchie & Olff 1999). In systems where both water and nutrients are non-limiting, plant competition is primarily for light. Plants invest in structural tissues to enhance their light competitive ability, and so dominant species tend to be of low-quality, used primarily by large-bodied bulk-feeding herbivores. Grazing by large-bodied herbivores on low-quality plants facilitates coexistence of both grazing-tolerant high-quality plants as well as small-bodied herbivores that feed on them. However, reductions in the numbers of large-bodied herbivores can lead to low-quality plants dominating the system, causing both high-quality plants as well as smaller-bodied herbivores to decline, and reducing overall rates of nutrient cycling. In summary, differential effects of herbivory on plant composition and subsequent ecosystem functioning can arise from differences in (1) the nature of limiting resources (e.g. water, nitrogen), which in turn defines plant characteristics and herbivore selectivity, (2) herbivory characteristics such frequency, intensity and timing, and (3) herbivore characteristics including foraging behaviour, herbivore diversity and herbivore body size dichotomy (Augustine & McNaughton 1998; Olff & Ritchie 1998; Ritchie *et al*. 1998; Ritchie & Olff 1999).

References

Agrawal, A.A. (2000) Overcompensation of plants in response to herbivory and the by-product benefits of mutualism. *Trends in Plant Science* 5, 309–313.

Agrawal, A.A. & Karban, R. (1999) Why induced defences may be favored over constitutive strategies in plants. In: *The Ecology and Evolution of Inducible Defences* (eds R. Tollrian & C.D. Harvell), pp. 45–61. Princeton University Press, Princeton, NJ.

Arnold, A.E., Mejia, L.C., Kyllo, D. *et al*. (2003) Fungal endophytes limit pathogen damage in a tropical tree. *Proceedings of the National Academy of Sciences USA* 100, 15649–15654.

Asquith, N.M., Wright, S.J. & Clauss, M.J. (1997) Does mammal community composition control recruitment in neotropical forests? Evidence from Panama. *Ecology* 78, 941–946.

Augustine, D.J. & McNaughton, S.J. (1998) Ungulate effects on the functional species composition of plant communities: herbivore selectivity and plant tolerance. *Journal of Wildlife Management* 62, 1165–1183.

Aunapuu, M., Dahlgren, J., Oksanen, T. *et al*. (2008) Spatial patterns and dynamic responses of arctic food webs corroborate the exploitation ecosystems hypothesis (EEH). *American Naturalist* 171, 249–262.

Bakker, E.S. and Olff, H. (2003) Impact of different-sized herbivores on recruitment opportunities for subordinate herbs in grasland. *Journal of Vegetation Science* 14, 465–474.

Bakker, E.S., Ritchie, M.E., Olff, H., Milchunas, D.G. & Knops, J.M.H. (2006) Herbivore impact on grassland plant diversity depends on habitat productivity and herbivore size. *Ecology Letters* 9, 780–788.

Belsky, A.J., Carson, W.P., Jensen, C.L. & Fox, G.A. (1993) Overcompensation by plants: herbivore optimization or red herring? *Evolutionary Ecology* 7, 109–121.

Bentley, B.L. & Johnson, N.D. (1991) Plants as food for herbivores: the roles of nitrogen fixation and carbon dioxide enrichment. In: *Plant–Animal Interactions: Evolutionary Ecology in Tropical and Temperate Regions* (eds P.W. Price, T.M. Lewinsohn, G.W. Fernandes & W.W. Benson), pp. 257–272. John Wiley & Sons, Ltd, New York, NY.

Beschta, R.L. & Ripple, W. J. (2009) Large predators and trophic cascades in terrestrial ecosystems of the western United States. *Biological Conservation* 142: 2401–2414.

Cebrian, J. (1999) Patterns in the fate of production in plant communities. *The American Naturalist* 154, 449–468.

Cebrian, J. & Lartigue, J. (2004) Patterns of herbivory and decomposition in aquatic and terrestrial ecosystems. *Ecological Monographs* 74, 237–259.

Chase, J.M., Leibold, M.A., Downing, A.L. & Shurin, J.B. (2000) The effects of productivity, herbivory, and plant species turnover in grassland food webs. *Ecology* 81, 2485–2497.

Crawley, M.J. (1997) Plant–herbivore dynamics. In: *Plant Ecology* (ed. M.J. Crawley), pp. 401–474. Blackwell Science, Oxford.

Crete, M. (1999) The distribution of deer biomass in North America supports the hypothesis of exploitation ecosystems. *Ecology Letters* **2**, 223–227.

Dormann, C.F., van der Wal, R. & Bakker, J.P. (2000) Competition and herbivory during salt marsh succession: the importance of forb growth strategy. *Journal of Ecology* **88**, 571–583.

de Mazancourt, C., Loreau, M. & Dieckmann, U. (2001) Can the evolution of plant defence lead to plant–herbivore mutualism? *The American Naturalist* **158**, 109–123.

Elmhagen, B., Ludwig, G., Rushton, S. P., Helle, P. & Lindén, H. (2010) Top predators, mesopredators and their prey: interference ecosystems along bioclimatic productivity gradients. *Journal of Animal Ecology* **79**, 785–794.

Faeth, S.H. (2002) Are endophytic fungi defensive plant mutualists? *Oikos* **98**, 25–36.

Gehring, C.A. & Whitham, T.G. (1994) Interactions between above-ground herbivores and mycorrhizal mutualists of plants. *Trends in Ecology & Evolution* **9**, 251–255.

Hanley, M.E. (1998) Seedling herbivory, community composition and plant life history traits. *Perspectives in Plant Ecology, Evolution and Systematics* **12**, 191–205.

Hartley, S.E. & Gange, A.C. (2009) Impacts of plant symbiotic fungi on insect herbivores: mutualism in a multitrophic context. *Annual Review of Entomology* **54**, 323–342.

Hartley, S.E. & Jones, C.G. (1997) Plant chemistry and herbivory, or why the world is green. In: *Plant Ecology*, 2nd edn (ed M.J. Crawley), pp. 284–324. Blackwell Science, Oxford.

Kauffman, M.J., Varley, N., Smith, D.W. *et al.* (2007) Landscape heterogeneity shapes predation in a newly restored predator–prey system. *Ecology Letters* **10**, 690–700.

Lennartsson, T., Nilsson, P. & Tuomi, J. (1998) Induction of overcompensation in the field gentian, *Gentianella campestris*. *Ecology* **79**, 1061–1072.

McNaughton, S.J. (1983a) Compensatory plant growth as a response to herbivory. *Oikos* **40**, 329–336.

McNaughton, S.J. (1983b) Physiological and ecological implications of herbivory. In: *Physiological Plant Ecology III: Responses to the Chemical and Biological Environment*, Vol. 12C (eds O.L. Lange, P.S. Nobel, C.B. Osmond & H. Ziegler), pp. 657–678. Springer-Verlag, Berlin.

McNaughton, S.J. (1985) Ecology of a grazing ecosystem: The Serengeti. *Ecological Monographs* **55**, 259–294.

McNaughton, S.J., Oesterheld, M., Frank, D.A. & Willliams, K.J. (1989) Ecosystem-level patterns of primary productivity and herbivory in terrestrial habitats. *Nature* **341**, 142–144.

Meyer, G.A. (1993) A comparison of the impacts of leaf- and sap-feeding insects on growth and allocation of goldenrod. *Ecology* **74**, 1101–1116.

Milchunas, D.G., Sala O.E. & Lauenroth, W.K. (1988) A generalized model of the effects of grazing by large herbivores on grassland community structure. *The American Naturalist* **132**, 87–106.

Moen, J. & Oksanen, L. (1991) Ecosystem trends. *Nature* **353**, 510.

Mortimer, S.R., van der Putten, W.H. & Brown, V.K. (1999) Insect and nematode herbivory below ground: interactions and role in vegetation succession. In: *Herbivores: Between Plants and Predators* (eds H. Olff, V.K. Brown & R.H. Drent), pp. 205–238. Blackwell Science, Oxford.

Mothershead, K. & Marquis, R.J. (2000) Fitness impacts of herbivory through indirect effects on plant–pollinator interactions in *Oenothera macrocarpa*. *Ecology* **81**, 30–40.

Oksanen, L. & Oksanen, T. (2000) The logic and realism of the hypothesis of exploitation ecosystems. *The American Naturalist* **155**, 703–723.

Oksanen, L., Fretwell, S.D., Arruda, J. & Niemela, P. (1981) Exploitation ecosystems in gradients of primary productivity. *The American Naturalist* **118**, 240–261.

Olff, H. & Ritchie, M.E. (1998) Effects of herbivores on grassland plant diversity. *Trends in Ecology & Evolution* **13**, 261–265.

Omacini, M., Chaneton, E.J., Ghersa, C.M. & Muller, C.B. (2001) Symbiotic fungal endophytes control insect host–parasite interaction webs. *Nature* **409**, 78–81.

Paige, K.N. & Whitham, T.G. (1987) Overcompensation in response to mammalian herbivory: the advantage of being eaten. *The American Naturalist* **129**, 407–416.

Pichersky, E. & Gang, D.R. (2000) Genetics and biochemistry of secondary metabolites in plants: an evolutionary perspective. *Trends in Plant Science* **5**, 439–445.

Power, M.E. (1992) Top-down and bottom-up forces in food webs: do plants have primacy? *Ecology* **73**, 733–746.

Proulx, M. & Mazumder, A. (1998) Reversal of grazing impact on plant species richness in nutrient-poor vs. nutrient-rich ecosystems. *Ecology* **79**, 2581–2592.

Riginos, C. & Grace, J.B. (2008) Savanna tree density, herbivores, and the herbaceous community: bottom-up vs. top-down effects. *Ecology* 89, 2228–2238.

Ripple, W.J. & Beschta, R.L. (2004) Wolves and the ecology of fear: can predation risk structure ecosystems? *BioScience* 54, 755–766.

Ritchie, M.E. & Olff, H. (1999) Herbivore diversity and plant dynamics: compensatory and additive effects. In: *Herbivores: Between Plants and Predators* (eds H. Olff, V.K. Brown & R.H. Drent), pp. 175–204. Blackwell Science, Oxford.

Ritchie, M.E., Tilman, D. & Knops, J.M.H. (1998) Herbivore effects on plant and nitrogen dynamics in oak savanna. *Ecology* 79, 165–177.

Rosenthal, J.P. & Kotanen, P.M. (1994) Terrestrial plant tolerance to herbivory. *Trends in Ecology & Evolution* 9, 145–148.

Saikkonen, K., Faeth, S.H., Helander, M. & Sullivan, T.J. (1998) Fungal endophytes: a continuum of interactions with host plants. *Annual Review of Ecology & Systematics* 29, 319–43.

Saikkonen, K., U. Ahonen-Jonnarth, A.M. Markkola *et al.* (1999) Defoliation and mycorrhizal symbiosis: a functional balance between carbon sources and below-ground sinks. *Ecology Letters* 2, 19–26.

Saikkonen, K., Saari, S. & Helander, M. (2010) Defensive mutualism between plants and endophytic fungi? *Fungal Diversity* 41, 101–113.

Sirotnak, J.M. & Huntly, N.J. (2000) Direct and indirect effects of herbivores on nitrogen dynamics: voles in riparian areas. *Ecology* 81, 78–87.

Stohlgren, T.J., Schell, L.D. & Heuvel, B.V. (1999) How grazing and soil quality affect native and exotic plant diversity in Rocky Mountain grasslands. *Ecological Applications* 9, 45–64.

van der Wal, R., van Lieshout, S., Bos, D. & Drent, R.H. (2000) Are spring staging Brent geese evicted by vegetation succession? *Ecography* 23, 60–69.

Wegener, C. & Odasz-Albrigtsen, A.M. (1998) Do Svalbard reindeer regulate standing crop in the absence of predators? A test of the 'exploitation ecosystems' model. *Oecologia* 116, 202–206.

Whitham, T.G., Maschinski, J., Larson, K.C. & Paige, K.N. (1991) Plant responses to herbivory: the continuum from negative to positive and underlying physiological mechanisms. In: *Plant–Animal Interactions: Evolutionary Ecology in Tropical and Temperate Regions* (eds P.W. Price, T.M. Lewinsohn, G.W. Fernandes & W.W. Benson), pp. 227–256. John Wiley & Sons, Ltd, New York, NY.

9

Interactions Between Higher Plants and Soil-dwelling Organisms

Thomas W. Kuyper and Ron G.M. de Goede
Wageningen University, The Netherlands

9.1 Introduction

After the discovery of mycorrhizal and nitrogen-fixing associations in the 1880s, early plant ecologists, notably Schimper, Warming, Clements and Braun-Blanquet, recognized the importance of soil organisms for plant community ecology. However, interest in these interactions gradually declined when vegetation ecology became increasingly descriptive in the formal recognition of plant community types. The study of these interactions became part of applied plant ecology (agronomy, forestry), and had a minor impact on the development of vegetation ecological theory. In the past few decades, however, plant ecologists rediscovered the importance of below-ground interactions (Crawley 1997). One explanation for a renewed interest in soil biota was the recognized inadequacy of niche theories that only looked at abiotic factors, because all terrestrial plants need the same suite of essential nutrients in relatively fixed quantities and ratios. It is now generally accepted that studies on interactions between plants in the absence of the soil community are unrealistic. A simple comparison between a plant species grown in sterile and in unsterile soil demonstrates the importance of such below-ground biotic interactions. Comparison of these plants shows that such interactions range from antagonistic to beneficial. Effects on the performance of individual species will influence the competitive or facilitative interactions between species and finally scale up to effects on plant community species composition.

This chapter mainly treats mutualistic and antagonistic interactions in the **rhizosphere** (the soil environment in the immediate vicinity of and affected by roots). It addresses different mechanisms by which soil biota affect plant species and communities and focuses on **microbiota** (bacteria, fungi, nematodes), especially because the occurrence and magnitude of feedbacks (see Section 9.5)

Vegetation Ecology, Second Edition. Eddy van der Maarel and Janet Franklin.
© 2013 John Wiley & Sons, Ltd. Published 2013 by John Wiley & Sons, Ltd.

depend more on the microbiota than on meso- and macrofauna. It is important to realise that the outcome of these mechanisms is context-dependent.

9.2 Ecologically important biota in the rhizosphere

9.2.1 Introduction

Nutrient limitation of primary production of vegetation is usually caused by a shortage of nitrogen or phosphorus. In tropical ecosystems with old weathered soils phosphorus is usually the primary limiting nutrient, whereas in temperate and boreal ecosystems on relatively young soils nitrogen limitation is more common. Mutualistic symbioses between micro-organisms and plant roots increase the possibilities to exploit these scarce resources; different root symbioses are dominant depending on which nutrient is limiting (see Section 9.7). The roots of the overwhelming majority of plant species are associated with mycorrhizal fungi that enhance the uptake of plant nutrients. Roots of plants of several families are specifically associated with various bacteria that can fix atmospheric nitrogen and convert it to mineral nitrogen compounds. Roots also attract pathogens, parasites and herbivores that use carbon and nutrients, resulting in net losses of resources.

While we classify rhizosphere organisms as mutualistic symbionts or antagonistic pathogens and parasites, it is important to realise that there is in fact a continuum in behaviour. The context of the environmental conditions influences costs and benefits of the symbiotic partners affecting plants. Klironomos (2003) tested the performance of 56 plant species from an old-field community in the presence of one arbuscular mycorrhizal fungus, *Glomus etunicatum*, which was isolated from the same site. The average response was a yield increment of +17%, but the variation between plant species ranged from −46% to +48%.

9.2.2 Mycorrhizal fungi

The roots of the overwhelming majority of higher plant species (>80%) are colonized by fungi that live in a mutualistic relationship, called **mycorrhiza**. In mycorrhizal symbiosis the plants provide carbon to the fungus, whereas the fungus provides essential nutrients to the plant, especially those of low mobility such as P and Zn, but also N, K, S, Mg, Ca and Cu. Mycorrhizas have also been implicated in other beneficial effects to plants, for example (i) improved water relations, (ii) increased protection against acidity, aluminium toxicity and heavy metals, and (iii) protection against root pathogens. Mycorrhizal symbiosis is therefore multifunctional (Newsham *et al.* 1995).

On the basis of the morphology of mycorrhiza, and plant and fungal taxa involved, mycorrhizal associations are divided in four broad categories: (i) arbuscular mycorrhiza, (ii) sheathing mycorrhiza, including ectomycorrhiza, ectendomycorrhiza, arbutoid mycorrhiza and monotropoid mycorrhiza, (iii) ericoid mycorrhiza, and (iv) orchid mycorrhiza. In this chapter we focus on the similarities between different mycorrhizal types rather than on their differences.

Regarding plant benefit, the emphasis is often on the increased biomass as compared to non-mycorrhizal plants. Such benefits are easily demonstrated when plants are grown singly in an experimental system. However, with increasing plant density (and increasing root density), mycorrhizal benefit declines. Plant benefit may also be expected to be most important in the seedling and establishment stages, and in the reproductive phase, notably regarding seed quality. Similarly, pathogens and root herbivores usually exert a larger impact on seedlings than on adult plants.

The mycorrhizal symbiosis also leaves legacies. Litter decomposition of arbuscular mycorrhizal plants is enhanced, but litter decomposition of ectomycorrhizal plants is retarded (see Section 9.8). Due to their high P-demand, almost all nitrogen-fixing plants develop mycorrhiza as well (see also Chapter 7).

9.2.3   Nitrogen-fixing bacteria

Among higher plants, two major types of symbiotic nitrogen-fixing associations are recognized: (i) the legume **symbiosis**, with bacteria collectively referred to as rhizobia; (ii) the actinorrhizal symbiosis of several trees and shrubs with actinobacteria belonging to the genus *Frankia*. The amounts of nitrogen fixed are variable. Highly productive, early-successional stands of nitrogen-fixing shrubs and trees, for instance *Acacia*, *Robinia* (**rhizobial symbiosis**), *Alnus*, *Hippophae* or *Myrica* (**actinorrhizal symbiosis**) can fix up to a few hundred kg-N·ha^{-1}·yr^{-1}. Such amounts are in excess of plant demand and uptake capacity (even though legumes have a nitrogen-demanding lifestyle; McKey 1994). In such cases the excess nitrogen is finally converted to nitrate, which subsequently leaches from the ecosystem. Nitrate loss is accompanied by acidification, the leaching of basic cations and lower phosphorus availability. During primary succession in Alaska, sites with *Alnus sinuata* are prone to rapid acidification, as a consequence of which *Alnus* disappears and conifers, notably *Picea sitchensis* and *Tsuga heterophylla*, establish. Due to this replacement, nitrogen-availability decreases and the subsequent build-up of recalcitrant litter further reduces nitrogen availability (Hobbie *et al.* 1998). The nitrogen-fixing *Alnus* thus acts as a driving force in primary succession.

Symbiotic nitrogen fixation is often prominent in early successional stages and declines in later successional stages. This pattern seems paradoxical as many late-successional ecosystems are still nitrogen-limited. The cost of resource acquisition explains the paradox. Nitrogen-fixing plants can acquire nitrogen from two sources, from the soil and through fixation of atmospheric nitrogen. Symbiotic nitrogen fixation costs about twice the amount of carbon as uptake of mineral nitrogen from the soil. Furthermore nitrogen-fixers have a high nitrogen-demand (or low nitrogen-use efficiency), and during succession such plants are outcompeted by plants with higher nitrogen-use efficiency. This replacement is more common in temperate than in tropical forests, where the higher nitrogen-availability (due to rapid decomposition and mineralization) allows potentially nitrogen-fixing legumes to maintain their nitrogen-demanding lifestyle.

9.2.4   Root-feeding soil fauna

Nematodes and insects are considered the most important root feeders. Earthworms sometimes feed on senescent roots, but they are better classified as saprotrophs. Based on feeding behaviour, root-dwelling insects can be classified as: (i) internal chewers that burrow into large roots or subterranean storage organs, (ii) external chewers that consume whole roots or graze on the root surface, and (iii) sap feeders that feed on the phloem or xylem through specific mouthparts (stylets) (Brown & Gange 1990). Like the sap-feeding insects, all plant-feeding nematodes have a hollow stylet that is used to suck vascular tissue or cytoplasm. Based on their feeding behaviour these nematodes can be classified as root hair or epidermal cell feeders, **ectoparasites**, **semi-endoparasites**, migratory **endoparasites**, and sedentary endoparasites (Yeates *et al.* 1993). The first two groups live in the rhizosphere and penetrate the root only with their stylet. The semi-endoparasites penetrate the root also with part of their body, whereas the endoparasites live (part of their life-cycle) inside plant roots. Migratory endoparasites move freely within the roots and can even exploit several host plants during their life-cycle. On the other hand, sedentary endoparasites affect the physiology of plant root cells thereby inducing the development of specific feeding cells that are used by the nematode to feed on. Besides their feeding relationship with plants some insects and nematodes act as disease vectors that contribute to the distribution of root pathogenic fungi and viruses.

9.2.5   Saprotrophic organisms

Other soil-dwelling animals, such as earthworms, enchytraeids, isopods, mites and springtails live on dead organic material or feed on the saprotrophic fungi and bacteria on these substrates. Because these animals can also feed on the mycelium of mycorrhizal fungi, they can impact on mycorrhizal functioning. Generally speaking, fungivorous animals prefer saprotrophic (and pathogenic) fungi over mycorrhizal fungi. Saprotrophic fungi and bacteria are the primary decomposers in ecosystems, with fungi being more common in more nutrient-poor and acidic sites where litter of lower decomposability (higher recalcitrance) is produced. Mycorrhizal fungi have no or very little saprotrophic activity.

9.3   The soil community as cause and consequence of plant community composition

The species composition of the plant community and that of the below-ground community are often correlated. This correlation raises the question to what extent the soil community is a cause or a consequence of plant community composition. This question is of major importance for restoration management, after cessation of agricultural practices. If suitable abiotic conditions have been created, but if the target vegetation has not established, should we then introduce the target plant species or should we introduce the soil biota?

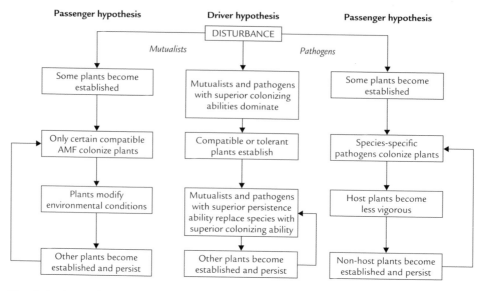

Fig. 9.1 A graphical model of two alternative mechanisms for compositional changes in the mutualistic or antagonistic soil community in interaction with compositional changes of the plant community. In the Driver Hypothesis changes in the soil community drive vegetation change, whereas in the Passenger Hypothesis changes in plant community composition result in changes in the soil community. AMF, arbuscular mycorrhizal fungi. (Modified after Hart *et al.* 2001.)

Hart *et al.* (2001) proposed a qualitative model to separate both mechanisms in the case of mycorrhizal fungi (Fig. 9.1 left). If mycorrhizal fungi are causes of vegetation dynamics (driver hypothesis), the presence of specific mycorrhizal fungi is required for the growth of specific plants. Plant species composition would then be a function of (and in principle predictable from) the presence of these fungi. If soil organisms are merely passive followers of vegetation dynamics (passenger hypothesis), specific plants are required to stimulate the growth of specific mycorrhizal fungi. The driver/passenger model is equally applicable to pathogenic organisms (Fig. 9.1 right).

In general, empirical tests for the driver/passenger hypothesis are scarce. Verschoor *et al.* (2002) investigated to what extent vegetation succession affected the root–parasitic nematode community, and to what extent that community accelerated or decelerated vegetation succession. They concluded that plant species-specific differences in tolerance to generalist root-feeding nematodes, rather than host selectivity of the nematodes, determine plant species replacement and hence plant species composition during succession. This example therefore supports the passenger hypothesis. Support for the passenger hypothesis was obtained for mycorrhizal fungi by Hausmann & Hawkes (2010). In their study, the order of plant establishment determined the assemblage of arbuscular mycorrhizal fungi in experimental grassland ecosystems. The authors noted that such priority effects could feed back to subsequent plant establishment.

However, feedback (mutual causation) does not get much attention in the driver/ passenger model, as it is monocausal and unidirectional. A further reason for the limited testing is that such models insufficiently take account of the dispersal limitations of plants and soil biota. Due to dispersal limitation neither plants nor soil biota seem sufficiently strong primary drivers.

9.4 Specificity and selectivity

9.4.1 Introduction

Interactions between plants and soil biota can be classified along a continuum from highly specific (private) to non-specific (shared). The gene-for-gene hypothesis, proposed for the co-evolutionary 'arms race' between plants and foliar pathogens, leads to conditions where only certain genotypes of the pathogen can colonize certain genotypes of a plant species. Such highly specific conditions are as yet unknown for rhizosphere organisms. However, selectivity seems to be common. A conceptual problem is that selectivity has different meanings and different underlying causes. Non-random association between plant and soil organisms under field conditions can be caused by joint abiotic preferences, by dispersal limitation, or by true selectivity. Shared preferences for the same abiotic factors was shown by Dumbrell *et al.* (2010) in a plant community along a pH gradient, where both plant and fungal species exhibited clear pH preferences. Next to pH, soil available phosphorus, nitrogen and C : N ratio of organic material are important niche axes for both arbuscular mycorrhizal fungi and plants, resulting in non-random association. True selectivity has been shown in the absence of major abiotic gradients. Vandenkoornhuyse *et al.* (2003) noted that co-occurring grass species harboured different mycorrhizal communities, and Helgason *et al.* (2007) showed non-random association between plants and fungi in a natural woodland. Only in the case of true selectivity may co-evolution between plants and their soil biota occur. This seems to be the case with myco-heterotrophic and mixotrophic plants (see Section 9.8). Selectivity can also be conceptualized as differential effects of the same soil organism on different plant species, as in the study by Klironomos (2003).

It has been argued that antagonistic associations show a larger degree of selectivity than mutualistic associations. The argument for that claim is that mutualistic symbionts that are adapted to rare partners or few species, gain smaller benefits than species that are more promiscuous. A counter argument could be that sharing symbionts leads to facilitation and hence increases interspecific competition. Empirical data (and subsequent theoretical models) suggest that in mutualisms selectivity is also widespread. The issue of specificity has been obscured by the fact that the species level has been considered as the relevant unit. As the number of worldwide species of rhizobia (itself a polyphyletic assemblage of two groups of proteobacteria, in all around 65 species) and of arbuscular mycorrhizal fungi (around 200 described species) is much lower than that of nitrogen-fixing legumes (16 000) or arbuscular mycorrhizal plants (more than 200 000), it is argued that specificity or selectivity must be low. However, genetic variation within both rhizobia and arbuscular mycorrhizal fungi suggests much more scope for selectivity on lower taxonomic levels. For ectomycorrhizal fungi, however, the situation is different with around 25 000 species of ectomycorrhizal fungi, associating with around 8000 ectomycorrhizal plant (mainly tree) species.

Shared mutualistic symbioses do also not necessarily equalize competitive abilities among plants; therefore shared symbioses can both promote and decrease floristic diversity (by reducing or enhancing competitive replacement), whereas private symbioses may also both increase and decrease differences between plant species. Whether such interactions increase or reduce plant species diversity depends primarily on the plant's responsiveness to the soil community. If dominant plants are more responsive to the mutualistic species or less negatively affected by the antagonistic members of the soil community, these species will increase in dominance and reduce plant species richness. If the dominant species are less responsive to mutualistic organisms or more affected by antagonistic species than the subordinate plant species, an increase in plant species richness will result (Urcelay & Díaz 2003).

9.4.2   Specificity of rhizobia

Only certain rhizobia are compatible, i.e. have the ability to induce nodulation and fix atmospheric nitrogen, with specific legumes. Legume selectivity is determined on the level of nodulation, not on the level of effective nitrogen fixation. Therefore, the legume × rhizobia association observed in the field is not the most productive one from the plant perspective. Local co-adaptation between rhizobia and legumes is also affected by plant neighbours. Various genotypes of *Trifolium repens* that co-occurred in a meadow were individually associated with (and adapted to) a specific neighbouring grass species (Expert *et al.* 1997). Chanway *et al.* (1989) showed that the compatibility between *Trifolium repens* and *Lolium perenne* depended on rhizobia: in their absence, the legume–grass compatibility was lost. Lafay & Burdon (1998) studied the diversity of rhizobia that nodulated on 32 native legume shrubs at 12 sites in south-eastern Australia. The occurrence of rhizobia on legume species was non-random, although distribution overlaps were common. True host specificity was not observed, while many legume species were selective. Part of this selectivity was due to the fact that several rhizobia were site-specific. Selectivity was observed for only three species – *Acacia obliquinervia*, *Goodia lotifolia* and *Phyllota phylicoides* – where the dominant rhizobium isolated was different from the most common rhizobium from the site. Specificity or selectivity of the actinorrhizal symbiosis is lower than that of the rhizobial symbiosis, but actinorrhizal plants in dry habitats are nodulated by only a limited number of *Frankia* strains (Benson & Dawson 2007).

9.4.3   Specificity of mycorrhizal fungi

Field studies from a diversity of habitats have now generally confirmed that associations between plant species and species of arbuscular mycorrhizal fungi are non-random. These observations conflict with the earlier claim that these fungi are generalists. However, it turned out that only a few species of arbuscular mycorrhizal fungi (in fact those that sporulate prolifically and therefore are most amenable to experiments under controlled conditions) are generalists, and most species show host selectivity. Johnson *et al.* (2010) showed geographic mosaics in the interaction between arbuscular mycorrhizal fungi and *Andropogon gerardii*. Local, co-evolved ecotypes of plants and fungi were more beneficial to the plant than combinations of plants or fungi that did not originate from the same habitat. They suggested that this geographic structure could have been driven by differential resource limitation by either nitrogen or phosphorus. Ji *et al.* (2010) equally showed selectivity of arbuscular mycorrhizal fungal assemblages (called ecological matching) and this was also driven by edaphic properties. Only in a few cases has true selectivity (co-adaptation and co-evolution, rather than joint adaptation to a major environmental factor) been conclusively shown (Öpik *et al.* 2010). Their meta-analysis also did not indicate a significant correlation between fungal species richness and plant species richness.

Functional selectivity (non-random benefit between plant and fungal species) has more often been shown. The relevant question for plant communities is whether such non-random combinations generally result in maximum plant benefit (as in the study by Johnson *et al.* 2010). Van der Heijden *et al.* (1998) compared the growth of 11 different plant species inoculated with four different species of arbuscular mycorrhizal fungi or with a mixture of these species. They noted that specific fungal species and the species mixture had significantly different effects on plant performance in some plant species but not in others. Plants that were more responsive to mycorrhizal fungi (compared to a non-mycorrhizal control) showed a higher mycorrhizal species sensitivity, variation in effect of the different fungal species. However, Klironomos (2003) was unable to confirm the positive relationship between mycorrhizal responsiveness and mycorrhizal species sensitivity. A positive relation between both parameters could imply that less abundant species (that are generally more responsive to mycorrhiza) have increased mycorrhizal species selectivity; if so, a large role for mycorrhizal fungal species composition in determining the success of rare species can be assumed. Alternatively, the positive relation between mycorrhizal responsiveness and mycorrhizal species sensitivity could be an artefact due to the very poor performance of responsive and sensitive plants with one non-beneficial fungus (an instance of the negative selection effect). In the study by van der Heijden *et al.* (1998) mycorrhizal species sensitivity was mainly due to the effects of one fungal species that on average was significantly less beneficial than the other fungal species. The study by Moora *et al.* (2004) in which a common species of *Pulsatilla* was compared with a rare one, indicated that the more common species (*P. pratensis*) benefitted more from its mycorrhizal fungus than the rare species (*P. patens*). Also the documented negative feedback driven by arbuscular mycorrhizal fungi (see Section 9.5.2) suggests that plants often do not associate with the fungus that is most beneficial to them.

9.4.4 Specificity of pathogens and root herbivores

Selectivity in soil-dwelling pathogens varies widely, from species being host-restricted to species with very wide host ranges. The pathogen *Phytophthora cinnamomi*, a species originally endemic to eastern Australia, has been introduced in many places around the world. As in its original area, the pathogen can cause rapid vegetation changes; within 5 years a closed *Eucalyptus* woodland with a dense understorey was transformed into an open woodland with an understorey dominated by the pathogen-resistant sedge *Lepidosperma concavum* (Weste 1981). Bishop *et al.* (2010) described how over 15 years in a *Banksia* woodland, where the pathogen was introduced, the dominant plant species *Banksia attenuata*, *B. ilicifolia* and *Daviesia flexuosa* strongly declined, while *Anarthria prolifera* significantly increased. The authors suggested that the strong species decline would make recovery of the original vegetation unlikely.

In other cases, root pathogens could drive cyclic succession. Conifers in the Pacific North-West are susceptible to the fungus *Phellinus weirii*, but susceptibility differs between conifer species. Pines (*P. contorta*, *P. monticola*) are much less affected than *Tsuga mertensiana*. The fungal pathogen therefore drives cyclic succession between pines and hemlock (Hansen & Goheen 2000).

Specific root herbivore–plant interactions result in similar outcomes. An example from natural vegetation in California is the caterpillar of the ghost moth *Hepialus californicus* that feeds on the roots of *Lupinus arboreus*. These caterpillars are largely monophagous and feed, when young, on the exterior of lupine roots, thereafter boring inside the roots. Field surveys showed a strong positive correlation between plant death rates and caterpillar densities inside roots, providing strong evidence that the caterpillar affected vegetation composition and dynamics (Strong 1999). The ghost moth itself is affected by an entomopathogenic nematode *Heterorhabdites marelatus*. As *L. arboreus* is a nitrogen-fixer, root herbivores and nematodes have cascading effects on ecosystem processes (Preisser 2003).

9.5 Feedback mechanisms

9.5.1 Introduction

Plant growth in its own soil ranges from significantly better to significantly worse compared to sterilized soil, demonstrating that plants build up their specific assemblage of mutualists and antagonists. But in order to translate the implication of such effects for vegetation processes, it is imperative to know how plant species perform in the soils that have been modified by another plant species, because it is differential performance of plants in their own ('home') and alien ('away') soil that determines the outcome of interactions between plants and their possible co-existence. Such studies can compare plant performance locally (home and away soils of neighbouring plant species) and over large geographical scales (for an understanding of the role that soil biota play in the success of invasive plants, see Section 9.6).

Following Bever *et al.* (1997), we can distinguish between **positive feedback** (when plants in their own soil perform better than in alien soil) and **negative feedback** (when the opposite happens). The impact of feedbacks has been repeatedly described (but without a formal theoretical framework) in the ecological literature. Negative feedbacks are implied in the Janzen-Connell hypothesis to explain maintenance of very high tree species richness in the tropics (negative density and distance dependence of seed and seedling performance) and the Red Queen hypothesis. Positive feedbacks have been described under concepts such as Pathogen Spillover or Local Accumulation of Pathogens. For invasive plants, where the native species is subject to negative feedback (replacement), but the invader escapes from that (and hence shows positive feedbacks), mechanisms such as Suppression of Mutualists and Enemy Release have been proposed.

Feedback is demonstrated in an experiment where two plant species are grown in two soils ('home' versus 'away'; own versus alien). Feedback occurs when the plant × soil biota interaction is a significant source of variation in an ANOVA. The magnitude of the feedback can be calculated as the natural log of the ratio (sum of the performance of both plants in their own soils divided by the sum of their performances in alien soil). This parameter is symmetrical around zero and allows comparison of the magnitude of feedbacks. A positive number indicates positive feedback, a negative number negative feedback. Positive and negative feedbacks are illustrated in Fig. 9.2 (Bever 2003). The figure shows that both mutualists and antagonists can generate positive and negative feedbacks. Note that in most experiments the combined feedback of mutualists and antagonists is recorded. Studies involving feedbacks with only subsets of the soil community (pathogen complexes, assemblages of mycorrhizal fungi) or even tests of effects of individual species are much less common.

9.5.2   Negative feedbacks

Negative feedbacks occur when plants create a rhizosphere that is less beneficial or more detrimental to conspecific than to heterospecific plants. Negative feedback can be easily envisaged for root pathogens. Van der Putten *et al.* (1993) described plant–soil biota interactions during primary succession in the coastal dunes of the Netherlands. The clonal species *Ammophila arenaria* plays an important role in early soil development by fixation of windblown beach sand. The species deteriorates, unless it extends its root system annually into a new layer of fresh windblown sand. In more fixed dunes, a species-specific pathogen complex will develop resulting in reduced growth and vigour of *A. arenaria*. Later successional species, on the other hand, are not or are less affected by this pathogen complex and will replace *A. arenaria*. Subsequently, in the rhizosphere of these later-successional plant species, species-specific pathogen complexes will develop that in their turn decrease the vigour of these plant species. Such pathogen complexes generate directional succession when the pathogens are persistent and plants are resistant to or tolerant of the pathogen complex of the previous species but not vice versa. If, however, the pathogen complex is not persistent, cyclic succession will result. Reciprocal transplant pot experiments with *A. arenaria* and the later successional graminoids *Festuca rubra* ssp. *arenaria* and *Carex arenaria*, showed that each of these species produced more biomass in pre-successional soil than in their own soil. The pathogen complex in this primary succession comprises mainly plant-feeding nematodes and pathogenic fungi.

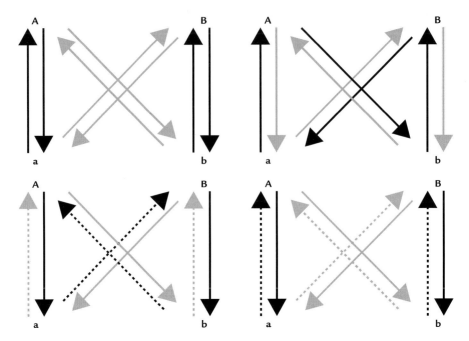

Fig. 9.2 Mechanisms of positive and negative feedbacks between plants (capital A and B) and the soil community (lower case a and b): upper left, positive feedback between plants and mutualists; upper right, negative feedback between plants and mutualists; lower left, positive feedback (indirect negative feedback) between plants and soil pathogens; lower right, negative feedback between plants and soil pathogens. Arrows, direction of effects; continuous lines, beneficial effects; dashed lines, harmful effects. Black lines indicate a stronger positive or negative effect than grey lines. Positive feedback occurs if either benefits for a plant and its mutualists are symmetrical, or if damage to a plant and benefits for pathogens are asymmetrical. Negative feedback occurs if damage to a plant and benefits for its pathogens are symmetrical, or if benefits for a plant and its mutualists are asymmetrical. (Modified after Bever 2003.)

However, so far, laboratory and field experiments in which field densities of specific ectoparasitic and endoparasitic nematodes and pathogenic fungi were added to sterilized soil, did not result in similar large growth reductions as found for field soils. This suggests that the organisms involved and their interactions have not yet been completely identified (de Rooij-van der Goes 1995). Negative feedbacks have been described from many plant communities, ranging from species-poor to very species-rich communities as in tropical rain forests.

Negative feedbacks between plant and soil communities have been proposed as a potential mechanism to maintain species richness, the Janzen–Connell hypothesis (Hyatt *et al.* 2003). For temperate grasslands, Petermann *et al.* (2008) stated that Janzen–Connell effects were sufficiently strong to maintain plant species diversity. Mangan *et al.* (2010) reported negative feedbacks among seedlings of six tropical rainforest tree species on Barro Colorado Island, Panama.

All species showed a negative feedback. Interestingly, the more common species were less sensitive to feedback, an observation that Klironomos (2002) had made before (see Section 9.5.5). Kiers *et al.* (2000) observed a negative feedback between two tropical rainforest species, *Dipteryx panamensis* and *Anacardium excelsum*. Both species grew better in soils with arbuscular mycorrhizal fungi obtained from the heterospecific tree than from a conspecific tree. Their data suggest that arbuscular mycorrhizal fungi generated negative feedback. However, their study included the effect of the whole rhizosphere community, not that of the mycorrhizal community only. The meta-analysis by Hyatt *et al.* (2003) suggested that the Janzen–Connell hypothesis was less often confirmed than originally proposed. The authors also showed that there was more support for the hypothesis in tropical forests than in temperate forests (for a possible explanation see Section 9.8).

An example of negative feedback generated by arbuscular mycorrhiza was provided by Bever (2002). Arbuscular mycorrhizal fungal species *Scutellospora calospora* formed more spores when associated with *Plantago lanceolata* than when associated with *Panicum sphaerocarpon*. But growth of *P. lanceolata* was less promoted by *S. calospora* than by the fungal species *Acaulospora morrowiae* and *Archaeospora trappei* that accumulated with *P. sphaerocarpon*. These latter fungal species stimulated growth of *P. lanceolata* more than that of *P. sphaerocarpon*. This interaction resulted in a decline in benefit received by *P. lanceolata*, allowing *P. sphaerocarpon* to increase. An increase in the grass species has an indirect beneficial effect on the growth of *P. lanceolata*, thereby contributing to the co-existence of both competing species.

Soil invertebrate fauna also generates feedback. Root-feeding nematodes and larvae of click-beetles selectively suppressed early-successional, dominant plant species and enhanced subordinate species and species of later succession stages, thereby increasing plant species diversity and accelerating plant succession (de Deyn *et al.* 2003).

In many cases, negative feedback is reciprocal, that is, both plants perform better in the alien soil than in their home field soil. For invasive plants, the indigenous species performs better in the alien soil, whereas the exotic plant is not affected by the soil community in its new habitat. The invader shows negative feedback in its native habitat, but positive feedback when introduced. *Prunus serotina* is a North American tree that has become invasive in Europe. In its original area the tree does not rejuvenate, or rarely does, under its own canopy, due to the build-up of specific pathogens (species of the oomycete *Pythium*); but in Europe these specific pathogens are absent and the species successfully regenerates under its own canopy. Consequently, where in America plants grow widely spaced, they may form dense stands in Europe (Reinhart *et al.* 2003). In sandy areas in the Netherlands, dense thickets of *P. serotina* threaten native vegetation in nature reserves. Removal of this plant is very difficult (see Chapter 13 for invasive species). For the genus *Acer* the same phenomenon was described, where a North American species escaped from negative feedback in Europe, while a European species escaped from negative feedback in North America (Reinhart & Callaway 2004). These examples fall under the class of Enemy Release. Suppression of Mutualists has been recorded for the invasive species *Alliaria petiolata* in North America, which suppresses both arbuscular and ectomycor-rhizal plants (see Section 9.6).

9.5.3 Positive feedbacks

Positive feedbacks occur when plants create their rhizosphere environment whereby they outperform other species. This feedback occurs if plants are associ-ated with rhizobia or mycorrhizal fungi from which they derive the largest benefits. (Note, however, that it is commonly assumed that plants select their most beneficial symbionts, but that this has hardly been demonstrated.) Positive feedback has been described for invasive ectomycorrhizal plants (*Pinus* in the southern hemisphere) and invasive actinorrhizal and rhizobial plants. On a more local scale, positive feedbacks are probably responsible for the development of monodominance by ectomycorrhizal trees in tropical forests (see Section 9.8).

Indirect positive feedbacks with pathogenic soil organisms can also occur. Olff *et al.* (2000), in a study of cyclic succession in grasslands in the Netherlands, found that this grazed vegetation consisted of shifting mosaics of patches where the grass *Festuca rubra* and the sedge *Carex arenaria* are dominant. In the rhizo-sphere of *C. arenaria* nematodes develop that reduce the performance of the sedge and allow the grass to become dominant. While the grass performed more poorly in non-sterile compared to sterile soils, the authors noted that the grass maintained the nematodes that are detrimental to the sedge, the enemies of its competitor (positive feedback). Nematode decline (which is responsible for maintenance of both species) was not driven by the grass, but by the addition of fresh (nematode-free) sand due to the digging activities of rabbits and ants. Addition of fresh sand could also have reduced arbuscular mycorrhiza (which is beneficial for the grass but not for the sedge), but that was not investigated. Mangla *et al.* (2008) demonstrated that the invasive plant *Chromolaena odorata* in India accumulated a generalist pathogen (*Fusarium* cf. *semitectum*) from which its competitors suffered from more than *Chromolaena* itself. This case falls in the class of Local Accumulation of Pathogens or Pathogen Spillover. Also the success of *Ammophila arenaria* in North America is due to its ability to accumulate pathogens in its rhizosphere from which it suffers less than its com-petitors. Interestingly, *A. arenaria* is invasive in several parts of the world. In New Zealand the species became a successful invader due to absence of specific pathogens (Enemy Release), but Enemy Release could not be demonstrated in North America (Beckstead & Parker 2003; Local Accumulation of Pathogens) and South Africa, where the species suffers from the pathogens from its native competitor *Sporobolus virginicus* (also a case of Local Accumulation of Patho-gens). Apparently, specific feedback mechanisms cannot be linked to plant species or plant traits (Inderjit & van der Putten 2010).

9.5.4 Feedbacks through saprotrophic organisms?

The same question can be addressed regarding decomposition: do plant species selectively enhance those saprotrophs that specialize in more rapid decomposi-tion of their own litter, thereby generating a positive feedback? To approach this

Plate 6.1 Application of a sheep dummy to collect retention data in a standardized way (e.g. number of seeds 'collected' in the fleece over a certain distance and at a distinct time in the season)

Vegetation Ecology, Second Edition. Eddy van der Maarel and Janet Franklin.
© 2013 John Wiley & Sons, Ltd. Published 2013 by John Wiley & Sons, Ltd.

Plate 6.2 Standardized measurement of the attachment capacity and the retention potential of seeds with a machine which simulates the movement of an animal. Demonstration by the first author. Fleeces/furs of different animals can be fixed (According to Tackenberg *et al.* 2006)

Plates 11.1 and 11.2 Jena experiment – an aerial view. One of the largest designed ecological experiments in the world. The species composition on individual plots is kept by weeding. **11.1** Survey. **11.2** Detail from another angle from the left part seen on Plate 11.1 (Photos courtesy of Winfried Voigt)

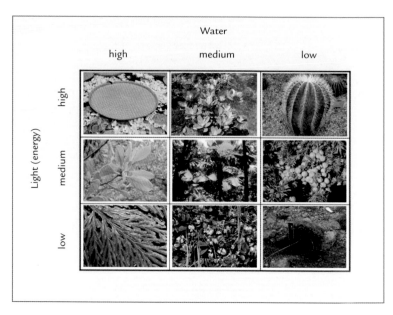

Plate 12.1 Different whole-plant PFT syndromes, subjectively positioned along gradients of light (energy) and moisture. Left to right: *Victoria regia* (Amazon basin); *Metrosideros* sp. (Philippines); *Echinocactus* sp. (Mexico); mangrove *Lumnitzera littorea* (Indomalesia); phanerophytic swamp fan palm, *Licuala ramsayi* (Tropical North Australia); *Juniperus communis* (Fennoscandia); fern *Selaginella* sp. (Indomalesia); *Vaccinium vitis-idaea* (boreal region); cushion plant, *Azorella macquariensis* (subantarctic Macquarie Island). Dominant functional traits are reflected in life-form, chlorophyll distribution, notably leaf size and inclination and green stem.

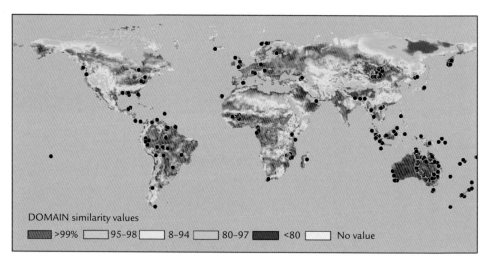

Plate 12.2 Distribution of 1066 VegClass sites (40 × 5 m transects) used in the recording of *modal* PFTs, taxa and vegetation structure. Degree of environmental coverage by all sites is indicated via a DOMAIN environmental similarity map (Carpenter *et al*. 1993) based on elevation, total annual precipitation, minimum temperature of coldest month and total annual actual evapotranspiration.

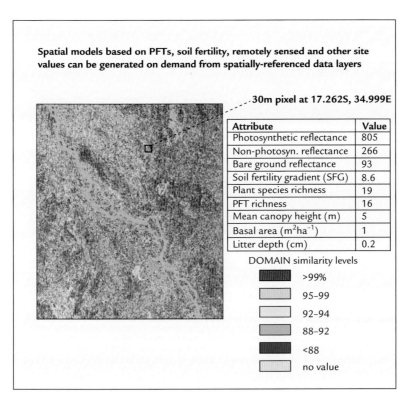

Spatial models based on PFTs, soil fertility, remotely sensed and other site values can be generated on demand from spatially-referenced data layers

30m pixel at 17.262S, 34.999E

Attribute	Value
Photosynthetic reflectance	805
Non-photosyn. reflectance	266
Bare ground reflectance	93
Soil fertility gradient (SFG)	8.6
Plant species richness	19
PFT richness	16
Mean canopy height (m)	5
Basal area (m^2ha^{-1})	1
Litter depth (cm)	0.2

DOMAIN similarity levels

>99%
95–99
92–94
88–92
<88
no value

Plate 12.3 DOMAIN spatial model of landscape similarity values in the lower Zambezi river basin, Mozambique, at 30-m pixel resolution using ordinated values of Landsat satellite imagery, *modal* PFT and plant species richness, PFT complexity, soil properties and vegetation structure. (From Gillison *et al.* 2012.).

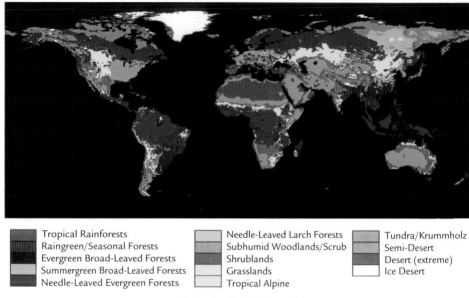

Tropical Rainforests	Needle-Leaved Larch Forests	Tundra/Krummholz
Raingreen/Seasonal Forests	Subhumid Woodlands/Scrub	Semi-Desert
Evergreen Broad-Leaved Forests	Shrublands	Desert (extreme)
Summergreen Broad-Leaved Forests	Grasslands	Ice Desert
Needle-Leaved Evergreen Forests	Tropical Alpine	

Climatically Estimated
PNV-Based World Terrestrial Biomes

Plate 15.1 World pheno-physiognomic vegetation pattern predicted from climate. The map shows the pheno-physiognomical vegetation types of Table 15.7 (groupings in left column), as predicted by climatic envelopes for the individual types (right column of Table 15.7). (From Box 1995b).

question the same methodology has to be applied. Ayres *et al.* (2009) demon-strated a positive feedback (a home field advantage), especially in forest ecosys-tems. The feedback was stronger when plants of different biomes (forest versus grassland) were compared, but the very small data set does not yet allow firm conclusions. Positive feedback on litter decomposition for three species of *Not-hofagus* in a mixed old-growth forest in South America was demonstrated by Vivanco & Austin (2008). According to Milcu & Manning (2011) this positive feedback is a function of litter quality, with low-quality litters showing stronger positive feedback. However, that proposal fits poorly with the litter-driven feed-back models for monodominant forests (see Section 9.7). A positive feedback through saprotrophic organisms implies that litter quality is not only an intrinsic property of that plant material, but is co-determined by the organisms that degrade it (Strickland *et al.* 2009).

Strong pleas for the importance of negative feedback have also been made. Mazzoleni *et al.* (2007) listed cases of negative plant–soil feedbacks that were driven by litter autotoxicity. In such cases, heterospecifc plants (provided they suffer less from litter toxicity; the authors claim that autotoxicity is more prevalent than generic phytotoxicity of decomposing litter) could replace plant species and thereby maintain diversity. We return to the issue of combina-tions of feedbacks by rhizosphere and saprotrophic organisms when discussing the origin and maintenance of monodominant forests in the tropics (see Section 9.8).

9.5.5   Evaluation

How strong is the evidence for positive and negative feedbacks structuring plant communities? A conceptual co-evolutionary framework was proposed by Thrall *et al.* (2007). The authors proposed the following hypotheses:

1   in species-poor vegetation, negative feedbacks are more likely than positive feedbacks;
2   in early stages (of primary succession), positive feedback is more likely than negative feedback;
3   under nutrient-poor conditions, positive feedback is more likely than nega-tive feedback;
4   introduced species become invasive due to the absence of a negative feedback in their introduced range.

Support for the fourth hypothesis has been given above. It seems that hypoth-eses 2 and 3 essentially refer to the same phenomenon. Kardol *et al.* (2006) showed that in a reversed succession (succession after agricultural cessation towards a more nutrient-poor and species-rich vegetation) negative feedbacks were important in the earlier stages and that gradually positive feedbacks became more prominent. Otherwise, these hypotheses have not yet been rigorously tested.

A meta-analysis by Kulmatiski *et al.* (2008) indicated that negative feedbacks were more common and of larger magnitude than positive feedbacks. This prevalence could be caused by a higher degree of specificity and selectivity by antagonists than by mutualists. The study by Fitzsimons & Miller (2010) in a tallgrass prairie in North America showed overall significant feedback, but a test of feedback that only involved the role of mycorrhizal fungi showed that the feedback effect was not significantly different from zero.

The ultimate question pertains to what extent negative and positive feedbacks explain plant species diversity. Positive feedbacks could augment local competitive ability, but this local success decreases species richness on small spatial scales (though not necessarily on larger scales, if dispersal limitation of the below-ground mutualists is important). Negative feedbacks allow co-existence of two plant species, but very few tests have been executed whether the mechanism is equally likely in the case of three or more interacting species. Negative feedbacks explain species richness when species rankings after accounting for feedback effects are intransitive; that is, every species must be competitively superior to all other species under one heterospecific plant species. That condition assumes long-term legacies of the soil-borne community over various generations of plants. It also assumes that the strength of the negative feedback is not stronger for rare species than for common species. However, the two studies carried out to date that investigated this property showed the opposite. Both Klironomos (2002) for grassland and Mangan *et al.* (2010) for tropical rainforest, observed that rare species were more sensitive to negative feedbacks; and suggested that the negative feedbacks were responsible for their rarity. It is therefore not very plausible that negative feedbacks provide the major explanation for the existence of species-rich forests, a conclusion that echoes an earlier conclusion by Hubbell (1980) that the Janzen–Connell mechanism could not explain co-existence of large species numbers except at unrealistically high and uniform spacing distances. Negative feedbacks are therefore more important in species-poor vegetation during primary succession on coastal dunes than in rainforests, as predicted by hypothesis 1 (Thrall *et al.* 2007).

9.6   Soil communities and invasive plants

The existence of soil community feedbacks is probably an important factor that determines why some exotic plants become invasive (see Chapter 13). Both pathogens and mutualists are involved. For pathogens, exotic plants could have been introduced without their below-ground pathogenic community (Enemy Release). Consequently, the general negative feedback between two plant species is broken if the exotic plant outperforms a native plant both in its 'own' soil and in the soil of the native plant (due to the build-up of a pathogenic soil community of the indigenous plant).

Exotic plant invasion has been observed for nitrogen-fixing symbioses. A good example of this is the success of *Myrica faya* in Hawaii. Following its introduction, soil characteristics changed and indigenous plants were unable to compete with exotics under conditions of higher nitrogen availability. Interestingly, *Myrica faya* also profits from enhanced seed dispersal by introduced birds, ultimately causing what has been described as invasional meltdown (Vitousek & Walker

1989). Invasional meltdown has also been proposed for rhizobial plants. Rodríguez-Echevarría (2010) described invasion by the Australian *Acacia longifolia* in the western Mediterranean. With the invasive tree, exotic rhizobia were introduced and these exotic rhizobia nodulated on native legumes and replaced the native rhizobia. Because the exotic rhizobia were less effective in promoting growth of the native legumes, the below-ground and above-ground invasion may have further cascading effects. For ectomycorrhizal fungi and their associated trees (especially the genus *Pinus*, but also *Eucalyptus*) invasional meltdown has also been reported, resulting in substantial ecosystem changes (Chapela *et al.* 2001). For arbuscular mycorrhizal fungi, invasional meltdown has not yet been mentioned. The opposite effect, Suppression of Mutualists, has been noted for *Alliaria petiolata*. This species has been introduced in North America, where it effectively suppresses both arbuscular and ectomycorrhizal plants (Stinson *et al.* 2006; Wolfe *et al.* 2008).

Feedbacks may be especially noticeable with invasive plants, due to a disruption of the plant and the below-ground organisms that co-evolved with them. However, in their new habitat antagonists may again gradually build up in the rhizosphere. Diez *et al.* (2010) described how negative feedbacks developed over time in plants that had colonized New Zealand for different time periods. Lankau (2011) recorded how soil microbial communities recovered in sites that were colonized a long time ago by *Alliaria petiolata*.

A major conclusion from these studies is that various feedback mechanisms are responsible for the success of invasive plants and that plant species and plant traits do not yet allow a predictive theory indicating the mechanism that successfully predicts whether a plant becomes invasive in a new habitat.

9.7 Mutualistic root symbioses and nutrient partitioning in plant communities

Next to feedbacks, soil biota could have a major impact on interactions between plant species and thereby influence plant community composition through a second mechanism – differential use of soil resources – particularly partitioning of various sources of nitrogen and phosphorus through mutualistic root symbioses (Bever *et al.* 2010). This mechanism could explain how plants, which basically need the same nutrients in relatively constant ratios, could co-exist. The clearest example is nitrogen fixation by rhizobia and certain actinobacteria. This mechanism explains the co-existence of legumes in species-rich vegetation and also provides a mechanism through which this increased diversity translates into higher plant productivity (positive selection effect). Could mycorrhizal associations also contribute to more narrow abiotic niches through differential access to different forms of nitrogen and phosphorus? In soils, nitrogen available to plants consists of the cation ammonium, the anion nitrate and dissolved organic nitrogen. It has been suggested that ectomycorrhizal and ericoid mycorrhizal plants have much larger access to organic nitrogen (a suggestion based on a presumed substantial saprotrophic capacity of these fungi). Support for that suggestion was found in different ^{15}N signatures of plants with different mycorrhizal habits. However, while differential signatures are consistent with, they are not a demonstration of access to different nitrogen pools with different ^{15}N signatures, because fractionation processes after uptake equally generate different signatures. Also, for phosphorus Turner (2008) hypothesized that plants with different kinds of mycorrhizal associations have access to different soil phosphorus pools. However, there is very little empirical support for these hypotheses.

Stoichiometric considerations (nitrogen-to-phosphorus ratio) could also result in increased niche differentiation between plants of different mycorrhizal types. Read (1991) noted that, alongside with the gradient where tropical ecosystems are usually P-limited and temperate and boreal ecosystems N-limited, there was a gradient with arbuscular mycorrhizal associations being prominent in the tropics and ectomycorrhizal and ericoid mycorrhizal associations being prominent in temperate and boreal regions. However, that rule has important exceptions, such as the occurrence of extensive tracts of ectomycorrhizal forests in the humid tropics and savanna (see Section 9.8).

This hypothesized mechanism raises the question of whether classifying plant species according to their mycorrhizal association adds a new dimension to classifications of plant functional types (Cornelissen *et al.* 2001). Such a classification links mycorrhizal type, plant functional type, litter quality, decomposition and mineralization characters, and N:P stoichiometry. A model is presented in Fig. 9.3. Ectomycorrhizal and ericoid mycorrhizal plants generally grow in soils and on humus profiles that are rich in organic matter and where decomposition and mineralization are hampered. The associated mycorrhizal fungi of these plants have the enzymatic ability to access some organic nutrients. On the other hand, arbuscular mycorrhizal fungi and plants have less access to these sources. Under external forcing (e.g. atmospheric nitrogen deposition), the higher availability of mineral nitrogen then allows arbuscular mycorrhizal plants

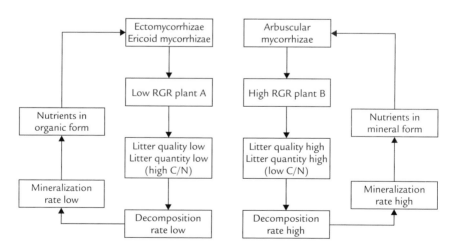

Fig. 9.3 Positive feedbacks between mycorrhizal type, plant functional type, quality of litter produced and the resulting decomposition and mineralization rates.

to outcompete ericoid mycorrhizal plants. This replacement has been observed in heathland in north-western Europe where arbuscular mycorrhizal grasses (*Deschampsia flexuosa, Molinia caerulea*) have replaced ericoid mycorrhizal shrubs (*Calluna vulgaris, Erica tetralix*). Because mycorrhizal type is part of a larger set of properties (mycorrhizal type correlates with litter decomposability), grass encroachment results in litters with higher decomposability and hence higher nutrient cycling, which enlarges the mineral nitrogen pool, resulting in a positive feedback.

A similar mechanism may be relevant for competition between ectomycorrhizal and arbuscular mycorrhizal trees in tropical forests. Plant ecologists do not only wish to explain the huge tree species diversity in most tropical forests (for which the Janzen–Connell hypothesis as an example of negative feedback has been considered a major mechanism), but also need to explain the existence of monodominance, in forests in which one species or a few related species (from the same family) dominate the canopy. Such monodominant forests occur in all continents: in south-east Asia forests are dominated by Dipterocarpaceae, in the Guineo-Congolian region forests by Caesalpinioideae (Fabaceae), and in tropical South America forests by *Dicymbe corymbosa* (also belonging to Caesalpinioideae). Apparently, soil properties do not differ between monodominant forests and adjacent patches of more species-rich forests. This has resulted in the suggestion that biotic factors (and especially below-ground factors) play a major role in causing monodominance (Peh *et al.* 2011).

Many, though not all, monodominant forest trees form ectomycorrhiza. This correlation has led to the hypothesis that the ectomycorrhizal habit is causally relevant for monodominance. Monodominance would be the result of positive feedbacks. Positive feedbacks occur when trees preferentially regenerate under their own canopy (because inoculum limitation of ectomycorrhizal fungi is more important than density-dependent mortality as implied in the Janzen–Connell hypothesis). A study by Norghauer *et al.* (2010) indicated that in an African rainforest, the arbuscular mycorrhizal *Oubangia alata* showed strong density dependence, whereas the ectomycorrhizal *Microberlinia bisulcata* was unaffected by density. Many of these ectomycorrhizal trees show mast fruiting, by which they satiate the seed predators. While density-dependent seed predation and mortality occurs, the massive seed production close to the trees (in combination with mycorrhizal inoculum limitation outside the crown circumference and an effective ectomycorrhizal network, see Section 9.8) would not result in over-compensation, which is necessary for negative feedback.

Interestingly, there is also some evidence that in these monodominant forests, the litter feedback is negative rather than positive (McGuire *et al.* 2010). Irrespective of litter quality, litter decomposes more slowly in the monodominant stands of *Dicymbe corymbosa* than in neighbouring more species-rich forest stands, suggesting that micro-organisms have been selected to slow down decomposition and mineralization processes (contrary to the positive litter feedback of Section 5.4). This phenomenon has been known in the literature as the Gadgil effect (Gadgil & Gadgil 1971) – the negative impact of ectomycorrhizal fungi on litter breakdown by saprotrophic fungi. The existence on the same soils of species-poor forests dominated by ectomycorrhizal trees next to species-rich forests dominated by arbuscular mycorrhizal trees provides important opportunities to test hypotheses of the role of soil organisms in determining plant community assembly and composition. A major question is whether it is coincidental that in ectomycorrhizal forests a positive rhizosphere feedback and a negative litter feedback co-occur, whereas the opposite is the case in forests with arbuscular mycorrhiza.

9.8   Mycorrhizal networks counteracting plant competition?

Laboratory experiments have shown that most plant × fungus combinations are functional. Apparently, specificity or selectivity of these mycorrhizal fungi is low. Such fungal species with low selectivity have the ability to connect different plant species through a common mycorrhizal network (CMN; Selosse *et al.* 2006). The existence of such links has been demonstrated repeatedly and it is clear that nutrients and carbon could move through these networks. But the ecological relevance (quantities moved, control over fluxes, implications for plant community dynamics) of these networks is still contentious.

For carbon fluxes, the consensus is that the quantities moved are ecologically unimportant and that the carbon remains under fungal control. One group of plants, known as mycoheterotrophic plants, form an exception. Mycoheterotrophic plants are achlorophyllous plants that share a small number of mycorrhizal fungi with other plant species in the same plant community. Such plants do not have the ability to photosynthesize and therefore there is a carbon flow from a photosynthetic mycorrhizal plant through the fungus to the mycoheterotrophic plant. Mycoheterotrophic plants associating with ectomycorrhizal trees belong to (i) Monotropoideae (Ericaceae), (ii) several orchid species (*Neottia nidus-avis, Corallorhiza, Epipogium*) and (iii) the moss *Cryptothallus mirabilis*. Mycoheterotrophic plants that associate with arbuscular mycorrhizal plants occur in various tropical ecosystems, and there is no physical connection between the mycoheterotrophic plant and the photosynthetic plant upon which it ultimately parasitizes for carbon. This lack of physical connection is a major difference with parasitic plants of the Orobanchaceae that directly attach themselves to the roots of green plants through specialized organs called haustoria. Next to mycoheterotrophic ectomycorrhizal plants, other plants of the same families have green leaves, but measurements of isotopic carbon (^{13}C) suggest that these plants (*Pyrola, Epipactis, Cephalanthera*) obtain part of their carbon from a neighbouring green plant (mixotrophy; Selosse & Roy 2009). Transfer of nitrogen and phosphorus in CMNs has also been reported, but again quantities are ecologically unimportant in most cases. It has been stated that source–sink dynamics regulate these fluxes, but evidence is mixed. A review by van der Heijden & Horton (2009) indicated a diversity of outcomes in terms of carbon flow to seedlings. For ectomycorrhizal plants, seedlings generally benefit from the presence of adults, but for arbuscular mycorrhizal plants there were equal numbers of positive and negative responses for seedlings.

Bever *et al.* (2010) mentioned that under the concept of CMN, two distinct ideas were placed. Networks not only allow transfer of nutrients and carbon, but also provide compatible fungal species that facilitate seedlings of the same or of other species to establish. For conspecifics, networks override the effects of negative distance and density dependence, as implied in the Janzen–Connell hypothesis. For heterospecifics, networks provide a clear mechanism for facilitation in plant communities.

9.9 Pathogenic soil organisms and nutrient dynamics

Below-ground herbivores can contribute to soil nutrient dynamics which in turn affect nutrient availability and plant productivity of the host, and possibly also of companion plant species. Low levels of root herbivory by plant-feeding nematodes result in leakage of nutrients from the damaged roots. This results in an increased supply of carbon and nutrients for microbial metabolism in the rhizosphere, affecting nutrient mineralization. A pulse labelling experiment with *Trifolium repens* that was colonized by field densities of clover cyst nematodes (*Heterodera trifolii*) showed a significant increase in the leakage of carbon from the roots and an increased microbial biomass (Bardgett *et al.* 1999a). Such observations are not restricted to endoparasites such as *H. trifolii* but are found also for semi-endoparasites, ectoparasites and epidermal and root-hair-cell feeding nematodes. Small migratory ectoparasites, such as species that feed on epidermal and root-hair cells, increase root exudation relatively more than other plant feeders, which may be explained by the ephemeral feeding behaviour leading to a relatively large number of minor damages in the roots (Bardgett *et al.* 1999b).

At present, the rate of this increased root exudation and its stimulating effect on nutrient mineralization (as a function of C : N ratio of exudates) relative to the total nutrient uptake by plants is unknown and still needs to be quantified (Frank & Groffman 2009). In a pot experiment where *T. repens* was grown together with *Lolium perenne*, the addition of clover cyst nematodes in field densities below the damage threshold for white clover resulted in a significant increase in root biomass of the host (141%) as well as of the neighbouring *L. perenne* (217%). Furthermore, nitrogen uptake, derived in pots with nematodes was higher than in pots were they were absent. Bardgett *et al.* (1999b) suggested that such increases in nutrient uptake and root growth of *L. perenne* may alter the competitive balance between the two species, most likely to the detriment of the nematode-infested white clover, thereby influencing plant community structure. Whether this is true for natural field situations remains to be investigated.

9.10 After description

During the early 2000s, ecologists have made substantial progress in identifying mechanisms through which soil biota *could* influence plant species and, through interactions between plant species, ultimately on plant communities. However, despite this progress a paradoxical situation remains. Several mechanisms have been identified; and it is virtually certain that these co-occur in plant communities. Furthermore, different mechanisms could result in similar outcomes, while the same mechanism could result in different outcomes, depending on the environmental matrix. This lack of one-to-one correspondence between mechanism and outcome suggests that mutualistic and antagonistic rhizosphere biota *could* have a major and diverse impact on plant species and plant community composition. But the extent to which this potential leads to actual determinants, and the relative importance of these mechanisms, in relation to other mechanisms, has only seldom been addressed. At present, support for enhanced niche differentiation through root mutualism is scant at best, except for nitrogen-fixing plants through rhizobial and actinorrhizal symbioses. Support for positive and negative feedbacks is well established. However, scaling up from pairwise interactions to species-rich communities is still a daunting task. In order to remain a dominant mechanism, feedbacks of three or more species should generate intransitivity (which requires very long-term soil legacies that have not yet been demonstrated) and different plant species should experience feedbacks of comparable strength (but in the only two studies negative feedback was stronger for rare species).

Two approaches have been applied in investigating the causes and consequences of the feedback processes, a whole-community approach (synthesis; black box approach) and an experimental approach (dissection; opening the black box) based on the putatively most important organisms. Basically the first approach involves the use of rhizosphere soil, in which different plants are grown. While such experiments have provided evidence for the operation of biotic feedbacks, they do not indicate the nature of the responsible organisms. In fact, plant responses do not even indicate to what extent pathogens and mutualists are responsible for the negative feedback.

Species-directed approaches can be successfully applied when a dominant species has been tentatively identified. In many cases, however, none of the antagonists is by itself powerful enough to provide the full explanation. This was observed in the studies of the effect of the soil community on the performance of *Ammophila arenaria* in coastal foredunes. Neither individual nematode species nor individual fungal species could drive the outcome under controlled conditions. In fact, only combinations of antagonistic fungi and nematodes, applied at unrealistically high densities, could replicate the results obtained in the field. Apparently, other biotic agents, interacting with the nematode and fungi, are involved too (de Rooij-van der Goes 1995). A multitude of as yet unidentified members of the soil community that all contribute to feedbacks, raises methodological questions about the sufficiency of experimental demonstration of feedback. To what extent do we need to experimentally replicate the field situation with the addition of various combinations of species groups in order to arrive at a sufficient grasp of the importance of the various mechanisms involved? A reductionistic approach will rapidly run the risk of getting too complicated because of higher order interactions.

Finally, soil-dwelling organisms do not act in isolation. Although the task to study the impact of groups of soil organisms and the interactions between the different groups on vegetation processes and plant communities is already Herculean, the picture that ultimately emerges from such studies is inevitably incomplete. The next step therefore should include interactions between bio-trophic (both mutualistic and antagonistic) and saprotrophic soil organisms, as many fungivorous and predatory soil animals prey on both biotrophic and sapro-trophic soil organisms. Furthermore, interactions and feedbacks between below-ground herbivores and pathogens and above-ground herbivore grazers and browsers (see Chapter 8) could further complicate the final outcome of these interactions on the plant community.

References

Ayres, E., Steltzer, H., Simmons, B.L. *et al.* (2009) Home-field advantage accelerates leaf litter decomposi-tion in forests. *Soil Biology & Biochemistry* **41**, 606–610.

Bardgett, R.D., Denton, C.S. & Cook, R. (1999a) Below-ground herbivory promotes soil nutrient transfer and root growth in grassland. *Ecology Letters* **2**, 357–360.

Bardgett, R.D., Cook, R., Yeates, G.W. & Denton, C.S. (1999b) The influence of nematodes on below-ground processes in grassland ecosystems. *Plant and Soil* **212**, 23–33.

Beckstead, J. & Parker, I.M. (2003) Invasiveness of *Ammophila arenaria*. Release from soil-borne patho-gens? *Ecology* **84**, 2824–2831.

Benson, D.R. & Dawson, J.O. (2007) Recent advances in the biogeography and genecology of symbiotic *Frankia* and its host plants. *Physiologia Plantarum* **130**, 318–330.

Bever, J.D. (2002) Negative feedback within a mutualism: host-specific growth of mycorrhizal fungi reduces plant benefit. *Proceedings of the Royal Society of London* **269**, 2595–2601.

Bever, J.D. (2003) Soil community feedback and the co-existence of competitors: conceptual frameworks and empirical tests. *New Phytologist* **157**, 465–473.

Bever, J.D., Westover, K.M. & Antonovics, J. (1997) Incorporating the soil community into plant popula-tion dynamics: the utility of the feedback approach. *Journal of Ecology* **85**, 561–573.

Bever, J.D., Dickie, I.A., Facelli, E. *et al.* (2010) Rooting theories of plant community ecology in microbial interactions. *Trends in Ecology and Evolution* **25**, 468–478.

Bishop, C.L., Wardell-Johnson, G.W. & Williams, M.R. (2010) Community-level changes in *Banksia* woodland following plant pathogen invasion in the Southwest Australian Floristic Region. *Journal of Vegetation Science* **21**, 888–898.

Brown, V.K. & Gange, A.C. (1990) Insect herbivory below ground. *Advances in Ecological Research* **30**, 1–58.

Chanway, C.P., Holl, F.B. & Turkington, R. (1989) Effect of *Rhizobium leguminosarum* biovar *trifolii* genotype on specificity between *Trifolium repens* and *Lolium perenne*. *Journal of Ecology* **77**, 1150–1160.

Chapela, I.H., Osher, L.J., Horton, T.R. & Henn, M.R. (2001) Ectomycorrhizal fungi introduced with exotic pine plantations induce soil carbon depletion. *Soil Biology and Biochemistry* **33**, 1733–1740.

Cornelissen, J.H.C., Aerts, R., Cerabolini, B., Werger, M.J.A. & van der Heijden, M.G.A. (2001) Carbon cycling traits of plant species are linked with mycorrhizal strategy. *Oecologia* **129**, 611–619.

Crawley, M.J. (ed.) (1997) *Plant Ecology*, 2nd edn. Blackwell Science, Oxford.

de Deyn, G.B., Raaijmakers, C.E., Zomer, H.R. *et al.* (2003) Soil invertebrate fauna enhances grassland succession and diversity. *Nature* **422**, 711–713.

de Rooij-van der Goes, P.C.E.M. (1995) The role of plant-parasitic nematodes and soil-borne fungi in the decline of *Ammophila arenaria* (L.) Link. *New Phytologist* **129**, 661–669.

Diez, J.M., Dickie, I., Edwards, G., Hulme, P.E., Sullivan, J.J. & Duncan, R.P. (2010) Negative soil feedbacks accumulate over time for non-native plant species. *Ecology Letters* **13**, 803–809.

Dumbrell, A.J., Nelson, M., Helgason, T., Dytham, C. & Fitter, A.H. (2010) Relative roles of niche and neutral processes in structuring a soil microbial community. *ISME Journal* **4**, 337–345.

Expert, J.M., Jacquard, P., Obaton, M. & Lüscher, A. (1997) Neighbourhood effect of genotypes of *Rhizobium leguminosarum* biovar. *trifolii*, *Trifolium repens* and *Lolium perenne*. *Theoretical and Applied Genetics* **94**, 486–492.

Fitzsimons, M.S. & Miller, R.M. (2010) The importance of soil microorganisms for maintaining diverse plant communities in tallgrass prairie. *American Journal of Botany* 97, 1937–1943.

Frank, D.A. & Groffman, P.M. (2009) Plant rhizospheric N processes: what we don't know and why we should care. *Ecology* 90, 1512–1519.

Gadgil, R.L. & Gadgil, P.D. (1971) Mycorrhiza and litter decomposition. *Nature* 233, 133.

Hansen, E.M. & Goheen, E.M. (2000) *Phellinus weirii* and other native root pathogens as determinants of forest structure and process in western North America. *Annual Review of Phytopathology* 38, 515–539.

Hart, M.M., Reader, R.J. & Klironomos, J.N. (2001) Life-history strategies of arbuscular mycorrhizal fungi in relation to their successional dynamics. *Mycologia* 93, 1186–1194.

Hausmann, N.T. & Hawkes, C.V. (2010) Order of plant host establishment alters the composition of arbuscular mycorrhizal communities. *Ecology* 91, 2333–2343.

Helgason, T., Merryweather, J.W., Young, J.P.W. & Fitter, A.H. (2007) Specificity and resilience in the arbuscular mycorrhizal fungi on a natural woodland community. *Journal of Ecology* 95, 623–630.

Hobbie, E.A., Macko, S.A. & Shugart, H.H. (1998) Patterns in N dynamics and N isotopes during primary succession in Glacier Bay, Alaska. *Chemical Geology* 152, 3–11.

Hubbell, S.P (1980) Seed predation and the co-existence of tree species in tropical forests. *Oikos* 35, 214–229.

Hyatt, L.A., Rosenberg, M.S., Howard, T.G. *et al.* (2003) The distance dependence prediction of the Janzen–Connell hypothesis: a meta-analysis. *Oikos* 103, 590–602.

Inderjit & van der Putten, W.H. (2010) Impacts of soil microbial communities on exotic plant invasions. *Trends in Ecology & Evolution* 25, 512–519.

Ji, B., Bentivenga, S.P. & Casper, B.B. (2010) Evidence for ecological matching of whole AM fungal communities to the local plant-soil environment. *Ecology* 91, 3037–3046.

Johnson, N.C., Wilson, G.W.T., Bowker, M.A., Wilson, J.A. & Miller, R.M. (2010) Resource limitation is a driver of local adaptation on mycorrhizal symbioses. *Proceedings of the National Academy of Sciences of the United States of America* 107, 2093–2098.

Kardol, P., Bezemer, T.M. & van der Putten, W.H. (2006) Temporal variation on plant-soil feedback controls succession. *Ecology Letters* 9, 1080–1088.

Kiers, E.T., Lovelock, C.E., Krueger, E.L. & Herre, E.A. (2000) Differential effects of tropical arbuscular mycorrhizal fungal inocula on root colonization and tree seedling growth: implications for tropical forest diversity. *Ecology Letters* 3, 106–113.

Klironomos, J.N. (2002) Feedback with soil biota contributes to plant rarity and invasiveness in communities. *Nature* 417, 67–70.

Klironomos, J.N. (2003) Variation in plant response to native and exotic arbuscular mycorrhizal fungi. *Ecology* 84, 2292–2301.

Kulmatiski, A., Beard, K.H., Stevens, J.R. & Cobbold, S.M. (2008) Plant-soil feedbacks: a meta-analytical review. *Ecology Letters* 11, 980–992.

Lafay, B. & Burdon, J.J. (1998) Molecular diversity of rhizobia occurring on native shrubby legumes in southeastern Australia. *Applied and Environmental Microbiology* 64, 3989–3997.

Lankau, R.A. (2011) Resistance and recovery of soil microbial communities in the face of *Alliaria petiolata* invasions. *New Phytologist* 189, 536–548.

Mangan, S.A., Schnitzer, S.A., Herre, E.A. *et al.* (2010) Negative plant-soil feedback predicts tree-species relative abundance in a tropical forest. *Nature* 466, 752–755.

Mangla, S., Inderjit & Callaway, R.M. (2008) Exotic invasive plant accumulates native soil pathogens which inhabit native plants. *Journal of Ecology* 96, 58–67.

Mazzoleni, S., Bonanomi, G., Gianno, F. *et al.* (2007) Is plant biodiversity driven by decomposition processes? An emerging new theory on plant diversity. *Community Ecology* 8, 103–109.

McGuire, K.A., Zak, D.R., Edwards, I.P., Blackwood, C.B. & Upchurch, R. (2010) Slowed decomposition is biotically mediated in an ectomycorrhizal, tropical rain forest. *Oecologia* 164, 785–795.

McKey, D. (1994) Legumes and nitrogen: the evolutionary ecology of a nitrogen-demanding lifestyle. *Advances in Legume Systematics* 5, 211–228.

Milcu, A. & Manning, P. (2011) All size classes of soil fauna and litter quality control the acceleration of litter decay in its home environment. *Oikos* 120, 1366–1370.

Moora, M., Öpik, M., Sen, R. & Zobel, M. (2004) Native arbuscular mycorrhizal fungal communities differentially influence the seedling performance of rare and common *Pulsatilla* species. *Functional Ecology* 18, 554–562.

Newsham, K.K., Fitter, A.H. & Watkinson, A.R. (1995) Multifunctionality and biodiversity in arbuscular mycorrhizas. *Trends in Ecology & Evolution* 10, 407–411.

Norghauer, J.M., Newbery, D.M., Tedersoo, L. & Chuyong, G.B. (2010) Do fungal pathogens drive density-dependent mortality in established seedlings of two dominant African rain-forest trees? *Journal of Tropical Ecology* 26, 293–301.

Olff, H., Hoorens, B., de Goede, R.G.M., van der Putten, W.H. & Gleichman, J.M. (2000) Small-scale shifting mosaics of two dominant grassland species: the possible role of soil-borne pathogens. *Oecologia* 125, 45–54.

Öpik, M., Vanatoa, A., Moora, M. *et al.* (2010) The online database MaarjAM reveals global and ecosystemic distribution patterns in arbuscular mycorrhizal fungi (Glomeromycota). *New Phytologist* 188, 223–241.

Peh, K.S.-H., Sonké, B., Lloyd, J., Quesada & Lewis, S.L. (2011) Soil does not explain monodominance in a Central African tropical forest. *PLoS ONE* 6, e16996.

Petermann, J., Fergus, A.J.F., Turnbull, L.A. & Schmid, B. (2008) Janzen–Connell effects are widespread and strong enough to maintain diversity in grasslands. *Ecology* 89, 2399–2406.

Preisser, E.L. (2003) Field evidence for a rapidly cascading underground food web. *Ecology* 84, 869–874.

Read, D.J. (1991) Mycorrhizas in ecosystems. *Experientia* 47, 376–391.

Reinhart, K.O. & Callaway, R.M. (2004) Soil biota facilitate exotic *Acer* invasions in Europe and North America. *Ecological Applications* 14, 1737–1745.

Reinhart, K.O., Packer, A., van der Putten, W.H. & Clay, C. (2003) Plant-soil biota interactions and spatial distribution of black cherry in its native and invasive ranges. *Ecology Letters* 6, 1046–1050,

Rodríguez-Echevarría, S. (2010) Rhizobial hitchhikers from Down Under: invasional meltdown in a plant-bacteria mutualism? *Journal of Biogeography* 37, 1611–1622.

Selosse, M.-A., Richard, F., He, X. & Simard, S.W. (2006) Mycorrhizal networks: *des liaisons dangereuses? Trends in Ecology & Evolution* 21, 621–628.

Selosse, M.-A. & Roy, M. (2009) Green plants that feed on fungi: facts and questions about mixotrophy. *Trends in Plant Science* 14, 64–70.

Stinson, K.A., Campbell, S.A., Powell, J.R. *et al.* (2006) Invasive plant suppresses the growth of native tree seedlings by disrupting belowground mutualisms. *PLoS Biology* 4: e140

Strickland, M.S., Osburn, E., Lauber, C., Fierer, N. & Bradford, M.A. (2009) Litter quality is in the eye of the beholder: initial decomposition rates as a function of inoculum characteristics. *Functional Ecology* 23, 627–636.

Strong, D.R. (1999) Predator control in terrestrial ecosystems: the underground food chain of bush lupin. In: *Herbivores: Between Plant and Predators* (eds H. Olff, V.K. Brown & R.H. Drent), pp. 577–602. Blackwell Science, Cambridge.

Thrall, P.H., Hochberg, M.E., Burdon, J.J. & Bever, J.D. (2007) Coevolution of symbiotic mutualists and parasites in a community context. *Trends in Ecology & Evolution* 22, 120–126.

Turner, B.L. (2008) Resource partitioning for soil phosphorus: a hypothesis. *Journal of Ecology* 96, 698–702.

Urcelay, C. & Díaz, S. (2003) The mycorrhizal dependence of subordinates determines the effect of arbuscular mycorrhizal fungi on plant diversity. *Ecology Letters* 6, 388–391.

van der Heijden, M.G.A. & Horton, T.R. (2009) Socialism in soil? The importance of mycorrhizal fungal networks for facilitation in natural ecosystems. *Journal of Ecology* 97, 1139–1150.

van der Heijden, M.G.A., Klironomos, J.N., Ursic, M. *et al.* (1998) Mycorrhizal fungal diversity determines plant biodiversity, ecosystem variability and productivity. *Nature* 396, 69–72.

van der Putten, W.H., van Dijk, C. & Peters, B.A.M. (1993) Plant-specific soil-borne diseases contribute to succession in foredune vegetation. *Nature* 362, 53–56.

Vandenkoornhuyse, P., Ridgway, K.P., Watson, I.J., Fitter, A.H. & Young, J.P.W. (2003) Co-existing grass species have distinctive arbuscular mycorrhizal communities. *Molecular Ecology* 12, 3085–3095.

Verschoor, B.C., Pronk, T.E., de Goede, R.G.M. & Brussaard, L. (2002) Could plant-feeding nematodes affect the competition between grass species during succession in grasslands under restoration management? *Journal of Ecology* 90, 753–761.

Vitousek, P.M. & Walker, L.R. (1989) Biological invasion by *Myrica faya* in Hawai'i: plant demography, nitrogen fixation, ecosystem effects. *Ecological Monographs* 59, 247–265.

Vivanco, L. & Austin, A.T. (2008) Tree species identity alters forest litter decomposition through long-term plant and soil interactions in Patagonia, Argentina. *Journal of Ecology* 96, 727–736.

Weste, G. (1981) Changes in the vegetation of sclerophyll shrubby woodland associated with invasion by *Phytophthora cinnamomi*. *Australian Journal of Botany* 29, 261–276.

Wolfe, B.E., Rodgers, V.L., Stinson, K.A. & Pringle, A. (2008) The invasive plant *Alliaria petiolata* (garlic mustard) inhibits ectomycorrhizal fungi in its introduced range. *Journal of Ecology* 96, 777–783.

Yeates, G.W., Bongers, T., de Goede, R.G.M., Freckman, D.W. & Georgieva, S.S. (1993) Feeding habits in soil nematode families and genera – an outline for soil ecologists. *Journal of Nematology* 25, 315–331.

10

Vegetation and Ecosystem

Christoph Leuschner

University of Göttingen, Germany

10.1 The ecosystem concept

The ecosystem concept was introduced by Tansley (1935) who stated that organisms cannot be separated from their environment if their ecology is to be understood. Nowadays, the ecosystem concept is one of the most influential ideas in contemporary ecology (Waring 1989). Modern definitions view the **ecosystem** as an energy-driven complex of the biological community (plants, animals, fungi and procaryotes) and its physical environment which has a limited capacity for self-regulation. Ecosystems may have fundamental principles in common with quantum mechanics (Kirwan 2008).

Organisms and their environment form complex biophysical systems with a system being defined as a set of elements (e.g. plants and climate factors) to which a relationship of cause and effect exists. Our understanding of complex systems is hindered by the fact that organisms and atmosphere, lithosphere (soil) and hydrosphere are linked by a multitude of interactions that can rarely by disentangled completely. Moreover, many interactions are non-linear, include feedback loops, or occur accidentally which makes prediction often difficult (Pomeroy *et al.* in Pomeroy & Alberts 1988).

The adoption of the ecosystem concept starts from the notion that we cannot understand important processes on community or landscape levels from a knowledge of the ecology and interaction of the community members alone. For example, we are not able to precisely predict the consequences for the ecosystem carbon balance of a doubling of atmospheric carbon dioxide if we refer to data on plant ecophysiology and population biology only (Körner 1996). Moreover, agriculture, forestry and water resources management often deal with ecosystem level processes such as nitrogen loss or groundwater recharge and their

Vegetation Ecology, Second Edition. Eddy van der Maarel and Janet Franklin.
© 2013 John Wiley & Sons, Ltd. Published 2013 by John Wiley & Sons, Ltd.

prediction. These goals require a shift in view from the organism and community levels to larger spatial and temporal scales.

The ecosystem approach adopts a system's perception of the living world. A hierarchy of biological organisation levels can be identified with a sequence from the biomolecule through the cell, organism, population, community and ecosystem to the landscape. Ecosystems are found at the top end of this gradient in biological complexity. To study complex environmental systems, 'hierarchy theory' has been developed which recognizes ecosystems as multiscale phenomena, ranging from the biochemical or organismic levels to the ecosystem level, and covering processes from seconds to thousands of years, with the different levels of organization being connected by asymmetric relationships (Allen & Starr 1982; O'Neill *et al.* 1986). Ecosystem studies based on hierarchy theory often use both 'bottom-up' and 'top-down' approaches. The former approaches attempt to assemble large-scale phenomena from smaller-scale components. The top-down approaches proceed in a reductionistic way by attempting to identify processes at lower levels that might cause observed ecosystem patterns.

The theory of complex systems states that each of the different organizational levels reveal 'emergent properties' (see Chapter 1) that cannot be predicted simply by adding up the properties of the next-lower level. The existence of emergent properties has been rejected by some ecologists (e.g. Harper 1982), but has been found by physicists to hold even at the subatomic level, and the notion of emergent properties has proved to be of practical use in ecosystem analysis. Higher levels often respond more slowly to disturbance and can buffer faster system dynamics at lower levels with the consequence that ecosystems exhibit a higher stability than do species populations because interactions and replacements do occur among organisms that dampen the rate of change at the ecosystem level. Moreover, a given property often behaves in a manner that is totally different from that of lower-level properties (Lenz *et al.* 2001). For example, negative effects of environmental factors on the growth and fitness of certain populations may have no effect or even a positive influence on productivity at the ecosystem level because the species of an ecosystem are partially redundant in their use of resources and decreases in one population can be over-compensated by the growth of other populations.

Evapotranspiration or primary productivity in a patch of vegetation are characteristic **emergent properties** at the ecosystem level that not only depend on the ecophysiological controls of water loss and carbon gain at the leaf and plant levels. At the ecosystem level, additional factors such as canopy structure ('roughness'), atmospheric turbulence or competition with neighbours are increasingly important in regulating evapotranspiration and primary productivity while ecophysiological factors often seem to lose their significance during the scaling-up process. Biodiversity (the number of species per plot) is another emergent property of ecosystems that can influence ecosystem functions, and may be indispensable for ecosystem persistence in a variable environment (Naeem 2002). However, the role of biodiversity in ecosystems can hardly be inferred from the species level. See also Chapter 11.

During the past 40 years, ecosystem level studies have helped to understand the regulation of energy and matter turnover in the biosphere, to comprehend the biological basis and major controls of productivity, and to increase yield in agriculture and forestry. More recently, ecosystem analysis has moved to a global perspective in order to predict how the biosphere and global element cycles will respond to an ever-growing human impact on the natural resources (see also Chapter 17).

10.2 The nature of ecosystems

A basic classification of ecosystems distinguishes between terrestrial, limnic and marine systems. A more detailed classification of terrestrial ecosystems is often based on vegetation types because plants are the main primary producers of organic matter, and they often define the spatial structure of ecosystems.

All ecosystems are open systems with respect to the exchange of energy, matter and organisms with their surroundings (Fig. 10.1). Indeed, ecosystems typically have no clearly defined boundaries, except (for example) small atoll islands, ponds and forest fragments that are isolated in the landscape. Instead, the

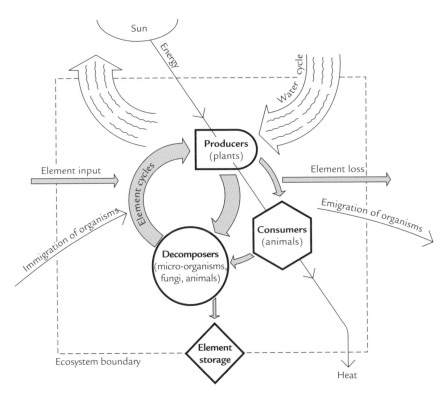

Fig. 10.1 The coupling of energy flow, and water and element cycling in ecosystems. Element cycles, which include those of carbon and nutrients, channel the bulk of matter from the producers (plants) to reducing organisms which decompose plant material and release carbon dioxide and nutrient ions to the environment.

boundaries of most terrestrial ecosystems may be defined by the purpose of study and, thus, are somewhat arbitrary. A useful criterion for defining ecosystem boundaries would be the homogeneous distribution in space of key processes such as water and nutrient fluxes. Plant ecologists usually delimitate ecosystems on the basis of vegetation structure and species composition, as they result from environmental gradients or management.

All ecosystems change in their structure and function over time. Change can be driven by external influences such as disturbance, or altered resource supplies. In many cases, however, change is the outcome of processes within the community. Individuals change with season and age, populations increase and decrease, the number of species present may vary, and fluxes of energy and matter change with season and community age. The most fundamental changes occur during succession when soil and microclimate of the ecosystem are altered (see Chapter 4). Different ecosystem properties can change at different rates. Consequently, static (time-independent) perceptions of communities and ecosystems may be oversimplifications which can lead to wrong conclusions. Hence modern ecosystem science concentrates on understanding the dynamic properties of ecosystems and the role ecosystems play in global biogeochemical cycles.

Ecosystems have a limited capacity for self-regulation. For example, gales and fires may episodically destroy large patches of temperate and boreal forest in North America and Europe. After century-long forest succession, stand structure, primary productivity and nutrient flux rates will eventually regain a state which is close to the pre-disturbance situation. Indeed, many structural and functional properties of ecosystems are restored rapidly during the process of succession whereas others, including species composition, can differ substantially in communities before and after perturbation.

How an ecosystem responds to disturbance is of crucial interest in an era with rapidly growing human impact on nature. A widely used term is **stability**, here in the sense of persistence of structural and functional attributes of ecosystems over time. A mature beech forest would be called stable if only minor changes in canopy structure, species composition and soil chemistry occur over 20 or 30 years. If ecosystem change is related to the degree of environmental change or disturbance intensity, the terms **resistance** and **resilience** are used. Ecosystems that show relatively small changes upon disturbance are said to be resistant. For example, *Fagus* forests are more resistant to disturbance by gales and catastrophic insect attacks than are *Picea* forests. A severe disturbance is required to change the state of beech ecosystems. Ecosystem characteristics that buffer against disturbance are large storage reservoirs for carbon and nutrients in long-lived stems and roots, and high turnover rates of nutrients with plant uptake and mineralization. On the other hand, resistant ecosystems often take a longer time to return to their initial condition following a severe perturbation than do less-resistant ecosystems.

Resilience expresses the speed at which an ecosystem returns to its initial state (Gunderson 2000). Resilient systems can be altered relatively easily but return to the pre-disturbance structure and function more rapidly; thus, their organisms are better adapted to tolerate disturbance. An example of a resilient ecosystem is subalpine alder (*Alnus* spp.) scrub that rapidly resprouts from stumps after mechanical disturbance by ice. Plant productivity and leaf area typically show a high resilience after disturbance, whereas plant biomass and species composition do not. The impact of timber extraction in tropical moist forests can be detected centuries later by an altered species composition although forest structure and productivity may have been restored. Ecosystems in cold environments typically have much lower resilience than those in warm climates, and species-rich eco-systems with high redundancy relative to productivity tend to be more resilient than species-poor ones (Johnson 2000). In any case, the capacity for self-regulation is an important ecosystem property which has implications for the restoration of damaged landscapes.

Self-regulation may also be relevant at higher levels of biological organization. The 'Gaia hypothesis' (Lovelock 1979) views the whole biosphere of the Earth as one, large ecosystem that evolved over geological time periods and has the capability for self-regulation (Gaia is the Greek goddess of the Earth). If photosynthesis did not exist, there would be much more carbon dioxide in the atmosphere and the surface temperature of the Earth would be much hotter than it presently is. According to this hypothesis, life acts as a stabilizing negative feedback system on the global climate by maintaining the oxygen and carbon dioxide levels within narrow limits. The hypothesis is both attractive and controversial because the nature of the suggested climate control mecha-nisms of the biota is not sufficiently understood, and there are doubts on the efficiency of biological control of atmospheric chemistry which has experi-enced large changes of CO_2 concentrations over the past million years (Watson *et al.* 1998).

10.3 Energy flow and trophic structure

10.3.1 Primary productivity

Solar radiation is the direct driving force behind the functioning of nearly all ecosystems on Earth. Only some specialist ecosystems in the deep sea or inside the Earth's mantle which are dominated by micro-organisms are maintained by geothermal energy and, thus, exist independently from the Sun's energy (Gold 1992). The quantity of solar energy reaching the surface of the Earth is about one-half (*c.* 47%) of the radiation flux at the top of the atmosphere ($5.6 \times 10^{24} \, J \cdot yr^{-1}$). The remainder is either absorbed by the molecules of the atmosphere and increases the temperature of the atmosphere, or is reflected back to space. There are large geographical differences in the annual input of solar radiation: while the tropical regions receive 50–$80 \times 10^8 \, J \cdot m^{-2} \cdot yr^{-1}$, the polar regions are only supplied with about half that amount. Temperate regions such as Great Britain and the north-eastern USA are somewhat intermediate with 40–$50 \times 10^8 \, J \cdot m^{-2} \cdot yr^{-1}$. Only a very small proportion of the solar radiation is captured through the carbon fixing activity of the primary producers, plants and other autotrophic organisms (photosynthetic bacteria and cyanobacteria). Typically, photosynthesis converts not more than 1–2% of the incoming solar radiation to chemical energy in carbohydrate compounds. This is because much of the solar energy is unavailable for use in carbon fixation by plants. In particular, only about 44% of incident solar radiation occurs at wavelengths suitable for photosynthesis (400–700 nm). However, even when this is taken into account, photosynthetic efficiency rarely exceeds 2 or 3%. The bulk of the radiation energy either fuels the water cycle (see Section 10.4.4) or simply warms the surface of plants and soil and, thus, drives the movement of air masses across continents and oceans.

The rate at which solar energy is used by the primary producers, to convert inorganic carbon into organic substances through the process of photosynthesis, is termed the **gross primary productivity** (GPP) of the ecosystem. It can be expressed either in units of energy (e.g. $J \cdot m^{-2} \cdot d^{-1}$) or of dry organic matter or carbon (e.g. $g \cdot m^{-2} \cdot d^{-1}$ or $mol \cdot m^{-2} \cdot d^{-1}$). Flows of energy and carbon are interrelated because organic matter contains roughly constant amounts of carbon (*c.* $0.48 g \cdot g^{-1}$), and the energy content of organic matter is reasonably estimated by a value of $18 kJ \cdot g^{-1}$.

Plants use a substantial part of the energy fixed in photosynthesis to support the synthesis of biological compounds and to maintain themselves, energy which is lost by respiration and is not available for plant growth. The difference between GPP and plant respiration (R) is termed the **net primary productivity** (NPP) of the ecosystem. It gives the production rate of new plant biomass plus the carbon lost by litter fall or consumed by herbivores. NPP is also a measure of how much 'food' is at maximum available for consumption by heterotrophic organisms which respire organic carbon back to CO_2. What is left after autotrophic (plants) and heterotrophic respiration (animals and microorganisms) is termed **net ecosystem productivity** (NEP). NEP can be positive, if ecosystems accumulate organic matter in biomass and soil, or negative, if decomposition processes (respiration) dominate.

10.3.2   Secondary productivity

Unlike chemical elements, energy does not circulate in the ecosystem but flows unidirectionally through it. The solar energy fixed in carbohydrates by the primary producers is transferred rapidly through several levels of heterotrophs by consumption and predation. Ultimately, all biological energy is converted to heat via respiration and in this way it leaves the system. Heterotrophs (animals, fungi, most bacteria) ingest autotroph or other heterotroph tissues to suit their own respiratory and tissue-building requirements. According to the type of food ingested, these organisms are termed primary, secondary and higher-level consumers. The productivity of primary consumers is always much less than that of the plants on which they feed because only a fraction of plant productivity is consumed by plant-eating herbivores, and much energy of the incorporated plant mass is lost to faeces and respiration, and thus does not add to herbivore productivity. The rate of biomass production by all heterotrophs is called the **secondary productivity** of the ecosystem.

10.3.3 Trophic levels and food chains

Ecosystems differ greatly in their trophic structure, i.e. the pattern of energy and matter flow through the different trophic levels of primary, secondary, tertiary and higher-level consumers. Energy and organic substances are transferred from one trophic level to another as living tissue (or bodies), dead tissue, faeces, particles of organic matter (POM) or dissolved organic matter (DOM). According to the preferred type of food, organisms can be grouped as herbivores (consumers of living plant tissue), carnivores (consumers of living animals; consumers of microbial biomass are termed microbivores), detritivores (consumers of dead tissue, POM or faeces, also called decomposers), and DOM feeders (consumers of dissolved organic substances). The distinction of producers, consumers and decomposers emphasizes the role organisms are playing in the assimilation and release of CO_2 and nutrients: producers (plants and certain micro-organisms) assimilate CO_2 and inorganic nutrients; consumers release CO_2; and decomposers (or mineralizers) release CO_2 and inorganic nutrients (Fig. 10.1). Both classifications are based on organism functions in the matter cycle but do not refer to taxonomic position (Pimm in Pomeroy & Alberts 1988).

The functional groups of organisms assemble into two principal food chains: (i) the live-consumer (or herbivory) chain with the sequence: herbivores – primary carnivores – secondary and higher-level carnivores; and (ii) the detritus chain with the sequence: detritivorous bacteria, fungi and animals – primary carnivores – secondary and higher-level carnivores. The former chain is based on living plant tissue; the latter uses dead tissue and bodies, POM and faeces (Fig. 10.2). In a number of ecosystems, a third type of food chain (iii) can be recognized which is based on dissolved organic matter (DOM) that feeds bacteria which themselves are consumed by carnivorous animals ('microbivores'). In the soils of terrestrial ecosystems, DOM originates from carbohydrates that are exudated from living plant roots or that are released during the decomposition process of soil organic matter. All three food chains can have quite a number of cross-links and most often form a complex food web rather than a simple chain.

Terrestrial ecosystems differ greatly with respect to the consumption efficiency (CE) of the herbivore community, i.e. the percentage of plant mass produced that is subsequently ingested by herbivores. CE is very low in many temperate forests (less than 5%; see the beech forest example in Fig. 10.3) where it is only significant in years of moth attack – occurring every 5–10 years in many temperate *Quercus* forests. Very high consumption efficiencies are characteristic for tropical grasslands where insects (e.g. locusts) and, in Africa, megaherbivores (among them antelopes and elephants) annually ingest from 20% to more than 50% of the plant mass produced (Lamotte & Bourlière 1983). Similarly high consumption efficiences are reached in fertilized pastures of the industrialized countries.

In all terrestrial ecosystems, a large amount of energy-rich plant material is not used by herbivores but dies without being grazed and thus supports a community of detritivorous animals, fungi and bacteria. Indeed, the world is green despite the activity of herbivores. This is a consequence of the fact that, in most

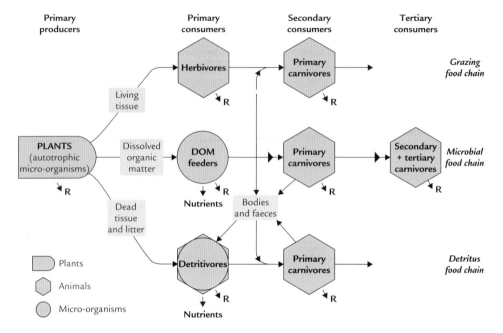

Fig. 10.2 Schematic diagram of the trophic structure of a community in which three prinicipal food chains with different substrates exist. The arrows represent fluxes of organic matter and associated energy. Minor fluxes ore omitted. Release of inorganic nutrients and respirative heat (R) is also indicated. DOM, dissolved organic matter.

terrestrial ecosystems, herbivores are effectively held in check by various mechanisms, among them an effective plant defence by secondary compounds such as lignin, growth-limiting nutrient concentrations in the herbivore's diet and control of herbivore population size by enemies and intraspecific competition (Hartley & Jones 1997). Increasing evidence is accumulating that apex consumers play important roles with respect to top-down forcing in ecosystems and that they are often also shaping the structure and composition of the vegetation (Ripple & Beschta 2003; Estes *et al.* 2011).

10.3.4   Decomposition of organic matter

Perhaps the most important role heterotrophic organisms play in the matter cycles of ecosystems is in decomposing organic substances. Decomposition is the gradual disintegration of dead organic material by both organisms (detritivorous animals, decomposing fungi and bacteria) and physical agencies. Fire can play an important role as a physical disintegrating agent in the dry regions of the world. Decomposition eventually leads to the breakdown of complex energy-rich molecules (such as carbohydrates, proteins and lipids) into carbon dioxide, water, inorganic nutrients and heat. This process – **mineralization** – regenerates nutrients for uptake by plants, fungi and micro-organisms. It has the important

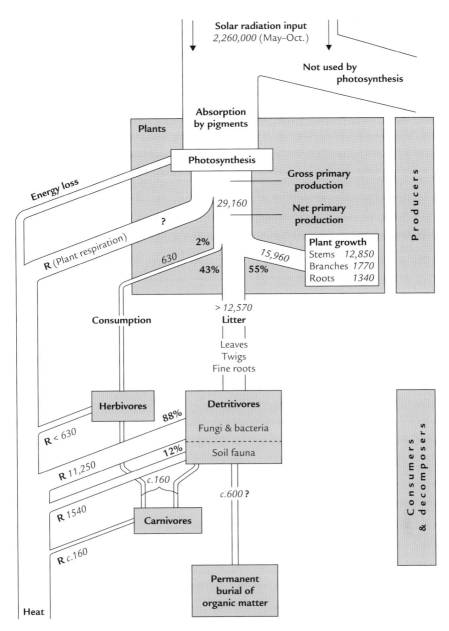

Fig. 10.3 Energy flow through a temperate broad-leaved forest ecosystem with dominating beech (*Fagus sylvatica*; Solling mountains, Germany). All fluxes are in kJ·m²·yr¹; they can be approximately expressed as fluxes of dry matter (g·m²·yr¹) by division with 20. During photosynthesis, an unknown part of the absorbed radiative energy is lost as heat, and through chlorophyll fluoresence and plant respiration. Note that energy flow from dead herbivores and carnivores to detritivores is omitted. (After sources in Ellenberg *et al.* 1986.)

consequence that nutrients immobilized in organic compounds are eventually released into the soil solution where plant roots and fungal hyphae can assimilate them. Detritus with higher nutrient concentrations tends to be decomposed faster. Decomposition and mineralization complete the carbon and nutrient cycles in ecosystems which started with the fixation of carbon dioxide and the assimilation of nutrients by primary producers.

Over decades and centuries, carbon fixation by primary producers and carbon release through the respiration of heterotrophic organisms will be balanced in the majority of terrestrial ecosystems. Notable exceptions are ecosystems that exist on wet soils or in cold climates where decomposition is hampered temporarily or permanently. Here, organic matter gradually accumulates in the soil and eventually may form peat and, in geological time spans, coal and oil. Carbon fixation with primary production and carbon release through heterotrophic respiration may also remain imbalanced in ecosystems that receive an input of organic matter from external sources. This occurs in riverine forests and coastal marshes where organic matter and nutrients are deposited during inundation.

Over periods of months to years, carbon gain and carbon loss may differ greatly in most terrestrial ecosystems because photosynthesis and heterotrophic respiration are controlled by different factors and, thus, fluctuate independently. In periods of maximum plant growth, when biomass rapidly increases, carbon fixation largely exceeds carbon release by respiration and the ecosystem will function as a net sink of carbon dioxide. This happens in spring in seasonal climates and, more generally, during the juvenile stages of plant life. In contrast, carbon losses will exceed carbon gain when plant productivity decreases with plant senescence as occurs in autumn and, more gradually, during the senescence phase of plant development (as in ageing forests) when decomposition of accumulated plant biomass dominates over production. However, there is growing evidence that temperate old-growth forests may function as carbon sinks even in their senescent stage, i.e. C-assimilation can dominate over respiration for long periods (Luyssaert *et al.* 2008).

10.3.5   Energy flow in a temperate forest

A good illustration of how energy flows through ecosystems is given by the example of the Solling forest, a temperate broad-leaved summer-green forest in central Germany (Fig. 10.3). A team of plant and animal ecologists, soil biologists, and micrometeorologists synchronously measured the fixation of solar radiation energy by trees and herbaceous plants, quantified the growth of plant leaves, branches, stems and roots, and studied food consumption, growth and respiratory activity of all major animal, fungal and bacteria groups in this ecosystem (Ellenberg *et al.* 1986). The primary production of this stand is provided nearly exclusively by European beech (*Fagus sylvatica*) that builds pure stands with an only a sparse herbaceous layer on the forest floor.

Net primary production ($29\,160\,\mathrm{kJ \cdot m^{-2} \cdot yr^{-1}}$) accounts for about 1.3% of the solar radiation input during the growing season (May–October). Gross primary production (which includes plant respiration) is estimated to be twice as high. The bulk of the incoming radiation energy (more than 80%) is consumed by the evaporation of water with the transpiration of leaves and rainfall interception of the canopy.

Of this net primary production, 55% is used in the growth of long-lived structural organs of beech (mainly stems, but also branches and large roots). These plant organs may be accessible by detritivorous organisms only after 100 or 200 years. Consequently, a substantial net accumulation of biomass still occurs in the 130-year-old Solling forest which indicates that this ecosystem actually functions as a net sink of carbon (Dixon *et al.* 1994). Other plant biomass fractions such as beech leaves, acorns and part of the twig and fine root fractions are turned over annually and, thus, represent plant litter which is the basis of the detritus food chain (43% of NPP). Although a number of herbivore populations reach high densities, leaf- and root-eating animals consume only a negligible fraction of NPP in this forest: *c.* 2% only of NPP is channelled through the live-consumer chain which starts with herbivore consumption on leaves and roots.

Shed leaves, twigs and acorns accumulate on the forest floor in autumn; dying roots represent an important additional litter source in the soil. Both components are decomposed by fungi, bacteria and soil animals that are present with high species numbers: more than 1500 soil animal species alone were counted on 1-ha plots in this forest. Decomposition is mainly carried out through the activity of fungi and bacteria which consume *c.* 88% of the energy contained in the detritus material. Soil animals are much less important in terms of energy flow but fulfil important roles in the decomposition process as shredders of dead tissue and by vertically mixing the soil organic layers. A small, chemically inert fraction of the plant litter resists the attack of the detritivorous organisms for years and decades, and appears as dark humic substances in the topsoil.

10.3.6   Global patterns of terrestrial primary productivity

Terrestrial net primary productivity of the globe is estimated at about 60×10^{15} g-C·yr^{-1} (Houghton & Skole 1990) which is in the same magnitude as the known world oil reserves (70×10^{15} g-C). NPP varies greatly across the ecosystems of the world (Fig. 10.4). The most productive terrestrial ecosystems of the globe are found among wetlands (i.e. swamps, marshlands and fens), tropical moist forests, and cultivated lands with typical NPP values in the range of 750–1300 g·m^{-2}·yr^{-1} of carbon fixed (1 g of dry phytomass contains *c.* 0.48 g of carbon). Low productivities dominate in desert, tundra and arctic ecosystems with 0–150 g-C·m^{-2}·yr^{-1} (circles in Fig. 10.5).

The principal factors determining NPP in terrestrial ecosystems are water availability and the temperature regime. A more thorough analysis shows that temperature seems to influence annual NPP mainly through the length of the growing season rather than through a direct dependence of annual mean temperature on productivity. Indeed, tundra, boreal and temperate ecosystems show remarkably high productivities in comparison to tropical moist forest if one considers the different lengths of the growing seasons in these four ecosystem types (1–3, 4–6, 6–8, and 12 months, respectively). Temperate forests reach equal, or even higher NPP rates than many tropical forests on a monthly basis,

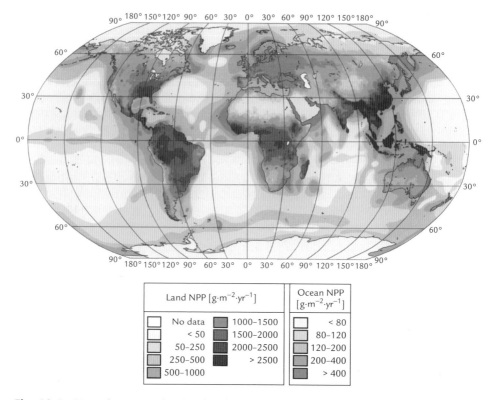

Fig. 10.4 Net primary production (NPP) of the biosphere. Note different scales for terrestrial and marine productivity and the generally smaller NPP in the oceans than on land. (Source: Berlekamp, Stegemann & Lieth, URL http://www.usf.uni-osnabrueck.de/~hlieth.)

while plant respiration and thus GPP are much lower. Minimum temperature, in particular the occurrence of frost, is another crucial factor for plant productivity.

On a global scale, light and nutrient availability are less important than water and temperature for the level of terrestrial NPP. However, nutrient availability often determines local differences in NPP among sites, for example in the tropical forests of Amazonia (Malhi *et al.* 2006).

10.3.7 Productivity and energy flow in different ecosystem types

In Fig. 10.6, general patterns of productivity and energy flow in four key ecosystems of the globe are compared, i.e. temperate deciduous forest, tropical moist forest, boreal coniferous forest, and temperate, summer-dry and winter-cold grassland. These ecosystems are dominated by very different functional groups of plants (summer-green broad-leaved trees, evergreen broad-leaved trees, evergreen needle-leaved trees, or grasses) with largely different GPP (which

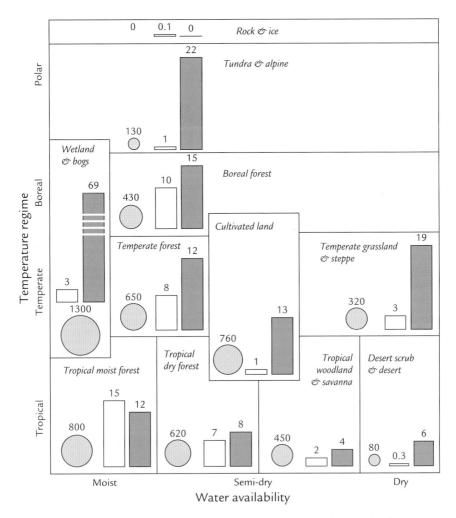

Fig. 10.5 Approximate stores of carbon and net primary production in the major biomes of the world. Soil organic matter (SOM, dark bars) and plant biomass (only above-ground; light bars) are given in kg-C·m^{-2}, NPP (circles) in g-C·m^2·yr^{-1} . The biomes or ecosystem types are arranged along axes of temperature and water availability – the two key factors that determine terrestrial productivity. (After data in Schlesinger 1997.)

is the annual total of photosynthesis) resulting from different physiological constitutions and also contrasting temperature, water and nutrient regimes. Boreal, temperate and tropical systems seem to differ mostly with respect to plant respiration, while NPP is not that different (see also Section 10.3.6). It is estimated that more than 65% of GPP is lost to the atmosphere by plant respiration in tropical moist forests. In comparison, plant respiration may consume a smaller proportion of GPP in the cooler temperate and boreal forests (Luyssaert *et al.* 2007; Ellenberg & Leuschner 2010).

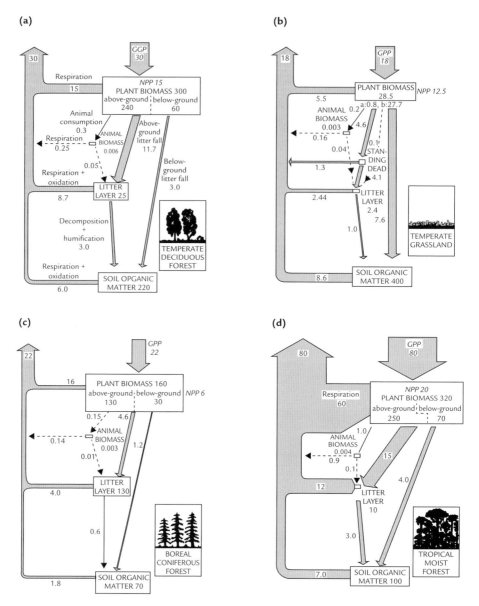

Fig. 10.6 Principal patterns of energy flow through four key ecosystems of the globe. Arrows and related numbers give fluxes of organic matter (in $Mg\text{-}dry\text{-}matter\cdot ha^{-1}\cdot yr^{-1}$), squares indicate pools of organic matter (in $Mg\cdot ha^{-1}$). The relative size of arrows and squares allows a comparison of the four ecosystems. (Data assembled from various sources: mostly Luyssaert *et al.* 2007; Ellenberg & Leuschner 2010; Schultz 2000; based partly on estimations; thus, only a rough picture is given.)

Plant biomass typically increases in the sequence temperate grassland – boreal forest – temperate forest – tropical forest (white rectangles at the top of Fig. 10.6), if plant mass is not reduced by frequent disturbances. In contrast, soil organic matter (SOM) in the mineral soil often is much higher in the grassland than in the three forest types (rectangles at the bottom). SOM pools are relatively small under many boreal coniferous and temperate broad-leaved forests because much plant litter accumulates on the forest floor and is not incorporated into the soil profile itself. The contrasting patterns of detritus storage in the four ecosystem types are primarily a consequence of rapid litter decomposition in the tropical moist forest where most detritus typically is decayed within weeks to several months, and thus is not accumulating on the forest floor. This contrasts with low decay rates in the dry and winter-cold grassland and the cold boreal forest. The size of SOM pools also depends on NPP and thus on the amount of litter that is annually supplied above- and below-ground: above-ground litter fall typically decreases in the sequence tropical forest – temperate forest – coniferous forest – temperate grassland. Below-ground litter from decaying roots, however, seems to be more important in the grassland than in the three forest ecosystems.

Animal biomass is negligible in comparison to plant biomass or soil organic matter in all four ecosystems although animals play important roles in many ecosystem functions such as soil formation, pollination and dispersal. Consumption of plant tissues by herbivorous animals is also quantitatively of minor importance in all three (boreal, temperate and tropical) forest ecosystems. Consumption efficiency is low in the temperate winter-cold grassland as well. All four ecosystems in Fig. 10.6 are characterized by a detritus food chain which, in terms of energy flow, is much more important than the live-consumer chain. This contrasts with intensive herbivore consumption in tropical grassland ecosystems (see Chapter 8).

10.4 Biogeochemical cycles

Nutrients and water tend to circulate along characteristic pathways in ecosystems, in marked contrast to energy. Energy is never recycled but flows through the ecosystem, being finally degraded to heat and lost from the system (Fig. 10.1). Carbon, nitrogen and phosphorus are the functionally most important chemical elements (besides hydrogen and oxygen) in plants, with nitrogen and phosphorus often limiting plant growth. They participate in biogeochemical cycles between living organisms and the environment with movement by wind in the atmosphere and by running water in soil, streams and ocean currents.

10.4.1 Carbon cycle

The cycle of carbon is predominantly a gaseous one which is driven by the two key processes photosynthesis and respiration (Fig. 10.7a). The only source of

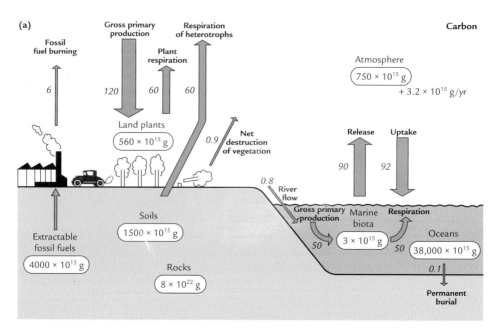

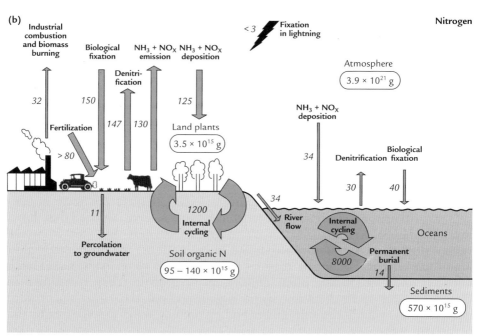

Fig. 10.7 The global cycles of (a) carbon and (b) nitrogen. The estimated sizes of pools (white boxes) is given in g-C or g-N, the transfers between compartments (arrows) in 10^{15} g-C·yr^{-1} and 10^{12} g-N·yr^{-1}, respectively. Internal cycling refers to plant uptake and release through decomposition. (Data from Schlesinger 1997; Ellenberg & Leuschner 2010; Jaffe 1992; redesigned after Schlesinger (1997) and updated.)

carbon available to land plants is carbon dioxide (CO_2) in the atmosphere, whereas aquatic plants can assimilate CO_2 and/or bicarbonate (HCO_3^-) dissolved in water. The large carbon stocks present in rocks (mainly as carbonates) are not available to plants. Large stores of inorganic and organic carbon also exist in the oceans.

In terrestrial ecosystems, considerable amounts of carbon are sequestred either in plant biomass, or in dead organic matter in the soil (SOM) or on the forest floor (litter layer). About 55% of the living plant biomass are found in the tropical moist forests (roughly $320\,Mg\text{-}DM\cdot ha^{-1}$); temperate forests store not much less (*c.* $300\,Mg\cdot ha^{-1}$) but occupy a smaller area on Earth due to the long history of forest destruction. The largest stocks of SOM are stored in wetland, mire and certain grassland ecosystems where detritus decomposition is inhibited by the lack of oxygen or by low temperatures (Fig. 10.7a).

10.4.2 Nitrogen cycle

The main reservoir of nitrogen is the atmosphere where the inert gas N_2 constitutes *c.* 78% by volume. Most nitrogen that is available to biota was originally derived from this atmospheric pool through nitrogen fixation, either by oxidation of N_2 through lightning in the atmosphere or by the activity of nitrogen-fixing micro-organisms on land and in the seas (Fig. 10.7b). Nitrogen fixers are found among free-living bacteria (e.g. *Azotobacter*, *Azotococcus*), cyanobacteria (e.g. *Nostoc*), and bacteria associated with plant roots (e.g. *Rhizobium*); this process is highly dependent on energy in terms of ATP. Plants with symbiotic nitrogen fixation include the legumes (family Fabaceae) and *Alnus* species (alders). Land plants are typically capable of acquiring NH_4^+, NO_3^- and organic N, where the amount of the N forms used mostly depends on their availability in the soil. However, certain plants show preferences either for NH_4^+ (mostly acid-tolerant species) or NO_3^- (some species of calcareous and/or base-rich soils).

Organic nitrogen (amino acids and oligopeptides) is an important nitrogen source for plants primarily in cold and acidic environments, such as arctic and boreal ecosystems, where symbiotic fungi (mycorrhiza) and bacteria shortcut the nitrogen cycle by absorbing organic nitrogen-rich compounds and eventually release ammonium that may be absorbed by plant roots or is exchangeably bound to soil clays and organic matter. NH_4^+ is oxidized by autotrophic soil bacteria (*Nitrosomonas*, *Nitrobacter*) to NO_3^- under a sufficiently high pH and the presence of oxygen. In acid soils, heterotrophic bacteria and certain fungi oxidize organic nitrogen either directly, or with NH_4^+ as an intermediate product, to nitrate (Killham 1990).

Microbial denitrification is a process in which the major end product is removed from the biological nitrogen cycle. Micro-organisms reduce nitrate to the gases N_2, nitrogen monoxide (NO) or dinitrogenoxide (N_2O) in the absence of oxygen, and thus lower the nitrogen load of ecosystems. Denitrification is an important process in wetland and mire ecosystems, but it also occurs in other terrestrial ecosystems such as forests after anthropogenic N-input. Undisturbed mature ecosystems typically have more or less equal inputs and outputs of nitrogen. Increased leaching of nitrate or other nutrients may serve as an indicator of instability in ecosystems (Likens & Bormann 1995; Brumme & Khanna 2009) which contrasts sharply with the nitrogen cycle in unpolluted temperate forests in Chile and Argentina where N is leached from the forest primarily as dissolved organic N (DON) (Perakis & Hedin 2002).

10.4.3   Phosphorus cycle

Besides nitrogen, phosphorus is the major limiting element for terrestrial primary productivity (Elser *et al.* 2007). In contrast to N, the cycle of P is a sedimentary one with no significant gaseous component. The main P-reservoir exists in rocks (apatite) which are slowly eroded, releasing P to ecosystems (Fig. 10.8a). The main flux of phosphorus in the global cycle is carried by streams to the oceans where P is eventually deposited in the sediments. Similar to N, there is an intensive internal cycling of P in ecosystems through plant uptake and assimilation, and subsequent release via microbial or fungal release. Unlike N, however, P is much less mobile in soils and is often immobilized by chemical and physical processes. Many plants depend on extra-cellular phosphatase enzymes that are mainly released by mycorrhizae to mobilize P. Mining of phosphate-rich rocks has greatly increased the availability of phosphorus in agricultural ecosystems, and has resulted in P-enrichment of adjacent limnic and marine systems.

10.4.4   Water cycle

The global water cycle moves much more substance than any other biogeochemical cycle on Earth. Receipt of water through rainfall is one of the key factors controlling primary productivity on land. Water exists in three different states, solid (ice), liquid and gas (water vapour). By far the largest reservoir of water are the oceans, which comprise *c.* 97% of the total amount of water on Earth. Another 2% is bound in polar ice caps and glaciers. Only a very small proportion, less than 1%, of the global water resource is available for the growth of terrestrial plants, in freshwater reservoirs in rivers, lakes and groundwater (Fig. 10.8b). Water vapour in the atmosphere constitutes only *c.* 0.08% of the total (Chahine 1992; Schlesinger 1997).

These pools are linked by flows of water with evaporation, transpiration, precipitation and overland flow. Water is transferred from land and ocean surfaces to the atmosphere through three processes, (i) evaporation from wet soils and water surfaces; (ii) transpiration from plant leaves; and (iii) evaporation of water from plant surfaces that is intercepted during rain. Among these three evapotranspiration components, only transpiration can be regulated by stomatal aperture according to plant demand. The energy required to move liquid water into the vapour phase is called the latent heat of vaporization and equals $44\,kJ\cdot mol^{-1}$ of water. This large amount of heat is consumed when water evaporates from vegetation, water tables or moist soil, and is regained during the process of condensation when clouds form. Therefore, the cycling of water on

(a)

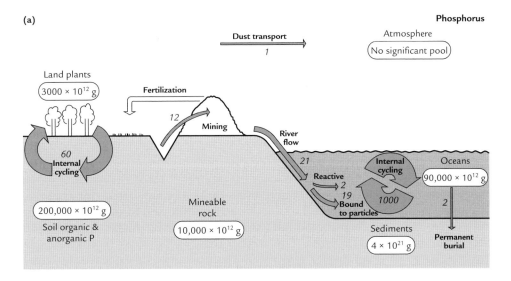

(b)

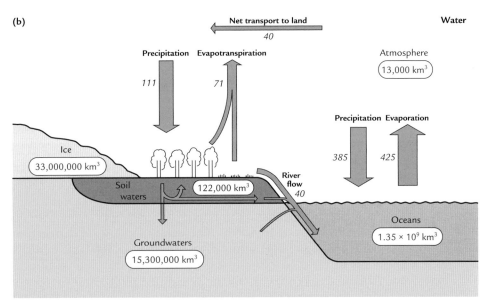

Fig. 10.8 The global cycles of (a) phosphorus and (b) water. Transfers between compartments are given in 10^{12} g-P·yr^{-1} and 10^{3} km^{3}-water·yr^{-1}, respectively. The net transport of rain water to land is the difference between evaporation and precipitation over the sea, and equals worldwide river flow which results from a precipitation surplus on land. (Redesigned after Schlesinger 1997 and updated.)

Earth is a highly effective means to distribute solar energy absorbed by land and water surfaces.

On the oceans, a surplus of evaporation over rainfall exists whereas the land surfaces, on average, receive more water through precipitation than they lose it through evapotranspiration. Consequently, an annual net transport of 40 000 km^3 of water occurs with cloud movement from the oceans to the land. An equal volume flows with rivers from the land to the oceans. River flow also carries the products of mechanical and chemical weathering to the sea and thus is an important agent in nutrient cycles.

Terrestrial vegetation can substantially modify the cycling of water. Vegetation types with large leaf areas such as tropical mountain forests may catch much more water through interception and re-evaporation of rainfall than, for example, a temperate shortgrass steppe. Thus, less water infiltrates into the subsoil under forests. In addition, deep-rooted tall forests typically have higher transpiration rates than low vegetation types. This further reduces the water that is available for a recharge of groundwater reservoirs. The principal consequences of large-scale forest destruction, as it occurs in the wet tropics, are a speed-up of erosion and soil nutrient impoverishment, and a lowered transpiration which may result in less cloud formation in the region.

10.4.5   Anthropogenic alterations of biogeochemical cycles

Major perturbations of the biogeochemical cycles have occurred on Earth with 30–50% of the land surface transformed by human action. In a few generations the world population will have exhausted the fossil fuels that were generated over several hundred million years. More than half of the accessible fresh water is already used by humans.

Rising atmospheric carbon dioxide concentrations are likely to cause global warming and reflect the human perturbation of the carbon cycle (IPCC 2007; also see Chapter 17). Compared to pre-industrial values, the atmospheric concentration of carbon dioxide (about 390 p.p.m.) has increased by more than 30%, that of methane by more than 100%. The recent large-scale destruction and burning of both tropical and boreal forests greatly reduces the large biomass and soil carbon pools of these ecosystems and increases the net flux of CO_2 to the atmosphere. This source represents about 25% of the global anthropogenic CO_2 emissions, while the remaining 75% relate to the burning of fossil fuel and industrial processes. Large stores of inorganic and organic carbon exist in the oceans which currently act as net sinks of carbon. While the oceans absorb part of the anthropogenic carbon dioxide emissions, they suffer from ongoing acidification due to the formation of carbonic acid, which has negative consequences for the marine biota.

Humans also have a dramatic impact on the nitrogen cycle. Today, more N enters the terrestrial ecosystems through the activity of humans (industrial N fixation for fertilizers, N release from fossil fuels, planting of legume crops) than is annually fixed by micro-organisms in all terrestrial ecosystems (Vitousek *et al.* 1997). Fertilization results in nitrogen enrichment (eutrophication) in terrestrial, limnic and marine ecosystems, and may lead to dramatic changes in plant species composition (Bobbink *et al*. 2010). Burning of fossil fuels and biomass, and the emission of ammonia (NH_3) and nitrogen oxides (NO_x) by modern agriculture increases the concentration of these N compounds in the atmosphere which are subsequently returned to surrounding ecosystems with rainfall deposition. Much of the N deposited to forest ecosystems is retained either through accumulation in the biomass, or, on acid soils, mainly through retention in growing humus stocks (Brumme & Khanna 2009; Ellenberg & Leuschner 2010). Several temperate forest ecosystems have already reached 'nitrogen saturation' which results in the increased transport of nitrate with infiltrating water from the soil to groundwater reservoirs, which are polluted. The mineral resources of P will be exhausted within the next 50–100 years and P-shortage may severely limit agricultural production in the future (Cordell *et al*. 2009). Despite efficient strategies for P-uptake, many temperate forests are increasingly limited by P-shortage due to high actual atmospheric N-inputs and intensification of forest use (Duquesnay *et al*. 2000; Elser *et al*. 2007). Humans release many toxic substances into the environment that often accumulate in organisms. Emissions of chlorofluorocarbon gases have led to the ozone hole over the Antarctic and would have destroyed much of the ozone layer if no international measures to end their emission had been taken (Ehhalt & Prather 2001).

The plundering of resources and the substantial alteration of the global biogeochemical cycles pose a serious threat to the functioning of most ecosystems on Earth. To preserve the biosphere with its indispensable life support functions is one of the great future tasks of humankind.

References

Allen, T.F.H. & Starr, T.B. (1982) *Hierarchy: Perspective for Ecological Complexity*. University of Chicago Press, Chicago. IL.

Bobbink, R., Hicks, K., Galloway, J. *et al*. (2010) Global assessment of nitrogen deposition effects on terrestrial plant diversity: a synthesis. *Ecological Applications* **20**, 30–59.

Brumme, R., Khanna, P.K. (eds.) (2009) *Functioning and Management of European Beech Ecosystems*. Ecological Studies 208. Springer Verlag, Berlin, Heidelberg.

Chahine, M.T. (1992) The hydrological cycle and its influence on climate. *Nature* **359**, 373–380.

Cordell, D., Drangert, J.-O. & White, S. (2009) The story of phosphorus: global food security and food for thought. *Global Environmental Change* **19**, 292–305.

Dixon, R.K., Brown, S., Houghton, R.A. *et al*. (1994) Carbon pools and flux of global forest ecosystems. *Science* **263**, 185–190.

Duquesnay, A., Dupouey, J.L., Clement, A., Ulrich, E. & Le Tacon, F. (2000) Spatial and temporal variability of foliar mineral concentration in beech (*Fagus sylvatica*) stands in northeastern France. *Tree Physiology* **20**, 13–22.

Ehhalt, D. & Prather, M. (2001) Atmospheric chemistry and greenhouse gases. In: *Climate Change 2001: the Scientific Basis. Contribution of Working Group I to the Third Assessment Report of the Intergovernmental Panel on Climate Change* (eds J.T. Houghton, Y. Ding, D.J. Griggs *et al*.), pp. 239–287. Cambridge University Press, New York, NY.

Ellenberg, H. & Leuschner, Ch. (2010) *Vegetation Mitteleuropas mit den Alpen in ökologischer, dynamischer und historischer Sicht*, 6th edn. Ulmer Verlag, Stuttgart.

Ellenberg, H., Mayer, R. & Schauermann, J. (1986) Ökosystemforschung. Ergebnisse des Sollingprojekts 1966–1986. Ulmer Verlag, Stuttgart.

Elser, J.J., Bracken, M.E.S., Cleland, E.E. *et al.* (2007) Global analysis of nitrogen and phosphorus limitation of primary producers in freshwater, marine and terrestrial ecosystems. *Ecology Letters* **10**, 1135–1142.

Estes, J.A., Terborgh, J., Brashares, J.S. *et al.* (2011) Trophic downgrading of planet Earth. *Science* **333**, 301–306.

Gold, T. (1992) The deep hot biosphere. *Proceedings of the National Academy of Sciences of the United States of America* **89**, 6045–6049.

Gunderson, L.H. (2000) Ecological resilience – in theory and application. *Annual Review of Ecology and Systematics* **31**, 425–439.

Harper, J.L. (1982) After description. In: *The Plant Community as a Working Mechanism* (ed. E.I. Newman), pp. 11–26. Blackwell, Oxford.

Hartley, S.E. & Jones, C.G. (1997) Plant chemistry and herbivory or why the world is green. In: *Plant Ecology*, 2nd edn (ed. M.J. Crawley), pp. 284–324. Blackwell, Oxford.

Houghton, R.A. & Skole, D.L. (1990) Carbon. In: *The Earth as Transformed by Human Action* (eds B.L. Turner, W.C. Clark, R.W. Kates, *et al.*), pp. 393–408. Cambridge University Press, Cambridge.

IPCC (2007) Contribution of Working Group I to the Fourth Assessment Report of the Intergovernmental Panel on Climate Change, 2007 (eds S. Solomon, D. Qin, M. Manning, *et al.*). Cambridge University Press, Cambridge.

Jaffe, D.A. (1992) The nitrogen cycle. In: *Global Biogeochemical Cycles* (ed. S.S. Butcher, R.J. Charlson, G.H. Orians & G.V. Wolfe), pp. 263–284. Academic Press, London.

Johnson, K.H. (2000) Trophic-dynamic considerations in relating species diversity to ecosystem resilience. *Biological Reviews* **75**, 347–376.

Killham, K. (1990) Nitrification in coniferous soils. *Plant and Soil* **128**, 31–44.

Kirwan, A.D. (2008) Quantum and ecosystem entropies. *Entropy* **10**, 58–70.

Körner, Ch. (1996) The response of complex multispecies systems to elevated CO_2. In: *Global Change and Terrestrial Ecosystems* (eds B. Walker & W. Steffen), pp. 20–42. Cambridge University Press, Cambridge.

Lamotte, M. & Bourlière, F. (1983) Energy flow and nutrient cycling in tropical savannas. In: *Tropical Savannas* (ed. F. Bourlière), pp. 583–603. Ecosystems of the World 13. Elsevier, Amsterdam.

Lenz, R., Haber, W. & Tenhunen, J.D. (2001) A historical perspective on the development of ecosystem and landscape research in Germany. In: *Ecosystem Approaches to Landscape Management in Central Europe* (eds J.D. Tenhunen, R. Lenz & R. Hantschel), pp. 17–35. Ecological Studies 147. Springer Verlag, Berlin.

Likens, G.E. & Bormann, F.H. (1995) *Biogeochemistry of a Forested Ecosystem*, 2nd edn. Springer Verlag, New York, NY.

Lovelock, J.E. (1979) *Gaia: A New Look at Life on Earth*. Oxford University Press, Oxford.

Luyssaert, S., Imglima, I., Jung, M. *et al.* (2007) CO_2 balance of boreal, temperate, and tropical forests derived from a global database. *Global Change Biology* **13**, 2509–2537.

Luyssaert, S., Schulze, E.-D., Boerner, A. *et al.* (2008) Old-growth forests as global carbon sinks. *Nature* **455**, 213–215.

Malhi, Y., Wood, D., Baker, T.R. *et al.* (2006) The regional variation of aboveground live biomass in old-growth Amazonian forests. *Global Change Biology* **12**, 1107–1138.

Naeem, S. (2002) Ecosystem consequences of biodiversity loss: the evolution of a paradigm. *Ecology* **83**, 1537–1552.

O'Neill, R.V., deAngelis, D.L., Waide, J.B. & Allen, T.F.H. (1986) *A Hierarchical Concept of Ecosystems*. Princeton University Press, Princeton, NJ.

Perakis, S.S. & Hedin, L.O. (2002) Nitrogen loss from unpolluted South American forests mainly via dissolved compounds. *Nature* **415**, 416–419.

Pomeroy, L.R. & Alberts, J.J. (1988) *Concepts of Ecosystem Ecology*. Ecological Studies 67. Springer Verlag, New York, NY.

Ripple, W.J. & Beschta, R.L. (2003) Wolf reintroduction, predation risk, and cottonwood recovery in Yellowstone National Park. *Forest Ecology and Management* **184**, 299–313

Schlesinger, W.H. (1997) *Biogeochemistry*, 2nd edn. Academic Press, San Diego, CA.

Schultz, J. (2000) *Handbuch der Ökozonen*. Ulmer Verlag, Stuttgart.

Tansley, A.G. (1935) The use and abuse of vegetation concepts and terms. *Ecology* **42**, 237–245.

Vitousek, P.M., Mooney, H.A., Lubchenco, J. & Melillo, J.M. (1997) Human domination of earth's ecosystems. *Science* **277**, 494–499.

Waring, R.H. (1989) Ecosystems: fluxes of matter and energy. In: *Ecological Concepts* (ed. J.M. Cherrett), pp. 17–41. Blackwell, Oxford.

Watson, R.T., Zinyowere, M.C. & Moss, R.H. (eds) (1998) *The Regional Impacts of Climate Change: An Assessment of Vulnerability*. Cambridge University Press, Cambridge.

11

Diversity and Ecosystem Function

Jan Lepš

University of South Bohemia, Czech Republic

11.1 Introduction

It is generally supposed that species diversity is important for the stability and proper functioning of ecosystems and for ecosystem services. Indeed, the Shannon formula (H') for species diversity was introduced to ecology as a stability index (MacArthur 1955). The relation between diversity and stability is complex. For instance, population outbreaks are more common in species-poor boreal regions than in species-rich tropical communities, or in species-poor agro-ecosystems and planted tree monocultures than in the species-rich natural communities. This has led to the 'diversity begets stability' statement. However, the causality of observed patterns could be reversed: the tropics are so rich in species because they have experienced long-term environmental stability, which enabled survival of many species, or even both; stability and diversity can be dependent on similar sets of external characteristics, so that they are just statistically correlated, without any direct causal relationship.

Since the 1990s, the global loss of biological diversity has become a major concern. Could indeed the decline of biodiversity impair the functioning of ecological systems? And do we have sound evidence of ecological consequences of declining biodiversity? These are matters of concern and controversy (e.g. Naeem *et al.* 1999; Wardle *et al.* 2000; Grace *et al.* 2007), but also of a growing consensus (e.g. Loreau *et al.* 2001; Hooper *et al.* 2005). Pimm (1984) and others showed that there are many aspects of stability and diversity. The term ecosystem functioning includes a variable set of characteristics. In natural ecosystems, diversity is a 'dependent variable', i.e. it is a result of evolutionary and ecological processes, which affect community composition and also ecosystem functioning.

Vegetation Ecology, Second Edition. Eddy van der Maarel and Janet Franklin.
© 2013 John Wiley & Sons, Ltd. Published 2013 by John Wiley & Sons, Ltd.

If we want to study diversity and its effects, we must be able to quantify diversity, and we need to understand the factors that affect diversity in nature. Then, we will have to quantify the ecosystem functions that are expected to be affected. Next we must find interrelationships and find ways to test the causality behind statistical relationships.

11.2 Measurement of species diversity

Ecologists use various terms for diversity: species diversity, ecological diversity, richness, and recently biodiversity and complexity. However, the concepts underlying these terms differ among ecologists, and also, various terms are sometimes used for the same concept.

11.2.1 Which organisms to include

In most studies, the community is defined taxonomically. Many descriptions of plant communities are restricted to vascular plants, while in some studies, where bryological expertise was available, bryophytes are also included. However, the diversity of vascular plants is not necessarily related to the diversity of all the plant species, and the diversity of all plant species is not necessarily a good indicator of the richness of the whole ecosystem. Another restriction concerns the lack of below-ground data, notably about the seed bank. This might cause some problems, because a seed is a substantial part of the species' life-cycle, particularly in arid systems. For the study of some processes (e.g. response to certain perturbations), the seed bank can be very important. There is no general rule for what should be included in the analysis. Decisions are often made on pragmatic grounds. They may be decisive for the type of relationships found.

11.2.2 Number of species and diversity

Let us imagine a plant community being analysed on a given location. Its species number is not sufficient to characterize the community diversity. Two communities with the same number of species may differ in the variation in species abundances. This leads to the distinction between two components of species diversity: species richness and evenness. However, species richness is often called diversity as well. Several diversity indices have been devised, the two most popular being the Shannon index and the reciprocal or complement of the Simpson dominance index.

Let us call species number S, define p_i as the proportion of the i-th species, i.e. $p_i = N_i/N$, where N_i is the quantity of the i-th species, usually its abundance or biomass, and $N = \Sigma\, N_i$, i.e. the total quantity of all the species. The Shannon index is then defined as

$$H' = -\sum_{i=1}^{S} p_i \log p_i \qquad (11.1)$$

The Shannon index is based on information theory; hence, $\log_2$ was used. Later also natural log and $\log_{10}$ have been used, but all three functions are called Shannon (or Shannon–Wiener, or Shannon–Weaver) index. Although nowadays natural log (ln) usually is used, it is necessary to indicate which logarithm was used to avoid confusion. The antilogarithm of H', i.e. $2^{H'}$, $e^{H'}$ or $10^{H'}$ for H' based on $\log_2$, natural log and $\log_{10}$, respectively, can also be used. This value can be interpreted as the number of species needed to reach diversity H', when the species are equally represented. The value of H' equals 0 for a monospecific community, and $\log S$ for a community of S equally represented species.

The second most frequently used diversity index is the reciprocal of the Simpson dominance index, $1/D$. The Simpson dominance index is defined as:

$$D = \sum_{i=1}^{S} p_i^2 \qquad (11.2)$$

As for antilog H', the minimum value of $1/D$ equals 1 for a monospecific community, and its maximum is S for a community of S equally represented species. Sometimes, $1 - D$ is used as measure of diversity; this value ranges from zero for monospecific communities to $1 - 1/S$ in cases of maximum evenness. If p_i is defined as the proportion of *individuals* in an indefinitely large community, then $1 - D$ is the probability that two randomly selected individuals will belong to different species.

Hill (1973) has shown that the common indices of diversity are related to each other (and to Rényi's definition of generalized entropy) and suggested a unifying notation. His general diversity index can be written as:

$$N_a = \left(\sum_{i=1}^{S} p_i^a \right)^{1/(1-a)} \qquad (11.3)$$

N_a is a general numerical diversity of 'order' a – which should not be confused with N_i, the quantity of the i-th species in the community! By increasing a, an increasing weight is given to the most abundant species. The following series arises (in some cases as a limit of equation 11.3):

$N_{-\infty}$ reciprocal of the proportion of the rarest species;
N_0 number of species;
N_1 antilog of H', the Shannon index (asymptotically);
N_2 reciprocal of the Simpson index, $1/D$;
N_∞ reciprocal of the proportion of the most abundant species, also known as the Berger–Parker Index.

Evenness is usually expressed as the ratio of the actual diversity and the maximum possible diversity for a given number of species. More complicated evenness indices were also suggested. However, the interpretation of evenness indices is sometimes problematic (e.g. Magurran 2004).

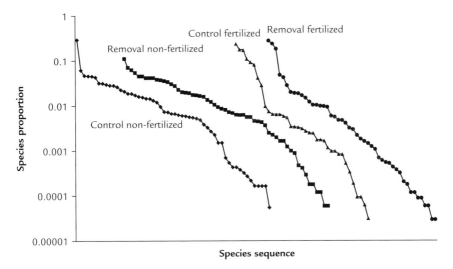

Fig. 11.1 Diversity–dominance curves for four plots in a wet oligotrophic meadow in Central Europe under different treatments (Lepš 1999), combining fertilization and removal of the dominant grass *Molinia caerulea*. Curves are based on pooled biomass values in three 0.5 × 0.5 m quadrats, 6 years after the start of the experiment. The values of the reciprocal Simpson index are (from left to right) 9.2, 22.6, 7.0 and 5.9; the values of the antilogarithm of *H'* are 19.7, 30.0, 9.6 and 9.7; the numbers of species are 54, 57, 37, and 47, respectively.

The variation in species quantities can also be expressed graphically, using so-called dominance–diversity curves (also called rank/abundance curves; see Whittaker 1975). Species are ranked from the most to the least abundant and the relative abundance (proportion of community biomass, or of the total number of individuals) is plotted on a logarithmic scale against species rank number. In this way, we obtain a decreasing curve, which varies in shape and length, and characterizes the community (Fig. 11.1). For other possibilities of graphical representation, see for example Hubbell (2001) and Magurran (2004). Sometimes, various species abundance models are fitted to the data, notably the geometric series, the log series, the lognormal distribution and the broken stick model (e.g. Whittaker 1975; Hubbell 2001). Their parameters are also used as diversity indices (Magurran 2004).

The shape of the dominance–diversity curve often varies in a predictable way along gradients, or among community types. In the example of Fig. 11.1 the four curves reflect the effects of fertilization and removal of the dominant grass *Molinia caerulea* in a yearly mown wet meadow. The slope of the curve is much steeper in the fertilized plots, reflecting a higher degree of dominance, and the curves are shorter, reflecting fewer species. The non-fertilized, non-removal plots are strongly dominated by *Molinia*, but the remaining species occur in rather equal proportions. Six years after removal, none of the remaining species had developed a strong dominance in the non-fertilized removal plots. Comparison of the curves with values of $1/D$ and H' shows that the reciprocal Simpson index is much more affected by the presence of the single dominant than the antilogarithm of H'.

In most plant communities, regardless of their species richness, the community consists of relatively few dominant species and many subordinate species, most of which have a low abundance – and consequently have a small effect on community productivity or nutrient cycling. In the example in Fig. 11.1, 90% of the biomass was made up by 27 of 57 species (47%) in the non-fertilized plots with the dominant removed, by 12/47 (25%) in removal/fertilization plots, 23/54 species (42%) in non-fertilized control plots, and 8/37 (21%) in fertilized control plots. Unlike the dominants, the low-abundant species will have a limited effect on ecosystem productivity or nutrient retention (the mass ratio hypothesis of Grime 1998). On the other hand, even low-abundant species can support populations of specialized herbivores, for example monophagous insects in a wet alder forest (Lepš et al. 1998). Such species may thus be crucial for the maintenance of diversity at higher trophic levels.

11.2.3   Spatial characteristics of diversity

The preceding section dealt with communities occupying a delimited area. However, we usually sample only part of a much larger community. With an increasing area sampled in the community, the number of species will normally increase, at a rate that varies among communities. The dependence of the number of species S found on the size of the investigated area A is described by the species–area relationship. Mainly two functions are considered: the power curve (Arrhenius model), usually written as $S = c \cdot A^z$ (and often fitted after log–log transformation $\log S = \log c + z \cdot \log A$) and the semi-logarithmic (Gleason model) curve $S = a + b \cdot \log(A)$; c, z, a and b are parameters estimated by the methods of regression analysis. The power curve starts in the origin – there are no species present at plot size zero; c is the species number in a plot of unit size; z measures the rate of increase: when doubling the plot size, the number of species increases 2^z times; z usually ranges from 0.15 to 0.3. According to the semi-logarithmic curve, a sample plot of unit size contains a species, and when doubling the area, $b.\log(2)$ new species are *added*. The number of species at zero area is not defined; actually, for very small plot sizes S would become negative. Note that both c and a depend on the units in which area is measured, whereas z and b do not. Theoretical arguments supporting either of the relationships were suggested. For example, the Arrhenius model was advocated by Preston (1962) on the basis of his analysis of abundance distributions (distribution of commonness and rarity in his words). When used for real data, neither of the two is consistently superior. Hence usually both are tried and the function best fitting the actual data is chosen. Functions with three parameters were also suggested, but are seldom used. See further Rosenzweig (1995).

Species-area curves are used on widely varying spatial scales, from within-community areas of square centimetres to whole continents. However, each curve should be interpreted solely in relation to the scale at which it was derived, and not for extrapolations. Indeed, it was shown (Rosenzweig 1995; Crawley & Harral 2001) that the slope of the relationship changes when based on different ranges of spatial scales. Lepš & Štursa (1989) showed that the estimate of the species number in the whole Krkonoše Mountains, as extrapolated from the within-habitat species area curve for mountain plains would be 30.3, and from avalanche paths 8225 species; the real value is *c.* 1220. Species–area relationships are governed by various mechanisms at various scales. At within-community scales, the increase of the number of sampled individuals is decisive, together with the ability of species to co-exist. The number of sampled individuals is negatively related to the mean size of an individual – a 1-m^2 plot may host thousands of individuals of tiny spring therophytes, but not a single big tree. With increasing area, the effect of environmental heterogeneity increases. This can be biotically generated heterogeneity – for example the variability between the matrix of dominant species and the gaps between them occupied by competitively inferior species – or small-scale heterogeneity in soil conditions at the within-habitat scale, or heterogeneity of habitats at the landscape scale. At continental scales the evolutionary differentiation between subareas starts to play a role. Fridley *et al.* (2006) demonstrated that similarly to the accumulation of species with increasing area, there is also a characteristic accumulation of species over time (i.e. when an identical plot is sampled repeatedly; see also the carousel model of van der Maarel & Sykes 1993) and that integration of these two processes can partially disentangle various mechanisms behind the species–area relationship.

To characterize the spatial aspects of diversity, the terms α or within-habitat diversity and β or between-habitat diversity are sometimes used. Whereas α-diversity can be measured by the number of species or any of the diversity indices within a limited area, β-diversity is characterized by differences between species composition in different (micro-)habitat types, or by species turnover along environmental gradients. A simple straightforward way for measuring β-diversity was suggested by Whittaker (1972; see Magurran 2004) as $\beta_w = S/\alpha - 1$, where *S* is the total number of species in the habitat complex studied (called sometimes γ-diversity) and α is the α-diversity, expressed as the mean number of species per fixed sample size. It would provide a good diversity estimate if we have a good estimate of *S*, the total number of species in the complex studied.

Usually, the number of all species in all quadrats is used as an estimate of *S*. This causes a problem: the mean number of species per quadrat is independent of the number of quadrats investigated, but the total number of species increases with the number of quadrats in the study, and thus β_w will increase with the number of quadrats used. A better approach to β-diversity is based on (dis)similarity measures. The distribution of (dis)similarity values between all pairs of samples is a good indication of β-diversity (Magurran 2004). We can base (dis)similarity measurements on both presence–absence and quantitative data.

As noted, the total richness of a community is usually not known because we are seldom able to investigate its entire distribution area. Usually a mean richness value can be obtained through analysis of sample plots of a size considered representative. Nevertheless, information on species accumulation by increasing the number of sampled plots (which is affected by β-diversity) can be used for estimation of the total species richness of the community by extrapolation, provided that the sample plots are distributed across the whole considered area. Various methods are available; see the free EstimateS software (Colwell 2009).

11.2.4 Species diversity, phylogenetic diversity and functional diversity

A community composed of four annuals will be less diverse from a functional point of view than a community composed of four species of different life-forms. This leads to the concept of functional diversity (Loreau 2000). Similarly, a community composed of four *Taraxacum* species is phylogenetically less diverse than one composed of four species from different genera. In several theories, the functional and phylogenetic differentiation within communities is more important than the plain number of co-existing species.

The traditional approach to functional diversity was based on the recognition of functional groups of species. Community diversity can be described in a hierarchical way – as diversity of functional groups, and as species diversity within functional groups. Similarly, phylogenetic diversity can be approached as diversity of genera, families, etc. The definition of functional group is crucial here, and there is a wide range of possible approaches (see Chapter 12). Clearly, by assigning individual species to usually broad functional types means a considerable loss of information. Recently, more quantitative approaches to functional and taxonomic diversity have been suggested. The most promising is the use of the Rao coefficient (e.g. see Botta-Dukát 2005; Lepš *et al.* 2006). In fact, it is a generalized form of the Simpson index of diversity (expressed as $1 - D$). Using the same notation as for diversity indices, with d_{ij} being the (functional or phylogenetic) dissimilarity of species i and j, the functional (phylogenetic) diversity (*FD*) has the form:

$$FD = \sum_{i=1}^{S}\sum_{j=1}^{S} d_{ij}p_i p_j \qquad (11.4)$$

By definition, $d_{ii} = 0$, i.e. dissimilarity of each species to itself is zero. If p_i is the proportion of individuals of species i in an infinitely large community, then FD is the expectation of dissimilarity of two individuals, randomly selected from the community. If $d_{ij} = 1$ for any pair of species (i.e. complete difference), then FD is the Simpson index of diversity $(1 - D)$, i.e. $1 - \sum_{i=1}^{S} p_i^2$ (see e.g. Botta-Dukát 2005 for details).

The main methodical decision is how to measure species dissimilarity (see e.g. Lepš *et al.* 2006 for discussion). For functional diversity, the dissimilarity measure is usually some multivariate metric (e.g. Gower distance) based on *functional traits*, i.e. species properties believed to be important for species function.

However, because we need to know the trait values for all (or the vast majority of) constituent species, the calculation is usually based on easy to measure 'soft' traits (Lavorel & Garnier 2002). These are usually morphological characteristics, supposed to be correlated with functional properties. This is often supported by available data (see the discussion of specific leaf area, seed mass and plant height in Westoby 1998). Various soft traits are included in databases (e.g. Klimešová & de Bello 2009), which often cover most of the species in an area. As to phylogenetic diversity, the dissimilarity can be based either on classical taxonomy, or, preferably, on phylogenetic analyses, often based on DNA sequences (obtained usually from GenBank, as in Cadotte *et al.* 2008). Alternatively, the functional or phylogenetic diversity can be based on a hierarchical classification of species (using cluster analysis for functional traits or phylogeny reconstruction), and express the distance using the topology of the trees, e.g. total phylogenetic branch lengths connecting species together (Cadotte *et al.* 2008). We are not yet able to measure all functional traits or gene sequences for all species in a community; consequently we have to rely on databases. Here we will have to choose between widely available but less 'functional' traits and more functional traits which we have to approximate. As to not available gene sequences, we need to find a reasonable estimation of the dissimilarity, and also cope with the situation that different genes were sequenced for different species.

Both functional and phylogenetic diversity can be partitioned into their components, particularly α- and β-diversity. By partitioning functional diversity, one can reveal trait convergence vs. divergence (de Bello *et al.* 2009), and suggest a mechanism of community assembly. If α-diversity is lower than expected under a null model (i.e. species in a sampling unit are functionally more similar than expected in a random selection from the species pool), this would indicate trait convergence, which can be explained by environmental filtering, but also by elimination of weak competitors in a highly productive environment. Trait divergence may be interpreted as support for the limiting similarity hypothesis, i.e. competitive exclusion of species that are too similar (see Section 11.3.3).

11.2.5 Intraspecific diversity

Each population is composed of different genotypes. The genotype composition depends on the mating system in the population, on the clonality of plants, and also on population size. Recent studies suggest that the fitness of a population and its ability to cope with environmental variability can be dependent on its genetic structure. Population decline is usually correlated with a loss of genotype diversity (Alsos *et al.* 2012).

11.3 Determinants of species diversity in the plant community

11.3.1 Two sets of determinants

Species occurrence in a community is a function of arriving at the site and coping with the conditions in the community. Species diversity in a plant community is thus determined by two sets of factors. The first is concerned with the species pool; the set of species propagules which is able to reach a site. The second comprises local ecological interactions; selecting species from the pool that are able to co-exist (Zobel 1992; Pärtel *et al.* 1996). In this 'community filter', both abiotic and biotic interactions operate. Abiotic conditions include physical conditions such as climate, soil, moisture, but also the disturbance regime (e.g. avalanches, fire). Biotic conditions include competitive relations, grazing pressure and effects of pathogens. In some cases, the absence of a species can be caused by the absence of a specialized dispersal agent, or absence of mycorrhizal fungi.

11.3.2 The species pool

The definition of species pool used here is broad; according to a narrower definition (e.g. Zobel *et al.* 1998) the pool will include species able to reach the site and survive. Recently, a conceptual synthesis was attempted by Vellend (2010), explaining community composition by four groups of processes: selection (deterministic fitness differences among species), drift (stochastic changes in species abundance), speciation, and dispersal. Dispersal is the basic factor influencing the composition of the species pool, whereas selection and probably also drift decide which species from the pool will finally form the community. Speciation, which is also affected by community processes, operates on a longer time scale and also affects the species pool.

The species pool is affected mostly by historical factors: the place where the species evolved, and whether they were able to migrate to a certain site. For example, many species migrated into boreal areas after the postglacial retreat of the ice sheets (Tallis 1991). The species pool is affected by the proximity of glacial refugia, and by migration barriers between the refugium and the site. The barriers are either physical (e.g. mountains), or biological. For example, the most important barrier for the dispersal of heliophilous mountain plants are forests in between the mountains, causing shade. The species pool is thus also affected by past and present competition (including competition that occurred on migration pathways). Also, postglacial micro-evolutionary processes modified species to make them better adapted to newly arising habitats, and new species also developed. Probably more species became adapted to postglacial habitats that were abundant (Taylor *et al.* 1990; Zobel 1992, Zobel *et al.* 2011).

For the sake of simplicity, the species pool is generally described as a fixed set of species. However, establishment of a single seed is highly improbable. The amount of seeds (or other propagules) needed for establishment of a viable population has to exceed a species-specific threshold. Not all populations are viable. Metapopulation theory (Hanski 1999) distinguishes source and sink populations; source populations are donors of propagules to other populations, sink populations are passive recipients of propagules. Sink populations, found in suboptimal habitats, need a constant influx of propagules from source populations to keep a stable population size (Cantero *et al.* 1999). Such 'transitional species' (Grime 1998) are probably not rare and may substantially increase the species richness of some communities. There is a mass effect occurring in the species pool: the probability that a species will pass through the community filter increases with the influx of propagules, which is related to its abundance in surrounding communities ('vicinism', see van der Maarel 1995). Another source of variation is in the dispersal ability of species. Good (e.g. anemochorous) dispersers can reach distant sites but many of their small propagules will be needed for a successful establishment, whereas bad (e.g. blastochorous) dispersers may need only a few propagules to establish, but will not reach far.

In relation to dispersal capacity, local and regional species pools are distinguished. This distinction is arbitrary, but can be useful when clearly defined. See also Chapter 6.

11.3.3   Species co-existence

Classical theory predicts that the number of co-existing species will not exceed the number of limiting resources. The competitive exclusion principle of Gause (see Chapter 7) states that two species cannot co-exist indefinitely in a homogeneous environment, if they are limited by the same resource. Nevertheless plant communities may consist of scores of species on a single square metre. This seems to contradict the competitive exclusion principle (Palmer 1994). There are many possible explanations for species co-existence. For example, Wilson (2011) counted 12 basic mechanisms suggested in the literature; he also noted that each realistic mechanism should include an 'increase-when-rare process'. Palmer (1994) suggested that mechanisms of species co-existence should be seen as a violation of assumptions of the competitive exclusion principle. The mechanisms are either equilibrium-based or not. Equilibrium-based explanations question the spatial homogeneity, i.e. species may use different parts of an existing resource gradient, or use resources in different ways: 'niche differentiation', for example different rooting depths, uses of light and phenologies. In order to co-exist, species should be functionally different (the limiting similarity concept; MacArthur & Levins 1967).

Non-equilibrium explanations challenge the assumption of permanence. If there is small-scale environmental variability and the rate of competitive displacement is low, competitive displacement may be prevented. For example, competitive hierarchies in grassland communities can change from year to year, depending on the weather (Herben *et al.* 1995). Recruitment of seedlings is more affected by heterogeneity, fluctuation and their interaction than the occurrence of established plants. The theory of the regeneration niche (Grubb 1977) assumes that co-existence is promoted through the differentiation of species requirements for successful germination and establishment. Many species are dependent on their recruitment in gaps in an otherwise closed canopy, in forests, or in grasslands. The gaps can be seen as highly variable resources; they differ not only in size, but also in the time of their creation – and plants differ in their seedling phenology (Kotorová & Lepš 1999). All this might lead to postponement of competitive exclusion and species co-existence. Indeed, the small-scale species composition in a community patch changes in time, whereas the species composition on a larger scale is fairly constant (a key element in the carousel model of van der Maarel & Sykes 1993; see Chapter 3).

The above explanations are partly based on the effects of organisms from higher trophic levels. In particular, pathogens and specialized herbivores may have greater effects on dominant species: the denser a host population, the higher the probability that a specialized herbivore or pathogen spreads in the population. This idea was behind the Janzen–Connell hypothesis (Janzen 1970), which explains the extraordinary diversity of tropical forests. No tree would become dominant, because specialized seed predators close to a parent tree would prevent establishment of seedlings around the parent tree. Although this hypothesis has not received sufficient empirical support, particularly concerning insect or vertebrate herbivores, similar mechanisms may support species diversity through specialized pathogens, particularly in soil, and not only in the tropics (Wills *et al.* 1997; Petermann *et al.* 2008).

Hubbell (2001), using mathematical models, demonstrated that species co-existence could be maintained under the species 'neutrality' hypothesis – i.e. when all the species have the same competitive abilities, in case there is some constant influx of new species (by immigration, or by speciation). However, because species differ in their competitive abilities, it is difficult to see how such neutral models can be ecologically realistic. Still, the role of 'lottery recruitment', implying that the identity of a species entering a gap is determined at random (which is one of the bases of Hubbell's model) is increasingly accepted. The chances to be the winner, however, differ among species, and a 'weighted lottery' (Busing & Brokaw 2002) is probably a more realistic model.

11.3.4 Distinguishing the effect of the species pool from local ecological interactions

The relative importance of historical factors (as reflected in the species pool) vs. that of local ecological factors is often discussed, but it is difficult to separate these effects, particularly because the actual species pool is also affected by local species interactions. A positive correlation between the actual species richness of a community and the number of species able to grow there has been demonstrated. However, the set of species able to grow in a habitat is determined by the species which actually occur in the communities, and hence by local ecological factors (Herben 2000). For example, calcareous grassland communities in Central Europe may be rich in species because there is a large pool of species adapted to these conditions. But it can also be argued that the large species pool is a consequence of the richness of calcareous grasslands, which is consequence of local ecological factors that promote species coexistence.

Probably the best way to separate the effects of local ecological interactions and general historical effects is to compare the patterns of species richness between geographical regions. Schluter & Ricklefs (1993) suggested a procedure for the decomposition of variance in species richness into parts attributable to habitat, geographical region and their interaction. The method is analogous to the decomposition of the sum of squares in two-way ANOVA. Repeated patterns in geographical regions differing in their history suggest the effects of local conditions, while differences indicate the effects of history.

Some patterns in species richness occur in various geographical regions; they are probably based on local mechanisms. For instance, tropical rainforests are always much richer in species than adjacent mangroves. This can be understood because of the physiologically extreme conditions in mangroves. On the other hand, mangroves in West Africa are poor in comparison with the richer mangroves of Malaysia. This difference may have historical reasons.

Experiments have shown that species which are missing in a community may be able to establish a viable population there, when their propagules are introduced. In this way we may get an indication whether limitation of diversity is related to species pool (dispersal) limitation, or to local ecological interactions, even though results of similar experiments must be interpreted with caution (Vítová & Lepš 2011). However, a successful experimental introduction should be followed by checking that none of the resident species was outcompeted from the community. An increasing species pool need not necessarily lead to increasing richness of a plant community. The introduction of a successful invasive species (i.e. increase of the species pool by a strong competitor) usually causes a reduction in diversity (see Chapter 13).

11.4   Patterns of species richness along gradients

11.4.1   Introduction

Ecologists have long since known that species richness of plant communities changes along environmental gradients in a predictable way (reviews in Huston 1994; Rosenzweig 1995). The decrease of species diversity from the equator to the poles is one of the most universal patterns in nature. This decrease does not only hold for species, but also for higher taxonomic levels (genera, families). Fossil records show that this pattern can be traced back at least to the Cretaceous (Crane & Lidgard 1989). At present, tropical rainforests are the richest plant communities on Earth at larger spatial scales; also, they are unsurpassed regarding their functional and phylogenetic diversity. Typical numbers of tall tree species are 100–300 ha^{-1}. For example, in the Lakekamu Basin alluvial plot in Papua New Guinea, 182 species belonging to 104 genera and 52 families were identified (Reich 1998). Typically, many species had a low abundance; 86 species (47%) were found with one single individual. There is little doubt that the high number of tropical species has historical reasons – the historically relatively stable environment minimizes extinction rates. Although glacial periods also affected the tropics, rainforest regions pertained through all the 'full-glacial' periods in the tropics of Africa, South America, South-east Asia and Oceania (Tallis 1991).

How are these hundreds of tree species able to co-exist? Many explanations have been suggested (e.g. reviewed by Hill & Hill 2001). The high photon flux enables the diversification of the tree canopy (emergent trees, several canopy layers), supporting niche differentiation. The decreasing species pool in forests further away from the tropics reflects both the historical reasons (e.g. the decreasing richness of genera and families), but also the increasing harshness of the environment (decreasing richness of life-forms). Nevertheless there are also extremely species-rich communities in various parts of the subtropical and temperate zones. For example, at finer scales temperate grasslands in various parts of the world, or even semi-deserts are among the most species-rich communities, with close to 100 vascular plant species per m² (e.g. Cantero *et al.* 1999). However, none of these communities is comparable to tropical rainforest for functional diversity and the diversity of higher taxonomic units, and also to species diversity at larger spatial scales.

Here, we will discuss the diversity response to productivity, and to disturbance. These two gradients are considered to be the most important axes determining the habitat templet (Grime 2001; Southwood 1988).

11.4.2   Relations between species richness and productivity

At the global scale, the productivity of terrestrial vegetation decreases from the equator to the poles, and species richness is positively correlated with productivity. At the local scale, however, unimodal (humped) relationships have often been found (Fig. 11.2). Meta-analyses of published studies (e.g. Mittelbach *et al.* 2001; Gillman & Wright 2006) have shown that unimodal relationships are common, but not ubiquitous. The validity of these meta-analyses have been questioned (see Forum in Ecology, *Ecology* vol. 91, e.g. Whittaker 2010 vs. Mittelbach 2010); the unimodal relationship is scale-dependent, i.e. it depends on the focal scale (size of plots included in the analyses) and also extent (total area, in which the samples were taken). The focal scale is particularly important because the shape of the species–area curve (the value of the z exponent in $S = c \cdot A^z$, see 11.2.3) often changes with the prevailing species strategy and with the size of the individuals of the constituent species, which in turn change with productivity or disturbance (Lepš & Štursa 1989). The productivity data (e.g. in $g \cdot m^{-2}$) should however be independent of plot size. Still, the pluriformity of the relationship species richness–productivity is clear; it seems that with increasing focal scale, the relationship changes from unimodal or negative to more positive. The situation is further complicated by selection of productivity measure (see various possibilities in Fig. 11.2), and also, by selection of community types (e.g. Gillman & Wright 2006 excluded from their meta-analysis all mown and grazed plots).

The impact of low productivity on richness can be adverse where the environment is so unproductive or otherwise extreme that no organism would survive. An increase in richness with increased productivity is then rather obvious. On the contrary, ecologists are puzzled by what happens at the other side of the hump (or in negative relationships): why does species richness decrease at high productivity levels.

Unimodal relationships between species diversity and standing crop, with a peak in richness at a moderate level and a decrease towards productive environments, have been found in many temperate grasslands, both natural and semi-natural (Fig. 11.2). A more rapid decline was also found in fertilization

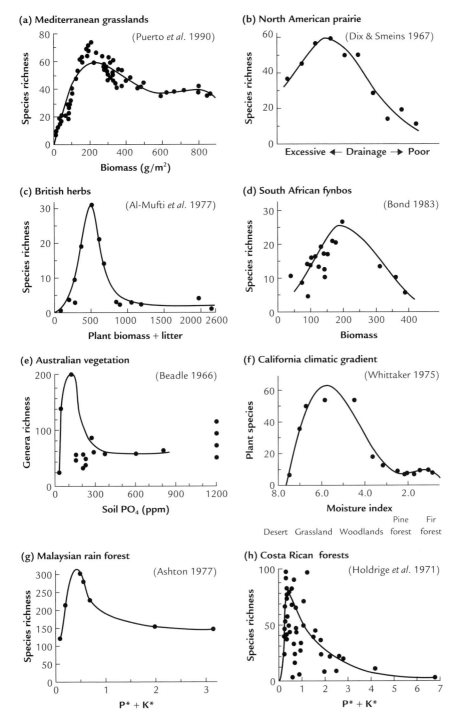

Fig. 11.2 Examples of unimodal relationships between species richness and measures of habitat productivity in plant communities. P* and K* are normalized concentrations of soil phosphorus and potassium, which were summed to give an index of soil fertility. (From Tilman & Pacala 1993, where also references to the original sources can be found.)

experiments (see Fig. 11.1). However, the unimodal relation was also found in woody vegetation (Fig. 11.2). Generally, eutrophication, an increased nutrient load, is considered one of the most important factors in the recent loss of diversity in European grasslands.

The reduction of species diversity in oligo- and mesotrophic grasslands and small sedge communities at increased nutrient levels may be caused by outcompetition of species by species increasing their growth rate faster (competitive displacement, Huston 1994). This has been confirmed by experiments where the faster growing species had been removed. Under increased soil productivity, competition for nutrients shifts to competition for light and the taller species take advantage. Competition for light is more asymmetric than below-ground competition. Soil heterogeneity, together with varying supply rates and varying rooting depth of plants may allow more niche differentiation and less asymmetric competition (Lepš 1999). Tilman & Pacala (1993) also suggested that the effective heterogeneity decreases when plant size increases.

11.4.3   Relations between species richness and disturbance

Similarly to the response to productivity, species richness also often exhibits a unimodal response along an axis of disturbance intensity, with the maximum found in the middle of the axis (or in the intermediate successional stages). The following discussion will be based on Grime's (2001) concept of disturbance: partial or complete destruction of plant biomass. Impacts of avalanches, fire, windstorms, but also grazing and mowing are all types of disturbance (and succession can be seen as a response in time since the last major disturbance event). There are at least three features that characterize the disturbance regime: *severity* (what proportion of biomass is destroyed), *frequency* (how often the disturbance occurs) and *spatial extent*. Again, it is easy to understand that at high disturbance levels, the species richness decreases with further increasing levels of disturbance, until no plant species will survive. The focus of attention is on diversity at medium disturbance levels, where the disturbance positively affects species richness.

The 'medium disturbance hypothesis' (Huston 1979), demonstrated that in systems where a competitively strong species prevails in the absence of disturbance, a medium frequency of disturbance leads to an increase in species richness, while under a higher frequency of disturbance, only fast growing species will survive. Huston (1994) demonstrated that the impact of medium disturbances depends on the system's productivity (i.e. on the growth rates of the prevailing species); in a more productive environment maximum diversity occurs at a higher disturbance level (Fig. 11.3).

As to possible mechanisms of response to disturbance, Huston (1979) showed that the destruction of a constant proportion of each species could postpone competitive exclusion. However, disturbance often harms the dominant species more, particularly those superior in competition for light, which leads to a 'increase-when-rare process' (Wilson 2011). The disturbance by mowing a grassland is more destructive to the taller species because a larger proportion of their biomass is removed (Klimešová *et al.* 2010). One of the effects is that

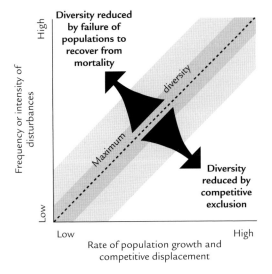

Fig. 11.3 Conceptual model of domains of the two primary processes that reduce species diversity. Diversity is reduced by competitive exclusion under conditions of high rates of population growth and competitive displacement and low frequencies and intensities of disturbance. Diversity is also reduced by failure of small and slowly growing populations to recover from mortality under conditions of low population growth rates and high frequencies and intensities of disturbance. Note that the frequency or intensity of disturbance supporting maximum diversity increases with population growth rate (i.e. with system productivity). (From Huston 1994.)

low-growing species are no longer outcompeted for light. An avalanche will destroy existing trees on its path and affect occurring shrubs, but it will usually not disturb the herb layer too much. Further, several forms of finer-scale disturbance may be spatio-temporally heterogeneous, which again promotes species co-existence. For some types of disturbance, e.g. windstorm damage, the spatial extent of the disturbance and the average time between two subsequent events are inversely related (single tree falls appear often, large windbreaks may happen only once in many decades). Medium disturbance leads to a mosaic community structure, with patches of various successional stages – and the resulting complex community is species-rich. In communities with many species dependent on regular seedling recruitment, disturbance provides the 'safe sites' for seedling recruitment.

Each community type has its typical disturbance regime. Changes in the intensity and type of a disturbance regime of an adapted community will often lead to a decrease in species richness: typical examples are fire suppression in North American forests (e.g. Hiers *et al.* 2000), and cessation of grazing and/or mowing in species-rich meadows (Lepš 1999).

The development of species diversity during a secondary succession often shows a similar pattern as described for the relation of diversity to productivity

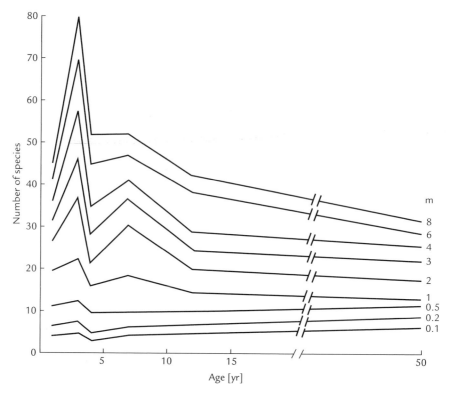

Fig. 11.4 Changes in species richness during an old-field succession, measured on various spatial scales. The numbers on the right side are sizes of sampling plots expressed as the lengths of the quadrat side. (From Osbornová *et al.* 1990.)

and disturbance: there is a rapid increase in species richness during the early years towards a maximum in intermediate stages, followed by a slow decrease. This is shown for an old-field succession (Fig. 11.4, which also elucidates the scale dependence). One may interpret this development as a response to the sudden drop in disturbance connected to the earlier management of the field. In the tropics, however, species richness usually increases steadily towards undisturbed mature forest.

11.5 Stability

11.5.1 Ecological stability

Ecologists have long believed that diversity begets stability (e.g. MacArthur 1955). On the other hand, May (1973) demonstrated that mathematical models predict a negative relationship between stability and complexity (including diversity). However, the results were based on unrealistic models: contrary to the model assumptions, ecological communities are far from random assemblages of species, and by analysing a linearized model close to system equilibrium one does not learn much about the many-sided behaviour of ecological systems. May's model demonstrated that the probability of the stability of an equilibrium of a randomly generated community matrix (in terms of Liapunov stability) decreases with the size of the matrix, i.e. with the number of species in the model community. Also, Liapunov stability as used in mathematics – and in models of theoretical ecology – is not an ideal reflection of what ecologists consider to be ecological stability (see Section 11.5.2). A positive effect of May's book was that ecologists realized that it is necessary to define clearly what ecological stability is and how we should measure it in real ecological systems, and also that the positive relationship between diversity and stability is not a necessity – should it be predicted by a model, then it depends on the model assumptions. In mathematical models, we have various analytical tools that enable the analysis of system equilibrium (equilibria), and its (their) stability. The only way to assess ecological stability in nature is to follow a real system trajectory in a 'state space' defined by selected measured variables such as total biomass, population sizes and rates of ecosystem processes. The evaluation of stability is then dependent on the variables selected for measurement, and on the length of the period and the frequency of the measurements. Regarding plant communities, we are usually mostly interested in species composition, total biomass and nutrient retention. These characteristics may behave independently; the total community biomass may be fairly constant while the species composition fluctuates, or the other way around.

11.5.2 Characteristics of ecological stability vs. non-stability

Various aspects of ecological stability are distinguished (e.g. Harrison 1979; Pimm 1984; Fig. 11.5). The first two concepts are based on system behaviour under 'normal conditions':

1 *Directional changes in the system state.* A lack of directional changes is usually interpreted as 'stability' (the system is considered to be in a state of 'equilibrium'); systems undergoing directional changes are called transient or unstable. This concept corresponds more or less to the existence of a stable equilibrium in mathematics. It is linked to that of succession – successional communities are by definition unstable, i.e. not in an equilibrium – but climax communities and also some 'blocked' successional stages are stable. A system may also be subjected to cyclic succession, as described by Watt (1947; see Chapter 4). This aspect of stability can only become clear after long-term analysis. Slow and small directional changes might be masked by random variability. This is why we will always use quotation marks when speaking about 'equilibrium' in real communities. The concept is also scale-dependent. Depending on spatial and temporal scales, even climax communities undergo local successional and cyclic changes, and

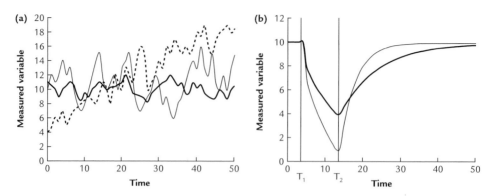

Fig. 11.5 Concepts of ecological stability. The measured variable is the choice of
the researcher (e.g. community biomass, photosynthesis rate, or population size).
(a) Community fluctuation in a constant environment. Broken line, unstable transient
community; full lines, communities in some steady state; heavy line, the more constant
(less variable) community. (b) Community stability when facing a perturbation
(sometimes called a stress period), which starts at T_1 and ends at T_2. Heavy line: the
community which is more resistant, but less resilient than the community indicated by
the light line. The time scale depends on the rate of ecosystem dynamics; for terrestrial
plants, it is usually measured in years. (Adapted from various sources.)

in sufficiently long-term perspective, the communities adapt to climate
changes.

2   *Temporal variation* (also indicated as variability) or, its opposite, *constancy*,
   determines how much the system fluctuates under 'normal conditions'.
   Standard measures of variability are used (e.g. standard deviation, SD, in a
   temporal sequence), usually standardized by the mean. For example, for total
   biomass, the coefficient of variation (CV = SD/mean) or the SD of log-
   transformed data would be appropriate measures of temporal variability.
   When the data are counts rather than a continuous variable, use of Lloyd's
   index of patchiness: $L = 1 + (SD/mean - 1)/mean$, will lead to a reasonable
   standardization by mean. When we are interested in species composition,
   the measured variable is multivariate; for the evaluation of such data, we
   should apply methods of multivariate analysis. For example, we can follow
   the community trajectory in ordination space, or measure the average (dis)
   similarity between subsequent measurements, or use multivariate analogs of
   variance, standardized by corresponding means.
      Ecological stability is often defined as the ability to remain in a state
   ('equilibrium') when facing some perturbation, and to return to the original
   state after the perturbation ceases. The next two characteristics are concerned with a response to external perturbation.

3   *Resistance*, the ability to resist a perturbation, and
4   *Resilience*, the ability to return to a pre-perturbation state. In both cases
   some period of 'normal conditions', i.e. some sort of equilibrium, is involved,

followed by a short period of limited perturbation. Each community exists in a variable environment, so that which variability is still 'normal' and which already means perturbation is quite arbitrary. Also, resistance and resilience should always be related to the perturbation under study.

Resistance is measured by the proximity to its original state of a system displaced by perturbation, i.e. by the similarity between the pre-perturbation and post-perturbation state. For example, during an extreme drought in 1976, young fallow decreased its standing crop by 64% in comparison with the 'normal' year 1975, but the old fallow only by 37% (Lepš *et al.* 1982, Fig. 11.6). So the old fallow was considered more resistant. When species composition of a community is concerned, we can use various (dis)similarity measures between the original and the perturbation state.

Resilience means the ability of a system to return to its original state after perturbation. It can, for instance, be measured as time, when the displacement caused by the perturbation has decreased to 50%. In many cases a return is neither smooth nor monotonous. Then, *ad hoc* measures of resilience have to be used, for example reduction of the displacement after a fixed period of time. The concepts of resistance and resilience can be applied to communities which are stable according to the first definition, i.e. being in 'equilibrium', a state towards which the system returns after perturbation. However, the concepts can also be applied to a successional community, provided that the rate of succession is much slower than the response to perturbation.

5   *Persistence.* In addition to these four frequently used aspects of stability there is *persistence*, defined as 'the ability of a system to maintain its population levels within acceptable ranges in spite of uncertainty of the environment' (Harrison 1979). Often the community is considered persistent when no species are lost during the observed time period.

After large or long-lasting perturbations, ecological systems may be too much damaged to recover, because species have become locally extinct, or the soil profile has been destroyed. These are examples of irreversible change. Thus, an important characteristic of stability is the range or/and the intensity of a perturbation from which the system is able to return to its original state.

For all aspects of variation, resistance and resilience, the temporal scale is very important. In comparative studies it might be more realistic to relate recovery time to the generation time of the constituent species.

Temporal variation, resistance and resilience reflect the community response to environmental fluctuation. An important part of the response by individual species is found in the physiological tolerance of their populations. In order to construct a realistic model, one would need to quantify the response of populations to environmental fluctuations (Yachi & Loreau 1999). Also, there is a physiological trade-off between growth rate and resistance to extreme events (MacGillivray *et al.* 1995); consequently the species that are highly resistant are usually not highly resilient. Similarly, communities composed of highly resistant species will not be highly resilient.

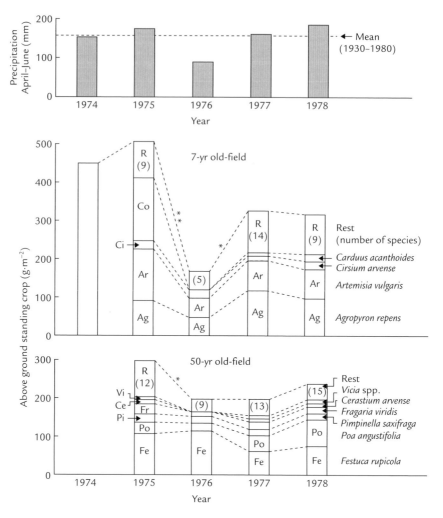

Fig. 11.6 Comparison of resistance and resilience of a 7-year old-field and a 50-year old-field. (a) Course of the spring precipitation in 1974–1978 suggesting that 1976 was an extreme year and can be considered as a 'perturbation period'. The decrease in the total productivity from 1975 to 1976 was considerably higher (and also more significant) in the younger field (so the younger field has a lower resistance). However, the younger field started earlier to return to the 'normal' state – so it has a higher resilience. Note that the characteristics were used for the successional stages; we expected that the successional development would be much slower than the response to drought. However, in the younger field, there is some decrease of productivity that should be taken into account – the standing crop never returned to the 1974 value in this plot. Differences between subsequent years were tested using the t-test ($* = P < 0.05$, $** = P < 0.01$). The number of species constituting the rest (in parentheses) has indicative meaning only. (From Lepš et al. 1982.)

11.6 On the causal relationship between diversity and ecosystem functioning

11.6.1 On correlations and causes

Not only does diversity change in a more or less predictable way along ecological gradients, this will also be the case for functional characteristics, such as primary productivity, nutrient retention or stability. Consequently, diversity and function will often be correlated. However, this does not necessarily imply a causal relationship. Both diversity and function can be dependent on the same set of environmental constraints. Also, diversity might be the consequence rather than the cause of stability, particularly on an evolutionary time scale.

11.6.2 Biodiversity experiments

Experiments have been carried out where community diversity, considered as an independent variable, is manipulated and the functional response, considered as a dependent variable, is measured. A significant statistical relationship is strong evidence for a causal relationship. This approach has a weak point: changing the diversity implies changing the species composition. However, as demonstrated (e.g. Lepš *et al.* 1982; MacGillivray *et al.* 1995; Rusch & Oesterheld 1997; Grime *et al.* 2000), the identity of the constituent species, and hence the plant functional types they belong to, is the basic determinant of ecosystem functioning. Whether it is possible to separate diversity and identity in such experiments, is a difficult and still debated question; maybe carefully designed experiments can provide some insight (Hooper *et al.* 2005).

Some of these biodiversity experiments comprise very extensive field experiments. One example is the 'Jena experiment', located near the German city Jena, jointly supervised by German and Swiss institutions, including ecologists from Jena. It includes 16 replicates of species richness 1, 2, 4 or 8 species, then 14 replicates of richness 16, and four replicates of a mixture of 60 species; each replicate comprises a 20×20 m plot, while there are also many additional 3.5×3.5 m plots, including monocultures of all constituent species. See Plates 11.1, 11.2 and Roscher *et al.* (2005). Other examples are Cedar Creek (Tilman *et al.* 1996), and multi-site European projects BIODEPTH (Hector *et al.* 1999), CLUE (van der Putten *et al.* 2000) and the experiments by the pan-European consortium (Kirwan *et al.* 2007).

How should such experiments be arranged? What are suitable methods for their analysis? And what are the lessons from their results for the functioning of real communities and ecosystems? Several analytical approaches are available, and they have various requirements on the experimental design. Consequently, the experimental design of a biodiversity experiment should ideally take the subsequent analytical tools into consideration. Further, irrespective of the analytical methods used, the species should as far as possible be represented equally at all richness levels, and individual richness levels should have replications differing in species composition.

A simple example may illustrate some of the problems. Three species, A, B and C are involved in an experiment on the effect of species richness (S) in mixtures on the final biomass yield (Y), which is often considered a parameter of 'ecosystem function'. When plants are grown from seed, biomass is a reasonable measure of productivity, and many of the functional characteristics (e.g. nutrient retention or CO_2 assimilation), are usually correlated with biomass and/or productivity. In that case, most of the reasoning presented below for biomass can be applied to some other ecosystem functions. By choosing productivity we can also rely on the large number of earlier experiments, both ecological and agronomical (e.g. Trenbath 1974; Austin & Austin 1980; Vandermeer 1989). But, we should be aware that most of the studies of 'effects of biodiversity' are based on simple measurements of the above-ground biomass, and it seems that these effects are too easily interpreted as effects on 'ecosystem functioning'. As noted by Srivastava & Vellend (2005), high productivity is not always a desirable property of an ecosystem, and so higher community above-ground biomass does not necessarily mean 'better ecosystem functioning' from the nature conservation point of view.

In our three species example, if all replications at $S = 1$ would be composed of species A, at $S = 2$ of mixtures of A and B, and at $S = 3$, of mixtures of A, B and C, the specific effect of species B would be indistinguishable from the increase of S from 1 to 2, and of species C from the increase of S from 2 to 3. This type of design, where the species composition is constant in all replications at a given species richness level, which form a subset of the composition at higher S-levels, was used in the pioneering Ecotron experiment (Naeem *et al.* 1994). The results were then heavily criticized (Huston 1997). In a much better design, the replications at $S = 1$, are monocultures of all three species A, B and C, which are equally replicated; at $S = 2$, all three possible pairs (i.e. AB, AC and BC) are equally replicated, so that we have three mixtures of two species; at $S = 3$ the (replicated) mixtures of all three species are included. In most experiments, a substitution design is used, leading to the replacement series of de Wit (1960), i.e. the total number of sown seeds is kept constant, and divided among the constituent species, which occur most often in equal proportions.

It is practical to have all the species combinations for mixtures up to say five species; however, we are usually not interested in the effects of diversity in five-species communities, but in considerably richer communities. Here, we will never have enough resources to include all possible species combinations for higher numbers of species; for $S = 10$ we would already have 45 possible two-species combinations, 120 three-species combinations, 210 four-species combinations, and so on. Usually, we are not even able to cover all possible richness values, so we select just some of them, and for each of them, select some species combinations (and here we need to care about equal representation of species in various richness levels). For practical reasons, the number of experimental units which we are able to handle is more limited if we require that not only the total 'ecosystem function', but also the contributions of individual species are determined.

In most similar studies, the final yield of the community is positively correlated with species richness (see the meta-analysis of Cardinale *et al.* 2007). Two main mechanisms are supposed to generate this relationship: the effect of the selection, sometimes called sampling or chance effect (Huston 1997; Aarssen 1997) and the effect of complementarity (which in some calculations also includes possible facilitation). We will use the simple three-species example from above to illustrate the selection effect (Fig. 11.7). Suppose A and B are small annual weeds (e.g. *Viola arvensis* and *Arabidopsis thaliana*) and C is a highly productive species (e.g. *Chenopodium album*). We expect that *Chenopodium* will dominate all mixtures where it is present, and consequently these communities will have a much higher biomass than the other ones. We also expect that – if the sowing density is not very low – *Chenopodium* will achieve a biomass in the mixtures which is close to its biomass in a monoculture, whereas the biomass of the other two monocultures, and the mixture of *Viola arvensis* and *Arabidopsis thaliana* will be very low. As the highly productive species (i.e. *Chenopodium*) is present in one third of the monocultures, in two thirds of the two-species mixtures and in all of the three-species mixtures, the average biomass will increase with species richness. Thus, we can expect a positive, highly significant regression of Y on S (the sampling or chance effect): by simply increasing the number of species we increase the chance that a productive species will be present. There was much controversy about this effect in the recent biodiversity debate. Loreau (2000) suggested that this is called the (positive) selection effect, to stress the fact that the most productive species has to prevail in the mixture to produce this effect, and this term is now often used. The average biomass increases as a consequence of the positive selection effect, but the selection effect itself is not sufficient for the biomass of the mixture to exceed the biomass of the most productive monoculture. When the mixture exceeds the biomass of its most productive constituent species monoculture, we speak of overyielding (Trenbath 1974). However, the term is also used in a much wider sense: for example Tilman (1999b) used the term for the situation where the productivity of a species in a mixture is higher than its yield in a monoculture divided by the number of species in a community. To avoid confusion, the term transgressive overyielding is used for the situation where the mixture is more productive than the most productive monoculture. Transgressive overyielding is strong evidence that there is more than a selection effect playing a role.

Mechanisms that potentially can (but need not) lead to transgressive over-yielding, are complementarity and facilitation. Complementarity means that various species are limited by different resources, or differ in the mode of use of a resource. Typical examples are the different rooting depths of species (Fig. 11.7), or the separation in time of species (e.g. spring vs. summer species). Complementarity is equivalent to niche differentiation – which, in equilibrium theory, is considered a necessary condition for species co-existence – and is probably very common in nature. A typical example of facilitation is the increase of soil nitrogen as a consequence of the presence of legumes, leading to the increased productivity of other species. Usually, on the basis of the final outcome, complementarity cannot be distinguished from facilitation; here we need knowledge of the biology of the constituent species and supplementary experiments focused directly on the mechanisms of interactions. Of the three mechanisms mentioned, only facilitation potentially can (but need not) lead to a

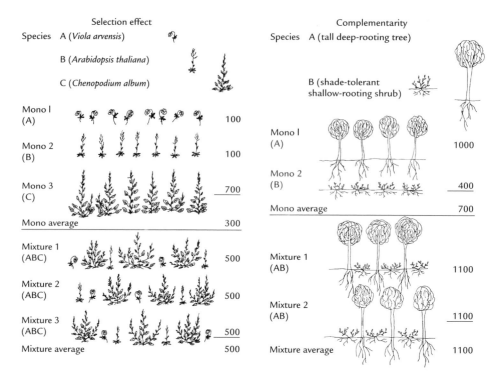

Fig. 11.7 Selection (sampling, chance) effect and complementarity affecting the final yield in biodiversity experiments. Comparison of three monocultures (Mono1, Mono2, . . .) with mixture of the three species is shown for selection effect, of two monocultures and their mixture, for complementarity. Mixture 1, Mixture 2, are replications of the mixture. Final biomass (in arbitrary units) is taken as the response variable. In the selection effect, the most productive species (such as *Chenopodium album*) prevails in the mixtures, and suppresses the less productive species. When sown in sufficient density, the mixture biomass approaches the biomass of the most productive species, but does not surpass it. In the complementarity effect, the species use resources in a different way, and consequently, the biomass of the mixture might (but need not) exceed that of the most productive monoculture. (Figure drawn by Eva Chaloupecká.)

situation, where a population in the mixture has a higher biomass than in the monoculture.

To evaluate the results of biodiversity experiments, various indices of biodiversity effects were suggested. The most common are the ratio of the mixture biomass to the biomass of its most productive species, characterizing the transgressive overyielding and the additive partitioning of the net effect to selection and complementarity effects suggested by Loreau & Hector (2001). This method is based on the idea of relative yield total (RYT, de Wit 1960); the net effect is the difference between the actual yield of the mixture and the average of monoculture yields of its constituent species (corrected for sowing proportions if the species are not sown in equal proportions). This value is partitioned into selection and complementarity effect on the basis of contributions of the individual species – selection effect is characterized by high covariance between a species monoculture yield and its deviation between realized and expected mixture yield (in other words, the selection effect is high if species with high monoculture yield prevail disproportionally in the mixture at the expense of species with low monoculture yield). Although highly intuitively appealing, the method of additive partitioning does not directly measure the mechanisms; these are just inferred from the productivity of individual species in mixtures. For example, the positive complementarity effect can be generated not only by real complementary use of resources, but also by facilitation (typically by the presence of a legume in the mixture), or by the fact that herbivores spread more easily (and so decrease the biomass more) in monocultures than in polycultures (see discussion in Trenbath 1974 or Vandermeer 1989). Simulations by Fibich & Lepš (2011) demonstrated that various shapes of dependence of final yield on sowing density in combination with the substitutive design used in these experiments can also generate non-zero values of these parameters in the absence of complementarity or facilitation. We should be aware that most of the recent conclusions about the mechanisms of biodiversity effects (e.g. in the meta-analysis of Cardinale *et al.* 2007 that the complementarity increases with duration of the experiment), are not based on direct measurement of mechanisms, but just on this additive partitioning of the net effect. Fox (2005) suggested a tripartite partitioning which is an extension of the method of Loreau & Hector (2001); it further divides the selection effect into two terms – dominance effect and trait-dependent complementarity. Other methods imply a direct application of classical statistical methods (general linear models, see Kirwan *et al.* 2007, 2009), where species identity in a mixture and species richness are used as sets of predictors of mixture performance (usually yield). In this way, the method should be able to separate the effect of species identity from the effect of species richness. A comprehensive review of the methods for the analysis of biodiversity experiments was recently published by Hector *et al.* (2009).

The meta-analysis of Cardinale *et al.* (2007) showed that mixtures were more productive than monocultures in 79% of the experiments. Similarly, species-rich communities are on average more efficient, for example in nutrient uptake (Tilman *et al.* 1996), or in overall catabolic activity of soil bacteria (Stephan *et al.* 2000). There is little doubt that this is a prevailing pattern; nevertheless, some studies found no effect of diversity on productivity (e.g. Kenkel *et al.* 2000). When the additive partitioning is used, the complementarity effect increases with the duration of the experiment (Cardinale *et al.* 2007). It also seems that this effect is saturating, i.e. it is most pronounced at low richness, but reaching an upper asymptote rather soon (Fig. 11.8). However, evidence that species-rich communities are more productive than the most productive monocultures or species-poor communities is mostly lacking (Cardinale *et al.* 2007 found transgressive overyielding in only 12% of experiments). This leaves room for contradictory interpretations (Garnier *et al.* 1997 vs. Loreau & Hector 2001). Fig. 11.8 gives an example of how different graphical presentations based on the same data might lead to different interpretations.

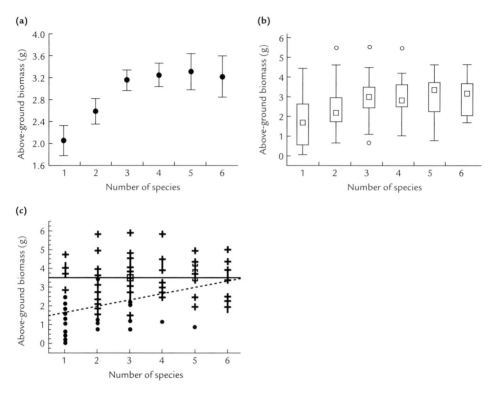

Fig. 11.8 The perception of the results of a biodiversity experiment can be affected by the way of statistical analysis and graphing (data from the low sowing density in the pot experiment of Špačková & Lepš (2001). (a) mean and standard error of mean showing that mean biomass increases with species richness. (b) Median values shown by squares, interquartil ranges by boxes, non-outlier extremes by whiskers, outliers by circles; outliers are more than 1.5× the interquartile range from the quartiles. Median values increase, minimum values increase as well, but the maximum is more or less independent of species richness. (c) Biomass value for each pot is shown separately. Data set divided into pots containing the most productive species, *Holcus lanatus*, (+, regression shown by full line) and those without this species (●, broken line). When the most productive species is absent, the average biomass is lower and increases with the number of species, as the probability that the second most productive species will be present increases.

Species richness is the directly manipulated variable in most biodiversity experiments. However, what should really matter for ecosystem functioning is the diversity of functional traits in a community (Loreau 2000); species richness is just a surrogate characteristic reflecting functional diversity. Indeed, some analyses show that functional or phylogenetic diversities are often better predictors of ecosystem function than the number of species (e.g. Lanta & Lepš 2006, Cadotte *et al.* 2008, 2009). Also, in the majority of experiments, the species are sown in equal proportions, and so attention is only paid to species richness.

Nevertheless, as shown by Kirwan *et al.* (2007), evenness can be the driving force of biodiversity effects.

The problem of biodiversity experiments is that in nature, species richness is basically a 'dependent variable', i.e. the result of ecological forces. Plant communities can be species-poor for three basic reasons: (i) lack of species in the species pool, i.e. of species able to reach the site, (ii) an extremely harsh environment (low productivity or high disturbance), and (iii) a highly productive environment, where competitive exclusion is fast. We can expect that ecological functioning of these three types of species-poor communities will be very different. The low diversity treatments in biodiversity experiments are achieved by the low number of species sown (often together with weeding), which corresponds to a lack of species in the species pool (Lepš 2004a). The experimental gradient in species richness is created by limiting the number of species allowed to enter the experimental plot, which might correspond to plant communities limited by the size of the species pool, but very probably not to communities where a highly productive environment leads to fast rates of competitive exclusion. The simulation study of Stachová & Lepš (2010) demonstrated that a pronounced increase in productivity with an increasing number of species in the community is expected only when the underlying richness gradient is caused by limitation of the species pool, i.e. limitation of number of species available at the site. This can explain why many patterns observed in biodiversity experiments are not confirmed in nature (e.g. the predicted positive correlation between species richness and productivity). Also, an important difference may exist between synthetic communities in biodiversity experiments and mature natural communities. Grace *et al.* (2007), on the basis of a structural equation modelling approach controlling for possible environmental effects, suggested that the influence of small-scale diversity on productivity in mature natural systems is weak.

11.6.3   Does diversity beget stability?

Like ecosystem functioning, ecological stability is mostly determined by the life histories of the prevailing species (Fig. 11.9). A community of cacti will be highly drought resistant, regardless of its species richness. However, when damaged, the recovery, depending on the resilience, will be slow, regardless of the species richness. Since the species (and life history) composition is determined by habitat characteristics, the latter are expected to be the main determinants of both species richness and stability.

Ecosystem functioning (energy flow and matter cycling) is dependent on a limited number of dominant species. The subordinate species will not be very important for the actual functioning of the community, but they might play an important role when the conditions change (Grime 1998). As far as the environment is variable, species richness might help to cope with these changes. MacArthur (1955), when proposing the Shannon index as an index of stability, suggested that the functional redundancy amongst species may increase the possibility that when a species fails to fulfil its role in the community, its function can be taken over by another species (risk spreading). Since then,

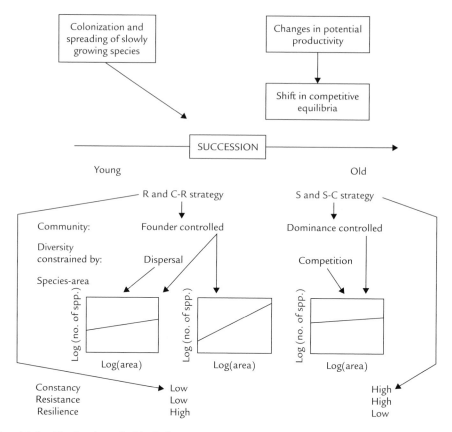

Fig. 11.9 Mechanisms behind changes in the species–area curve and stability characteristics during secondary succession. (Based on an old-field succession in Central Europe: Lepš *et al.* 1982; Lepš & Štursa 1989; Osbornová *et al.* 1990.).

several mechanisms have been proposed, some with rather complicated mathematical models; however, these are mostly variations on the original idea of MacArthur.

Doak *et al.* (1998) suggested that aggregate community characteristics such as total biomass should be less variable as an effect of statistical averaging. It follows from basic probability laws that the coefficient of variation (CV; see section 11.5.2) of the sum of *independent* random variables should generally decrease with the number of variables included (Doak *et al.* 1998). According to Tilman (1999b), this decrease will depend on (i) the way the variance is scaled with the mean, and (ii) the independence of the variables. The stabilizing effect would weaken when the more abundant species are less variable, i.e. have a lower CV than the less abundant species, which is often the case (Lepš 2004b). In that study, the CV of the dominant species (*Molinia caerulea*) was, in an unfertilized semi-natural meadow, smaller than the CV of the whole community, including *Molinia*. This suggests that the ability to attain dominance might be dependent on similar traits as is the ability to maintain a constant biomass over time. However, the negative correlation between average biomass and CV is usually not strong enough to fully compensate for the averaging effect. The assumed independence of the variables may be even more important: biomass variation over time will not be damped by diversity when the different species involved respond to environmental variation in a concordant way. However, only a perfect positive correlation could counteract the averaging effect, but this is rather unlikely. Species are different to the extent that they will respond in different ways to environmental variation (which corresponds to MacArthur's idea that when one species fails, it can be replaced by another species). Moreover, due to the effect of interspecific competition in the community, decline of one species can enable an increase of its competitor, leading to negative correlation. However, in a grassland community study where environmental variation was restricted to weather fluctuations, species responded concordantly and thus were positively correlated (Lepš 2004b).

Yachi & Loreau (1999) used a different mathematical model to describe risk spreading (called 'insurance hypothesis' by them), which is based on the differences in species responses to environmental variation. However, the more different species responses are, the lower will be the correlation of species abundances in a variable environment (as in Doak *et al.* 1998). Community resistance will thus be higher in species-rich communities. We can also expect a higher resilience, because there is a greater possibility that a fast regenerating species forms part of the community which can compensate for the decline of other species. Consequently, under conditions of natural environmental fluctuations, community biomass will be less variable in species-rich communities. However, the species composition will change; consequently, species richness does not support compositional stability (Tilman 1999b). Data on the variability in the BIODEPH experiment support this pattern – the total community biomass CV decreased, but the population biomass CV increased with species richness of a community (Hector *et al.* 2010). Should a relative constancy of ecosystem function be achieved by the internal substitution of species, then the substituting species should have a similar effect on ecosystem functioning (as suggested by MacArthur 1955), but differ in response to the environmental variation. Lavorel & Garnier (2002) argued that the insurance effect will be dependent on a decoupling between 'response traits', i.e. traits that determine the species response to environmental variation, and 'effect traits', i.e. traits that determine the species effects on ecosystem function.

In comparisons between habitats in nature, much depends on how the richness differences originated. Species richness may vary along environmental or successional gradients. Community stability (relative constancy in biomass) seems to be determined by the impact of species life history rather than species richness as such. For an old-field succession (Lepš *et al.* 1982; Figs 11.6, 11.9) community productivity decreased, while strategy types according to Grime 2001 shifted from R-strategy via C-R strategy to S-strategy. This change was related to a change in the species–area curve. The S-strategists which grow in number are able to co-exist on small areas; in the stage dominated by C-R strategists, a mosaic of diverse species-poor patches are found which at larger spatial scales are richer in species (Lepš & Štursa 1989; Osbornová et al. 1990). The C-R dominated stage is more productive, less resistant and more resilient than the S-strategist stage.

Resistance to invasion of alien species can be seen as a special case of stability. Species-rich communities were traditionally considered more resistant to invasions of exotic species (Elton 1958). This statement was partly based on a comparison between species-rich tropical forests and species- poorer extra-tropical communities (comparison subject to the effects of confounding factors). A theoretical explanation here is that in a species-rich community there are less 'empty niches' available for possible newcomers. Indeed, when species richness was manipulated, species-poor communities were shown to be more susceptible to invasions (Naeem et al. 2000). However, empirical support from observational data is not unequivocal (Rejmánek 1996). Undisturbed tropical forests are both extremely rich in species and highly resistant to invasions; on the other hand, some extra-tropical centres of diversity such as the South African Cape Floral Region are very vulnerable to plant invasions. Some of the factors promoting species co-existence (e.g. repeated disturbance in African fynbos) can also promote invasions. Consequently, rather than species richness per se, the factors determining species richness are also important for invasibility.

It seems that there is a difference between communities that are species-poor because of the harsh environment or strong competition on one hand and communities where low diversity is a consequence of limited size of the species pool (e.g. as on islands) on the other hand. Only the latter type is more vulnerable to invasions, as predicted by biodiversity experiments. This corresponds well to the larger invasibility of island ecosystems (Rejmánek 1996; see Chapter 13).

11.6.4   Biodiversity experiments, real consequences of species losses and conservation consequences

Some biodiversity experiments and the biodiversity debate were encouraged by the reality of the global decline of diversity. Will the loss of species impair the functioning of ecosystems? Will biodiversity experiments help predict changes in ecosystems? And do they provide directions for conservation efforts?

In biodiversity experiments, the set of sown species (in fact, the species pool) is manipulated; hence it is considered as an independent variable (predictor of ecosystem functioning). Not all the species usually survive to form the actual community, but data on the resulting species richness are seldom reported. If the actual pool in individual experimental units is a random subset of some larger species pool of the whole experiment (e.g. a random selection of species used in experiments or simulations), the realized richness is positively correlated with the size of the species pool (Stachová & Lepš 2010). If new species are added to an existing species pool, and the new species differ in their traits from those in the original species pool, the actual result need not always be an increase in realized species richness, similarly the experimental removal of a species from a community (and so a factual reduction of the species pool available there) can result in an increase of actual species richness in a community. In a long-term multisite study (Lepš et al. 2007), high and low richness meadow species mixtures were sown in a newly abandoned field (thus enhancing the species pool), the plots were not weeded and were left to colonize naturally. The sown meadow species were different from the pool of natural colonizers (mostly competitively weak species). After 10 years, the productivity generally decreased (as expected from biodiversity experiments) in the order: high richness mixture > low richness mixture > unsown control plots. Nevertheless, the unsown (low productive) control exhibited the highest actual number of species, because in the sown plots the competitive exclusion of naturally colonizing species was faster. As a result, the productivity was positively related to the number of sown species, but not to the realized species richness.

We should be aware that there are important differences between the gradient of species richness created by biodiversity experiments and a sequence created by the loss of species in nature. The equal representation of species on all the diversity levels, important for disentangling the effects of species diversity from the effect of species composition, corresponds to the situation when species are lost from the community at random, irrespective of their traits. In nature, however, the species that are lost from the communities are not a random subset of their species (Lepš 2004a, Srivastava & Vellend 2005). If we want to construct a realistic scenario of species loss, we need to identify the expected sequence of species to be lost, probably according to their traits. In this case, however, we will not get the effect of species richness *per se*, but the expected effect of the loss of particular species (i.e. those that we consider candidates for extinction). For example, in Central Europe, the most endangered species are those of nutrient-poor habitats (which are usually less productive), while the non-endangered species are more productive (Lepš 2004a). In Central European grasslands, species loss is mostly the result of agricultural intensification leading to increased productivity.

In natural communities, species are usually not lost at random (Lepš 2004a; Srivastava & Vellend 2005), but as a result of many specific factors, some of them being species-specific, and some not. Species-specific factors are usually direct and negative in their effect; a typical example is the introduction of a new pathogen or a specialized herbivore. Also, human exploitation is often species-specific, for example selective logging or the collection of plants for pharmaceutical use. The species affected by species-specific negative effects are often selected independently of their function in a community and independently of their effect traits, which also means independently of their competitive strength. Dutch elm disease (*Ceratocystis ulmi*) eliminated *Ulmus* from part of the European forest; species of the genus *Ulmus*, but no other functionally similar species were affected; similarly the decline of *Gentiana pannonica* in the Bohemian forest in the first half of the 20th century was caused by selective digging of its roots for a local liqueur – their functionally analogous species were not affected. Both cases are examples of the decoupling of the response and effect traits (Lavorel & Garnier 2002); the species lost can be functionally replaced by other species from the community that were not affected and the chance that an 'appropriate' species will be present increases with diversity. In those cases, the lessons from the biodiversity experiments are relevant. Indeed, in the case of elimination of elms from part of European forests, their functional role was taken over by other tree species in mixed forests, and the general functioning of the respective eco-systems did not change considerably. In contrast, elimination of *Picea abies* in Central European mountain forests, where the species was a single dominant tree with no functional analogues, led to tremendous changes in the whole eco-system, regardless of whether the spruce was planted or indigenous, and regard-less of whether the spruce dieback was caused by emissions (acid rain) or by a bark beetle outbreak.

When environmental conditions (e.g. productivity or disturbance regime) are changed, many if not all species are affected simultaneously. Typical examples are land-use changes, changes in nature management, large-scale pollution (e.g. nitrogen deposition) or climate change. Under such circumstances, some species are eliminated or at least negatively affected, while other species may benefit, or invade the community undergoing change. Regardless of the final net change in species richness, the most pronounced effect is the change in life history spectra, which will probably overrule any possible diversity effect (Srivastava & Vellend 2005). Functionally similar species will be affected in similar ways, and so there is only a small chance that lost species will be replaced by functionally analogous species. The species will be outcompeted. The change in the produc-tivity or in the disturbance regime will affect species according to the traits that are important for competitive strength, which are usually also important for primary productivity and other ecosystem functions; there is no or slight decou-pling of response and effect traits and species richness has a small stabilizing effect. Although changes in environmental conditions can lead to both increase and decline of species richness, most of the recent changes result in a net decline of species richness. Typical examples are: the recent loss of species due to eutrophication; where few productive species prevail in a community; excluding less competitive species; and loss of species due to the abandonment of previ-ously extensively managed grasslands (Bakker 1989) – cessation of regular mowing or grazing leads to extinction of many species, usually less productive, weak competitors. The serious loss of biodiversity in European meadows is partially caused by the increasing nutrient load, which leads to an increased productivity. Conservationists in several European countries have tried to per-suade farmers to keep productivity of species-rich grasslands low in order to keep diversity high. Under those circumstances, the use of the argument based on biodiversity experiments that keeping diversity high might be eco-nomical because of increased productivity (as suggested by Tilman 1999a) is counter-productive.

Whatever the impact of the loss of species on community functioning, the identity of lost species is probably more important than their number (Aarssen 2001). The loss of any species means that the functional properties of a com-munity is impaired to some extent. As Stampfli & Zeiter (1999) showed for an abandoned formerly managed meadow, the loss of species cannot easily be reversed by the re-introduction of mowing, because the species lost would not return to the earlier state, because the propagules are no longer available. Stamp-fli & Zeiter (2010) also found that the productivity of these less species-rich meadows is lower than that of the meadow in its original state. The ability of a community to respond to environmental change could be a function of the richness of the species pool rather than the number of species already present in the community. With the exception of species with a permanent seed bank, and species with long-distance dispersal mechanisms, the species pool is determined by species growing in nearby communities in the landscape (Cantero *et al.* 1999). From this point of view, the simultaneous loss of species in the landscape (which we recently observed in various types of previously species-rich grasslands in Europe) would have serious consequences, not envisaged by small-scale biodiversity experiments.

Both species gains and species losses are considered negatively by conservationists. Gaining a new species by an alien invasion or by expansion of the original area of distribution is in fact an increase of a community species pool, which can, however, have detrimental effects on the native biota (Wardle *et al.* 2011; Chapter 13). As noted by Wardle *et al.* (2011), research on the effects of species gains and species losses has developed largely independently from each other; however, they have a common basis: for the functioning of the new community, it is important which traits are gained/lost in the process (in comparison with the traits of the other species in a community). Thus, for the community to function, trait composition is much more important than the number of species.

The relationship between biodiversity and ecosystem functioning is not only of academic interest, but also has important consequences for environmental policy. Research efforts on this topic resulted in several books providing new syntheses (e.g. Kinzig *et al.* 2001; Naeem *et al.* 2009), and also attempts to reconcile contrasting interpretations (Loreau *et al.* 2002; Loreau 2010). This chapter has concentrated on the study of vascular plant communities (where we expect the competition to be the main interspecific interaction). However, vascular plants (and their diversity) are not the only ecosystem component determining ecosystem functioning; particularly the linkage to below-ground components is also of basic importance (Wardle 2002; see Chapter 9).

Acknowledgements

The chapter is to a large extent based on experiences gained in the European projects TERI-CLUE (ENV4-CT95-0002) and TLinks (EVK2-CT-2001-00123). I am grateful to Marcel Rejmánek and Eddy van der Maarel for invaluable comments on earlier drafts of the chapter.

References

Aarssen, L.W. (1997) High productivity in grassland ecosystems: effected by species diversity or productive species? *Oikos* **80**, 183–184.

Aarssen, L.W. (2001) On correlation and causation between productivity and species richness in vegetation: predictions from habitat attributes. *Basic and Applied Ecology* **2**, 105–114.

Alsos, I.G., Ehrich, D., Thuiller, W. *et al.* (2012) Genetic consequences of climate change for northern plants. *Proceedings of the Royal Society Series B. Biological Sciences*, doi: 10.1098/rspb.2011.2363.

Austin, M.P. & Austin, B.O. (1980) Behaviour of experimental plant-communities along a nutrient gradient. *Journal of Ecology* **68**, 891–918.

Bakker, J.P. (1989) *Nature Management by Grazing and Cutting.* Kluwer, Dordrecht.

Botta-Dukát, Z. (2005) Rao's quadratic entropy as a measure of functional diversity based on multiple traits. *Journal of Vegetation Science* **16**, 533–540.

Busing, R.T. & Brokaw, N. (2002) Tree species diversity in temperate and tropical forest gaps: the role of lottery recruitment. *Folia Geobotanica* **37**, 33–43.

Cadotte, M.C., Cardinale, B.J. & Oakley T.H. (2008) Evolutionary history predicts the ecological impacts of species extinction. *Proceedings of the National Academy of Sciences of the United States of America* **105**, 17012–17017.

Cadotte, M.W., Cavender-Bares, J., Tilman, D. & Oakley, T.H. (2009) Using phylogenetic, functional and trait diversity to understand patterns of plant community productivity. *PLoS ONE* **4**, e5695.

Cantero, J.J., Pärtel, M. & Zobel, M. (1999) Is species richness dependent on the neighbouring stands? An analysis of the community patterns in mountain grasslands of central Argentina. *Oikos* **87**, 346–354.

Cardinale, B.J., Wright, J.P., Cadotte, M.W. *et al.* (2007) Impacts of plant diversity on biomass production increase through time because of species complementarity. *Proceedings of the National Academy of Sciences of the United States of America* **104**, 18123–18128.

Colwell, R.K. (2009) *EstimateS: Statistical estimation of species richness and shared species from samples.* Version 8.2. User's Guide and application published at: http://purl.oclc.org/estimates.

Crane, P.R. & Lidgard, S. (1989) Angiosperm diversification and paleolatitudinal gradients in cretaceous floristic diversity. *Science* **246**, 675–678.

Crawley, M.J. & Harral, J.E. (2001) Scale dependence in plant biodiversity. *Science* **291**, 864–868.

de Bello, F., Thuiller, W., Lepš, J. *et al.* (2009) Partitioning of functional diversity reveals the scale and extent of trait convergence and divergence. *Journal of Vegetation Science* **20**, 475–486.

de Wit, C T. (1960) On competition. *Verslagen Landbouwkundig Onderzoek Wageningen* **66.8.**

Doak, D.F., Bigger, D., Harding, E.K. *et al.* (1998) The statistical inevitability of stability–diversity relationships in community ecology. *The American Naturalist* **151**, 264–276.

Elton, C.S. (1958) *The Ecology of Invasions by Animals and Plants.* Methuen, London.

Fibich, P. & Lepš, J. (2011) Do biodiversity indices behave as expected from traits of constituent species in simulated scenarios? *Ecological Modelling* **222**, 2049–2058.

Fox, J.W. (2005) Interpreting the 'selection effect' of biodiversity on ecosystem function. *Ecology Letters*, **8**, 846–856.

Fridley, J.D., Peet, R.K., van der Maarel, E. & Willems, J.H. 2006. Integration of local and regional species–area relationships from space–time species accumulation. *The American Naturalist* **168**, 133–143.

Garnier, E., Navas, M.L., Austin, M.P., Lilley, J.M. & Gifford, R.M. (1997) A problem for biodiversity–productivity studies: how to compare the productivity of multispecific plant mixtures to that of monocultures? *Acta Oecologica* **18**, 657–670.

Gillman, L.N. & Wright S. D. 2006. The influence of productivity on the species richness of plants: a critical assessment. *Ecology* **87**, 1234–1243.

Grace, J.B., Anderson, T.M., Smith, M.D. *et al.* (2007) Does species diversity limit productivity in natural grassland communities? *Ecology Letters* **19**, 680–689.

Grime, J.P. (1998) Benefits of plant diversity to ecosystems: immediate, filter and founder effects. *Journal of Ecology* **86**, 902–910.

Grime, J.P. (2001) *Plant Strategies, Vegetation Processes, and Ecosystem Properties.* John Wiley & Sons, Ltd, Chichester.

Grime, J.P., Brown, V.K., Thompson, K. *et al.* (2000) The response of two contrasting limestone grasslands to simulated climate change. *Science* **289**, 762–765.

Grubb, P.J. (1977) The maintenance of species richness in plant communities: the importance of the regeneration niche. *Biological Reviews of the Cambridge Philosophical Society* **52**, 107–145.

Hanski, I. (1999) *Metapopulation Ecology.* Oxford University Press, Oxford.

Harrison, G.W. (1979) Stability under environmental stress: resistance, resilience, persistence, and variability. *The American Naturalist* **113**, 659–669.

Hector, A., Schmid, B. & Beierkuhnlein, C. *et al.* (1999) Plant diversity and productivity in European grasslands. *Science* **286**, 1123–1127.

Hector, A., Bell, T., Connolly, J. *et al.* (2009) The analysis of biodiversity experiments: from pattern toward mechanism. In: *Biodiversity, Ecosystem Functioning, and Human Wellbeing: An Ecological and Economic Perspective* (eds S. Naeem, D.E. Bunker, M. Hector, M. Loreau & C. Perrings), pp. 94–104. Oxford University Press, Oxford.

Hector, A., Hautier, Y., Saner, P. *et al.* (2010) General stabilizing effects of plant diversity on grassland productivity through population asynchrony and overyielding. *Ecology* **91**, 2213–2220.

Herben, T. (2000) Correlation between richness per unit area and the species pool cannot be used to demonstrate the species pool effect. *Journal of Vegetation Science* **11**, 123–126.

Herben, T., Krahulec, F., Hadincová, V. & Pecháčková, S. (1995) Climatic variability and grassland community composition over 10 years – separating effects on module biomass and number of modules. *Functional Ecology* **9**, 767–773.

Hiers, J.K., Wyatt, R. & Mitchell, R.J. (2000) The effects of fire regime on legume reproduction in longleaf pine savannas: is a season selective? *Oecologia* **125**, 521–530.

Hill, M.O. (1973) Diversity and evenness: a unifying notation and its consequences. *Ecology* **54**, 427–432.

Hill, J.L. & Hill, R.A. (2001) Why are tropical rain forests so species rich? Classifying, reviewing and evaluating theories. *Progress in Physical Geography* **25**, 326–354.

Hooper, D.U., Chapin III, F. S., Ewel, J.J. *et al.* (2005) Effects of biodiversity on ecosystem functioning: a consensus of current knowledge *Ecological Monographs* **75**, 3–35.

Hubbell, S.P. (2001) *The Unified Neutral Theory of Biodiversity and Biogeography*. Princeton University Press, Princeton, NJ.

Huston, M.A. (1979) A general hypothesis of species diversity. *The American Naturalist* **113**, 81–101.

Huston, M.A. (1994) *Biological Diversity. The Co-existence of Species on Changing Landscapes*. Cambridge University Press, Cambridge.

Huston, M.A. (1997) Hidden treatments in ecological experiments: re-evaluating the ecosystem function of biodiversity. *Oecologia* **110**, 449–460.

Janzen, D.H. (1970) Herbivores and the number of tree species in tropical forests. *The American Naturalist* **110**, 501–528.

Kenkel, N.C., Peltzer, D.A., Baluta, D. & Pirie, D. (2000) Increasing plant diversity does not influence productivity: empirical evidence and potential mechanisms. *Community Ecology* **1**, 165–170.

Kinzig, A.P., Pacala, S.W. & Tilman, D. (eds) (2001) *The Functional Consequences of Biodiversity. Empirical Processes and Theoretical Extensions*. Princeton University Press, Princeton, NJ.

Kirwan, L., Luescher, A., Sebastia, M.T. *et al.* (2007) Evenness drives consistent diversity effects in intensive grassland systems across 28 European sites. *Journal of Ecology* **95**, 530–539.

Kirwan, L., Connolly, J., Finn, J. A. *et al.* (2009) Diversity–interaction modeling: estimating contributions of species identities and interactions to ecosystem function. *Ecology* **90**, 2032–2038.

Klimešová, J. & de Bello, F. (2009). CLO-PLA: the database of clonal and bud bank traits of Central European flora. *Journal of Vegetation Science* **20**, 511–516.

Klimešová, J., Janeček, Š., Bartušková, A., Lanta, V. & Doležal, J. (2010) How is regeneration of plants after mowing affected by shoot size in two species-rich meadows with different water supply? *Folia Geobotanica* **45**, 225–238.

Kotorová, I. & Lepš, J. (1999) Comparative ecology of seedling recruitment in an oligotrophic wet meadow. *Journal of Vegetation Science* **10**, 175–186.

Lanta, V. & Lepš, J. (2006) Effect of functional group richness and species richness in manipulated productivity–diversity studies: a glasshouse pot experiment. *Acta Oecologica* **29**, 85–96.

Lavorel, S. & Garnier, E. (2002) Predicting changes in community composition and ecosystem functioning from plant traits: revisiting the Holy Grail. *Functional Ecology* **16**, 545–556.

Lepš, J. (1999) Nutrient status, disturbance and competition: an experimental test of relationships in a wet meadow. *Journal of Vegetation Science* **10**, 219–230.

Lepš, J. (2004a) What do the biodiversity experiments tell us about consequences of plant species loss in the real world? *Basic and Applied Ecology* **5**, 529–534.

Lepš, J. (2004b) Variability in population and community biomass in a grassland community affected by environmental productivity and diversity. *Oikos* **107**, 64–71.

Lepš, J. & Štursa, J. (1989) Species–area relationship, life history strategies and succession – a field test of relationships. *Vegetatio* **83**, 249–257.

Lepš, J., Osbornová, J. & Rejmánek, M. (1982) Community stability, complexity and species life-history strategies. *Vegetatio* 50, 53–63.

Lepš, J., Spitzer, K. & Jaroš, J. (1998) Food plants, species composition and variability of the moth community in undisturbed forest. *Oikos* 81, 538–548.

Lepš J., de Bello, F., Lavorel, S. & Berman, S. (2006) Quantifying and interpreting functional diversity of natural communities: practical considerations matter. *Preslia* 78, 481–501.

Lepš, J., Doležal, J., Bezemer, T. M. *et al.* (2007) Long-term effectiveness of sowing high and low diversity seed mixtures to enhance plant community development on ex-arable fields in five European countries. *Applied Vegetation Science* 10, 97–110.

Loreau, M. (2000) Biodiversity and ecosystem functioning: recent theoretical advances. *Oikos* 91, 3–17.

Loreau, M. (2010) *From Populations to Ecosystems: Theoretical Foundations for a New Ecological Synthesis*. Monographs in Population Biology. Princeton University Press, Princeton, NJ.

Loreau, M. & Hector, A. (2001) Partitioning selection and complementarity in biodiversity experiments. *Nature* 412, 72–76.

Loreau, M., Naeem, S., Inchausti, P. *et al.* (2001) Biodiversity and ecosystem functioning: current knowledge and future challenges. *Science* 294, 804–808.

Loreau, M., Naeem, S. & Inchausti, P. (eds) (2002) *Biodiversity and Ecosystem Functioning. Synthesis and Perspectives*. Oxford University Press, Oxford.

MacArthur, R.H. (1955) Fluctuations of animal populations and a measure of community stability. *Ecology* 36, 533–536.

MacArthur, R.H. & Levins, R. (1967). The limiting similarity, convergence and divergence of coexisting species. *American Naturalist* 101, 377–385.

MacGillivray, C.W., Grime, J.P., Band, S.R. *et al.* (1995) Testing predictions of the resistance and resilience of vegetation subjected to extreme events. *Functional Ecology* 9, 640–649.

Magurran, A.E. (2004) *Measuring Biological Diversity*. Blackwell Publishing, Oxford.

May, R.M. (1973) *Stability and Complexity in Model Ecosystems*. Princeton University Press, Princeton, NJ.

Mittelbach, G.G. (2010) Understanding species richness–productivity relationships: the importance of meta-analyses. *Ecology* 91, 2540–2544.

Mittelbach, G.G., Steiner, C.F., Scheiner, S.M. *et al.* (2001) What is the observed relationship between species richness and productivity? *Ecology* 82, 2381–2396.

Naeem, S., Thompson, L.J., Lawler, S.P., Lawton, J.H. & Woodfin, R.M. (1994) Declining biodiversity can alter the performance of ecosystems. *Nature* 368, 734–737.

Naeem, S., Chapin III, F.S., Costanza, R. *et al.* (1999) Biodiversity and ecosystem functioning: maintaining natural life support processes. *Issues in Ecology* 4, 1–14.

Naeem, S., Knops, J.M.H., Tilman, D. *et al.* (2000) Plant diversity increases resistance to invasion in the absence of covarying extrinsic factors. *Oikos* 91, 97–108.

Naeem, S., Bunker, D.E., Hector, M., Loreau, M. & Perrings, C. (eds) (2009) *Biodiversity, Ecosystem Functioning, and Human Wellbeing: An Ecological and Economic Perspective*. Oxford University Press, Oxford.

Osbornová, J., Kovářová, M., Lepš, J. & Prach, K. (eds) (1990) *Succession in Abandoned Fields. Studies in Central Bohemia, Czechoslovakia*. Geobotany 15. Kluwer, Dordrecht.

Palmer, M.W. (1994) Variation in species richness – towards a unification of hypotheses. *Folia Geobotanica & Phytotaxonomica* 29, 511–530.

Pärtel, M., Zobel, M., Zobel, K. & van der Maarel, E. (1996) The species pool and its relation to species richness: evidence from Estonian plant communities. *Oikos* 75, 111–117.

Petermann, J.S., Fergus, A.J., Turnbull, L.A. & Schmid, B. (2008) Janzen–Connell effects are widespread and strong enough to maintain diversity in grasslands. *Ecology* 89, 2399–2406.

Pimm, S.L. (1984) The complexity and stability of ecosystems. *Nature* 307, 321–326.

Preston, F.W. (1962) The canonical distribution of commonness and rarity: Part I. *Ecology* 43, 185–215.

Reich J.A. (1998) Vegetation Part 1: A comparison of two one-hectare tree plots in the Lakekamu basin. In: *A Biological Assessment of the Lakekamu Basin, Papua New Guinea*. RAP Working Papers 9 (ed. A.L. Mack), pp. 25–35. Conservation International, Washington, DC.

Rejmánek, M. (1996) Species richness and resistance to invasions. In: *Biodiversity and Ecosystem Processes in Tropical Forests*. Ecological Studies 122. (eds G.H. Orians, R. Dirzo & J.H. Cushman), pp. 153–172. Springer, Berlin.

Roscher, C., Temperton, V.M., Scherer-Lorenzen, M. *et al.* (2005) Overyielding in experimental grassland communities – irrespective of species pool or spatial scale. *Ecology Letters* 8, 419–429.

Rosenzweig, M.L. (1995) *Species Diversity in Space and Time*. Cambridge University Press, Cambridge.

Rusch, G.M. & Oesterheld, M. (1997) Relationship between productivity, and species and functional group diversity in grazed and non-grazed Pampas grassland. *Oikos* 78, 519–526.

Schluter, D. & Ricklefs, R.E. (1993) Convergence and the regional component of species diversity. In: *Species Diversity in Ecological Communities. Historical and Geographical Perspectives* (eds R.E. Ricklefs & D. Schluter), pp. 230–240. The University of Chicago Press, Chicago, IL.

Southwood, T.R.E. (1988) Tactics, stategies and templets. *Oikos* 52, 3–18.

Špačková, I. & Lepš, J. (2001) Procedure for separating the selection effect from other effects in diversity–productivity relationship. *Ecology Letters* 4, 585–594. [Name of first author erroneously spelled Spaékova in the journal.]

Srivastava, D.S. & Vellend, M. (2005) Biodiversity–ecosystem research: Is it relevant to conservation? *Annual Reviews of Ecology and Evolution* 36, 267–294.

Stachová, T. & Lepš, J. (2010) Species pool size and realized species richness affect productivity differently: a modeling study. *Acta Oecologica* 36, 578–586.

Stampfli, A. & Zeiter, M. (1999) Plant species decline due to abandonment of meadows cannot easily be reversed by mowing. A case study from the southern Alps. *Journal of Vegetation Science* 10, 151–164.

Stampfli, A. & Zeiter, M. (2010) Der Verlust von Arten wirkt sich negativ auf die Futterproduktion aus. *Agrarforschung Schweiz* 1, 184–189.

Stephan, A., Meyer, A.H. & Schmid, B. (2000) Plant diversity affects culturable soil bacteria in experimental grassland communities. *Journal of Ecology* 88, 988–998.

Tallis, J.H. (1991) *Plant Community History. Long-term Changes in Plant Distribution and Diversity*. Chapman and Hall, London.

Taylor, D.R., Aarssen, L.W. & Loehle, C. (1990) On the relationship between r/K selection and environmental carrying-capacity – a new habitat templet for plant life-history strategies. *Oikos* 58, 239–250.

Tilman, D. (1999a) Diversity and production in European grasslands. *Science* 286, 1099–1100.

Tilman, D. (1999b) The ecological consequences of changes in biodiversity: a search for general principles. *Ecology* 80, 1455–1474.

Tilman, D. & Pacala, S. (1993) The maintenance of species richness in plant communities. In: *Species Diversity in Ecological Communities. Historical and Geographical Perspectives* (eds R.E. Ricklefs & D. Schluter), pp. 13–25. The University of Chicago Press, Chicago, IL.

Tilman, D., Wedin, D. & Knops, J. (1996) Productivity and sustainability influenced by biodiversity in grassland ecosystems. *Nature* 379, 718–720.

Trenbath, B.R. (1974). Biomass productivity of mixtures. *Advances in Agronomy* 26, 177–210.

van der Maarel, E. (1995). Vicinism and mass effect in a historical perspective. *Journal of Vegetation Science* 1, 135–138.

van der Maarel, E. & Sykes, M.T. (1993) Small-scale plant-species turnover in a limestone grassland – the carousel model and some comments on the niche concept. *Journal of Vegetation Science* 4, 179–188.

Vandermeer, J. (1989) *Ecology of Intercropping*. Cambridge University Press, Cambridge.

van der Putten, W.H., Mortimer, S.R., Hedlund, K. *et al.* (2000) Plant species diversity as a driver of early succession in abandoned fields: a multi-site approach. *Oecologia* 124, 91–99.

Vellend, M. (2010) Conceptual synthesis in community ecology. *The Quarterly Review of Biology* 85, 183–206.

Vítová, A. & Lepš, J. (2011) Experimental assessment of dispersal and habitat limitation in an oligotrophic wet meadow. *Plant Ecology* 212, 1231–1242.

Wardle, D.A. (2002) *Communities and Ecosystems. Linking the Aboveground and Belowground Components*. Princeton University Press, Princeton, NJ.

Wardle, D.A., Huston, M.A., Grime, J.P. *et al.* (2000) Biodiversity and ecosystem functioning: an issue in ecology. *Bulletin of the Ecological Society of America* 81, 235–239.

Wardle, D.A., Bardgett, R.D., Callaway, R.M. & van der Putten, W.H. (2011) Terrestrial ecosystem responses to species gains and losses. *Science* **332**, 1273–1277.

Watt, A.S. (1947) Pattern and process in the plant community. *Journal of Ecology* **35**, 1–22.

Westoby M. (1998) A leaf–height–seed (LHS) plant ecology strategy scheme. *Plant and Soil* **199**, 213–227.

Whittaker, R.H. (1972) Evolution and measurement of species diversity. *Taxon* **21**, 213–251.

Whittaker, R.H. (1975) *Communities and Ecosystems*, 2nd edn. Macmillan, New York, NY.

Whittaker R.J. (2010) Meta-analyses and mega-mistakes: calling time on meta-analysis of the species richness–productivity relationship. *Ecology* **91**, 2522–2533.

Wills, C., Condit, R., Foster, R.B. & Hubbell, S.P. (1997) Strong density- and diversity-related effects help to maintain tree species diversity in a neotropical forest. *Proceedings of the National Academy of Sciences of the United States of America* **94**, 1252–1257.

Wilson, J.B. (2011) The twelve theories of co-existence in plant communities: the doubtful, the important and the unexplored. *Journal of Vegetation Science* **22**, 184–195.

Yachi, S. & Loreau, M. (1999) Biodiversity and ecosystem productivity in a fluctuating environment: The insurance hypothesis. *Proceedings of the National Academy of Sciences of the United States of America* **96**, 1463–1468.

Zobel, M. (1992) Plant-species co-existence – the role of historical, evolutionary and ecological factors. *Oikos* **65**, 314–320.

Zobel, M., van der Maarel, E. & Dupré, C. (1998) Species pool: the concept. Its determination and significance for community restoration. *Applied Vegetation Science* **1**, 55–66.

Zobel, M., Otto, R., Laanisto, L. *et al.* (2011) The formation of species pools: historical habitat abundance affects current local diversity. *Global Ecology and Biogeography* **20**, 251–259.

12

Plant Functional Types and Traits at the Community, Ecosystem and World Level

Andrew N. Gillison

Center for Biodiversity Management, Queensland, Australia

12.1 The quest for a functional paradigm

Eugenius Warming's insightful comment (1909) that we are '. . . yet far distant from the oecological interpretation of various growth-forms' still applies in a world where increased pressure on global resources and rapid environmental change generate questions that remain unsolvable through time-honoured methodologies. It is here that functional ecology can play an important role in helping to better understand ecosystem dynamics through a more detailed analysis of form, function and plant–environment interaction. Criteria for functional classifications vary. Lavorel *et al.* (1997) propose four main types of functional classifications of plant species: (1) **emergent groups** – groups of species that reflect natural correlations of biological attributes; (2) **strategies** – species within a strategy have similar attributes interpreted as adaptations to particular patterns of resource use; (3) **functional types** – species with similar roles in ecosystem processes that respond in similar ways to multiple environmental factors; and (4) specific **response groups** – containing species that respond in similar ways to specific environmental factors. To these may be added specific **effect groups** – containing species that influence ecosystem performance either directly or indirectly (Díaz *et al.* 2002; Lavorel *et al.* 2007). Each of these is discussed further in this chapter. Advances in functional ecology show significant gains in the quality of baseline data and readily classifiable functional types where cause and effect relationships can be demonstrated between the biophysical environment and readily measureable, non-phylogenetic, morphological and physiological adaptations of plants. Despite progress, the successful identification, measurement and testing of plant functional characteristics underpin the quest for a functional paradigm where the classification and application of entities such as **plant functional types (PFTs)** and related 'functional traits' play a central role.

Vegetation Ecology, Second Edition. Eddy van der Maarel and Janet Franklin.
© 2013 John Wiley & Sons, Ltd. Published 2013 by John Wiley & Sons, Ltd.

The plethora of definitions (Web Resource 12.4) highlights the uncertainty surrounding the meaning of 'functional' type, its component traits and whether functional types actually exist beyond the minds of ecologists. If PFTs and traits are to be useful, we need to know which functional traits are reliable predictors of species abundances, biodiversity or demographic change and whether functional traits can be used, for example, to assess and monitor vegetation change. Related questions concern the genetic basis for functional traits and their connections with phylogeny (Kooyman *et al.* 2011). A key requirement is to establish a robust, scientific basis for the generalization of ecological strategies based on functional traits and to demonstrate their applicability across ecological scales. This chapter addresses the evolution of the concept of plant function, the development of plant functional typology and includes case studies that illustrate the current and potential use of PFTs and functional trait-based approaches at the community, ecosystem and world level.

12.2 Form and function: evolution of the 'functional' concept in plant ecology

Early physiognomic-structural classification systems were designed primarily to communicate and compare vegetation physiognomy or appearance rather than function. Until the mid to late 19th century, physiognomic types were the primary descriptive units of a plant community and vegetation of a specific region (Du Rietz 1931). Then, during the late 19th century Eugenius Warming (1895, 1909) first attempted to arrange higher plants into biological groups – the early **epharmonic life-form** (the adaptive form) – a precursor to subsequent classifications of life-forms by others. During this period, Christen Raunkiær (1934) constructed a **life-form** ('*livsform*') classification system based on the position of the perennating organ during the most unfavourable season. Following Raunkiær, Fosberg (1967) argued a case for a functional classification based on dynamic rather than static vegetation descriptors – an approach developed by Gillison (1981) who combined modified Raunkiærean life-form criteria with adaptive photosynthetic leaf-stem attributes and above-ground rooting systems as a basis for classifying **whole-plant PFTs**.

12.3 The development of functional typology

There is a clear need to clarify and unify concepts surrounding PFTs (Gitay & Noble 1997; Semenova & van der Maarel 2000). The following sections summarize some key aspects of cross-related terms such as **guilds, growth-forms, life-forms, plant strategies, functional types** and **functional traits**.

12.3.1 Guilds

The term guild has important connotations for functional typology and emerged as an English translation of '*Genossenschaften*' applied by Schimper (1903) to plant types that depend on others for support (lianes, epiphytes, parasites, sapro-phytes; see also Simberloff & Dayan 1991). When applied to plants the term is frequently equivalent to **functional group** or **functional type** (e.g. Shugart 1997). Boutin & Keddy (1993) define plant guild composition according to functional traits and emphasize that there can be major obstacles when using guild classi-fications built on broad resource criteria where, for example, an entire com-munity may be included as a single guild (cf. Harper 1977; Grubb 1977). On the other hand, attributes of dispersal, establishment and growth were used to construct a guild hierarchy for the conterminous vegetation of the USA (Johnson 1981). The literature reveals other classificatory diversions such as '**functional guilds**' (Condit *et al.* 1996; Gitay *et al.* 1999) '**structural guilds**' (Gitay *et al.* 1999), '**management guilds**' (Verner (1984) and '**functional cliques**' (Yodzis 1982). Few protocols exist for the objective recognition of guilds. Recent usage in functional typology suggests there is much to support the view of Hawkins & MacMahon (1989) that the guild concept is a useful but artificial construct of the minds of ecologists.

12.3.2   Life-forms and growth-forms

Confusion surrounds the meaning and utility of these two widely used terms. Initial applications of growth-form expressed as physiognomy (appearance) and structure were largely developed for phytogeographical purposes. In functional typology, terminological clarity and ease of interpretation of results are manda-tory in today's demand for fast-paced, cost-effective methodology. In this respect Raunkiær's life-form terminology remains a clear winner (see also Floret *et al.* 1987). Attempts to expand Raunkiær's system, for example that of Mueller-Dombois & Ellenberg (1974), failed to capture the interest of practitioners who seek simpler and more readily quantifiable variables with improved return for effort. Unless otherwise indicated, in this chapter '**life-form**' follows Raunkiær (*sensu stricto*) with '**growth-form**' applied as a purely physiognomic descriptor.

12.3.3   Plant functional types and groups

With some exceptions (Vitousek & Hooper 1993; Cramer 1997; Hunt *et al.* 2004) most authors treat **types** and **groups** synonymously (e.g. Gitay & Noble 1997; Reich *et al.* 2003). For the purposes of this chapter, PFTs are considered synonymous with plant functional groups and include closely related entities such as the '**plant functional response type (PRT)**' of Louault *et al.* (2005). Fig. 12.1 provides a spatio-temporal context for measureable PFTs and functional traits above genetic and molecular level. The simplest definition of a PFT is that of Elgene Box (1996) 'PFTs are functionally similar plant types.' PFT definitions can vary according to whether the response to an environment or the effects on an ecosystem, singly or both, are intended. Smith *et al.* (1992) defined PFTs as 'sets of species showing similar responses to the environment and similar effects on ecosystem functioning,' a theme echoed by others (Díaz & Cabido 1997, 2001; Lavorel & Garnier 2002). PFTs are often regarded as trait assemblages or **trait syndromes** (Plate 12.1 shows nine different whole-plant PFT syndromes,

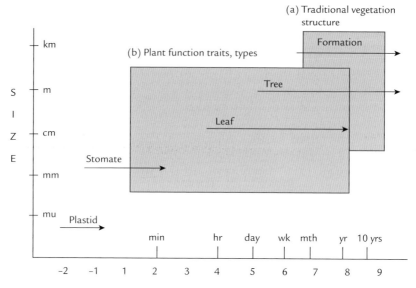

Fig. 12.1   Approximate log response time (s) of above-ground plant elements including spatio-temporal domains of PFT and individual trait sensitivity (a) Formation class and (b) generalized zone of plant functional classifications. (Adapted from Gillison 2002.)

subjectively positioned along gradients of light (energy) and moisture: *Victoria regia* (Amazon basin); *Metrosideros* (Phillipines); *Echinocactus* (Mexico); mangrove *Lumnitzera littorea* (Indomalesia): palm *Licuala ramsayi* (North Australia); *Juniperus communis* (Fennoscandia); *Selaginella* (Indomalesia); *Vaccinium vitis-idaea* (boreal); cushion plant *Azorella macquariensis* (subantarctic).) (Skarpe 1996; McIntyre & Lavorel 2001).

Reich *et al.* (2003) arbitrarily defined four different **functional groupings** expanded in Table 12.1 to include two additional groups. The first and most traditional grouping is based on discrete, typically qualitative individual traits. These include, for example, key ancestral or evolutionary criteria (conifer/ angiosperm, monocot/dicot), photosynthetic pathways (C_3/C_4) and seasonality (evergreen/deciduous). The second group is based on taxon position along a continuum of quantitative values for a shared trait such as leaf life-span, seed size, net photosynthetic capacity (A_{max}), or others. The third group is based on suites or syndromes of coordinated quantitative traits (Westoby *et al.* 2002; Wright *et al.* 2004; Hummel *et al.* 2007). The fourth represents a class of traits commonly recorded as ordered or multistate variables such as leaf size class. The fifth uses *post hoc* classification schemes to group plant species based on their responses to specific environmental factors (Gillison 1981; Lavorel *et al.* 1997; Garnier *et al.* 2007). This grouping is based on integrated whole-plant behaviour and outcomes and includes traditional classifications exemplified by shade and drought tolerance as well as plant strategy concepts such as the C-S-R triangle

Table 12.1 Different kinds of functional groupings.[a]

Basis	Trait (examples)
1 Qualitative, discrete trait	Dicot/monocot, woody/not, N-fixer/not, C_3/C_4, conifer/angiosperm, evergreen/deciduous
2 Relative value of quantitative, continuous trait	SLA, A_{max}, leaf life-span, height, seed mass, basal area, hydraulic conductance
3 Quantitative, suite of continuous traits	Leaf-trait syndrome, root-trait syndrome, seed trait syndrome
4 Qualitative suite of ordinal or multistate traits	Leaf (size class, inclination, phenology), plant inclination, canopy structure
5 Qualitative or quantitative; integrated response based mainly on functional strategies	Shade tolerance, drought tolerance, C-S-R scheme, LHS, LES, functional *modus*, optical spectra, predictive PFTs
6 Qualitative or quantitative; integrated effect of combined traits	Life-form, growth-form, litter structure and chemistry, species richness, herbivore palatability, pathogen defence, flammability, leaf, root and stem leachates

[a]Modified from Reich *et al.* (2003).

(Grime 1977) and the LHS approach (Westoby 1998). Growing evidence suggests that intraspecific as well as interspecific functional variability can influence community dynamics and ecosystem functioning across a range of ecological scales (Albert *et al.* 2010). The sixth group is therefore based on the concept that additional whole-plant behaviour can influence ecosystem process (see also Table 12.2) and includes life-form and growth-form.

12.3.4 Functional traits

Definitions. According to McGill *et al.* (2006) 'trait' refers to 'A well-defined, measurable property of organisms, usually measured at the individual level and used comparatively across species'. A **'functional trait'** on the other hand may be 'Any measurable feature at the individual level affecting its fitness directly or indirectly' (Albert *et al.* 2010). Apart from an emphasis on **'fitness'** (Violle *et al.* 2007; Vandewalle *et al.* 2010), functional traits may be characterized additionally by their **adaptive** or **strategic significance** (Semenova & van der Maarel 2000; Ackerly *et al.* 2000; Reich *et al.* 2003; Lavorel *et al.* 2007), growth and/or survival (Lusk *et al.* 2008), their combinatory role in forming a PFT (van der Maarel 2005) or their influence on 'organismal performance' (McGill *et al.* 2006). Functional traits can be further described according to 'biological function' (Gaucherand & Lavorel 2007; Aubin *et al.* 2009) or their perceived causal connection to **'response'** or **'effect'** in or on ecosystems (Díaz & Cabido 2001; Lavorel & Garnier 2002; Garnier *et al.* 2004, 2007; Violle *et al.* 2007) (see also Web Resource 12.2). For this chapter I define a **functional trait** as 'any measureable plant trait with potential to influence whole-plant fitness'.

Table 12.2 PFT and trait indicators of terrestrial ecosystem processes and properties.

Ecosystem process, properties	Laboratory[a]	Field
Productivity (NEP. NPP, SANPP)	SLA, LAI, LDMC, LNC, mycorrhizal diversity	Life-form, growth-form, canopy height, cover %, basal area all woody plants, root type and depth, bryophyte, lichen cover-abundance of types
Carbon assimilation and investment	Leaf N, P, photosynthetic light response curves, stomatal conductance, SLA, SLW, LAI, optical type, photosynthetic pathway (C_3, C_4, CAM), RCC	Life-form, growth-form, relative growth rate (RGR), leaf phenology, leaf type (e.g. needle- vs. broad-leaf, hardness, color), green stem, water storage, root diameter, bryophyte, lichen cover-abundance of types
Respiration, decomposition	LDMC, leaf N, P, base content, phenolics, SLA, mycorrhizae, stomatal and stem lenticel conductance	Leaf (size, inclination), phenology, litter depth and type, RGR, RNC, SRL, leaf turnover rate, bryophyte, lichen cover-abundance of types
Water use, evapotranspiration, drought resilience, hydrology	Stomatal conductance, xylem water potential, WUE, SLA	Life-form, growth-form, plant height, basal area, diameter increment, leaf (size, inclination, palisade distribution, phenology, succulence), green stem, root (type, depth, architecture), bryophytes, lichens
Tolerance to flooding, tidal movements, salinization, etc.	SLA, LWC	Adventitious rooting, salt glands, propagule dispersal, lenticels, succulence. Life-form, growth-form, tree height, leaf (succulence, inclination, thickness, palisade distribution) furcation index, green stem, known photosynthetic pathways (e.g. CAM), biocrusts (bryophytes, lichens)
Nutrient stocks, N mineralization, soil fertility	Leaf dry matter content; leaf N, P, mycorrhizal fungi, LDMC, SLA	Trait size, growth rate, litter depth, bryophyte, lichen composition, Mean canopy height, basal area, lichen cover-abundance, functional *modi*
Stress-tolerance, ruderal pioneers (C-S-R); Disturbance (loss of plant biomass and species)	LCC, LDMC, SLA, Succulence index, Standing biomass, resprouting abillity, vegetative spread	Clonality, canopy height, necromass persistence. Stem and canopy structure, basal area, furcation index, life-form, growth-form, seed dormancy, seed dispersal, species: functional *modi* to richness ratio, bryophytes, lichens

Table 12.2 *(Continued)*

Ecosystem process, properties	Laboratory[a]	Field
Response to grazing, herbivore resistance	SLA, palatability, standing biomass. Leaf N, P, phenolics, RTD	Canopy height, cover %, Vegetation structure, life-form, growth-form, species turnover
Response to fire	Flammability	Life-form, resprouting capacity, seed bank availability and persistence, seedling establishment, bark types, leaf volatiles, fuel load, clonal regeneration
Competition for light	SLA, RGR (seedling)	Plant height, leaf size, type, inclination, diameter, tiller increment, vegetative regeneration, seed size, type
Biodiversity	Functional diversity, functional complexity, mycorrhizal fungi	Functional types (e.g. *modi* this chapter), species richness, species composition, bryophyte, lichen cover-abundance, composition

[a]Includes instrumentation used to measure gas fluxes, xylem water potential, light dynamics, etc. in the field. Abbreviations: A_{max}, net photosynthetic capacity; PNUE, photosynthetic energy use efficiency; NEP = net ecosystem productivity; SANPP, specific annual net primary productivity; NPP, net primary productivity; LAI, leaf area index; SLA, specific leaf area; LCC, leaf carbon content; LDMC, leaf dry matter content; LNC, leaf nitrogen content; LWC, leaf water content; N, nitrogen; P, phosphorus; C, carbon; RCC, root construction cost; RGR, relative growth rate; RNC, root N concentration; SRL, specific root length; RTD, root tissue density; WUE, water use efficiency. (See Web Resource 12.5 for units used by different authors.)

Attributes and elements. Gillison (1981) applied a systematic approach to trait terminology in which a **plant functional attribute** or PFA is defined as 'any plant feature that responds in a demonstrable and predictable way with a change in the physical environment'. For PFT classification, Gillison & Carpenter (1997) use a hierarchical system whereby the lowest ranking **plant functional elements** (PFEs) (e.g. microphyll leaf size) are used to quantify PFAs at the next (class) level that, together with other PFAs, are then used to construct **whole-plant PFTs** according to specific assembly rules (Table 12.3, Section 12.4.5, Web Resource 12.7). A similar concept is described by Skarpe (1996), while van der Maarel (2005) considers PFAs to be different expressions of a trait that should rather be called 'states'. Although many ecologists frequently distinguish between '**soft traits**' (easy to measure) and '**hard traits**' (difficult to measure), I agree with Violle *et al.* (2007) that there is little evidence to support such distinction.

Table 12.3 Plant functional attributes and elements used to construct *modal* PFTs.

Attribute	Element	Description	
[Photosynthetic envelope]			
Leaf size	nr	no repeating leaf units	
	pi	picophyll	<2 mm^2
	le	leptophyll	2–25
	na	nanophyll	25–225
	mi	microphyll	225–2025
	no	notophyll	2025–4500
	me	mesophyll	4500–18200
	pl	platyphyll	18200–36400
	ma	macrophyll	36400–18 × 10^4
	mg	megaphyll	>18 × 10^4
Leaf inclination	ve	vertical	>30° above horizontal
	la	lateral	±30° to horizontal
	pe	pendulous	>30° below horizontal
	co	composite	
Leaf chlorotype	do	dorsiventral	
	is	isobilateral or isocentric	
	de	deciduous	
	ct	cortic	(photosynthetic stem)
	ac	achlorophyllous	(without chlorophyll)
Leaf morphotype	ro	rosulate or rosette	
	so	solid 3-D	
	su	succulent	
	pv	parallel-veined	
	fi	filicoid (fern)	(Pteridophytes)
	ca	carnivorous	(e.g. *Nepenthes*)
[Supporting vascular structure]			
life-form	ph	phanerophyte	
	ch	chamaephyte	
	hc	hemicryptophyte	
	cr	cryptophyte	
	th	therophyte	
	li	liane	
Root type	ad	adventitious	
	ae	aerating	(e.g. pneumatophore)
	ep	epiphytic	
	hy	hydrophytic	
	pa	parasitic	

12.4   Plant strategies, trade-offs and functional types

12.4.1   On plant strategies

'Plant strategy' is usually taken to mean a combination of plant characteristics that best maximize trade-offs in resource allocation patterns in order to achieve maximum growth rate, maximum size and maximum age along with the plant's growth response to different combinations of light and water availability (*cf.* Smith & Huston 1989). Strategy differentiation among species contributes to the maintenance of diversity and thus ecosystem performance (Kraft *et al.* 2008) and understanding plant ecological strategies is a fundamental aim of ecological research. When ecologically important plant traits are correlated they may be said to constitute an ecological 'strategy' dimension when matched against trade-offs in investment (Westoby *et al.* 2002; Wright *et al.* 2007). According to Craine (2009) all seed-plant diversity can be collapsed onto four central resource **strategy axes** – strategies for low nutrients, low light, low water and low CO_2 – with modifications for increases in resource supply. For practical purposes, the challenge is to identify the most parsimonious factors among whole-plant PFTs and individual traits that best explain causal links with such strategies. The functional significance of leaf traits within the context of the entire plant is highlighted where plant responses to environmental adversity require coordinated responses of both whole plant traits and leaf traits alike (Bonser 2006). Within the broad constraints of resource acquisition, four **axes of specialization** are considered pivotal to plant strategies (Westoby *et al.* 2002; Lavorel *et al.* 2007). These involve trade-offs between (1) specific leaf area (SLA) and leaf life-span (LLS), (2) seed mass and fecundity, (3) plant height at maturity (H) and shading, water use and response to disturbance, and (4) leaf size (LS) and twig size (TS). This framework has contributed to two key strategy models (LHS and LES; see Sections 12.4.3, 12.4.4).

Not all trade-offs are above-ground. Investment trade-offs between specific root length (SRL) (ratio of root length to root biomass) and root nitrogen and lignin concentrations indicate covarying plant response (e.g. potential growth rate) along environmentally limiting gradients for overall plant growth (Comas & Eissenstat 2002; Craine & Lee 2003; Craine *et al.*, 2005). Root structural and anatomical traits known to constrain RGR(max) and H(max) have potential links with hydraulic conductance, support and longevity (Hummel *et al.* 2007) and exert a feedforward effect on stomatal conductance. In many circumstances the functional significance of leaf traits can parallel that of root traits (Craine *et al.* 2005).

Among the more significant plant ecological strategies involving PFTs and individual traits is the **'resource-ratio'** model of Tilman (1982, 1985) (see also Clark *et al.* 2007) that views the spatial heterogeneity of resources as a selective force for optimal foraging in chronically unproductive habitats. Tilman's model requires precise ordering of trade-offs, for example between life history and competitive ability in which data for multiple co-existing species ability may be limiting (Pierce *et al.* 2005). The **'vital attribute'** strategy of Noble & Slatyer (1980) based on the residence time of specific life history traits following disturbance is theoretically insightful but limited in practice. Rather like the CSR strategy discussed in the next section, the well-known **r-K** model of MacArthur & Wilson (1967), while conceptually useful, also has methodological limitations in complex vegetational successional sequences and in isolated, floristically poor communities such as oceanic islands. Less widely established strategies are reviewed elsewhere (Westoby 1998; Lavorel *et al.* 2007).

Preceding the above and persisting remarkably through time is Raunkiær's (1934) life-form model. Raunkiær defines life-form theoretically as 'The sum of the adaptation of the plant to the climate' (Du Rietz 1931) but practically chooses one of the most fundamental adaptations as a base for his systems of life-forms – the survival of the perennating organ during the most unfavourable season. Although based primarily on sensitivity to winter temperatures, Raunkiær's strategy can be applied equally to 'unfavourableness' under other periodic and even episodic, thermal, light and moisture regimes including flood, fire and strong winds. It can be argued that, as a plant ecological strategy, Raunkiær's system is consistent with a theoretical trade-off of carbon investment per individual against tissue loss and reproductive and regenerative capacity under regimes of cyclic environmental extremes. Thus a gradient can be shown to exist between a preponderance of woody phanerophytes in 'optimal' environments with corresponding decreases towards less optimal habitats accompanied by increasing relative percentage of structurally reduced chamaephytes, geophytes and hemicryptophytes. Four strategies described here include leaf-based features and reflect a move beyond the more loosely defined adaptive or 'epharmonic' (cf. van der Maarel 1980, 2005; Floret *et al.* 1987) Raunkiærean descriptors towards more detailed evidence of cause and effect between functional traits and environment.

12.4.2   The C-S-R strategy

Other than Raunkiær's life-form model, the most widely known plant strategy is the C-S-R model of Grime (1977, 1979). CSR theory aims to describe the key mechanisms underlying vegetation processes and considers the interaction between competition (limitations to biomass production imposed by other species), stress (direct limitations to biomass production imposed by the environment) and disturbance (biomass removal or tissue destruction) in shaping phenotype. According to CSR theory, characteristic developmental traits are inherent to competitor (C), stress-tolerator (S) and ruderal (R) strategists, with apparent intermediate strategies (Caccianiga *et al.* 2006). Crucially, the CSR model suggests that stress and sporadic resource availability favour conservative phenotypes (Pierce *et al.* 2005). While theoretical support for CSR is derived from extensive studies in the UK, mainly on herbaceous vegetation, methodological limitations have precluded its application in other countries especially in species-rich, structurally and functionally complex woody vegetation. A partial solution to the methodological impasse (Hodgson *et al.* 1999; Hunt *et al.* 2004) is to allocate a functional type to an unknown subject using a few, simple predictor variables. Traits such as leaf weight (leaf dry matter content) can be statistically coupled with productivity traits that, for example, are relevant to S-type (slow-growing, stress-tolerant species of chronically unproductive habitats). An ordination of these more readily measureable traits then allows the taxa under study to be placed within CSR coordinate space.

The CSR triangle defines the axes with reference to concepts, for which there is no simple protocol for positioning species beyond the reference data sets within the scheme, and consequently benefits of global comparison have not materialized (Westoby 1998). Methodological and theoretical limitations are clearly apparent where, under studies of grazing impact and shoreline successional sequences, CSR types are not readily applicable (Oksanen & Ranta 1992; Ecke & Rydin 2000; Moog *et al.* 2005). Other problems with the CSR format have been noted elsewhere (Austin & Gaywood 1994; Onipchenko *et al.* 1998; Körner & Jeltsch 2008). With some exceptions (e.g. Cerabolini *et al.* 2010; Kılınç *et al.* 2010) and despite improved numerical procedures, the capacity of CSR theory to predict variation in species composition along environmental gradients worldwide remains problematic.

12.4.3 The Leaf-Height-Seed (LHS) strategy

A more parsimonious approach using a 'core' set of more readily measureable functional traits based on specific Leaf area, mature plant Height and Seed mass (the LHS system of Westoby 1998) represents a significant breakthrough in quantifying plant responses to the environment, with a capacity for general application. The LHS system represents a tightly defined functional concept using orthogonal (functionally independent) traits and as such indicates a paradigmatic shift towards the understanding and application of plant functional traits. As described by Westoby (1998), the LHS plant ecology strategy scheme employs three axes: SLA (light-capturing area deployed per dry mass allocated), height of the plant's canopy at maturity, and seed mass, in which the strategy of a species is described by its position in the volume formed by the three axes. The advantages of the LHS scheme can be understood by comparing it to Grime's CSR scheme, over which it has some significant advantages. Whereas certain elements of the CSR scheme (e.g. the C–S dimension) are overtly conceptual, and as such present methodological limitations (Westoby 2007), these limitations are essentially overcome by the more readily quantifiable LHS application to any vascular plant species in any terrestrial environment. Nonetheless, the advantage of the axes defined through a single readily-measured variable needs to be weighed against the disadvantage that single plant traits may not capture as much strategy variation as CSR's multi-trait axes (Westoby 1998).

12.4.4 The Leaf Economics Spectrum (LES) strategy

There are some common trends and linkages between the LHS strategy and the LES scheme proposed by Wright *et al.* (2004) which describes, at global scale, a universal spectrum of leaf economics consisting of key chemical, structural and physiological properties. The spectrum reflects a quick-to-slow return gradient on investments of nutrients and dry mass in leaves. Unlike several other strategies it is essentially independent of growth-form, plant functional type or biome. Functional linkages between leaf traits and net photosynthetic rate investigated by Shipley *et al.* (2005) provide a mechanistic explanation for the empirical trends relating leaf form and carbon fixation, and predict that SLA and leaf N must be quantitatively coordinated to maximize C fixation thus lending validity to the LES scheme. (See further Section 12.11.)

12.4.5 The Leaf–Life-form–Root (LLR) strategy

The LLR approach considers ways in which multiple traits can be used to construct PFTs via an assembly system that addresses whole-plant performance. This is achieved in part by coupling photosynthetic traits with life-form and readily observable rooting structures. When coupled with additional information that describes stand structure, the LLR methodology facilitates comparative analysis across a range of environmental scales (Fig. 12.1) (Gillison 1981, 2002). The LLR strategy complements significant gaps in the CSR, LHS and LES systems that otherwise exclude important photosynthetic traits such as leaf inclination (Falster & Westoby 2003; Posada *et al.* 2009), leaf phyllotaxis or insertion pattern such as rosettes (Withrow 1932; Lavorel *et al.* 1998, 1999a, 1999b; Díaz *et al.* 2007a; Ansquer *et al.* 2009; Bernhardt-Römermann *et al.* 2011a) and woody green-stem photosynthesis, all of which are noted plant adaptations to irradiance, nutritional and water availability.

As discussed earlier, one strategy that has stood the test of time is the Raunkiærian life-form system, partly because it is built on a fundamental survival adaptation to cyclic environmental and edaphic (nutritional) extremes and because of its sheer simplicity. On the other hand, in its basic form, the life-form model ignores photosynthetic traits. To help redress this issue Gillison (1981) devised a whole-plant classification system based on plant functional attributes in which a plant individual is classified as a '**functionally coherent unit**' composed of a photosynthetic 'envelope' supported by a modified Raunkiærean life-form and an aboveground rooting system – presented here as the' Leaf-Life-form-Root' or LLR spectrum. The LLR asserts that a single attribute, such as leaf size class, takes on increased functional significance when combined with leaf-inclination and other morphological (e.g. dorsiventral) and temporal (e.g. deciduous) descriptors of photosynthetic tissue. In this case the photosynthetic attributes describe a 'functional leaf' that includes any part of the plant (including the primary stem cortex) capable of photosynthesis. For convenience, and to indicate the unique type of PFT, specific LLR combinations are termed functional *modi* (from the Latin '*modus*' mode or manner of behaviour) (see also the '*modality*' of Violle *et al.* 2007). This initial model (Gillison 1981) was the first coordinated use of PFAs to relate *modal* PFTs to environmental conditions (Fig. 12.2). The method was later formalized (Gillison & Carpenter 1997) using an assembly rule set and syntactical grammar to construct *modal* PFTs based on 36 plant functional elements (PFEs) (Table 12.3). In this method, a typical PFT *modus* for an individual

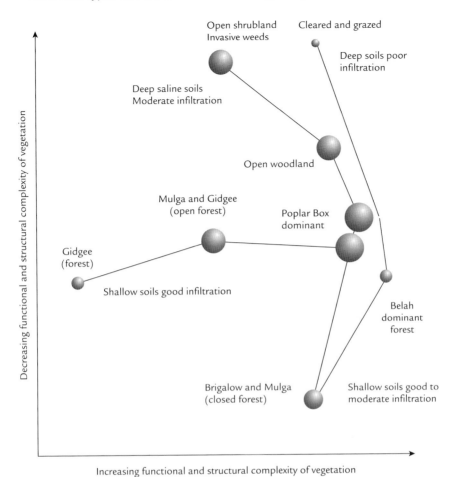

Fig. 12.2   Minimum spanning ordination (Gillison 1978) of plant functional attributes in ten 40 × 5 m transects (globes) mapped against soil depth, infiltration capacity and salinity. The x-axis indicates complexity in leaf size, inclination and phyllodes. The y-axis indicates decreasing functional complexity through decreasing phanerophytes, increasing cryptophytes, and dorsiventral leaves. The z-axis (visualized through decreasing size of the globes) represents mainly a response to vegetation structure (max height, canopy cover %). (Adapted from Gillison 1981.)

of *Acer palmatum* might be a mesophyll (**me**) size class with pendulous (**pe**), dorsiventral (**do**), deciduous (**de**) leaves with green-stem (cortex) (**ct**) photosynthesis attached to a phanerophyte (**ph**), the resulting *modal* PFT combination being **me-pe-do-de-ct-ph**. Within the same species on the same or other site, variation in any one functional element (e.g. a leaf size class), results in a new *modus* thereby facilitating further comparison of intraspecific as well interspecific variability within a described habitat. Using the public domain VegClass software package (Gillison 2002), quantitative and statistical comparisons within and between species and plots are facilitated via predetermined lexical distances between different PFTs (Gillison & Carpenter 1997). The system comprises many-to-many mapping whereby more than one *modal* PFT can be represented within a species and vice versa. While 7.2 million combinations are theoretically possible, a data set compiled from 1066 field sites worldwide (Plate 12.2) indicates the 'real' number of unique *modal* PFTs approximates 3500 for the world's estimated 300 000 vascular plant species.

At a global scale, Plate 12.1 illustrates an arrangement of whole-plant LLR functional syndromes arranged along two key environmental gradients or axes (irradiance and moisture; see also Lavers & Field 2006). Syndromes of this kind are readily described according to the *modal* schema. In the same way that the LES LMA varies with rainfall and temperature, preliminary results from a global survey illustrate how both *modal* PFTs and PFEs covary with global environmental gradients of rainfall and total annual actual evapotranspiration (Figs 12.3, 12.4).

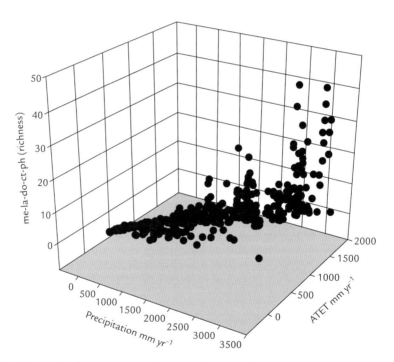

Fig. 12.3 Example of environmentally covarying distribution pattern of plants possessing the *modal* PFT combination me-la-do-ct-ph representing mesophyll (me), laterally inclined (la) dorsiventral (do) (hypostomatous) leaves with a photosynthetic stem cortex (ct), supported by a phanerophyte (ph). Covariates are mean annual rainfall and total annual actual evapostranspiration. Circles are records from 1066 (40 × 5 m) transects.

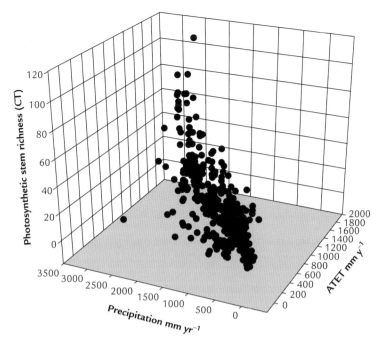

Fig. 12.4 Example of how a single PFE, representing a photosynthetic primary stem cortex (ct) covaries with mean annual rainfall and total annual actual evapotranspiration. Circles are records from 1066 (40 × 5 m) transects.

12.5   The mass ratio hypothesis

The mass ratio hypothesis (MRH) of Grime (1998) predicts that the effect of species or groups of species on ecosystem properties will depend on their proportional abundance in a community. The hypothesis is well supported by empirical evidence (Díaz *et al.* 2007b; Mokany *et al.* 2008) and implies that the ecosystem function is determined to a large extent by the trait values of the dominant contributors to the plant biomass. According to the MRH, ecosystem properties should be predictable from the community weighted mean of traits with proven links with resource capture, usage and release at the individual and ecosystem levels. Díaz *et al.* (2007c) alluded to overwhelming evidence that the more abundant traits are major drivers of short-term ecosystem processes and their feedbacks onto global change drivers. Garnier *et al.* (2004) found support for the MRH where ecosystem-specific net primary productivity, litter decomposition rate and total soil carbon and nitrogen varied significantly with field age, and with community-weighted functional leaf traits SLA, LDMC and leaf N. On the other hand, McLaren & Turkington (2010) show that the effects of losing a functional group do not depend solely on the group's dominance and that functional group identity plays a critical role in determining the effects of diversity loss.

12.6　Functional diversity and complexity

Measures of functional equivalence between many traits lack consensus as do measures of functional redundancy (see Web Resource 12.1.1). Similar debate surrounds measures of functional diversity (FD) that comprises the *kind, range* and *relative abundance* of functional traits present in a given community. There is, however, increasing evidence that FD can be a better predictor of ecosystem functioning than the number of species or the number of functional groups (Díaz & Cabido 2001; Lepš *et al.* 2006; Petchey & Gaston 2006; Villéger *et al.* 2008). To this end, Mayfield *et al.* (2006) further attach an abundance measure distinguishing '**functional composition**' as the identity and abundance of trait states found from a trait in a community. For rangeland studies in Australia, Walker *et al.* (1999) use two **functional attribute diversity** measures: FAD1: the number of different attribute combinations that occurs in the community that must be equal to or less than the number of species – a feature found to be questionable on ecological grounds (Mayfield *et al.* 2005; Villéger *et al.* 2008).

　　To counter the problem that a single measure of FD such as Euclidean distance (the FAD2 of Walker *et al.* 1999; Flynn *et al.* 2009) limits ecological interpretation, Mason *et al.* (2005) propose three additional indices: (a) the amount of niche space filled by species in the community (**functional richness**); (b) the evenness of abundance distribution in filled niche space (**functional evenness**); and (c) the degree to which abundance distribution in niche space maximizes divergence in functional characters within the community (**functional divergence**) (but see also Villéger *et al.* 2008; Bernhardt-Römermann *et al.* 2011b). A pervasive problem in estimating FD is the need to take into account multiple traits that can occur within and between species. To this end the Rao quadratic entropy index (The FDq of Botta-Dukát 2005) fulfils all *a priori* criteria identified by Mason *et al.* (2003, 2005) and according to Botta-Dukát (2005) surpasses other proposed indices, because it includes species abundances and more than one trait (see also de Bello 2012). This is similar to the inverse of Simpson's D index (1-D) used in the VegClass system (Gillison 2002) where species numbers are measured against counts of *modal* PFTs. Nonetheless difficulties remain in allocating standardized measures of different traits identified by different workers (Villéger *et al.* 2008) and in estimating distance measures between traits and combinations used to describe PFTs.

　　A very different approach (Gillison 2002; Gillison *et al.* 2012) explores descriptors of functional complexity and diversity based on *modal* PFTs. First, a minimum spanning tree (MST) (cf. Villéger *et al.* 2008) is used to calculate the total '*functional distance*' that represents a potentially useful measure of '**plant functional complexity**' (PFC) as distinct from 'diversity' *per se* (see Web resource 12.1.2) Dendrograms (*sensu* Petchey & Gaston 2002) or MST lengths are not, strictly speaking, measures of ecological diversity (Magurran 2004), hence the preferred alternative use of 'complexity'. As a measure of *modal* PFT complexity, PFC value can be a useful additional measure of biodiversity in discriminating for example, between two communities that may share the same number of PFTs, but otherwise differ in PFT composition as indicated by a PFC value. Second, whereas the estimation of species diversity relies on individual abundance counts per species, a **'plant functional diversity'** analogue can be estimated using the number of species per PFT instead, to compute three commonly used ecological diversity indices such as Fisher's *alpha* (α), Shannon-Wiener (*H'*) and Simpson's (dominance). A summary of different global vegetation types (Table 12.4) illustrates how PFC and FD values derived from *modal* PFTs vary with vegetation type. By implication, the alternative measurement of the number of species per PFT elevates the application (and testing) of the **mass ratio hypothesis** to another level as the focus changes from dominant species to dominant PFTs.

12.7 Moving to a trait-based ecology – response and effect traits

Whole-plant trait combinations or PFT syndromes facilitate a more holistic perspective of plant-environment interaction than their disaggregated, singular traits such as leaf size or plant height. This advantage is offset by difficulties in deciding how and why trait syndromes should be constructed and how and at what scales traits either singly or combined, interact within and between individuals and with the biophysical environment. Recent progress in formulating plant functional strategies through combinations of independently functioning (orthogonal) traits (12.4) is being increasingly complemented by parallel research that focuses on readily quantifiable, core functional traits. While a common functional thread links both trait syndromes and single traits in the study of plant-environment interaction, the following sections focus on how trait-centred aspects of plant functional ecology may complement the study of PFTs.

PFTs have been variously defined according to their response to environmental conditions or their effect on dominant ecosystem processes (cf. Díaz & Cabido 1997; Díaz Barradas *et al.* 1999). In similar vein, functional traits (FTs) may be described according to 'effect' (Díaz & Cabido 2001; Garnier *et al.* 2004; Violle *et al.* 2007) or 'response' (Garnier *et al.* 2007) (Further definitions of traits and trait types can be found in Web Resource 12.4, 12.5). The following subsections discuss these traits.

12.7.1 Response traits

Disturbance. Discrimination between response and effect phenomena in functional types and traits is obscured by complex feedback and feedforward systems. A comprehensive summary of response and effect phenomena by Lavorel *et al.* (2007) cross-links whole-plant and individual leaf, stem and belowground traits as well as regenerative traits based on trait responses to four classes of environmental change or 'environmental filters'; plant competition and plant defense against herbivores and pathogens (biological filters) and plant effects on biogeochemical cycles and disturbance regimes. Plant ecological strategies are inevitably

Table 12.4 Examples of *modal* PFT diversity and complexity indices across a range of global vegetation types.

Vegetation type	Country/region	Site ID	Spp	Modi	Spp/modi	Fisher's alpha	Shannon index	Simpson index	PFC
Rainforest, Broadleaf lowland	Indonesia/Sumatra	Tesso Nilo 2	202	73	2.77	38.49	3.59	3.11	370
Savanna Sub-Sahelian, open woodland	Africa/ Cameroon	Cameroon 17	45	41	1.10	153.35	0.027	2.91	439
Alpine meadow	Bhutan	Mt Jhomolari 7	34	24	1.42	36.34	3.01	3.33	84
Tundra	Kamchatka (Russia)	Mutrousky pass K01	28	16	1.25	15.51	2.60	3.17	90
Heath – coastal sandy	Australia/temperate/ Mediterranean	Hamelin Bay WA08	27	25	1.08	162.09	0.043	3.00	233
Woodland, Miombo	Africa/Malawi	Malawi01	24	20	1.20	56.46	0.056	2.75	159
Desert, hot, dry	United Arab Emirates/ desertic	Dubai01	22	15	1.47	20.76	0.095	3.60	91
Pasture 20yr	South America/ Perú/ Amazon Basin	Pucallpa PUC05	21	14	1.50	18.36	0.093	3.04	74
Conifer forest on sand	North America/USA	Ocala Florida Nam02	18	13	1.23	24.03	0.080	3.23	120
Deciduous broad-leaved forest, old growth	South America/ Argentina	Tierra del Fuego 01	10	8	1.25	18.57	0.140	2.57	48
Mangrove, Sonneratia	Indonesia/Sumatra	Jambi AHD05	6	6	1.00	166.75	0.167	3.11	61
Steppe	Outer Mongolia	Baatsaagan Nuur 24	6	4	1.5	1.24	5.24	3.28	30

Spp, species richness; *Modi*, richness of *modal* PFTs; Sp/*modi*, species/*modi* ratio; Fisher's alpha (*modi*) according to the logarithmic series; Shannon, Shannon–Wiener index (*modi*); Simpson dominance index (*modi*); PFC, plant functional complexity Index. For diversity indices see Magurran (2004). All data recorded from 40 × 5m transects using the uniform VegClass protocol (Gillison 2002). (See also Web Resource 12.6 for an extended list.)

connected with trait response and effect and include a variety of stressor axes among them disturbance and resource availability. Disturbance – *defined here as loss of tissue or taxa* – may result from natural phenomena, land use and other human-related activities. The intermediate disturbance hypothesis (IDH) asserts that along a disturbance gradient, highest species richness and diversity will occur at intermediate rather than extreme levels of disturbance (Connell 1983). Despite some evidence to the contrary, the IDH has general empirical support (Sheil & Burslem 2003; Bongers *et al.* 2009) with emerging implications for functional ecology. Bernhardt-Römermann *et al.* (2011a) for example, show that management treatments with intermediate disturbance regimes maximize biomass yields in temperate environments. Because of the implications for response and effect dynamics and because the IDH has received little attention thus far, Table 12.5 indicates trends in functional response along gradients of both disturbance and resource availability (see also Lavorel *et al.* 2007). Disturbance effects related to recolonization also reflect phylogenetic patterning in tropical and subtropical vegetation when examined using LHS, LES type functional traits (Kooyman *et al.* 2011). Some key elements of disturbance are described here.

Grazing. Investigations into response-based traits concern grazing dynamics are derived mainly from northern (American and European) temperate and Mediterranean grasslands. Differing levels of reporting and conclusions are a consequence of different investigators applying different techniques in different environments. Overall trait response to environmental gradients such as grazing intensity is not necessarily linear (Saatkamp *et al.* 2010) and reports vary as to response to palatability (Jauffret & Lavorel 2003), xeromorphy (Navarro *et al.* 2006) and mediation by climate (de Bello *et al.* 2005). Among the most commonly reported adaptive responses is that of phyllotaxy (arrangement of leaves along a stem) that directly influences the efficiency of light interception in rosette plants – an effect that greatly diminishes when leaves are vertically displaced by elongated internodes (Niklas 1988; Ackerly 1999). Rosettes harvest low-intensity rains and fogs and large succulent leaf rosettes are a characteristic life-form in many arid and semi-arid areas where large numbers of rosette species suggest a close relationship between form and environment (Martorell & Ezcurra 2002). The extent to which climate, soil and grazing individually influence rosette morphology as a functional trade-off to maximize carbon fixing is not clear, although rosette frequency is widely regarded as a common indicator trait of response to grazing (Lavorel *et al.* 1997, 2007; Díaz *et al.* 2007a; Klimešová *et al.* 2008; Ansquer *et al.* 2009; Bernhardt-Römermann *et al.* 2011a, b).

Other characteristics such as species richness and diversity, plant height, vegetative spread, canopy structure, leaf 'toughness', leaf mass, life history and seed mass, are linked with community response to grazing across continents (Lavorel *et al.* 1998; Díaz *et al.* 2001; Cingolani *et al.* 2005; Louault *et al.* 2005 and others). Debate centres around the utility of single versus multiple traits. Within temperate European grasslands, although plant height is a significant predictor of management impact, the existence of other important plant traits led Klimešová *et al.* (2008) to conclude that single traits cannot be the only basis for predicting vegetation changes under pasture management and that a functional analysis of

Table 12.5 Functional response trends along resource and disturbance gradients.

Functional scaling	High RA, low disturbance	High RA, high to intermediate disturbance	Low RA, high to intermediate disturbance	Low RA, low disturbance
Whole-plant				
Rosette crown	*	**	***	***
Geophytes	*	**	***	***
Liane form	**	***	**	*
Epiphyte	***	**	**	***
Biomass	***	***	**	*
Modal PFTs				
me-la-do-ph	***	***	**	*
pl-la-do-ct-ph	**	***	**	—
pi-ve-is-ro-ch	—	*	**	***
(see Table 12.3)				
Stem				
GSP (ct)	*	***	**	**
Height	***	***	**	*
Basal area	***	***	*	*
Specific density	**	*	**	***
Succulence	*	*	**	***
Shoot:root ratio	***	***	**	*
Aerial roots	**	***	*	*
Leaf				
SLA, LMA	**	***	**	*
C:N ratio	**	*	**	***
LDMC	**	*	**	***
CAM pathway	—	*	**	***
N & P content	**	***	**	*
Secondary metabolites	**	***	***	**
Tensile strength	**	*	**	***
life-span	**	**	**	***
Isostomatous	*	*	**	***
Ve or Pe incl.	*	**	**	***
Litter				
Decomposition rate	**	***	*	*
Fungal (mycorrhizal)	**	***	*	*
Below-ground roots				
SRL	***	***	**	*
Regenerative mode				
Clonality	*	**	***	**
Resprouting	*	***	**	*
Seed mass	**	***	**	*
Ecosystem performance				
Species richness	**	***	**	*
Modal PFT richness	**	***	**	*
Species:*modal* PFT ratio	***	**	**	*
PFC	***	***	**	*

RA, resource availability (light, moisture, nutrients); table excludes response to seasonality and thermal gradients; GSP, green-stem photosynthesis (*modal* element ct); CAM, Crassulacean Acid Metabolism (photosynthetic pathway). Isostomatous, stomata on both sides of leaf; Pe, pendulous inclination; Ve, vertical inclination (Table 12.3); SRL, specific root length; Number of * indicates relative increase in trait response. With the exception of below-ground traits, all or most traits are readily measureable. List is restricted to dry land, non-immersive, terrestrial vascular plants.

the trade-off between multiple key traits is needed. Gradients of grazing intensity are commonly associated with soil properties and functional traits, especially SLA (Ceriani *et al.* 2008; Rusch *et al.* 2009) but at global scale this relationship may be more strongly influenced by climate (Ordoñez *et al.* 2009). Because plant functional type classifications and response rules are frequently specific to regions with different climate and herbivory history, there is a need for more comprehensive studies of ecosystem dynamics at landscape level.

Fire. Regeneration strategies of woody plant species subject to recurrent fire vary between regions (Lloret & Montserrat 2003; Pausas *et al.* 2004; Lavorel *et al.* 2007; Müller *et al.* 2007). Among the primary functional traits that facilitate persistence following crown fire are resprouting capacity and the ability to retain a viable seed bank. Different combinations of these two traits have been preferentially selected in floras with different evolutionary histories. In Australian heathlands for example, the proportion of resprouters and non-resprouters is relatively even, compared with other fire-prone ecosystems, although post-fire obligate resprouters (resprouters without a seed bank) are almost absent. In the Mediterranean basin, most resprouters are obligate, while in California, shrub resprouters are evenly segregated among those having propagules that persist after fire (facultative species) and those without propagule persistence capacity (obligate resprouters). Species with neither persistence mechanism are rare in most fire-prone shrublands (Pausas *et al.* 2004; Lavorel *et al.* 2007), but other significant functional types such as obligate seeders are important components that, together with resprouters, may be adversely affected by short-term alterations to any long-standing fire regime (Regan *et al.* 2010). The highly dynamic nature of fire-prone ecosystems especially towards the lower latitudes, suggests that organization of PFTs and their assemblages is continually mediated by high environmental stochasticity. Limited evidence for deterministic relationships such as between herbivory and fire in savanna (van Langevelde *et al.* 2003) is affected by mainly stochastic phenomena (Jeltsch *et al.* 1996; D'Odorico *et al.* 2006; Keith *et al.* 2007; Regan *et al.* 2010) that typically operate in fire-prone graminoid and heathland ecosystems. In fire-prone wet heathlands of southeastern Australia, not all species within a PFT follow the predicted direction of change (Keith *et al.* 2007).

Land-use change. Future global change scenarios for terrestrial ecosystems suggest that land-use change will probably have the largest effect, followed by climate change, nitrogen deposition, biotic exchange and elevated carbon dioxide concentration (Sala *et al.* 2000; Bakker *et al.* 2011). Within landscapes, fire, grazing and land-use history are key determinants of vegetation performance. However, discrimination between their differential effects is complex as shown by studies across agricultural landscape mosaics in different countries. An analysis of species data and life history traits across northern temperate forested landscapes (Verheyen *et al.* 2003) showed that different groups of species respond to land-use change according to distinguishable trait syndromes. Simulations of different CSR-type PFT performance under fragmented landscapes (Körner & Jeltsch 2008) suggest that seed-based dispersal traits and PFTs play critical roles in vegetation performance. However, in other areas these can be mediated both by the level of disturbance and resource supply (Kleyer 1999) and land-use change may be more likely to affect community assembly processes than species *per se* (Mayfield *et al.* 2010). Bernhardt-Römermann *et al.* (2011a) found that vegetation resistance to disturbance within several European landscapes was related to the occurrence of species with traits selected by a history of intensive land use (smaller leaf size, rosette plant form) and local environmental conditions, whereas vegetation resilience was associated with ecosystem properties that facilitate higher growth rates. Liira *et al.* (2008) on the other hand showed that functional group composition and plant species richness are driven mainly by habitat patch availability and habitat quality. These diverse findings indicate that mechanistic responses to disturbance under different drivers (land use or climate) may depend on historical context.

Trait response in tropical forests. Studies in functional ecology are primarily focused in northern temperate regions and then, mostly in grasslands. For this reason, many hypotheses generated in temperate biomes remain to be tested in humid tropical forests where species and functional richness may increase by an order of magnitude. With some exceptions, hypotheses derived from the temperate zone concerning resource-acquisition trade-offs between traits and light, soil nutrients and disturbance tend nonetheless to apply in the tropics. In dipterocarp dominated forests, species distributions and traits show a significant response to soil nutrient gradients, in line with a generally consistent global pattern where rich-soil specialists have larger leaves, higher SLA, leaf N and P, and lower N:P ratios (Paoli 2006). As with some temperate region studies, research in tropical forests (Poorter & Bongers 2006; Markesteijn *et al.* 2007) indicate a similar pattern of correspondence between leaf trait values and growth, survival and light requirements. Compared to temperate regions and despite some exceptions (Gillison 2002), most studies in tropical forests rely on single rather than trait syndromes (Guehl *et al.* 1998; Kariuki *et al.* 2006; Kooyman & Rossetto 2008; Maharjan *et al.* 2011).

Climate. The analysis of plant functional response to climate is confounded by variation in land use, available nutrients, scale of analysis and the nature of the functional types used in the investigation. While a review of literature on plant response to climate is beyond the scope of this chapter (see Chapters 15 and 17), certain inferences can be drawn from studies of trait and whole-plant PFT response at several integrated levels, albeit with some confusing outcomes. At one level, simulated climate change impact on PFTs (Esther *et al.* 2010) suggests that responses are determined by specific trait characteristics and that community patterns can exhibit often complex responses to climate change. For example, while an increase in annual rainfall can cause an increase in the numbers of dispersed seeds for some PFTs, but decreased PFT diversity in the community, a simulated decrease in rainfall can reduce the number of dispersed seeds and diversity of PFTs. It can be concluded that, at this level, PFT interactions and regional processes must be considered when assessing how local community structure will be affected by environmental change (Esther *et al.* 2010). A climatic gradient may dominate and thus confound otherwise predictive functional traits related to grazing in the Mediterranean region (de Bello *et al.* 2005). In

Patagonia, Jobbágy & Sala (2000) demonstrated a differential effect of precipitation on functional type (grass and shrub) ANPP that shifted from precipitation alone to precipitation and temperature when the temporal scale of analysis changed from annual to seasonal. At a subregional level, leaf size class, leaf type, leaf longevity, photosynthetic pathway and rooting depth along a savanna transect in Southern Africa (Skarpe 1996) were strongly associated with total annual precipitation, precipitation of the wettest month, a moisture index and temperature of the coldest month. However, Maharjan *et al.* (2011) found that in West Africa, shade tolerance and drought resistance were the main strategy axes of variation, with wood density and deciduousness emerging as the best predictor traits of species position along the rainfall gradient.

Theoretical models of optimal, adaptive responses of leaf 'shape' size to irradiance also show a divergence in outcomes (Parkhurst & Loucks 1972; Givnish & Vermeij 1976; Shugart 1997). Givnish (1988) demonstrated how effective light compensation points can maximize tree heights as a function of irradiance, and that shade tolerance, in turn, is a consequent function of tree height. In practice however, simple models of this kind may mislead where complex cascade effects in response trade-offs to irradiance need to be considered. Apart from the influence of seasonal irradiance, rainfall seasonality has a profound influence on vegetation and traits associated with trade-offs between carbon investment and water use efficiency. In certain seasonal forest types (Enquist & Enquist 2011), climate may outweigh disturbance as a driver in ecosystem performance.

12.7.2 Effect traits

Effects on ecosystem properties and services. The Millennium Ecosystem Assessment synthesis (2005) covers *provisioning services* such as food, water, timber and fibre; *regulating services* such as the regulation of climate, floods, disease, wastes and water quality; *cultural services* such as recreational, aesthetic and spiritual benefits; and *supporting services* such as soil formation, photosynthesis and nutrient cycling. While these services are strongly affected by abiotic drivers and direct land-use effects, they are also modulated by community FD (Lloret & Montserrat 2003; Díaz *et al.* 2007a, b). Analyses of ecosystem services using plant functional variation across landscapes offer a powerful approach to understanding fundamental ecological mechanisms underlying ecosystem service provision, and trade-offs or synergies among services (Lavorel *et al.* 2011). On the negative side, univariate investigations of the response–effect relationships between functional traits and ecosystem performance show no coherent solution as yet to the search for a generic methodology or a unified syndrome of traits that can be applied worldwide. A significant contribution to solving this problem is a framework proposed by Díaz *et al.* (2007b) based on the way in which FD response to land-use change alters the provision of ecosystem services important to local stakeholders. Other workers (e.g. Quetiér *et al.* 2007), argue that because PFTs relate to universal plant functions of growth (e.g., light and nutrient acquisition, water-use efficiency) and persistence (e.g. recruitment, dispersal, defence against herbivores, and other disturbances), they have the potential to couple community structure to ecosystem functions. These authors also show that, at least for subalpine European grasslands, plant traits and PFTs are effective predictors of relevant ecosystem attributes for a range of ecosystem services including provisioning (fodder), cultural (land stewardship), regulating (landslide and avalanche risk), and supporting services (plant diversity). Leaf traits such as leaf nitrogen content (LNC) for example, are markers of plant nutrient economy (Wright *et al.* 2004) and are associated with faster nutrient cycling at the ecosystem level (see also Sierra 2009).

Single versus multiple trait effects. Emergent properties of vegetation are affected by interacting plant traits and trait syndromes but as yet, little is known about their degree of influence. Reduction of covarying traits to a minimum set of key predictors can suffice for monitoring effects of land-use change on ecosystem behaviour (cf. Ansquer *et al.* 2009; Falster *et al.* 2011). A novel theoretical framework, the **functional matrix** (Eviner & Chapin 2003; Eviner 2004) describes the relationship between ecosystem processes and multiple traits, treating traits as continuous variables, and determining if the effects of these multiple traits are additive or interactive. PFT assemblages or 'trait syndromes' thus described, provide a means of moving forward from individual to composite sets of traits. In this context, PFTs based on morphology have the potential to link ecophysiological traits with ecosystem processes relevant at large scales (Chapin *et al.* 1993) and to offer alternatives to species representing ecosystem structure (Smith *et al.* 1997). Supporting evidence from Aguiar *et al.* (1996) suggests that changes in PFT composition (diversity) – especially growth-form – independent of changes in biomass, affect ecosystem functioning.

12.8 Plant functional types and traits as bioindicators

Bioindicators are a well-established feature of modern ecology and are commonly used to assess and monitor the status of the biophysical environmental with regard to acid rain, pollutants, landscape rehabilitation, contamination and the like. For assessing and monitoring biodiversity, surrogate measures include a wide range of environmental units or arbitrary ecosystem 'types' or combinations of both (Oliver *et al.* 2004; Carmel & Stroller-Cavari 2006; Grantham *et al.* 2010). Because most definitions of biodiversity are impractical for operational purposes, I propose an operational definition of biodiversity as '*The number and composition of all recordable species and functional types and traits in any given area*'. This definition provides for the extension beyond the species as the most common currency of biodiversity, to include traits and trait syndromes that, as with Linnean species, are gene-based.

12.8.1 Species versus PFTs and functional traits as bioindicators

Taxon-based bioindicators are common surrogates for other taxa and more often an expression of taxonomic richness in biodiversity conservation; however, their application is not without debate (Lawton *et al.* 1998; Lewandowski *et al.* 2010; Lindenmayer & Likens 2011). Recent findings (Sætersdal & Gjerde 2011) point to a general failure of surrogate species or other taxonomic levels in conservation planning – an outcome that suggests functional rather than species-based measures of complementarity may be more appropriate for such purposes. Vandewalle *et al.* (2010) propose standard procedures to integrate different components of species-based functional traits into biodiversity monitoring schemes across trophic levels and disciplines where the development of indicators using functional traits could complement, rather than replace, existing biodiversity monitoring. The frequent use of confamilial or congeneric 'means' or species data, often from wide-ranging locations (Jackson *et al.* 1996; Duru *et al.* 2010; Moles *et al.* 2011; Ordoñez *et al.* 2010), runs the risk of misleading matches as illustrated, for example, by phenotypic plasticity within certain species of arctic or boreal *Salix* (Argus 2004) or tropical Rubiaceae (e.g. *Psychotria,* A.N. Gillison pers. obs.).

Despite the potential utility of traits and trait syndromes in biodiversity conservation, field-validated research is surprisingly sparse. The few examples available suggest only that at broad scale there is predictive potential between plant functional group composition and species richness and landscape 'patch' habitat availability and quality (e.g. Liira *et al.* 2008; Lavorel *et al.* 2011). A general review of PFTs and traits as biodiversity indicators is beyond the scope of this chapter. Three case studies below summarize outcomes from rapid biodiversity and land-use surveys in Sumatra, Indonesia and Mato Grosso, Brazil, using trait syndromes recorded as *modal* PFTs via the VegClass field recording protocol (Gillison 2002).

12.8.2 Regional biodiversity signatures and predictive functional traits

Functional 'signatures' can describe a quantitative profile of community–environment interaction. They can be derived by a variety of means such as a spreadsheet tool for calculating functional signatures for herbaceous vegetation within the context of the C-S-R system of plant functional types (Hunt *et al.* 2004), or spectral signatures for functional types from satellite imagery (Kooistra *et al.* 2007). *Modal* PFTs can be used as reliable indicators for plant species richness in different countries (Fig. 12.5) where, once baseline surveys have been conducted, plant species richness can be estimated from species-independent counts of unique *modal* PFTs, usually with a high degree of confidence. This can be useful where species richness is required for conservation purposes and especially so where species identification is difficult – as can be the case in poorly known areas. In Fig. 12.5, differences in the regression slope between Sumatra and Brazil may represent evolutionary and other historical differences in the species pool, suggesting differential species : PFT 'signatures'. The ratio of species to *modal* PFTs can vary predictably along resource availability and disturbance gradients (Gillison 2002) reflecting strategies such as LES and LHS. Certain faunal groups also exhibit a close relationship with the species : PFT ratio; changes in termite species richness along a Sumatran land-use intensity gradient are significantly correlated with plant species richness and *modal* PFT richness (Fig. 12.6a,b). However, the correlation becomes appreciably linear when termite species richness is regressed against the species : PFT ratio (Fig. 12.6c). As discussed in the foregoing, functional complexity (PFC) provides an additional measure of biodiversity. In Sumatra, PFC values were significantly related to the

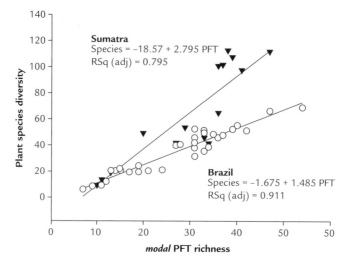

Fig. 12.5 Different regional 'signatures' in species to *modal* PFT ratios along land use intensity gradients and vegetation mosaics in Sumatra, Indonesia (triangles) and Brazil (circles) may reflect evolutional separation of floras and functional characteristics. Data points are 40 × 5 m transects.

distribution of certain invertebrate and vertebrate fauna (see Fig. 12.7 for mammals) and, as with termites, reflects variation in habitat as indicated by vegetation and land use and indirectly, availability of food and shelter resources (Gillison *et al.* in press).

12.9 Environmental monitoring

Research outcomes from the past two decades identify potential PFT and trait-based indicators for monitoring the effects of environmental change on bio-physical resources, although very few are actually taken up by management. A method proposed by Hodgson *et al.* (1999; see also Cerabolini *et al.* 2010) for assessing and monitoring change in CSR characteristics is based on long-term monitoring (1958 to date) of permanent plots in Northern England. Most potential indicators, such as CSR types, are based on herbaceous communities (Lavorel *et al.* 1998, 1999a, b; Díaz *et al.* 2007a; Jauffret & Lavorel 2003; Lavorel *et al.* 1997, 2007; Ansquer *et al.* 2009; Bernhardt-Römermann *et al.* 2011a, b) and are unlikely to apply as readily in non-herbaceous biomes. For monitoring purposes, criteria and indicators should target key ecosystem drivers with a focus on the most parsimonious sets of indicators that can be readily measured in a repeatable way by different observers. A move away from species to complementary, functional trait-based indicators is advocated by Vandewalle *et al.* (2010). Detailed studies in different environments suggest minimal indicator groups can be selected from trait syndromes where convergence between plant traits simplifies their monitoring (Ansquer *et al.* 2009).

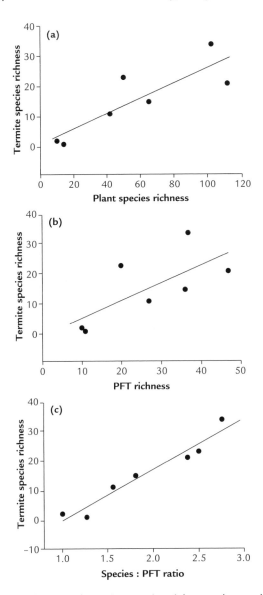

Fig. 12.6 Improved prediction of termite species richness along a land use intensity gradient in Sumatra when ratio of plant species richness to *modal* PFT richness is applied: (a) plant species richness; (b) *modal* PFT richness; (c) plant species richness:*modal* PFT richness. (Adapted from Gillison *et al.* 2003; Bardgett 2005.)

Rapid, repeatable, cost-effective assessment and monitoring methods are a central goal for environmental monitoring. Community-aggregated (i.e. weighted according to the relative abundance of species) functional traits (Garnier *et al.* 2004), while potentially useful in herbaceous assemblages are unlikely to be effective in botanically poorly known, structurally complex vegetation. As also pointed out by Gaucherand & Lavorel (2007), a standardized population-centred

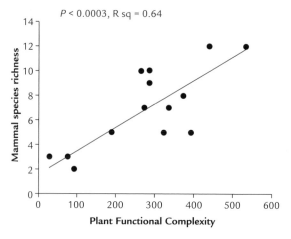

Fig. 12.7 Plant functional complexity (PFC) and mammal species richness in Jambi Sumatra, Indonesia. Dots are 40 × 5 m transects recorded along a land use intensity gradient. (Adapted from Gillison 2000.)

method for measuring species traits already exists (Cornelissen *et al.* 2003), but requires substantial labour and reliable botanical knowledge. An alternative low-cost approach using a rapid **trait-transect** combined with a minimal readily measureable set of traits (Gaucherand & Lavorel 2007) was found to be just as effective for the rapid assessment of functional composition in herbaceous communities. When combined with baseline ground data using rapid survey techniques, recent developments in satellite and airborne imagery are already delivering the next generation of environmental monitoring tools (White *et al.* 2000; Asner *et al.* 2005) that may be usefully combined with dynamic vegetation modelling (Kooistra *et al.* 2007).

In the lower Zambezi valley of Mozambique, Gillison *et al.* (2011) combined an LLR strategy with gradient-based rapid-survey using remote-sensing technology to explore linkages between biodiversity and agricultural productivity along multiple biophysical gradients. Baseline ground data were obtained using gradsects and the VegClass system to sample vascular plant species, *modal* PFTs and their PFEs, vegetation structure, soil properties and land-use characteristics along an inland-to-coast 450-km corridor. Landsat 7 satellite imagery was used to map photosynthetic and non-photosynthetic vegetation and bare substrate along each gradsect. Highly significant correlations between single and combined sets of plant, soil and remotely sensed variables permitted spatial extrapolation of biodiversity and soil fertility throughout a regional land-use mosaic (Plate 12.3) that at 30-m grid resolution, provides a rich source of spatially co-referenced data for management purposes.

12.10 Trait-based climate modelling

A wide range of models is now available for modelling vegetation response to climate change (Chapter 15) across biomes. In the absence of any global 'functional' (i.e. physiological) plant types, many of these are based on ecophysi-ognomic growth-form such as 'Evergreen broad-leaved laurophyll tree' (Box 1981, 1996; Box & Fujiwara 2005). To simulate vegetation responses to past and future climate change, well-known mechanistic vegetation models (see Chapters 15 and 17) are BIOME4 (Kaplan *et al.* 2003) and LPJ (Lund-Potsdam-Jena; Sitch *et al.* 2003). These models simulate the distribution of plant func-tional types (e.g. grass, evergreen needle-leaved trees), which can be combined to represent biomes and habitat types (Hickler *et al.* 2004; Broennimann *et al.* 2006; USGS 2010). Harrison & Prentice (2003) used similar growth-form in conjunction with BIOME4 to simulate climate and CO_2 controls on global pal-aeovegetation distribution at the last glacial maximum. They concluded that more realistic simulations of glacial vegetation and climate will need to take into account the feedback effects of these structural and physiological changes on the climate. Peppe *et al.* (2011) have since shown that, compared to ecophysiogno-mic growth-form, the inclusion of leaf traits that are functionally linked to climate improves palaeoclimate reconstructions.

A fundamental difficulty with 'biome' models is that they are rarely structur-ally monotypic (i.e. pure grass or pure trees) representing instead, a mix of many growth-forms or PFTs. To address this problem Oleson *et al.* (2010) describe a Community Land Model (CLM), a land surface parameterization of two other models – the Community Atmosphere Model (CAM4.0) and the Community Climate System Model (CCSM4.0). In the CLM, vegetation is not represented as biomes (e.g. savanna) but rather as patches of PFTs (e.g. grasses, trees) (Bonan *et al.* 2002). The PFT (broadly defined for example, as '*broad-leaved evergreen tree*') determines plant physiology while community composition (i.e. the PFTs and their areal extent) and vegetation structure (e.g. height, leaf area index) constitutes direct input to each grid cell for each PFT. This allows the model to interface with models of ecosystem processes and vegetation dynamics where each PFT is defined by a variety of optical (reflectance, transmittance), morpho-logical (e.g. leaf habit, stem type) and physiological (photosynthetic) parameters. In the same context Laurent *et al.* (2004) refined vegetation simulation models in order to apply global scale modelling to regional scale. From a bioclimatic analysis of 320 taxa they produced a series of **Bioclimatic Affinity Groups** (BAGs) based on growth-form that could be shown to correspond with different geo-graphical ranges and climatic tolerances.

The question of which level of PFT sensitivity is most appropriate for model-ling climate change impact has engendered much discussion where there is now a tendency to argue a case for finer scale interactive levels (Esther *et al.* 2010; Peppe *et al.* 2011). Here the use of spatially continuous distributions of coexist-ing PFTs may be a necessary step to link climate and ecosystem models (Bonan *et al.* 2002). Studies of functional traits such as phenology (Arora & Boer 2005), LLS, LMA, photosynthetic capacity, dark respiration and leaf N and P concentra-tions, as well as leaf K, photosynthetic N-use efficiency (PNUE), and leaf N:P ratio (Wright *et al.* 2005) show that at the global-scale, quantification of relation-ships between plant traits may be fundamental for parameterizing vegetation-climate models. Imaging spectroscopy is now capable of delivering full optical spectra (400–2500 nm) of the global land surface on a monthly time step and can be used to estimate (i) fractional cover of biological materials, (ii) canopy water content, (iii) vegetation pigments and light-use efficiency, (iv) plant functional types, (v) fire fuel load and fuel moisture content, and (vi) disturbance occurrence, type and intensity (Asner *et al.* 2005).

12.11   Scaling across community, ecosystem and world level

Here, 'scaling' refers to (1) dependent variation of an organism's form or function (e.g. body mass) in which the most commonly used scaling equation is the power function, and (2) the use of empirical models of trait-based, response–effect interactions that can be scaled up from local to global scale. In the former case, metabolic scaling theory is proposed by Enquist *et al.* (2007) as a basis for constructing a general quantitative framework to incorporate additional leaf-level trait scaling relationships such as LES, and hence integrate functional trait spectra with theories of relative growth rate.

Scales can range from biome, landscape and forest canopy, to leaves and their components (Fig. 12.1). The LES has helped defuse some of the earlier pessimism (He *et al.* 1994) that the effect of scaling on measures of biological diversity is non-linear and that heterogeneity increases with the size of the sampling units so that fine-scale information is lost at a broad scale. By itself, LES describes biome-invariant scaling functions for leaf functional traits that relate to global primary productivity and nutrient cycling. Similar scaling analogues have been proposed for wood (Chave *et al.* 2009) and leaf venation (Blonder *et al.* 2011). Although each of these strategies reveals scaling trends along global environmental gradients, all are concerned primarily with bivariate rather than multivariate relationships and as such may oversimplify significant plant–environment interaction at a variety of environmental scales (see also Shipley 2004; Grueters 2009). Serial dependency (where trait A depends on B for its existence but not *vice versa*) is also an issue when generating biologically realistic models (Shipley 2004) and in ensuring parsimony when selecting functional traits for generalizable scaling purposes. Both the LHS and LES avoid this issue by selecting traits with independent (orthogonal) functional axes, a feature demonstrated to a lesser extent by LLR, although the combinatory nature of the functional traits used in the LLR encapsulates vegetation features from leaf to structural formation level (Fig. 12.1).

Leaves represent a common starting point for upscaling to whole-plant properties that, according to Wardle *et al.* (1998), have the potential to manifest themselves over much larger scales (see also Read *et al.* 2006). The use of plant functional traits, rather than species and other taxa, to generalize complex community dynamics and predict the effects of environmental changes has been referred to as a 'Holy Grail' in ecology (Lavorel & Garnier 2002; Lavorel *et al.* 2007; Suding & Goldstein 2008). In this context, Kooistra *et al.* (2007) argue that PFTs should be adopted for global-scale modelling. An approach devised by Falster *et al.* (2011) uses a structured trait, size and patch model of vegetation dynamics based on four key traits (leaf economic strategy, height at maturation, wood density and seed size) that allows scaling up from individual-level growth processes and probabilistic disturbances to landscape-level predictions. Varying suites of traits have been suggested for scaling up from organ to ecosystem such as the 'functional markers' of Garnier *et al.* (2004) using three leaf traits, SLA, LDMC and N concentration. Again, field-testing of these and other scaling-up approaches at global scale (Körner 1994; Hodgson *et al.* 1999; Niinemets *et al.* 2007) remains a challenge as does common agreement on sampling protocols (Kattge *et al.* 2011) (see Web Resource 12.2).

12.12   Discussion

Despite a lack of consensus regarding the theory and practice of identifying appropriate PFTs and functional traits, it would be wrong to say that plant functional ecology is in a state of undisciplined chaos. Particularly in the past two decades, much has been achieved in the way of new insights and technology. Nonetheless, the search for a functional paradigm and the recent explosion of literature surrounding trait-based ecology reflects as much the multi-faceted interests of investigators as that of the entire research agenda of ecology itself (cf. Westoby 1998). In the move towards a more synthetic and more predictive science of ecology, the search for a single, comprehensive yet relatively parsimonious, plant functional classification remains as yet, an elusive Holy Grail (cf. Lavorel & Garnier 2002; Lavorel *et al.* 2007; Suding & Goldstein 2008). While quantification and synthesis remain a central focus, emerging theory related to the stoichiometry and metabolic scaling of functional traits and types (Web Resource 12.3) is also assisting in the move towards an improved understanding of plant functional interaction with global change.

Acknowledgements

I am indebted to Brian Enquist, Peter Reich, Ian Wright and others for their insightful comments and to the editors Janet Franklin and Eddy van der Maarel for their extraordinary patience and painstaking editing.

References

Ackerly, D.D. (1999) Self-shading, carbon gain and leaf dynamics: a test of alternative optimality models. *Oecologia* **119**, 300–310.

Ackerly, D.D., Dudley, S.A., Sultan, S.E. *et al.* (2000) The evolution of plant ecophysiological traits: recent advances and future directions. *Bioscience* **50**, 979–995.

Aguiar, M.R., Paruelo, J.M., Sala, O.E. & Lauenroth, W.K. (1996) Ecosystem responses to changes in plant functional type composition: an example from the Patagonian steppe. *Journal of Vegetation Science* **7**, 381–390.

Albert, C.H., Thuiller, W., Yoccoz, N.J. *et al.* (2010) A multi-trait approach reveals the structure and the relative importance of intra- *vs.* interspecific variability in plant traits. *Functional Ecology* **24**, 1192–1201.

Ansquer, P., Duru, M., Theau, J.P. & Cruz, P. (2009) Convergence in plant traits between species within grassland communities simplifies their monitoring. *Ecological Indicators* **9**, 1020–1029.

Argus, G.W. (2004) A guide to the identification of Salix (willows) in Alaska, the Yukon Territory and adjacent regions. Workshop on willow identification. http://137.229.141.57/wp-content/uploads/2011/02/GuideSalixAK-YT11May05.pdf (accessed 4 July 2011).

Arora, V.K. & Boer, G.J. (2005) A parameterization of leaf phenology for the terrestrial ecosystem component of climate models. *Global Change Biology* **11**, 39–59.

Asner, G.P., Knox, R.G., Green, R.O. & Ungar, S.G. (2005) Mission concept for the National Academy of Sciences decadal study: the Flora mission for ecosystem composition, disturbance and productivity. http://pages.csam.montclair.edu/~chopping/rs/FLORA_NRCDecadalSurvey_2005.pdf (accessed 25 May 2012).

Aubin, I., Ouellette, M.H., Legendre, P., Messier, C. & Bouchard, A. (2009) Comparison of two plant functional approaches to evaluate natural restoration along an old-field–deciduous forest chronosequence. *Journal of Vegetation Science* **20**, 185–198.

Austin, M.P. & Gaywood, M.J. (1994) Current problems of environmental gradients and species response curves in relation to continuum theory. *Journal of Vegetation Science* **5**, 473–482.

Bakker, M.A., Carreño-Rocabado, G. & Poorter L. (2011) Leaf economics traits predict litter decomposition of tropical plants and differ among land use types. *Functional Ecology* **25**, 473–483.

Bardgett, R.D. (2005) *The Biology of Soil: a Community and Ecosystem Approach.* Oxford University Press, Oxford.

Bernhardt-Römermann, M., Gray, A., Vanbergen, A.J. *et al.* (2011a) Functional traits and local environment predict vegetation responses to disturbance: a pan-European multi-site experiment. *Journal of Ecology* **99**, 777–787.

Bernhardt-Römermann, M., Römermann, C., Sperlich, S. & Schmidt, W. (2011b) Explaining grassland biomass – the contribution of climate, species and functional diversity depends on fertilization and mowing frequency. *Journal of Applied Ecology* **48**, 1088–1097. doi: 10.1111/j.1365-2664.2011.01968.x

Blonder, B., Violle, C., Bentley, L.P. & Enquist, B.J. (2011) Venation networks and the origin of the leaf economics spectrum. *Ecology Letters* **14**, 91–100.

Bonan, G.B., Levis, S., Kergoat, L. & Oleson, K.W. (2002) Landscapes as patches of plant functional types: an integrating concept for climate and ecosystem models. *Global Biogeochemical Cycles* **16** (2), 1021. doi:10.1029/2000GB001360

Bongers, F., Poorter, L., Hawthorne, W.D. & Sheil, D. (2009) The intermediate disturbance hypothesis applies to tropical forests, but disturbance contributes little to tree diversity. *Ecology Letters* **12**, 798–805.

Bonser, S.P. (2006) Form defining function: interpreting leaf functional variability in integrated plant phenotypes. *Oikos,* **114**, 187–190.

Botta-Dukát, Z. (2005) Rao's quadratic entropy as a measure of functional diversity based on multiple traits. *Journal of Vegetation Science* **16**, 533–540.

Boutin, C. & Keddy, P.A. (1993) A functional classification of wetland plants. *Journal of Vegetation Science* **4**, 591–600.

Box, E.O. (1981) *Macroclimate and plant forms: an introduction to predictive modeling in phytogeography.* Dr W. Junk, The Hague.

Box, E.O. (1996) Plant functional types and climate at the global scale. *Journal of Vegetation Science* **7**, 309–320.

Box, E.O. & Fujiwara, K. (2005) Vegetation types and their broadscale distribution. In: *Vegetation Ecology* (ed. E. van der Maarel), pp. 106–128. Blackwell Publishing UK, Oxford.

Broennimann, O., Thuiller, W., Hughes, G. *et al.* (2006) Do geographic distribution, niche property and life form explain plants' vulnerability to global change? *Global Change Biology* **12**, 1079–1093.

Caccianiga, M., Luzzaro, A., Pierce, S., Ceriani, R.M. & Cerabolini, B. (2006) The functional basis of a primary succession resolved by CSR Classification. *Oikos* **112**, 10–20.

Carmel, Y. & Stroller-Cavari, L. (2006) Comparing environmental and biological surrogates for biodiversity at a local scale. *Israel Journal of Ecology and Evolution* **52**, 11–27.

Carpenter, G., Gillison, A.N. & Winter, J. (1993) DOMAIN: a flexible modelling procedure for mapping potential distributions of plants and animals. *Biodiversity Conservation* **2**, 667–680.

Cerabolini, B.E.L., Brusa, G., Ceriani, R.M., de Andreis, R., Luzzaro, A. & Pierce, S. (2010) Can CSR classification be generally applied outside Britain? *Plant Ecology* **210**, 253–261.

Ceriani, R.M., Pierce, S. & Cerabolini, B. (2008) Are morpho-functional traits reliable indicators of inherent relative growth rate for prealpine calcareous grassland species? *Plant Biosystems* **142**, 60–65.

Chapin, F.S. III., Autumn, K. & Pugnaire, F. (1993) Evolution of suites of traits in response to environmental stress. *The American Naturalist* **142**, S78–S92.

Chave J., Coomes D., Jansen S. *et al.* (2009) Towards a worldwide wood economics spectrum. *Ecology Letters* **12**, 351–366.

Cingolani, A.M., Posse, G. & Collantes, M.B. (2005) Plant functional traits, herbivore selectivity and response to sheep grazing in Patagonian steppe grasslands. *Journal of Applied Ecology* **42**, 50–59.

Clark, J.S., Dietze, M., Chakraborty, S. *et al.* (2007) Resolving the biodiversity paradox. *Ecology Letters* **10**, 647–659.

Comas, L.H. & Eissenstat, D.M. (2002) Linking root traits to potential growth rate in six temperate tree species. *Oecologia* **132**, 34–43.

Condit, R., Hubbell, S.P. & Foster, R.B. (1996). Assessing the response of plant functional types to climatic change in tropical forests. *Journal of Vegetation Science* **7**, 405–416.

Connell, J. (1983) On the prevalence and relative importance of interspecific competition: evidence from field experiments. *The American Naturalist* **122**, 661–696.

Cornelissen, J.H.C., Lavorel, S., Garnier, E. *et al.* (2003) A handbook of protocols for standardised and easy measurement of plant functional traits worldwide. *Australian Journal of Botany* **51**, 335–380.

Craine, J.M. (2009) *Resource Strategies of Wild Plants*. Princeton University Press, Princeton, NJ.

Craine, J.M. & Lee, W.G. (2003) Covariation in leaf and root traits for native and non-native grasses along an altitudinal gradient in New Zealand. *Oecologia* **134**, 471–478.

Craine, J.M., Lee, W.G., Bond, W.J., Williams, R.J. & Johnson, L.C. (2005) Environmental constraints on a global relationship among leaf and root traits of grasses. *Ecology* **86**, 12–19.

Cramer, W. (1997) Using plant functional types in a global vegetation model. In: *Plant Functional Types: Their Relevance to Ecosystem Properties and Global Change* (eds T.M. Smith, H.H. Shugart & F.I. Woodward), pp. 271–288. Cambridge University Press, Cambridge.

de Bello, F. (2012) The quest for trait convergence and divergence in community assembly: are null-models the magic wand? *Global Ecology and Biogeography* **21**, 312–317.

de Bello, F., Lepš , J. & Sebastià, M.-T. (2005) Predictive value of plant traits to grazing along a climatic gradient in the Mediterranean. *Journal of Applied Ecology* **42**, 824–833.

Díaz Barradas, M.C., Zunzunegui, M., Tirado, R., Ain-Lhout, F. & García Novo, F. (1999) Plant functional types and ecosystem function in Mediterranean shrubland. *Journal of Vegetation Science* **10**, 709–716.

Díaz, S. & Cabido M. (1997) Plant functional types and ecosystem function in relation to global change. *Journal of Vegetation Science* **8**, 463–474.

Díaz, S. & Cabido, M. (2001) Vive la différence: plant functional diversity matters to ecosystem processes. *Trends in Ecology and Evolution* **16**, 646–655.

Díaz, S., Lavorel, S., McIntyre, S. *et al.* (2007a) Plant trait responses to grazing – a global synthesis. *Global Change Biology* **13**, 313–341.

Díaz, S., Lavorel, S., de Bello, F. *et al.* (2007b) Incorporating plant functional diversity effects in ecosystem service assessments. *Proceedings of the National Academy of Sciences of the United States of America* **104**, 20684–20689.

Díaz, S., Lavorel, S., Chapin, F.S. III *et al.* (2007c) Functional diversity – at the crossroads between ecosystem functioning and environmental filters. In: *Terrestrial Ecosystems in a Changing World* (eds J.G. Canadell, D. Pataki & L. Pitelka), pp. 81–90. Springer-Verlag, Berlin.

Díaz, S., McIntyre, S., Lavorel, S. & Pausas, J.G. (2002) Does hairiness matter in Harare? Resolving controversy in global comparisons of plant trait responses to ecosystem disturbance. *New Phytologist* **154**, 1–14.

Díaz, S., Noy-Meir, I. & Cabido, M. (2001) Can grazing response of herbaceous plants be predicted from simple vegetative traits? *Journal of Applied Ecology* **38**, 497–508.

D'Odorico, P., Laio, F. & Ridolfi, L. (2006) A Probabilistic analysis of fire-induced tree-grass coexistence in savannas. *The American Naturalist* **167**, E79–E87.

Du Rietz, G.E. (1931) Life-forms of terrestrial flowering plants. *Acta Phytogeographica Suecica* 3, 1–95.

Duru, M., Ansquer, P., Jouany, C., Theau, J.P. & Cruz, P. (2010) Comparison of methods for assessing the impact of different disturbances and nutrient conditions upon functional characteristics of grassland communities. *Annals of Botany* 106, 823–831.

Ecke, F. & Rydin, H. (2000) Succession on a land uplift coast in relation to plant strategy theory. *Annales Botanici Fennici* 37, 163–171.

Enquist, B.J. & Enquist, C.A.F. (2011) Long-term change within a Neotropical forest: assessing differential functional and floristic responses to disturbance and drought. *Global Change Biology* 17, 1408–1424.

Enquist, B.J., Kerkhoff, A.J., Huxman, T.E. & Economow, E.P. (2007) Adaptive differences in plant physiology and ecosystem paradoxes: insights from metabolic scaling theory. *Global Change Biology* 13, 591–609.

Esther, A., Groeneveld, J., Enright, N.J. *et al.* (2010) Sensitivity of plant functional types to climate change: classification tree analysis of a simulation model. *Journal of Vegetation Science* 21, 447–461.

Eviner, V.T. & Chapin, F.S. III (2003) Functional matrix: a conceptual framework for predicting multiple plant effects on ecosystem processes. *Annual Review of Ecology, Evolution and Systematics* 34, 455–485.

Eviner, V.T. (2004) Plant traits that influence ecosystem processes vary independently among species. *Ecology* 85, 2215–2229.

Falster, D.S. & Westoby, M. (2003) Leaf size and angle vary widely across species: what consequences for light interception? *New Phytologist* 158, 509–525.

Falster, D.S., Bränstrom, Å., Dieckmann, U. & Westoby, M. (2011) Influence of four major plant traits on average height, leaf-area cover, net primary productivity, and biomass density in single-species forests: a theoretical investigation. *Journal of Ecology* 99, 148–164.

Floret, C., Galan, N.J., Le Floc'h, E., Orshan, G. & Romane, F. (1987) Local characterization of vegetation through growth forms: Mediterranean *Quercus ilex* coppice as an example. *Vegetatio* 71, 3–11.

Flynn, D.F.B., Gogol-Prokurat, M., Nogeire,T. *et al.* (2009) Loss of functional diversity under land use intensification across multiple taxa. *Ecology Letters* 12, 22–33.

Fosberg, F.R. (1967) A classification of vegetation for general purposes. In: *Guide to the Checklist for I.B.P. areas.* I.B.P. Handbook, Number 4 (ed G.F. Peterken), pp. 73–120. Blackwell Scientific, Oxford, UK.

Garnier, E., Cortez, J., Billès, G. *et al.* (2004) Plant functional markers capture ecosystem properties during secondary succession. *Ecology* 85, 2630–2637.

Garnier, E., Lavorel, S., Ansquer, P. *et al.* (2007) Assessing the effects of land-use change on plant traits, communities and ecosystem functioning in grasslands: a standardized methodology and lessons from an application to 11 European sites. *Annals of Botany* 99, 967–985.

Gaucherand, S. & Lavorel, S. (2007) New method for rapid assessment of the functional composition of herbaceous plant communities. *Austral Ecology* 32, 927–936.

Gillison, A.N. (1978). Minimum spanning ordination: a graphic-analytical technique for three-dimensional ordination display. *Australian Journal of Ecology* 3, 233–238.

Gillison, A.N. (1981) Towards a functional vegetation classification. In: *Vegetation Classification in Australia.* (eds A.N. Gillison & D.J. Anderson), pp. 30–41. Commonwealth Scientific and Industrial Research Organization and the Australian National University Press, Canberra.

Gillison, A.N. (2000). Summary and overview. In: *Above-ground Biodiversity assessment Working Group Summary Report 1996–99 Impact of Different Land Uses on Biodiversity* (Coordinator A.N. Gillison), pp. 19–24. Alternatives to Slash and Burn project. ICRAF, Nairobi.

Gillison, A.N. (2002) A generic, computer-assisted method for rapid vegetation classification and survey: tropical and temperate case studies. *Conservation Ecology* 6, 3. [online]: http://www.consecol.org/vol6/iss2/art3 (accessed 25 May 2012).

Gillison, A.N. & Carpenter, G. (1997) A generic plant functional attribute set and grammar for dynamic vegetation description and analysis. *Functional Ecology* 11, 775–783.

Gillison, A.N., Jones, D.T., Susilo, F.-X. & Bignell, D.E. (2003) Vegetation indicates diversity of soil macroinvertebrates: a case study with termites along a land-use intensification gradient in lowland Sumatra. *Organisms, Diversity and Evolution* 3, 111–126.

Gillison A.N., Asner, G.P., Fernandes, E.C. *et al.* (2011) Biodiversity in changing environments: new approaches to integrated ground and satellite baseline surveys. (submitted for publication).

Gillison, A.N., Bignell, D.E., Brewer, K.R.W. *et al.* (2012) Plant functional types and traits as biodiversity indicators for tropical forests: two biogeographically separated studies including birds, mammals and termites (Unpublished as at 17 May 2012 – submitted for publication)

Gillison, A.N., Babu, M.M., Williams, A.C. *et al.* (in press) Low-cost, high-return methodology for rapid biodiversity assessment: a case study from the Eastern Himalayas. In: *Sustainable Development: Asia-Pacific Perspectives* (ed. P.S. Low), Chapter 36. Cambridge University Press.

Gitay, H & Noble, I.R. (1997) What are functional groups and how should we seek them? In: *Plant Functional Types: Their Relevance to Ecosystem Properties and Global Change* (eds T.M. Smith, H.H. Shugart & F.I. Woodward), pp. 3–19. Cambridge University Press, Cambridge.

Gitay, H., Noble, I.R. & Connell, J.H. (1999) Deriving functional types for rain-forest trees. *Journal of Vegetation Science* **10**, 641–650.

Givnish, T.J. (1988) Adaptation to sun and shade, a whole-plant perspective. *Australian Journal of Plant Physiology* **15**, 63–92.

Givnish, T.J. & Vermeij, G.J. (1976) Sizes and shapes of liane leaves. *The American Naturalist* **100**, 743–778.

Grantham, H.S., Pressey, R.L., Wells, J.A. & Beattie, A.J. (2010) Effectiveness of biodiversity surrogates for conservation planning: different measures of effectiveness generate a kaleidoscope of variation. *PLoS ONE* **5** (7), e11430 1–12.

Grime, J.P. (1977) Evidence for the existence of three primary strategies in plants and its relevance to ecological and evolutionary theory. *The American Naturalist* **111**, 1169–1194.

Grime, J.P. (1979) *Plant Strategies and Vegetation Processes.* John Wiley & Sons, Ltd, Chichester.

Grime, J.P. (1998) Benefits of plant diversity to ecosystems: immediate, filter and founder effects. *Journal of Ecology* **86**, 902–910.

Grubb, P.J. (1977) The maintenance of species-richness in plant communities: the importance of the regeneration niche. *Biological Reviews* **52**, 107–145.

Grueters, U. (2009) The universal, individual-based model. http://uibm-de.sourceforge.net/04a3d89c5a0f60e01/04a3d89c5a0fbcb09/index.php (accessed 28 May 2011).

Guehl, J.M., Domenach, A.M., Bereau, M. *et al.* (1998) Functional diversity in an Amazonian rainforest of French Guyana: a dual isotope approach (δ5N and δ3C). *Oecologia* **116**, 316–330.

Harper, J.L. (1977) *The Population Biology of Plants.* Academic Press, London.

Harrison S.P. & Prentice, I.C. (2003) Climate and CO_2 controls on global vegetation distribution at the last glacial maximum: analysis based on palaeovegetation data, biome modelling and palaeoclimate simulations. *Global Change Biology* **9**, 983–1004.

Hawkins, C.P. & MacMahon, J.A. (1989) Guilds: the multiple meanings of a concept. *Annual Review of Entomology* **34**, 423–451.

He, F., Legendre, P. & Bellehumeur, C. (1994) Diversity pattern and spatial scale: a study of a tropical rain forest of Malaysia. *Environment and Ecological Statistics* **1**, 265–286.

Hickler, T., Smith, B., Sykes, M.T. *et al.* (2004) Using a general vegetation model to simulate vegetation dynamics in northeastern U.S.A. *Ecology* **85**, 519–530.

Hodgson, J.G., Wilson, P.J., Hunt, R., Grime, J.P. & Thompson, K. (1999) Allocating C-S-R plant functional types: a soft approach to a hard problem. *Oikos* **85**, 282–294.

Hummel, I., Vile, D., Violle, C. *et al.* (2007) Relating root structure and anatomy to whole plant functioning: the case of fourteen herbaceous Mediterranean species. *New Phytologist* **173**, 313–321.

Hunt, R., Hodgson, J.G., Thompson, K. *et al.* (2004) A new practical tool for deriving a functional signature for herbaceous vegetation *Applied Vegetation Science* **7**, 163–170.

Jackson, R.B., Canadell, J., Ehleringer, J.R. *et al.* (1996) A global analysis of root distribution for terrestrial biomes. *Oecologia* **108**, 389–411.

Jauffret, S. & Lavorel, S. (2003) Are plant functional types relevant to describe degradation in arid, southern Tunisian steppes? *Journal of Vegetation Science* **14**, 399–408.

Jeltsch, F., Milton, S.J., Dean, W.R.J. & Van Royen, N. (1996) Tree spacing and coexistence in semi-arid savannas. *Journal of Ecology* **84**, 583–595.

Jobbágy, E.G. & Sala, O.E. (2000) Controls of grass and shrub aboveground production in the Patagonian steppe. *Ecological Applications* **10**, 541–549.

Johnson, R.A. (1981) Application of the guild concept to environmental impact analysis of terrestrial vegetation. *Journal of Environmental Management* 13, 205–222.

Kaplan, J.O., Bigelow, N.H., Prentice, I.C. *et al.* (2003) Climate change and Arctic ecosystems: 2. Modeling, paleodata-model comparisons, and future projections. *Journal of Geophysical Research* 108 (D19), 8171.

Kariuki, M., Rolfe, M., Smith, R.G.B., Vanclay, J.K. & Kooyman, R.M. (2006) Diameter growth performance varies with species functional-group and habitat characteristics in subtropical rainforests. *Forest Ecology and Management* 225, 1–14.

Kattge, J., Díaz, S., Lavorel, S. *et al.* (2011) TRY – a global database of plant traits. *Global Change Biology* 17, 2905–2935. doi:10.1111/j.1365-2486.2011.02451.x

Keith, D.A., Holman, L., Rodoreda, S., Lemmon, J. & Bedward, M. (2007) Plant functional types can predict decade-scale changes in fire-prone vegetation. *Journal of Ecology* 95, 1324–1337.

Kilinç, M., Karavin, N. & Kutbay, H.G. (2010) Classification of some plant species according to Grime's strategies in a *Quercus cerris* L. var. *cerris* woodland in Samsun, northern Turkey. *Turkish Journal of Botany* 34, 521–529.

Kleyer, M. (1999) Distribution of plant functional types along gradients of disturbance intensity and resource supply in an agricultural landscape. *Journal of Vegetation Science* 10, 697–708.

Klimešová, J., Latzel, V., de Bello, F. & van Groenendael, J.M. (2008) Plant functional traits in studies of vegetation changes in response to grazing and mowing: towards a use of more specific traits. *Preslia* 80, 245–253.

Kooistra, L., Sanchez-Prieto, L., Bartholomeus, H.M. & Schaepman, M.E. (2007) Regional mapping of plant functional types in river floodplain ecosystems using airborne imaging spectroscopy data In: *Proceedings of the 10th International Symposium on Physical Measurements and Spectral Signatures in Remote Sensing (ISPMSRS'07)*, Vol. XXXVI, Part 7/C50 (eds M. Schaepman, S. Liang, N. Groot & M. Kneubühler), pp. 291–296. ISPRS, International Society for Photogrammetry and Remote Sensing, Davos.

Kooyman, R. & Rossetto, M. (2008) Definition of plant functional groups for informing implementation scenarios in resource-limited multi-species recovery planning. *Biodiversity and Conservation* 217, 2917–2937.

Kooyman, R., Rossetto, M., Cornwell, W. & Westoby, M. (2011) Phylogenetic tests of community assembly across regional to continental scales in tropical and subtropical rain forests *Global Ecology and Biogeography* 20, 707–716.

Körner, Ch. (1994) Scaling up from species to vegetation: the usefulness of functional groups. In: *Biodiversity and Ecosystem Function* (eds E.-D. Schulze & H.A. Mooney), pp. 117–140. Springer-Verlag, Berlin.

Körner, K. & Jeltsch, F. (2008) Detecting general plant functional type responses in fragmented landscapes using spatially-explicit simulations. *Ecological Modelling* 210, 287–300.

Kraft, N.J.B., Valencia, R. & Ackerly, D.D. (2008) Functional traits and niche-based tree community assembly in an Amazonian forest. *Science* 322, 580–582.

Laurent, J.-M., Bar-Hen, A., François, L., Ghislain, M. & Cheddadi, R. (2004) Refining vegetation simulation models: From plant functional types to bioclimatic affinity groups of plants. *Journal of Vegetation Science* 15, 739–746.

Lavers, C. & Field, R. (2006) A resource-based conceptual model of plant diversity that reassesses causality in the productivity–diversity relationship. *Global Ecology and Biogeography* 15, 213–224.

Lavorel, S. & Garnier, E. (2002) Predicting changes in community composition and ecosystem functioning from plant traits: revisiting the Holy Grail. *Functional Ecology* 16, 545–556.

Lavorel, S., McIntyre, S., Landsberg, J. & Forbes, T.D.A. (1997) Plant functional classifications: from general groups to specific groups based on response to disturbance. *Trends in Ecology and Evolution* 12, 474–478.

Lavorel, S., Touzard, B., Lebreton, J-D. & Clément, B. (1998) Identifying functional groups for response to disturbance in an abandoned pasture. *Acta Oecologica* 19, 227–240.

Lavorel, S., McIntyre, S. & Grigulis, K. (1999a) Plant response to disturbance in a Mediterranean grassland: How many functional groups? *Journal of Vegetation Science* 10, 661–672.

Lavorel, S., Rochette, C. & Lebreton, J.-D. (1999b) Functional groups for response to disturbance in Mediterranean old fields. *Oikos* 84, 480–498.

Lavorel, S., Díaz, S., Cornelissen, J.H.C. *et al.* (2007) Plant functional types: are we getting any closer to the Holy Grail? In: *Terrestrial Ecosystems in a Changing World* (eds J.G. Canadell, D. Pataki & L. Pitelka), pp. 149–160. The IGBP Series. Springer-Verlag, Berlin & Heidelberg.

Lavorel, S., Grigulis, K., Lamarque, P. *et al.* (2011) Using plant functional traits to understand the landscape distribution of multiple ecosystem services. *Journal of Ecology* 99, 135–147.

Lawton, J.H., Bignell, D.E., Bolton, B. *et al.* (1998) Biodiversity inventories, indicator taxa and effects of habitat modification in tropical forest. *Nature* 391, 72–76.

Lepš, J., de Bello, F., Lavorel, S. & Berman, S. (2006) Quantifying and interpreting functional diversity of natural communities: practical considerations matter. *Preslia* 78, 481–501.

Lewandowski, A.S., Noss, R.F. & Parsons, D.R. (2010) The effectiveness of surrogate taxa for the representation of biodiversity. *Conservation Biology* 24, 1367–1377.

Liira, J., Schmidt, T., Aavik, T. *et al.* (2008) Plant functional group composition and large-scale species richness in European agricultural landscapes. *Journal of Vegetation Science* 19, 3–14.

Lindenmayer, D.B. & Likens, G.E. (2011) Direct measurement versus surrogate indicator species for evaluating environmental change and biodiversity loss. *Ecosystems* 14, 47–59.

Lloret, F. & Montserrat, V. (2003) Diversity patterns of plant functional types in relation to fire regime and previous land use in Mediterranean woodlands. *Journal of Vegetation Science* 14, 387–398.

Louault, F., Pillar, V.D., Aufrère, J., Garnier, E. & Soussana, J.-F. (2005) Plant traits and functional types in response to reduced disturbance in a semi-natural grassland. *Journal of Vegetation Science* 16, 151–16.

Lusk, C.H., Reich, P.B., Montgomery, R., Ackerly, D.D. & Cavender-Bares, J. (2008) Why are evergreen leaves so contrary about shade? *Trends in Ecology & Evolution* 23, 299–303.

MacArthur, R. & Wilson, E.O. (1967) *The Theory of Island Biogeography*, Princeton University Press, Princeton, NJ.

Magurran, A.E. (2004) *Measuring Biological Diversity*. Blackwell Publishing, Oxford.

Maharjan, S.K., Poorter, L., Holmgren, M. *et al.* (2011) Plant functional traits and the distribution of West African rainforest trees along the rainfall gradient. *Biotropica* 43, 552–561. doi: 10.1111/j.1744-7429.2010.00747.x

Markesteijn, L., Poorter, L. & Bongers, F. (2007) Light-dependent leaf trait variation in 43 tropical dry forest tree species. *American Journal of Botany* 94, 515–525.

Martorell, C. & Ezcurra, E. (2002) Rosette scrub occurrence and fog availability in arid mountains of Mexico. *Journal of Vegetation Science* 13, 651–662.

Mason, N.W.H., MacGillivray, K., Steel, J.B. & Wilson, J.B. (2003) An index of functional diversity. *Journal of Vegatation Science.* 14, 571–578.

Mason, N.W.H., Mouillot D., Lee, W.G & Wilson, J.B. (2005) Functional richness, functional evenness and functional divergence: proposed primary components of functional diversity. *Oikos* 111, 112–118.

Mayfield, M.M., Ackerly, D.D. & Daily, G.C. (2006) The diversity and conservation of plant reproductive and dispersal functional traits in human-dominated tropical landscapes. *Journal of Ecology*, 94, 522–536.

Mayfield, M.M., Boni, M.F., Daily, G.C. & Ackerly, D. (2005) Species and functional diversity of native and human-dominated plant communities. *Ecology* 86, 2365–2372.

Mayfield, M.M., Bonser, S.P., Morgan, J.P. *et al.* (2010) What does species richness tell us about functional trait diversity? Predictions and evidence for responses of species and functional trait diversity to land-use change. *Global Ecology and Biogeography* 19, 423–431.

McGill, B.J., Enquist, B.J., Weiher, E. & Westoby, M. (2006) Rebuilding community ecology from functional traits. *Trends in Ecology & Evolution* 21, 178–185.

McIntyre, S. & Lavorel, S. (2001) Livestock grazing in sub-tropical pastures: steps in the analysis of attribute response and plant functional types. *Journal of Ecology* 89, 209–226.

McLaren, J.R. & Turkington, R. (2010) Ecosystem properties determined by plant functional group identity. *Journal of Ecology* 98, 459–469.

Millennium Ecosystem Assessment (2005) *Millennium Ecosystem Assessment, Ecosystems and Human Well-being: Synthesis*. Island Press, Washington, DC.

Mokany, K., Ash, J. & Roxburgh, S. (2008) Functional identity is more important than diversity in influencing ecosystem processes in a temperate native grassland. *Journal of Ecology* 96, 884–893.

Moles, A.T., Bonser, S.P., Poore, A.G.B., Wallis, I.R. & Foley, W.J. (2011) Assessing the evidence for latitudinal gradients in plant defence and herbivory. *Functional Ecology* 25, 380–388.

Moog, D., Kahmen, S. & Poschlod, P. (2005) Application of *CSR*- and *LHS*-strategies for the distinction of differently managed grasslands. *Basic and Applied Ecology* 6, 133–143.

Müller, S.C., Overbeck, G.E., Pfadenhauer, J. & Pillar, V.D. (2007) Plant functional types of woody species related to fire disturbance in forest-grassland ecotones. *Plant Ecology* 189, 1–14.

Mueller-Dombois & Ellenberg, H. (1974) *Aims and Methods of Vegetation Ecology.* The Blackburn Press, Caldwell, NJ. (2002 Reprint by John Wiley & Sons, Ltd).

Navarro, T., Alados, C.L. & Cabezudo, B. (2006) Changes in plant functional types in response to goat and sheep grazing in two semi-arid shrublands of SE Spain. *Journal of Arid Environments* 64, 298–322.

Niinemets, Ü., Portsmuth, A., Tema, D. *et al.* (2007) Do we underestimate the importance of leaf size in plant economics? Disproportional scaling of support costs within the spectrum of leaf physiognomy. *Annals of Botany* 100, 283–303.

Niklas, K.J. (1988) The role of phyllotactic pattern as a 'developmental constraint' on the interception of light by leaf surfaces. *Evolution* 42, 1–16.

Noble, I.R. & Slatyer, R.O. (1980) The use of vital attributes to predict successional sequences in plant communities subject to recurrent disturbance. *Vegetatio* 43, 5–21.

Oksanen, L. & Ranta, E. (1992) Plant strategies along mountain vegetation gradients: a test of two theories. *Journal of Vegetation Science* 3, 175–186.

Oleson, K.W., Lawrence, D.M., Bonan, G.B. *et al.* (2010) *Technical Description of version 4.0 of the Community Land Model (CLM).* Climate and Global Dynamics Division National Center For Atmospheric Research. Technical Note NCAR/TN-478+STR.

Oliver, I., Holmes, A., Dangerfield, J.M. *et al.* (2004) Land systems as surrogates for biodiversity in conservation planning. *Ecological Applications* 14, 485–503.

Onipchenko, V.G., Semenova, G.V. & van der Maarel, E. (1998) Population strategies in severe environments: alpine plants in the northwestern Caucasus. *Journal of Vegetation Science* 9, 27–40.

Ordoñez, A., Wright, I.J. & Olff, H. (2010) Functional differences between native and alien species: a global-scale comparison. *Functional Ecology* 24, 1353–1361.

Ordoñez, J.C., van Bodegom, P.M., Witte, J.-P.M. *et al.* (2009) A global study of relationships between leaf traits, climate and soil measures of nutrient fertility. *Global Ecology and Biogeography* 18, 137–149.

Paoli, G.D. (2006) Divergent leaf traits among congeneric tropical trees with contrasting habitat associations on Borneo. *Journal of Tropical Ecology* 22, 397–408.

Parkhurst, D.F. & Loucks, O.E. (1972) Optimal leaf size in relation to environment. *Journal of Ecology* 60, 505–537.

Pausas, J.G., Bradstock, R.A., Keith, D.A., Keeley, J.E. & Network tGF (2004) Plant functional traits in relation to fire in crown-fire ecosystems. *Ecology* 85, 1085–1100.

Peppe, D.J., Royer, D.L., Cariglino, B. *et al.* (2011) Sensitivity of leaf size and shape to climate: global patterns and paleoclimatic applications. *New Phytologist* 190, 724–739.

Petchey, O.L. & Gaston, K.J. (2002) Functional diversity (FD), species richness, and community composition. *Ecological Letters* 5, 402–411.

Petchey, O.L. & Gaston, K.J. (2006) Functional diversity: back to basics and looking forward. *Ecology Letters* 9, 741–758.

Pierce, S., Vianelli, A. & Cerabolini, B. (2005) From ancient genes to modern communities: the cellular stress response and the evolution of plant strategies. *Functional Ecology* 19, 763–776.

Poorter L. & Bongers F. (2006) Leaf traits are good predictors of plant performance across 53 rain forest species. *Ecology* 87, 1733–1743.

Posada, J.M., Lechowicz, M.J. & Kitajima, K. (2009) Optimal photosynthetic use of light by tropical tree crowns achieved by adjustment of individual leaf angles and nitrogen content. *Annals of Botany* 103, 795–805.

Quetiér, F., Lavorel, S., Thuillier, W. & Davies, I. (2007) Plant-trait-based modeling assessment of ecosystem service sensitivity to land-use change. *Ecological Applications* 17, 2377–2386.

Raunkiær, C. (1934) *The Life Forms of Plants and Statistical Plant Geography.* Clarendon Press, Oxford.

Read, C., Wright, I.J. & Westoby, M. (2006) Scaling-up from leaf to canopy-aggregate properties in sclerophyll shrub species. *Austral Ecology* **31**, 310–316.

Regan, H.M., Crookston, J.B., Swab, R., Franklin, J. & Lawson, D.M. (2010) Habitat fragmentation and altered fire regime create trade-offs for an obligate seeding shrub. *Ecology* **91**, 1114–1123.

Reich, P.B., Wright, I.J., Cavender-Bares, J. *et al.* (2003) The evolution of plant functional variation: traits, spectra, and strategies. *International Journal of Plant Sciences* **164**, S3: *Evolution of functional traits in plants*, pp. S143–S164.

Rusch, G. M., Skarpe, C. & Halley, D. J. (2009) Plant traits link hypothesis about resource-use and response to herbivory. *Basic and Applied Ecology* **10**, 466–474.

Saatkamp, A., Römermann, C. & Dutoit, T. (2010) Plant functional traits show non-linear response to grazing. *Folia Geobotanica* **45**, 239–252.

Sætersdal, M. & Gjerde, I. (2011) Prioritising conservation areas using species surrogate measures: consistent with ecological theory? *Journal of Applied Ecology.* doi: 10.1111/j.1365–2664.2011.02027.x

Sala, O.E., Chapin, F.S., Armesto, J.J. *et al.* (2000) Global biodiversity scenarios for the year 2100. *Science* **287**, 1770–1774.

Schimper, A.F.W. (1903) *Plant-geography Upon a Physiological Basis* (Translated by Fisher, W.R.) (eds P. Groom & I.B. Balfour). The Clarendon Press, Oxford.

Semenova, G.V. & van der Maarel, E. (2000) Plant functional types: a strategic perspective. *Journal of Vegetation Science* **11**, 917–922.

Sheil, D. & Burslem, D.F.R.P (2003) Disturbing hypotheses in tropical forests. *Trends in Ecology & Evolution* **18**, 18–26.

Shipley, B. (2004) Analysing the allometry of multiple interacting traits. *Perspectives in Plant Ecology, Evolution and Systematics* **6**, 235–241.

Shipley, B., Vile, D., Garnier, E., Wright, I.J. & Poorter, H. (2005) Functional linkages between leaf traits and net photosynthetic rate: reconciling empirical and mechanistic models. *Functional Ecology* **19**, 602–625.

Shugart, H.H. (1997) Plant and ecosystem functional types. In: *Plant Functional Types: Their Relevance to Ecosystem Properties and Global Change* (eds T.M. Smith, H.H. Shugart & F.I. Woodward), pp. 20–43. Cambridge University Press, Cambridge.

Sierra, C. (2009) Plant functional constraints on foliar N:P ratios in a tropical forest landscape. *Nature Precedings*: hdl:10101/npre.2009.3185.1

Simberloff, D. & Dayan, T. (1991) The guild concept and the structure of ecological communities. *Annual Review of Ecology and Systematics* **22**, 115–143.

Sitch, S., Smith, B., Prentice, I.C. *et al.* (2003) Evaluation of ecosystem dynamics, plant geography and terrestrial carbon cycling in the LPJ Dynamic Global Vegetation Model. *Global Change Biology* **9**, 161–185.

Skarpe, C. (1996) Plant functional types and climate in southern African savanna. *Journal of Vegetation Science* **7**, 397–404.

Smith, T. & Huston, M. (1989) A theory of spatial and temporal dynamics of plant communities. *Vegetatio* **83**, 49–69.

Smith, T.M., Shugart, H.H. & Woodward, F.I. (1997) Preface. In: *Plant Functional Types: Their Relevance to Ecosystem Properties and Global Change* (eds T.M. Smith, H.H. Shugart & F.I. Woodward), pp. xii–xiv. Cambridge University Press, Cambridge.

Smith, T.M., Shugart, H.H., Woodward, F.I. & Burton, P.J. (1992) Plant functional types. In: *Vegetation Dynamics and Global Change* (eds A.M. Solomon & H.H. Shugart), pp. 272–292. Chapman & Hall, New York, NY.

Suding, K.S. & Goldstein, L.J. (2008) Testing the Holy Grail framework: using functional traits to predict ecosystem change. *New Phytologist* **180**, 559–562.

Tilman, D. (1982) *Resource Competition and Community Structure*. Princeton University Press, Princeton NJ.

Tilman, D. (1985) The resource-ratio hypothesis of plant succession. *The American Naturalist* **125**, 827–852.

USGS (2010) Exploring Future Flora, Environments, and Climates Through Simulations (EFFECTS). Ecosystem responses to climate change. Geology and Environment Change Science Center. http://esp.cr.usgs.gov/info/effects/responses.html (accessed 25 May 2012).

van der Maarel, E. (1980) Epharmony and bioindication of plant communities. In: *Epharmonie* (eds O. Wilmans & R. Tüxen), pp. 7–17. Cramer, Vaduz.

van der Maarel, E. (2005) Vegetation ecology – an overview. In: *Vegetation Ecology* (ed. E. van der Maarel), pp. 1–51. Blackwell Publishing, Oxford.

van Langevelde, F., van de Vijver, C.A.D.M., Kumar, L. *et al.* (2003) Effects of fire and herbivory on the stability of savanna ecosystems. *Ecology* 84, 337–350.

Vandewalle, M., de Bello, F., Berg, M. *et al.* (2010) Functional traits as indicators of biodiversity response to land use changes across ecosystems and organisms. *Biodiversity and Conservation* 19, 2921–2947.

Verheyen, K., Honnay, O., Motzkin, G., Hermy, M. & Foster, D.R. (2003) Response of forest plant species to land-use change: a life-history trait-based approach. *Journal of Ecology* 91, 563–577.

Verner, J. (1984) The guild concept applied to management of bird populations. *Environmental Management* 8, 1–14.

Villéger, S., Mason, N.W.H. & Mouillot, D. (2008) New multidimensional functional diversity indices for a multifaceted framework in functional ecology. *Ecology* 89, 2290–2301.

Violle, C., Navas, M.-L., Vile, D. *et al.* (2007) Let the concept of trait be functional! *Oikos*, 116, 882–892.

Vitousek, P.M. & Hooper, D.U. (1993) Biological diversity and terrestrial ecosystem biogeochemistry. In: *Biodiversity and Ecosystem Function*. (eds E.-D. Schulze & H.A. Mooney), pp. 3–14. Springer-Verlag, Berlin.

Walker, B., Kinzig, A. & Langridge, J. (1999) Plant attribute diversity, resilience, and ecosystem function: the nature and significance of dominant and minor species. *Ecosystems* 2, 95–113.

Wardle, D.A., Barker, G.M., Bonner, K.I. & Nicholson, K.S. (1998) Can comparative approaches based on plant ecophysiological traits predict the nature of biotic interactions and individual plant species effects in ecosystems? *Journal of Ecology* 86, 405–420.

Warming, E. (1895) *Plantesamfund. Grundtræk af den økologiske Plantegeografi* [Plant communities. Basics of Ecological Plant Geography]. P.G. Philipsens Forlag, Kjøbenhavn.

Warming, E. (1909) *Oecology of Plants – An Introduction to the Study of Plant Communities* (Translated By M. Vahl, P. Groom & I.B. Balfour). 2nd edn 1925. Clarendon Press, Oxford.

Westoby, M. (1998) A leaf–height–seed (LHS) plant ecology strategy scheme. *Plant and Soil* 199, 213–227.

Westoby, M. (2007) Generalization in functional plant ecology: the species-sampling problem, plant ecology strategy schemes, and phylogeny. In: *Handbook of Functional Plant Ecology*, 2nd edn *(eds* F.I. Pugnaire & F. Valladares), pp. 685–703. CRC Press, Boca Raton, FL.

Westoby, M., Falster, D.S., Moles, A.T., Vesk, P.A. & Wright, I.J. (2002) Plant ecological strategies: some leading dimensions of variation between species. *Annual Review of Ecology and Systematics* 33, 125–159.

White, M.A., Asner, G.P., Nemani, R.R., Privette, J.L. & Running, S.W. (2000) Measuring fractional cover and leaf area index in arid ecosystems: digital camera, radiation transmittance, and laser altimetry methods. *Remote Sensing and Environment* 74, 45–57.

Withrow, A.P. (1932) Life forms and leaf size classes of certain plant communities of the Cincinnati region. *Ecology* 13, 12–35.

Wright, I.J., Ackerly, D.D., Bongers, F. *et al.* (2007) Relationships among key dimensions of plant trait variation in seven Neotropical forests. *Annals of Botany* 99, 1003–1015.

Wright, I.J., Reich, P.B., Cornelissen, J.H.C. *et al.* (2005) Assessing the generality of global leaf trait relationships. *New Phytologist* 166, 485–496.

Wright, I.J., Reich, P.B., Westoby, M. *et al.* (2004) The worldwide leaf economics spectrum. *Nature* 428, 821–827.

Yodzis, P. (1982) The compartmentation of real and assembled ecosystems. *The American Naturalist* 120, 551–570.

13

Plant Invasions and Invasibility of Plant Communities

Marcel Rejmánek[1], David M. Richardson[2] and Petr Pyšek[3]

[1]University of California Davis, USA
[2]Stellenbosch University, South Africa
[3]Academy of Sciences of the Czech Republic, Czech Republic

13.1 Introduction

Historically, plant taxa have always been migrating and spreading. Colonization of deglaciated areas has been very well illustrated by many examples. For obvious reasons, less documented are plant migrations via the Bering landbridge and the Central American landbridge. Occasional long-distance dispersal events have been fundamental for assembling the floras of many islands. For example, while New Zealand is often characterized as a sort of living museum of late Gondwanan vegetation, most of the predecessors of the New Zealand flora arrived by long-distance dispersal. Transoceanic dispersal events have been apparently more frequent than we thought only 10 years ago. Nevertheless, we should note that associated time scales have been enormous: thousands to millions of years.

Currently, however, the rate of human-assisted migrations (i.e. invasions *sensu* Pyšek *et al.* 2004) of plants is several orders of magnitude higher. In California, for example, more than 1300 alien plant species, introduced either intentionally or accidentally, have established self-sustaining populations over the past 250 years. About half of them are spreading to some extent. Throughout the three million year history of the Galápagos Islands, only one new vascular plant species arrived with birds or sea currents approximately every 10000 years on average. Over the past 470 years, however, the human-assisted introduction rate has been about 1.2 established species per year – about 13000 times the background rate (Tye 2006). In light of these numbers, and for other reasons discussed later, human-mediated plant invasions are radically different from natural long-distance dispersal events.

Most human-introduced species stay in disturbed areas or are incorporated into resident plant communities and have no noticeable or measurable impact. A small percentage of introduced plants do have substantial environmental and/ or economic impacts. This is the main reason for the explosion of research interest in biological invasions.

Three basic questions arise:

1   What kind of ecosystems are more (or less) likely to be invaded by alien plants?
2   What kind of plants are the most successful invaders and under what circumstances?
3   What is the impact of plant invaders?

13.2   Definitions and major patterns

Unlike **natives** (taxa that evolved in the region or reached it without help from humans from another area where they are native), **aliens** ('exotic', 'introduced', or 'non-native') owe their presence to the direct or indirect activities of humans. Most aliens occur only temporarily and are not able to persist for a long time without human-assisted input of propagules; these are termed **casual**. **Naturalized taxa** form sustainable populations without direct human help but do not necessarily spread; the ability to spread characterizes their subset termed **invasive** taxa. This distinction is critical because not all naturalized taxa reported in floras and checklists are invasive. Not all naturalized plant taxa, and not even all invaders, are harmful – the last-mentioned should rather be called alien **weeds**, alien **pest plants** or **transformer species** (Richardson *et al.* 2000b; Pyšek *et al.* 2004). It is important to stress that the ecological definition of 'invasive' that we advocate is not universally accepted. For example, managers, particularly in the USA, define as invasive only those alien taxa that cause environmental or economic damage.

Weeds comprise both native and alien species and the relative contribution of alien species in weed floras varies across the world. Most weedy taxa in Europe, Malaysia, Mexico and Taiwan are native, whereas in Australia, Chile, Hawaii, New Zealand and South Africa weed floras are overwhelmingly dominated by non-natives. There may be inherent differences in invasibility of different parts of the world. Uneven representation of alien, mostly naturalized, plant species in regional floras along the Pacific shore of the Americas illustrates this point (Fig. 13.1). These differences are certainly partly due to the history of human colonization and trade. Nevertheless, similar patterns can be recognized on other continents (Rejmánek 1996; Lonsdale 1999). For instance, areas with mediterranean-type climates (with the exception of the Mediterranean Basin itself) seem to be more vulnerable, and the tropics probably more resistant, to plant invasions. This should not be generalized, however. Savannas and especially disturbed deforested areas in the Neotropics are very often dominated by African grasses such as *Hyparrhenia rufa*, *Melinis minutiflora* and *Urochloa mutica*, while similar tropical habitats in Africa and Asia are dominated by

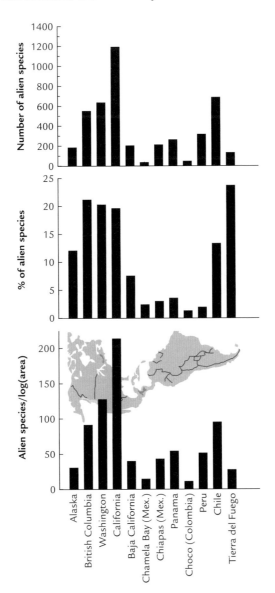

Fig. 13.1 Total number of alien plant species, percentage of alien plant species, and number of alien plant species per log(area) along the Pacific coast of Americas. 'Alien species' here are plants growing in individual areas without cultivation. Not all of them are fully naturalized and even fewer are invasive. Nevertheless, numbers of naturalized and invasive species are proportional to numbers of 'alien species' in this diagram. Primary data or references are in Kartesz & Meacham (1999) and Vitousek *et al.* (1997).

Neotropical woody plants, e.g. *Lantana camara* and *Opuntia* spp. (Foxcroft *et al.* 2010). The absolute number of alien species, therefore, is not necessarily the best indicator of ecosystem invasibility, at least at this scale (Stohlgren *et al.* 2011). Undisturbed tropical forests, however, harbour only very small numbers of alien plant species and most of them do not spread beyond trails and gaps. It is probably not the extraordinary species diversity of tropical forests that is important but simply the presence of fast-growing multilayered vegetation that makes undisturbed tropical forests resistant to invasions (Rejmánek 1996).

At the regional scale, enormous differences in the presence and abundance of invaders among different communities (ecosystems) within one area seem to be the rule. An overview is now available for several areas in Europe (Fig. 13.2, Table 13.1). Alien species are concentrated mostly in vegetation of deforested

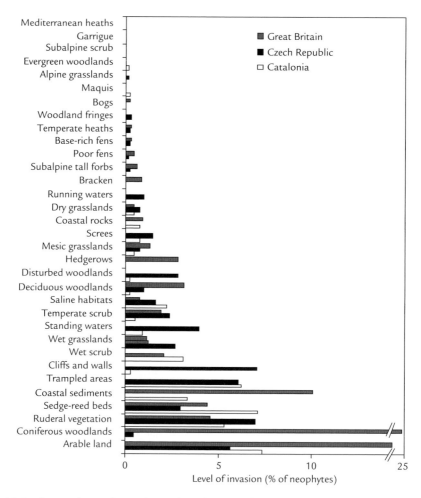

Fig. 13.2 Proportions of neophytes (species introduced after AD 1500) occurring in vegetation plots in different habitats in Catalonia, Czech Republic, and Great Britain. (Based on Chytrý *et al.* 2008.)

Table 13.1 Numbers of alien species, classified according to the time of introduction into archaeophytes and neophytes, in representative vegetation alliances of the Czech Republic.

Vegetation group[a]	No. of archaeophytes	No. of neophytes	% of invasive among neophytes
Ruderal vegetation			
Sisymbrion officinalis (tall-herb communities of annuals on nitrogen-rich mineral soils)	96	106	9.4
Aegopodion podagrariae (nitrophilous fringe communities)	16	76	36.8
Arction lappae (nitrophilous communities of dumps and rubbish tips)	36	45	31.1
Balloto–Sambucion (shrub communities of ruderal habitats)	18	34	41.2
Matricario–Polygonion arenastri (communities of trampled sites)	20	20	15.0
Potentillion anserinae (communities of salt-rich ruderal habitats)	12	20	10.0
Convolvulo–Agropyrion (communities of field margins and disturbed slopes)	24	16	31.3
Onopordion acanthii (thermophilous communities of village dumps and rubbish tips)	34	8	12.5
Weed communities of arable land			
Veronico–Euphorbion (weed communities of root crops on basic soils)	47	28	21.4
Panico–Setarion (weed communities of root crops on sandy soils)	28	15	40.0
Caucalidion lappulae (thermophilous weed communities on base-rich soils)	79	11	0.0
Aphanion (weed communities on acid soils)	41	8	12.5
Sherardion (weed communities of cereals on medium base-rich soils)	47	7	14.3
Grasslands			
Arrhenatherion (mesic *Arrhenatherum* meadows)	15	56	25.0
Festucion valesiacae (narrow-leaved dry grasslands)	12	12	0.0

(Continued)

Table 13.1   (Continued)

Vegetation group[a]	No. of archaeophytes	No. of neophytes	% of invasive among neophytes
Bromion erecti (broad-leaved dry grasslands)	6	8	0.0
Nardion (subalpine Nardus grasslands)	0	1	0.0
Helianthemo cani–Festucion pallentis (rock-outcrop vegetation with Festuca pallens)	2	0	–
Forests			
Alnion incanae (ash-alder alluvial forests)	4	15	40.0
Carpinion (oak-hornbeam forests)	6	14	14.3
Chelidonio–Robinion (plantations of Robinia)	5	10	60.0
Genisto germanicae–Quercion (dry acidophilous oak forests)	1	11	36.4
Tilio–Acerion (ravine forests)	5	8	37.5
Luzulo–Fagion (acidophilous beech forests)	0	4	50.0
Quercion pubescenti-petraeae (thermophilous oak forests)	1	2	0.0
Quercion petraeae (acidophilous thermophilous oak forests)	0	2	50.0
Salicion albae (willow-poplar forests of lowland rivers)	0	2	50.0
Alnion glutinosae (alder carrs)	0	2	0.0
Fagion (beech forests)	0	1	100.0
Betulion pubescentis (birch mire forests)	0	0	–
Piceion excelsae (spruce forests)	0	0	–
Aquatic and wetland vegetation			
Lemnion minoris (macrophyte vegetation of naturally eutrophic and mesotrophic still waters)	0	3	0.0
Cardamino–Montion (forest springs without tufa formation)	0	2	50.0
Phragmition (reed beds of eutrophic still waters)	1	1	0.0
Magnocaricion elatae (tall-sedge beds)	0	1	0.0
Nanocyperion flavescentis (annual vegetation on wet sand)	1	0	–

[a]Within each vegetation group, alliances are ranked according to the decreasing total number of alien species.
Data from Pyšek et al. (2002b).

mesic habitats with frequent disturbance (Pyšek *et al*. 2002a, b). Native forests generally harbour a low number and proportion of both archaeophytes (intro-duced before 1500) and neophytes (introduced later); alien species are com-pletely missing from many types of natural vegetation (e.g. bogs, natural *Picea abies* forest), and are rare in many natural herbaceous communities. Herbaceous communities of extreme habitats and/or with strong native clonal dominants (*Nanocyperion flavescentis*, *Phragmition*, *Nardion*) seem to be most resistant to invasions of both archaeophytes and neophytes. In general, Californian lowland communities (Fig. 13.3) are more invaded than corresponding communities in Europe. However, there are some important similarities. Open and disturbed communities are more invaded, while undisturbed forests are less invaded. It is important to stress, however, that the actual level of invasion may be mostly correlated with, but need not necessarily always correspond to, invasibility (see Section 13.3) of particular communities or habitats. To determine the invasibility of different communities, we need to factor out the effects of confounding vari-ables such as propagule pressure and climate on the level of invasion (Chytrý *et al*. 2008; Eschtruth & Battles 2011).

Data from California (Fig. 13.3) suggest that the proportions of alien species numbers are reasonably well correlated with their dominance (cover). This is probably attributable to a simple sampling effect: with an increasing proportion of alien species, there is an increasing chance that one or more of them will dominate the community. While there seems to be a general agreement between the proportion of alien species numbers and their actual importance (cover and biomass), some exceptions are very noteworthy. Whereas the number of alien species in European *Chelidonio–Robinion* woodland is not exceptionally high (Table 13.1), the dominant *Robinia pseudoacacia* is an alien tree from North America. On the other hand, while there are many alien species in some grass-land communities (*Festucion valesiaceae*, *Bromion erecti*), the dominants are all native and aliens are rarely invasive.

13.3 Invasibility of plant communities

Can we say anything conclusive about differences in invasibility (susceptibility to invasions) of particular ecosystems? Analyses of ecosystem invasibility based just on one-point-in-time observations (*a posteriori*) are usually unsatisfactory (Rejmánek 1989; Chytrý *et al*. 2008). In most cases we know nothing about the quality, quantity and regime of introduction of alien propagules. Never-theless, available evidence indicates that only a few non-native species invade successionally advanced plant communities (Rejmánek 1989; Meiners *et al*. 2002). Here, however, the quality of common species pools of introduced alien species – mostly rapidly growing and reproducing *r*-strategists – is probably an important part of the story. These species are mostly not shade-tolerant and many of them are excluded during the first 10 or 20 years of uninterrupted secondary succession (Fig. 13.4), or over longer periods of primary successions. However, some *r*-strategists are shade-tolerant, for example *Acer platanoides*,

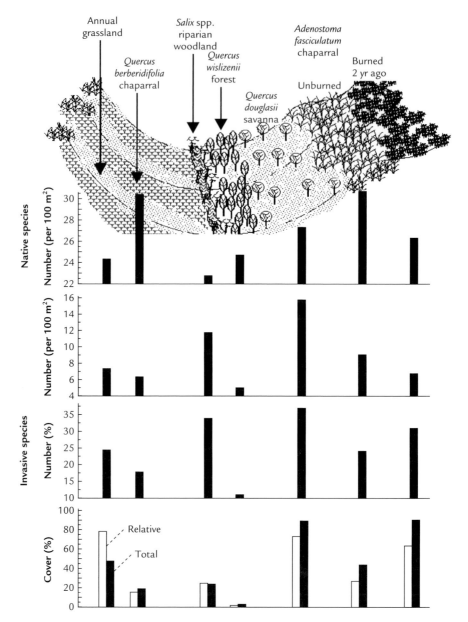

Fig. 13.3 Native and invasive species in seven plant communities of the Stebbins Cold Canyon Reserve, North Coast Ranges, California (150–500 m a.s.l.). Each column represents a mean from three 100-m^2 plots. 'Relative cover' of invaders is their cover with respect to the cumulative vegetation cover in all strata (herbs, shrubs, and trees). Comparing means for individual vegetation types, the only significant correlation is between percentage of invasive species and total cover of invasive species ($r = 0.75$; $n = 7$; $p = 0.05$). (M. Rejmánek, unpublished data.)

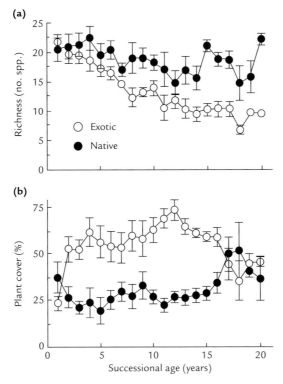

Fig. 13.4 Effect of time since abandonment on the mean species richness (a) and cover (b) of native and alien (non-native) species over 20 years of old-field succession in Argentina (Tognetti *et al.* 2010). Data points show means and vertical bars show SE for plots reaching a given age in different years. Decline of the mean percentage of alien species richness is even more dramatic. Mean relative cover of alien species usually temporarily increases during the first 10 years of succession. See also Meiners *et al.* (2002), Rejmánek (1989), and Schmidt *et al.* (2009).

Alliaria petiolata, Microstegium vimineum and *Triadica sebifera* (= *Sapium sebiferum*). Such species can invade successively advanced plant communities and, therefore, represent a special challenge to managers of protected areas (Martin *et al.* 2009).

Plant communities in mesic environments seem to be more invasible than those in extreme terrestrial environments (Rejmánek 1989). Xeric environments are much less favourable for germination and seedling survival of many introduced species (abiotic resistance) and wet terrestrial habitats do not provide resources – mainly light – for invaders because of fast growth and high competitiveness of resident species (biotic resistance). We have to be cautious, however, in interpretating these patterns. When the 'right' species are introduced, even ecosystems that have been viewed as invasion-resistant for a long time may turn out to be susceptible, for instance the Mojave and Sonoran deserts are facing invasions following introductions of *Brassica tournefortii* and *Pennisetum ciliare*.

Open water is notoriously known as vulnerable to invasions of all kinds of non-native aquatic plants. Disturbance, nutrient enrichment, slow recovery rate of resident vegetation and fragmentation of successionally advanced communities generally promote plant invasions (Rejmánek 1989; Hobbs & Huenneke 1992; Cadenasso & Pickett 2001; but see Moles *et al.* 2012). In addition, increasing CO_2 levels will probably accelerate invasions in arid ecosystems (Smith *et al.* 2000; Dukes *et al.* 2011).

A general theory of invasibility was put forward by Davis *et al.* (2000): intermittent resource enrichment (eutrophication) or release (due to disturbance) increases community susceptibility to invasions. Invasions occur if/when this situation coincides with the availability of suitable propagules. The larger the difference between gross resource supply and resource uptake, the more susceptible the community to invasion. This was anticipated by Vitousek & Walker (1987) (Fig. 13.5) and expressed more rigorously by Shea & Chesson (2002). Davis & Pelsor (2001) experimentally manipulated resources and competition in an herbaceous community to show that fluctuations in resource availability of as little as one week in duration could greatly enhance plant invasion success (survival and cover of alien plants) up to one year after such events. Not all field experiments, however, support this theory (Walker *et al.* 2005; Maron & Marler 2008).

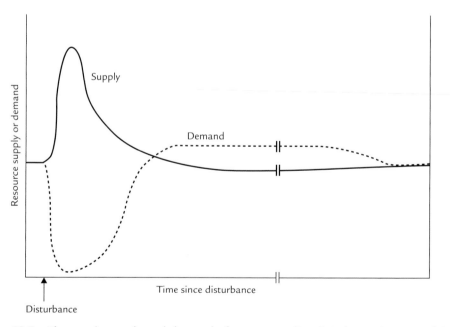

Fig. 13.5 Changes in supply and demand of resources after disturbance in terrestrial ecosystems. Resource availability is generally at its maximum shortly after disturbance, although conditions of bare ground can inhibit seedling establishment in some sites. (Modified from Vitousek & Walker 1987.)

Experiments on invasibility of different types of ecosystems have been gaining momentum in recent years (Fargione *et al.* 2003; Roscher *et al.* 2009; Petermann *et al.* 2010). The notion that there is a causal connection between invasibility of a plant community and the number of species present in that community (biotic invasions' resistance due to species richness) is usually attributed to Charles Elton (Fridley 2011). However, Crawley *et al.* (1999), Davis *et al.* (2000), and Schamp & Aarssen (2010), among others, pointed out that there is not necessarily any unambiguous relationship between these two phenomena. Other studies show that such a relationship exists: positive at the landscape scale (e.g. Stohlgren *et al.* 1999; Davies *et al.* 2011) and negative at scales usually of $1\,m^2$ or smaller (neighbourhood scales). This fact is sometimes called the 'invasion paradox' (Fridley *et al.* 2007). Many recent, well-designed experimental studies confirmed a negative relationship between resident plant species richness and invasibility in small (Fargione & Tilman 2005; Maron & Marler 2008; Fig. 13.6) and even in somewhat larger plots ($4\,m^2$; Petermann *et al.* 2010). Kennedy *et al.* (2002) concluded that in herbaceous communities, neighbourhood species richness (within 5–15 cm radius) represents 'an important line of defence against the spread of invaders'. Hubbell *et al.* (2001) found that in an undisturbed forest in Panama, neighbourhood species richness (within 2.5–50 m radius) had a weak but significantly negative effect on focal tree survival. Is there a generalization emerging from studies on neighbourhood scales? This would not be surprising as vascular plants are sedentary organisms and actual interactions are occurring among neighbouring individuals. The most plausible explanation of low invasibility of highly diverse communities at this scale is not the effect of diversity *per se*, but rather species complementarity in the use of resources and their uniformly low levels in high-diversity communities (Tilman 2004).

In this context, it is not surprising that some studies concluded that it is not necessarily the diversity of taxa, but that of functional groups (guilds; see Chapter 12), that makes communities in small plots more resistant to invasion (Symstad 2000; Lanta & Lepš 2008; Hooper & Dukes 2010). On the other hand, dominant species identity (Emery & Gross 2007) and/or intraspecific genetic diversity of dominant species may also contribute to invasion resistance (Crutsinger *et al.* 2008).

The experimental studies just mentioned usually relate the number of resident plant species to the number and/or abundance of alien plant species that establish or become invasive. But, the diversity of organisms at other trophic levels in the receiving environment may well be as important as, if not more important than, the number of plant species. We can expect that diverse assemblages of mutualists (pollinators, seed dispersers, microbiota that form symbioses with plant roots) would promote invasibility (Simberloff & Von Holle 1999; Richardson *et al.* 2000a). Experiments by Klironomos (2002) on species from Canadian old-fields and grasslands showed that rare species of native plants accumulate soil plant pathogens rapidly, while invasive species do not. This plant-soil feedback and similar findings of other authors (Callaway *et al.* 2004; Inderjit & van der Putten 2010) have potentially important consequences for community invasibility.

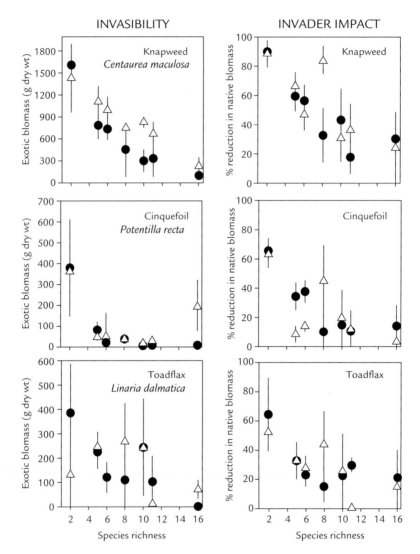

Fig. 13.6 Effect of native species richness on invasibility (left panel), quantified as mean (±SE) above-ground biomass of invaded non-natives, and invader impact (right panel), quantified as mean (±SE) per cent reduction in above-ground biomass of native plants. Circles, dry treatments; triangles, wet treatments. (Maron & Marler 2008.)

When introduced outside of their native territories, plants are often liberated from their enemies, including soil pathogens. This should be a clear advantage that would make natives and aliens, at least temporarily, different. However, the evidence is mixed. The recent meta-analysis by Chun *et al.* (2010) showed that non-native plant species may not always experience enemy release and that enemy release may not always result in greater plant performance. In their meta-analysis of the role of biotic resistance in invasions of non-native plants, Levine *et al.* (2004) concluded that biotic interactions rarely enable communities to resist invasion, although they do very often constrain the abundance of invasive species once they have successfully established.

A conceptual cause–effect diagram (Fig. 13.7) captures all the fundamental components of the current debate on the issue of invasibility. The fact that both invasibility and species diversity of residents are regulated in a similar way by the same set of factors – (micro)climate, spatial heterogeneity, long-term regime of available resources – explains why there are so many reports of a positive correlation between numbers of native and non-native species when several different communities or areas are compared (Tilman 2004; Davies *et al.* 2005; Stohlgren *et al.* 2006). Fast post-disturbance recovery of residents (native and already established non-native species) may be a key factor making the wet tropics more resistant to plant invasions – measured as the number of invading species per log(area) (Rejmánek 1996).

However, there is very likely one extra factor that is currently poorly understood: the historical and prehistoric degree of exposure of resident taxa to other biota (Fig. 13.7). Is this why islands are more vulnerable and Eurasia least vulnerable to invasions? Is instability of so many artificial monocultures a result of the 'lack of any significant history of co-evolution with pests and pathogens' (May 1981)? Actual species richness may not be as important as the complexity of assembly history. In addition to mathematical models and computer simulations (Law 1999), relevant experiments with plant communities will have to be conducted to resolve this question. Artificial experimental plant communities that are so often used for invasibility experiments have a clear advantage of homogeneous substrata and microclimates. However, assembly processes in such communities are very short and/or artificially directed via arbitrary species pool selection, weeding, reseeding, etc. The existence of well-established associations and the fact that plant species are combined in highly non-random patterns within their natural communities (Gotelli & McCabe 2002) indicate that historical assembly processes cannot be substituted by arbitrary mixtures of species. In this context, the size and composition of alien species pools (Fig. 13.7) play an important role. Such pools determine the traits and identity of invading alien species, as well as the composition of all communities in the landscape.

First, the size: size of the species pool ultimately determines the range of trait variation of the available species. It is more likely that some better competitors and species better adapted to the local environment will exist in large species pools. Island communities represent limited samples of potential species matching their habitats. Therefore, invasibility of islands should be studied in terms of the differences in species pools, not local differences in the species richness of invaded communities (Herben 2007). Also, it has been proposed that species in larger and diverse regions are 'more advanced' by a greater diversity and intensity of competition to match a wider range of both abiotic and biotic challenges. Floras consisting of such species should be less invasible because of their greater evolutionary advancement (Fridley 2011). This seems to be in agreement with a recent analysis of plant communities in the Netherlands which showed

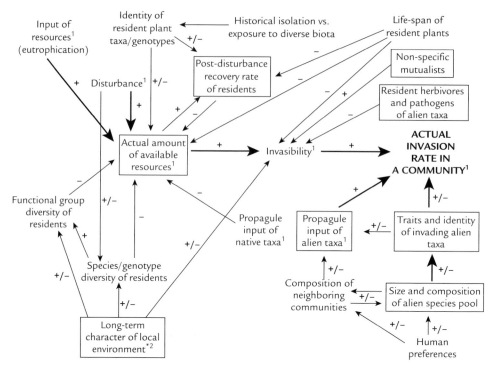

Fig. 13.7   Causal relationships between factors and processes that are assumed responsible for invasions of alien species into plant communities. The most important relationships are indicated by thick arrows. *, spatial heterogeneity, (micro)climate, and long-term regime of available resources and toxic compounds. Time scale: 1, days–years; 2, years–centuries. The key components are in boxes.

that phylogenetically less diverse communities are invaded by more alien species (Gerhold *et al.* 2011).

Second, phylogenetic relatedness clearly matters. The recent study by Davies *et al.* (2011) demonstrated that at both small (16 m^2) and large (10 816 m^2) scales, native and alien plant species in Californian grasslands are more distantly related than expected from a random assemblage model. Alien species closely related to those already present are excluded. Therefore, even communities that appear unsaturated still can be structured by biotic resistance. This is in agreement with the so-called Darwin's naturalization hypothesis: introduced species that are phylogenetically distant from their recipient communities should be more successful invaders than closely related introduced species (Rejmánek 1999; Proches *et al.* 2008; Parker *et al.* 2012).

Third, introduced species are not random samples from donor floras. Particularly, intentional introductions are heavily biased toward potentially invasive species: '. . . a useful exotic pasture species is almost certain to become a weed in some circumstances' (Lonsdale 1994). Fast-growing plantation trees with short juvenile periods and ornamental woody species with showy fleshy fruits are other examples. Moreover, due to the nursery's cultivation experience, plant species that have been sold more recently are more likely to naturalize than those sold earlier (Pemberton & Liu 2009). Obviously, biases in introduction pools make often an *a priori* trait difference between introduced non-native and native species.

Finally, longevity/persistence of resident plants is a distinct component of resistance to invasions (Von Holle *et al*. 2003), especially in forest communities, resulting in 'biological inertia', including allelopathic chemicals produced by living or dead residents. This is essentially identical to the idea proposed by Bruun & Ejrnaes (2006) that a community's invasibility is positively influenced by the turnover rate of reproductive genets in the community, which they call the 'community-level birth rate'.

13.4 Habitat compatibility

The identity of non-native taxa (Fig. 13.7) is important for two reasons. First, they may or may not survive and reproduce in habitats where they are introduced. Second, they may or may not spread and become invasive. Recipient habitat compatibility is usually treated as a necessary condition for all invasions. The match of primary (native) and secondary (adventive) environments of an invading taxon is not always perfect but usually reasonably close (e.g. Hejda *et al*. 2009b; Petipierre *et al*. 2012). In North America, for example, latitudinal ranges of naturalized European plant species from the Poaceae and Asteraceae are on average 15° to 20° narrower than their native ranges in Eurasia and North Africa (Rejmánek 2000). These differences essentially reflect the differences in the position of corresponding isotherms and major biomes in Eurasia and North America. Knowledge of species or genotype tolerance limits (Richards & Janes 2011) and habitat compatibility is essential for predictions of the potential distribution of invasive species (Gallien *et al*. 2010; Franklin 2010).

Major discrepancies between primary and secondary ranges have been found for aquatic plants where secondary distributions are often much less restricted than their primary distributions. Vegetative reproduction of many aquatic species seems to be the most important factor. Obviously, secondary ranges, if already known from other invaded continents, should be used in any prediction of habitat compatibility.

As for plants introduced (or considered for introduction) from Europe, several useful summaries of their 'ecological behaviour' are available. The combination of Ellenberg indicator values (Ellenberg *et al*. 1992) with Grime's functional types (strategies) (Grime *et al*. 2007) especially can be a powerful tool for predictions of habitat compatibility of European species. The strength of affiliation with phytosociological syntaxa is well known for almost all European taxa. Environmental conditions (climate, soil, disturbance, management) of all syntaxa are available and potential habitat compatibility of taxa can be extracted from the European literature. Also, knowledge of this 'phytosociological behaviour'

of taxa allows predictions about compatibility with analogous (vicarious) vegetation types, even if these will not always be correct.

'Open niches', habitats that can support life-forms that are not present in
local floras for historical and/or evolutionary reasons, deserve special attention.
Dramatic invasions have occurred in such habitats, e.g. *Ammophila arenaria*
(a rhizomatous grass) in coastal dunes in California, *Lygodium japonicum*
(a climbing fern) in bottomland hardwoods from Louisiana to Florida, *Acacia*
and *Pinus* species in South African fynbos shrublands, many Cactaceae species
in arid regions of the Palaeotropics, *Rhizophora mangle* (mangrove) in treeless
coastal marshes of Hawaii, and the tree *Cinchona pubescens* (Rubiaceae)
in mountain shrublands on Santa Cruz Island, Galápagos. The explanation
of such invasions is confirmed by experiments showing that the competitive
inhibition of invaders increases with their functional similarity to resident
abundant species (Fargione *et al.* 2003; Hooper & Dukes 2010; Petermann
et al. 2010).

13.5   Propagule pressure and residence time

Invasions result from an interplay between habitat compatibility and propagule
pressure (Fig. 13.7). This is illustrated by the invasion dynamics of the New
Zealand tree *Metrosideros excelsa* (Myrtaceae) in South African fynbos (details
in Richardson & Rejmánek 1998). Multiple regression of the number of
Metrosideros saplings on a potential seed rain index (PSRI) and soil moisture
revealed that, in this case, both factors are about equally important (Fig.
13.8). This example shows that classification of habitats or communities into
'invasible' and 'non-invasible' cannot be absolute in many situations. Habitats
that are currently unaffected (or only slightly affected) by plant invasions may
be deemed resistant to invasion. However, as populations of alien plants build
up and propagule pressure increases outside or within such areas, invasions could
well start or increase (Foster 2001; Duncan 2011; but see Nunez *et al.* 2011).
Estimates of propagule pressure are essential for distinguishing between the
extent of invasion and invasibility of biotic communities (Eschtruth & Battles
2011). A highly relevant aspect is the propagule pressure of native species: if
propagules of natives are not available, as for instance around abandoned fields
in California, the 'repairing' function of ecological succession (Fig. 13.4) does
not work.

Residence time – the time since the introduction of a taxon to a new area –
represents another dimension of propagule pressure. As we seldom know exactly
when taxa are introduced, we use 'minimum residence time' (MRT) based on
herbarium specimens or reliable records. Nevertheless, the number of discrete
localities of naturalized species is significantly positively correlated with MRT
(Fig. 13.9). There is usually longer MRT for naturalized species compared with
casual species and even longer MRT for invasive species. However, transformers
may have their MRT even shorter than other invasive species (Fig. 13.10). MRT
is an important factor explaining the extent of invasion of alien plants at a
regional scale (Wilson *et al.* 2007; Pemberton & Liu 2009).

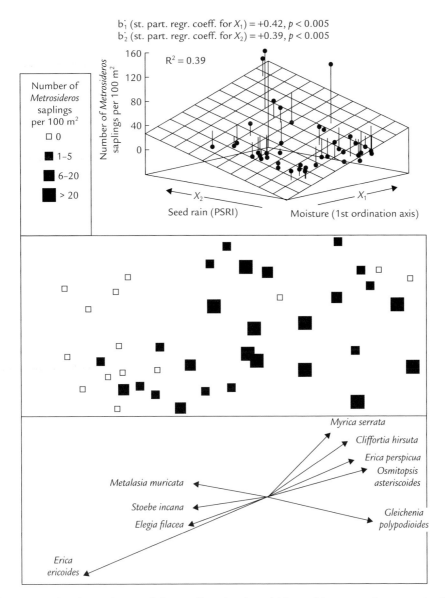

b_1' (st. part. regr. coeff. for X_1) = +0.42, $p < 0.005$
b_2' (st. part. regr. coeff. for X_2) = +0.39, $p < 0.005$

Number of
Metrosideros
saplings
per 100 m^2

☐ 0

■ 1–5

■ 6–20

■ > 20

$R^2 = 0.39$

X_2

X_1

Seed rain (PSRI)

Moisture (1st ordination axis)

Myrica serrata

Cliffortia hirsuta

Erica perspicua

Osmitopsis asteriscoides

Metalasia muricata

Stoebe incana

Gleichenia polypodioides

Elegia filacea

Erica ericoides

Fig. 13.8 The dependence of the sapling density of *Metrosideros excelsa* on potential seed rain index (PSRI) and moisture in fynbos of the Western Cape, South Africa. PSRI = SUM(1/d_i), where d_i is distance to the i-th mature tree in metres within the radius 300 m. The first ordination axis (below) serves as a surrogate for moisture gradient. Standardized partial regression coefficients of the multiple regression are almost identical. Therefore, both independent variables – environment and propagule pressure – are equally important in this case. (M. Rejmánek & D.M. Richardson, unpublished data.)

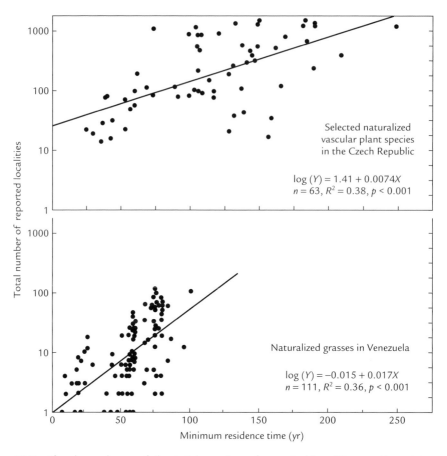

Fig. 13.9 The dependence of the total number of reported localities on the minimum residence time (years since the first record) of selected naturalized species in the Czech Republic and Venezuela. (P. Pyšek & M. Rejmánek, unpublished data.)

13.6 What are the attributes of successful invaders?

The identity of introduced species certainly matters (Fig. 13.7). One of the basic questions is whether some taxa are more invasive than others and if so, which biological attributes are responsible for that difference. The current consensus is that plant species are not equal in their invasiveness; however, different biological attributes may be important in different life-forms of plants and in different environments. Many other factors, namely propagule pressure (introduction effort) and residence time, often mask differences in invasiveness that are due to biological attributes.

Several prerequisites and stages of biological invasions are usually recognized: (1) selection of species and genotypes → (2) transport → (3) introduction → (4) establishment (consistent reproduction) = naturalization → (5) spread (invasion

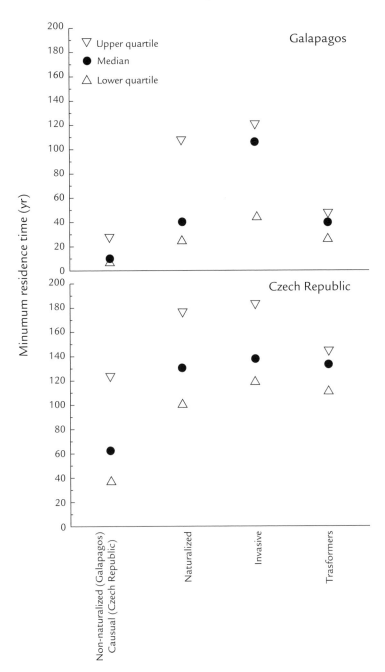

Fig. 13.10 Medians and interquartile ranges of minimum residence times for alien plant species categories in Galapagos and the Czech Republic. The subset of 'Invasive' is excluded from 'Naturalized', and the subset 'Transformers' is excluded from 'Invasive'. (Based on Trueman *et al*. 2010 and P. Pyšek, unpublished data.)

sensu stricto) → (6) environmental and/or economic impact. Obviously, very different factors may be important at each stage (Dawson *et al.* 2009). The first three steps entail intentional or unintentional human assistance. The remaining steps are spontaneous but may still be assisted by human activities. The first three steps determine the species pool of potential invaders. Species that are invasive may be introduced due to different selection processes operating during these stages. Here we will focus on step 5 (spread), which implicitly includes step 4 (reproduction). It is conceptually useful to distinguish between steps 4 and 5, but they are tightly interconnected. For a species to be invasive, it has to reproduce (establish), successfully disperse (spread) and reproduce (establish) again in new locations, and so on.

Extrapolations based on previously documented invasions are fundamental for predictions in invasion ecology. With the development of relevant databases – see, for example, Richardson & Rejmánek (2011) for invasive trees and shrubs – this approach should lead to immediate rejection of imports of many taxa known to be invasive in similar habitats elsewhere (prevention) and prioritized control of those that are already established. Such transregional, taxon-specific extrapolations are very useful in many situations, but our lack of mechanistic understanding makes them intellectually unsatisfying. Understanding how and why certain biological characters promote invasiveness is extremely important, since even an ideal whole-Earth database will not cover all (or even most) potentially invasive taxa. In New Zealand, for example, Williams *et al.* (2001) reported that 20% of the alien weedy species collected for the first time in the second half of the 20th century had never been reported as invasive outside New Zealand.

Basic taxonomic units used in plant invasion ecology are usually species or, much less often, subspecific taxa. However, genera are certainly worth considering. Plant species belonging to genera notoriously known for their invasiveness or 'weediness' (e.g. *Amaranthus, Cuscuta, Echinochloa, Ehrharta, Myriophyllum*) should all be treated as high risk. However, a continuum from invasive to non-invasive species is also common in some genera (*Acer, Amsinckia, Centaurea, Eichhornia, Pinus*). Which pattern is more typical should be rigorously tested. Naturally, attention has been paid to taxonomic patterns of invasive plants. In terms of relative numbers of invasive species, some plant families are consistently over-represented: Amaranthaceae, Brassicaceae, Chenopodiaceae, Fabaceae, Gramineae, Hydrocharitaceae, Papaveraceae, Pinaceae, and Polygonaceae. Among large families, the only conclusively under-represented one is Orchidaceae (Daehler 1998; Pyšek 1998).

Assuming abiotic environment compatibility, five biological attributes are, to different degrees, responsible for invasiveness of all kinds of organisms: (a) population fitness homeostasis, (b) population fitness, (c) minimum generation time, (d) rate of population expansion, and (e) organismal competitiveness and/or self-suitable modification of the environment (Fig. 13.11).

The relative importance of these attributes varies depending on the amount of critical resources, disturbance regimes and spatial heterogeneity of the environments. Their components are not necessarily compatible and may be important under different circumstances. For example, the ability to use available

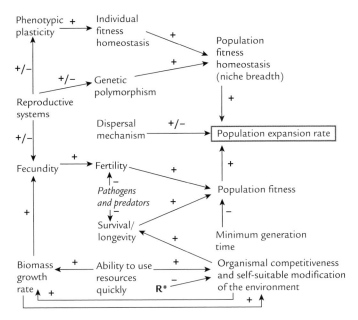

Fig. 13.11   Positive (+) and negative (−) causal relationships among biological attributes responsible for species invasiveness. R* is the level to which the amount of the available form of the limiting resource is reduced by a monoculture of a species once that monoculture has reached equilibrium (i.e. once it has attained its carrying capacity). (Modified from Rejmánek 2011.)

resources quickly is important in disturbed habitats, while the ability to reduce the amount of critical resources (lower R*) is important when invading successionally advanced communities. Also, short minimum generation time (positively influencing fitness) is usually associated with short longevity (negatively influencing fitness).

(a) *Population fitness homeostasis* (PFH) means consistent fitness at a population level over a broad range of environments. PFH is determined by individual fitness homeostasis and genetic polymorphism. Individual fitness homeostasis (IFH), or Herbert Baker's (1965) *general purpose genotype*, is the ability of an individual to maintain consistent fitness across a range of conditions through phenotypic plasticity. Phenotypic plasticity is responsible for both IFH and PFH of many plant invaders with little or no genetic diversity (e.g. *Alternanthera philoxeroides* in Asia, *Arundo donax* and *Hieracium aurantiacum* in North America, *Clidemia hirta* and *Pennisetum setaceum* in Hawaii). However, our current understanding of the role of phenotypic plasticity is far from conclusive (Davidson *et al.* 2011). On the other hand, there is abundant evidence for local adaptations through selection acting on population genetic diversity of introduced plant species (e.g.

Escholzia californica in Chile, *Hypericum perforatum* in North America, *Phyla canescens* in Australia). In this context, polyploidy, as a source of genetic diversity, can be a particularly important factor (te Beest *et al.* 2012). One important source of genetic diversity within invasive species is their repeated introduction from multiple sources (Novak 2011). Multiple introductions often transform among-population variation in native ranges to within-population variation in introduced areas. High PFH of a species translates into its broader ecological niche. It is reasonable to expect that a wide native habitat range of a species is a good indicator of its high PFH and therefore high invasiveness.

(b) *Actual level of population fitness in particular environments* is the key component of all invasions. Unfortunately, fitness quantified as finite rate of population increase (λ) is only rarely properly measured and comparisons of fitness between invasive and non-invasive species are almost non-existent. In an exceptional study, Burns (2008) found that invasive plant species in the family Comelinaceae had significantly larger λ values than non-invasive ones, but only under high-nutrient conditions. More often, fitness is just estimated on the basis of its components: fertility or fecundity. A positive correlation between individual plant biomass and seed production per plant is one of the most robust generalizations of plant ecology. Therefore, higher values of relative growth rates (RGR) in plants may often indicate higher fitness and invasiveness (Grotkopp *et al.* 2002, 2010). The recent meta-analysis of all available studies by van Kleunen *et al.* (2010) revealed that both growth rates and fitness-related attributes are significantly higher for invasive plant species when compared with either non-invasive or native plant species. However, there are trade-offs between biomass growth rate and survival – another component of fitness. There are both benefits and costs to fast living. For example, because RGR of plants is usually negatively related to water-use efficiency, fast growth is not the best strategy for perennial plant invaders in arid environments. Based on their studies in resource-poor habitats in Hawaii, Funk & Vitousek (2007) showed that invasive plant species were generally more efficient than native species at using limited light, water and nitrogen.

Last but not least, fecundity depends on reproductive systems. The consistent production of offspring in new environments is usually associated with rather simple or flexible breeding systems. For example, rare and endangered taxa in the genus *Amsinckia* (e.g, *A. furcata*, *A. grandiflora*) are heterostylic, while derived invasive taxa (*A. lycopsoides*, *A. menziesii*) are homostylic and self-compatible. Self-pollination has been consistently identified as a mating strategy in colonizing species. Nevertheless, not all sexually reproducing successful invaders are selfers.

Vegetative reproduction can compensate more than sufficiently for sexual reproduction in some invasive plant species. Water hyacinth (*Eichhornia crassipes*) and infertile hybrid giant salvinia (*Salvinia molesta*) are well-known examples. The ability to allocate energy to different modes of reproduction depending on environmental conditions is one type of phenotypic plasticity and increases IFH and PFH. Apomictic plants (like dandelions) have an advantage, at least initially, as a single individual can establish a population (Koltunow *et al.* 2011).

(c) *Short minimum generation time*, also called juvenile period, is an obvious advantage for invasive species. Not surprisingly, substantial proportions of non-native floras in temperate zones are annual species. Short minimum generation time is usually a prominent attribute used for identification of (potentially) invasive woody species. Invasiveness of woody taxa in disturbed landscapes is associated with short juvenile period (<10 years), small seed mass (<50 mg), and short intervals between large seed crops. Differences between invasive and non-invasive pine (*Pinus*) species served as the first illustration of such regularities (Fig. 13.12; Rejmánek & Richardson 1996). The three attributes, listed above, contribute, directly or indirectly, to higher values of three parameters critical for population expansion: net reproduction rate, reciprocal of mean age of reproduction and variance of the marginal dispersal density. For wind-dispersed seeds, the last parameter is negatively related to terminal velocity of seeds, which is positively related to $\sqrt{\text{(seed mass)}}$. Because of the trade-off between seed number and mean seed mass, small-seeded taxa usually produce more seeds per unit biomass. Invasions of woody species with very small seeds (<3 mg), however, are limited to wet and preferably mineral substrates (Fig. 13.13). Based on invasibility experiments with herbaceous species, it seems that somewhat larger seeds (3–10 mg) extend species habitat compatibility (Burke & Grime 1996). As seed mass seems to be positively correlated with habitat shade, large-seeded aliens may be more successful in undisturbed, successionally more mature plant communities.

(d) *Fast dispersal of propagules* is another crucial component of plant invasiveness. Rate of dispersal always depends on two species-specific characteristics: fertility and efficiency of dispersal mechanism. This is also the substance of Fisher–Kolmogorov's classic formulation of population rate of expansion of the population front (how many metres a constant population density can propagate in one dimension in one year) in a homogeneous environment: $2\sqrt{(rD)}$, where r is the intrinsic rate of population increase (fertility minus mortality, i.e., individual/individual/year) and D is the diffusion coefficient ($m^2\,yr^{-1}$). The first term is directly related to population fitness: $r = \ln \lambda$.

The most important long-distance dispersal agents for plants are people, other vertebrates (mostly birds), water, and wind. Plants have many different adaptations or preadaptations for dispersal by these vectors. Plant species with seeds without any dispersal-promoting appendages are usually less invasive (*Eucalyptus* spp.). However, because increasing volumes of soil are moved around by people (in topsoil, in mud on cars, with horticultural stock), plant species with numerous, dormant, soil-stored seeds are preadapted for this kind of dispersal (Hodkinson & Thompson 1997; Von der Lippe & Kovarik 2007).

Seed dispersal by vertebrates is responsible for the success of many invaders in disturbed as well as 'undisturbed' habitats (Aslan & Rejmánek 2010;

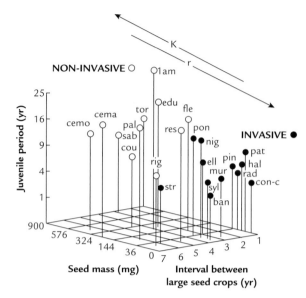

Fig. 13.12 Distribution of 23 frequently cultivated *Pinus* species in a space created by three biological variables critical in separating invasive and non-invasive species. The *K-r* selection continuum running from the upper left to the lower right corner of the diagram also represents the direction of the discriminant function (*Z*) separating non-invasive and invasive *Pinus* species. $Z = 23.39 - 0.63\sqrt{M} - 3.88\sqrt{J} - 1.095S$, where *M* = mean seed mass (in milligrams), *J* = minimum juvenile period (in years), and *S* = mean interval between large seed crops (in years). Pine species with positive *Z* scores are invasive and species with negative *Z* scores are non-invasive. Species abbreviations: ban, *banksiana*; cema, *cembra*; cemo, *cembroides*; con, *contorta*; cou, *coulteri*; edu, *edulis*; ell, *elliotii*; eng, *engelmannii*; fle, *flexilis*; hal, *halepensis*; lam, *lambertiana*; mur, *muricata*; nig, *nigra*; pal, *palustris*; pat, *patula*; pin, *pinaster*; pon, *ponderosa*; rad, *radiata*; res, *resinosa*; sab, *sabiniana*; str, *strobus*; syl, *sylvestris*; tor, *torreyana*.

Richardson & Rejmánek 2011). Even some very large-seeded alien species like mango (*Mangifera indica*) or avocado (*Persea americana*) can be dispersed by large mammals. Assessment of whether there is an opportunity for vertebrate dispersal is an important component of the screening procedure for woody plants (Fig. 13.13).

(e) *Undisturbed (natural and semi-natural) plant communities* in mesic environments are more likely invaded by tall plant species. The most prominent examples are new, taller, life-forms (*Acacia* spp. and *Pinus* spp. in South African fynbos, *Cinchona pubescens* in shrub and fern/grassland communities of the Galapagos highlands). Undisturbed plant communities in *semi-arid* habitats seem to be invasible especially by environmentally compatible species that rapidly develop deep root systems (e.g. *Bromus tectorum* or *Centaurea solstitialis*). In short, in undisturbed plant communities, efficient competitors for limiting resources will very likely be successful invaders and the worst environmental weeds. Theoretically, given a set of R_i^* values

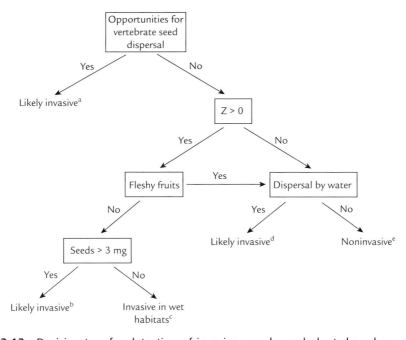

Fig. 13.13 Decision tree for detection of invasive woody seed plants based on values of the discriminant function **Z***, seed mass values and presence or absence of opportunities for vertebrate dispersal (derived from Table 6.1 in Rejmánek *et al.* 2005). ***Z** = 23.39 − 0.63$\sqrt{M}$ − 3.88$\sqrt{J}$ − 1.09S, where M = mean seed mass (in milligrams), J = minimum juvenile period (in years), and S = mean interval between large seed crops (in years). This discriminant function (Z) was derived on the basis of differences between invasive and non-invasive pine (*Pinus*) species. Positive **Z** indicates invasive species; negative **Z** indicates non-invasive species. The function was later successfully applied to other gymnosperms and, as a component of broader frameworks, to woody angiosperms. 'Opportunities for vertebrate seed dispersal' mean that plant species are producing fruits attractive for vertebrates, usually fleshy fruits or nuts, and that at least one member of the local vertebrate fauna is can serve as a dispersal agent.
[a]Examples of invasive species in this group are many fleshy-fruiting species with small seeds: *Berberis* spp. *Clidemia hirta, Lantana camara, Ligustrum* spp., *Lonicera* spp., *Muntingia calabura, Passiflora* spp., *Pittosporum undulatum, Psidium guajava, Rosa* spp., *Rubus* spp., *Solanum* spp. Species with seeds possessing large arils (*Acacia saligna*) or with seeds coated with a wax (*Triadica sebifera*) are dispersed by birds. Even some large-seeded species may be dispersed by some vertebrates: *Pinus pinea* and *Melia azedarach* in South Africa, *Olea europaea* in Australia, *Juglans regia* and *Quercus rubra* in Europe, *Mangifera indica* in the Neotropics, and *Persea americana* in Galapagos.
[b]Mainly wind- and ant-dispersed species, e.g. *Acer platanoides, Ailanthus altissima, Clematis vitalba, Cryptomeria japonica, Cytisus scoparius, Pinus radiata, Pseudotsuga menziesii, Robinia pseudoacacia, Tecoma stans, Ulex europaeus.*
[c]Examples of these are *Alnus glutinosa* and *Salix* spp. in New Zealand, *Eucalyptus camaldulensis* in South Africa, *Melaleuca quinquenervia* in southern Florida, *Tamarix* spp. in the south-western US, and *Baccharis halimifolia* in Australia. If species in this category reproduce only by seeds, they need wet mineral substrata for their establishment. Some species in this category can also propagate vegetatively: viable branches of *Salix* spp. and *Populus* spp. can be dispersed by water in streams and rivers over a long distance.
[d]*Nypa fruticans* spreads along tidal streams in Nigeria and Panama, *Thevetia peruviana* can be dispersed over short distances by rain-wash in Africa.
[e]Examples of non-invasive species are *Aesculus hippocastanum, Araucaria araucana, Bertholletia excelsa, Camellia* spp., *Fagus* spp., *Pinus lambertiana, Tilia* spp. Some fleshy-fruiting species with $Z > 0$ may be locally non-invasive if opportunities for vertebrate dispersal are not present: *Acca sellowiana, Rhaphiolepis indica, Pyrus calleryana,* and *Nandina domestica* are frequently cultivated but non-invasive species in California because very few vertebrates eat their fruits; *N. domestica,* however, is dispersed by birds and water in the south-eastern USA.

(R_i^* is a level of resource below which an *i*-th species cannot survive), for a pool of potential invaders, it should be possible to predict the average likely success of each invading species in undisturbed communities (Tilman 1999; Shea & Chesson 2002). However, if seasonality, senescence, or even very low levels of natural disturbance allow establishment of shade-intolerant taxa that are taller than resident vegetation at maturity, then such taxa can still be very successful and influential invaders in spite of their high R^* for light.

The ability to use available resources quickly is an attribute of many successful plant invaders in disturbed habitats. Obviously, there is a trade-off between this kind of strategy and possession of low R^*. Whether some species can quickly use resources and also reduce their levels below those tolerable by resident species remains to be seen. Such species would be the most successful invaders.

Recently, there has been a renewal of interest in the role of allelopathy in plant invasions. It seems that some chemical substances released from the living or decaying biomass of non-native species can inhibit the growth of native plants and/or soil micro-organisms. This can increase the invasiveness of such species. However, with the exception of some consistent effects (e.g. juglone released by walnuts, *Juglans* spp.), results are highly inconsistent, depending on climate and soil properties (Blair *et al.* 2006; Inderjit *et al.* 2006; Callaway 2011). Allelo-pathic substances are potentially more influential in soils with low organic content and in habitats with low precipitation.

In general, reducing the amount of critical resources below the level needed by resident species or release of chemicals inhibiting growth of residents by non-native species are examples of 'niche constructions' accelerating plant invasions, particularly in undisturbed environments. Some invasive grasses (e.g. *Andropogon gayanus, Bromus tectorum, Hyparrhenia rufa*) can initiate and maintain a positive grass-fire feedback and transform whole ecosystems to their benefit (Foxcroft *et al.* 2010; Mack 2011).

Long-term population invasiveness, however, does not depend only on organismal anatomical or physiological properties treated above, but on relationships between population fitness values of invaders and residents and the degree of niche overlap between invaders and residents (Fig. 13.14). As Chesson (1990) and more recently MacDougall *et al.* (2009) showed, there are essentially three possible invasion outcomes for all possible combinations of niche and fitness differences: (1) when fitness of residents > fitness of invader and niche overlap is large, residents will repel the potential invader; (2) when there is either no difference in fitness, or niche overlap is small, invader and residents can co-exist; (3) when fitness of invader > fitness of residents and niche overlap is large, the invader can exclude residents. High PFH may contribute to the third outcome. In general, successful invasion can result from either fitness differences that favour the dominance of invader, or niche differences that allow the invader to establish despite lower population fitness. However, the outcomes of invasion will differ. Only the former leads to displacement of resident species. The latter leads to co-existence and not to local extinctions of residents. This model

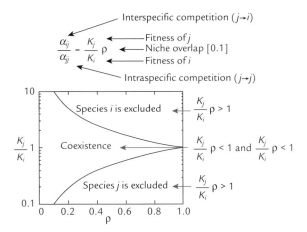

Fig. 13.14 According to Robert MacArthur's consumer-resource model, the ratio of interspecific (between species, a_{ij}) and intraspecific (within species, a_{jj}) competitive effects can be expressed in terms of population fitness (k_j and k_i) of species j and i and resource-use (niche) overlap (ρ) between those species. Species j competitively excludes species i if the ratio is greater than one. In general, when $\rho = 1$ (niche overlap is complete), the species with the larger population fitness excludes the other. However, niche overlap less than 1 constrains the fitness differences compatible with coexistence. This model provides a theoretical framework for potential outcomes of interactions between invading and resident species. (Derived from Chesson 1990; Chesson & Kuang 2008.)

explicitly connects two major topics of invasion biology that are often treated independently: species invasiveness and invasibility of biotic communities. Even though quantifications of both fitness and niche overlap are far from simple measurements, this model provides a useful theoretical framework that will very likely guide research on biological invasions in years to come.

13.7 Impact of invasive plants, justification and prospects of eradication projects

Many invasive taxa have transformed the structure and function of ecosystems by, for example, changing disturbance- or nutrient-cycling regimes (Ehrenfeld 2010). In many parts of the world, impacts have clear economic implications for humans, for example as a result of reduced stream flow from watersheds in South African fynbos following alien tree invasion, or through disruption to fishing and navigation after invasion of aquatic plants such as *Eichhornia crassipes*.

It is important to stress, however, that the impacts of invasive plants on biodiversity are generally less dramatic than the impacts of non-native pathogens, herbivores or predators. It seems that most naturalized/invasive plant species have hardly any detectable effect on biotic communities (Williamson & Fitter

1996; Meiners *et al.* 2001). There are at least 3000 naturalized plant species in North America and more than 1000 of them are invasive. However, not a single native plant species is known to have been driven to extinction due to interactions with alien plants alone. Even on islands, where numbers of non-native plant species are often increasing exponentially, extinctions of native plant species cannot be attributed to plant invasions *per se* (Sax *et al.* 2002). Also, the often reported correlation between numbers of native and non-native plant species on the landscape scale can be interpreted as a lack of mechanisms for competitive exclusion of native plants by non-native ones. Nevertheless, we should be careful with conclusions – many invasions are quite recent and extinction takes a long time.

While there has been substantial progress in understanding the plant attributes responsible for or, at least, correlated with successful reproduction and the spread of invasive plant species, our ability to predict their impacts, or even measure their impact using standardized methods, is still very rudimentary. This fact is very important in the context of the ongoing discussion about the possible overestimation of negative impacts of non-native species (Simberloff *et al.* 2011). Several meta-analyses of published data on the ecological impacts of invasive plant species have been published recently (e.g. Powell *et al.* 2011; Vilà *et al.* 2011). In general, they conclude that many alien plants have a statistically significant negative effect on native plant abundance, fitness and diversity. However, at least 80% of over 1000 field studies included in these meta-analyses were based on a 'space-for-time-substitution' approach. Particular examples of results obtained this way are presented in Fig. 13.15 and Table 13.2. However, without pre-invasion data from the invaded and non-invaded sites, conclusions may be

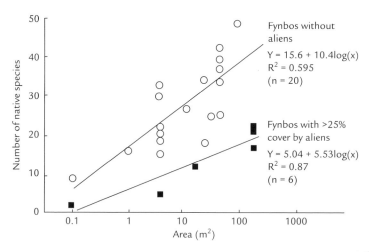

Fig. 13.15 Species–area relationships for native vascular plant species in South African fynbos areas densely infested (squares) by alien woody plants and in uninfested areas (circles). Elevations of the two regression lines are significantly different (*p* < 0.001). Sources of the data are acknowledged in Richardson *et al.* (1989).

Table 13.2   Impact of 12 invasive plant species on species richness of invaded plots.

Species	Cover range (%)	Species number[a]		Impact (%)
		Uninvaded	Invaded	
Fallopia sachalinensis	70–100	13.3 ± 4.9	1.8 ± 1.6	86.4***
F. japonica	100	12.1 ± 3.5	3.3 ± 2.8	73.0**
F. x bohemica	40–100	14.8 ± 7.3	5.4 ± 5.0	65.9*
Heracleum mantegazzianum	90–100	16.7 ± 4.5	7.4 ± 3.1	52.6**
Rumex alpinus	75–100	12.6 ± 2.5	7.7 ± 2.4	39.1***
Aster novi–belgii	60–90	14.1 ± 4.8	8.9 ± 6.3	38.7
Helianthus tuberosus	50–100	12.7 ± 6.5	8.0 ± 4.9	33.7
Rudbeckia laciniata	80–100	10.6 ± 2.6	6.9 ± 3.0	29.8
Solidago gigantea	70–100	16.4 ± 6.7	12.0 ± 6.3	25.5
Imperatoria ostrunthium	50–80	14.3 ± 5.6	9.9 ± 2.6	21.4
Lupinus polyphyllus	60–95	21.1 ± 2.3	16.4 ± 3.8	21.2
Impatiens glandulifera	60–90	10.9 ± 1.8	9.5 ± 2.6	12.3

[a]Species numbers are expressed as mean ± SD per $16\,m^2$, $n = 10$, $*p < 0.05$, $**p < 0.01$, $***p < 0.001$).
From Hejda *et al.* (2009a).

misleading. For example, invaded sites that have lower species richness than non-invaded sites in the post-invasion condition may suggest that non-native species negatively affected diversity of native species. An alternative interpretation is that invaded sites could have had lower species richness than the non-invaded ones prior to invasion. This is possible if, for example, invaded sites had lower habitat heterogeneity and/or other environmental conditions that limit numbers of both native species and non-native species. Another possibility is that non-native species invaded less rich sites because of lower biotic resistance. Thus, one cannot determine whether the non-native species really had a negative impact on diversity of native species.

Although the time approach (comparisons of sites in pre- and post-invasion situations) is apparently the only option for resolving the above limitations and serves the purpose of measuring the real impact of non-native species, it can nonetheless also produce mistaken conclusions. Without data from equivalent non-invaded habitats in pre- and post-invasion situations, one may not estimate the direction of the effects of non-native species, nor their magnitudes. Such sources of confusion (see also Thiele *et al.* 2010) could be resolved by testing the effects of non-native species through experiments in conditions that are as realistic as possible.

Competition experiments that are usually limited just to pairs of species represent one option (Vilà & Weiner 2004). Responses to invaders in multispecies communities can be evaluated in invader addition experiments (Maron & Marler 2008; see Fig. 13.6), invader removal experiments (Schutzenhofer & Valone 2006) and experiments where passive colonization of invader monocultures is analysed (Hovick *et al.* 2011). Preferably, in all situations the multiple mechanisms of impacts of invasive species should be anticipated and systematically tested (Bennett *et al.* 2011). Demographic matrix models are an increasingly standard method for quantitative evaluation of invader's impacts on endangered plant species (Thomson 2005).

Invasiveness and impact are not necessarily positively correlated. Some fast-spreading species, such as *Aira caryophyllea* or *Cakile edentula*, exhibit little (if any) measurable environmental or economic impact. On the other hand, some relatively slowly spreading species (e.g. *Ammophila arenaria* or *Robinia pseudoacacia*) may have far-reaching environmental effects (stabilization of coastal dunes in the first case and nitrogen soil enrichment in the second).

There is a need for universally acceptable, and objectively applicable, procedures for the assessment of influential invasive plant taxa within given regions, or globally. Some attempts in this direction (Magee et al. 2010; Thiele *et al.* 2011) are more promising than others. A potentially useful term to use in this regard is 'transformer species' (Richardson *et al.* 2000b). Such species, comprising perhaps only about 10% of invasive species, have profound effects on biodiversity and clearly demand a major allocation of resources for containment/control/eradication. Several categories of transformers may be distinguished.

1 Excessive users of resources: water – *Tamarix* spp., *Acacia mearnsii*; light – *Pueraria lobata* and many other vines, *Heracleum mantegazzianum*, *Rubus armeniacus*; water and light – *Arundo donax*; light and oxygen – *Salvinia molesta*, *Eichhornia crassipes*; high leaf area ratio, LAR, of many invasive plants is an important prerequisite for excessive transpiration; *Andropogon gayanus* inhibits soil nitrification and thereby depletes total soil nitrogen from nitrogen-poor soils and promotes fire-mediated nitrogen loss;

2 Donors/enhancers of limiting resources: nitrogen – *Acacia* spp., *Lupinus arboreus*, *Morella* (*Myrica*) *faya*, *Robinia pseudoacacia*, *Salvinia molesta*; phosphorus – *Buddleja davidii*, *Centaurea maculosa*, *Solidago gigantea*;

3 Fire promotors/suppressors: promotors – *Andropogon gayanus*, *Bromus tectorum*, *Melaleuca quinquenervia*; suppressors – *Mimosa pigra*;

4 Sand stabilizers: *Ammophila* spp., *Elymus* spp.;

5 Erosion promotors: *Andropogon virginicus* in Hawaii, *Impatiens glandulifera* in Europe;

6 Colonizers of intertidal mudflats – sediment stabilizers: *Spartina* spp., *Rhizophora* spp.;

7 Litter accumulators: *Centaurea solstitialis*, *Eucalyptus* spp., *Lepidium latifolium*, *Pinus strobus*, *Taeniatherum caput-medusae*;

8 Soil carbon storage modifiers: promotor – *Andropogon gayanus*; suppressor – *Agropyron cristatum*;

9 Salt accumulators/redistributors: *Mesembryanthemum crystallinum*, *Tamarix* spp.

The potentially most important transformers are taxa that add a new function, such as nitrogen fixation, to the invaded ecosystem (Vitousek & Walker 1989). Many impacts, however, are not so obvious. For example, invasive *Lonicera* and *Rhamnus* change the vegetation structure of the forest, and *Lythrum salicaria* and *Impatiens glandulifera* can have negative impacts on the pollination and reproductive success of co-flowering native plants (Grabas & Laverty 1999; Chittka & Schürkens 2001). A meta-analysis recently published by Morales and Traveset (2009) demonstrated the predominant detrimental impact of alien plants on the pollination and reproduction of natives. Morover, hybridization with native congeners may be the most important permanent impact of some invaders (Mercure & Bruneau 2008; Hall & Ayers 2009).

In attempting to quantify the value of ecosystem services of South African fynbos systems and the extent to which these values are reduced by invasions, Higgins *et al.* (1997) showed that the cost of clearing alien plants was very small (<5%) as compared to the value of the services provided by these ecosystems. Their conclusion was that pro-active management could increase the value of these ecosystem services by at least 138%. The most important ecosystem service was water, and much work has been done on developing models for assessing the value (in monetary terms) of allocating management resources to clearing invasive plants from fynbos watersheds.

It follows from the discussion on impacts of non-native plants that careful prioritization is needed before starting often very expensive and time-consuming eradication projects. Maintenance of biodiversity is dependent on the maintenance of ecological processes. Our priority should be the protection of ecological processes. Attempts to eradicate widespread invasive species, especially those that do not have any documented environmental impacts (including suppression of rare native taxa), may be not only useless but also a waste of time and resources. Non-native taxa with large-scale environmental impacts (transformers) are usually obvious targets for control and eradication. But when is complete eradication a realistic goal?

There are numerous examples where small infestations of invasive plant species have been eradicated. There are also several encouraging examples where widespread alien animals have been completely eradicated. Can equally widespread and difficult alien plants also be eradicated? On the basis of a unique data set on eradication attempts by the California Department of Food and Agriculture on 18 species and 53 separate infestations targeted for eradication in 1972–2000 (Table 13.3), it is shown that professional eradication of non-native

Table 13.3 Areas of initial gross infestations (at the beginning of eradication projects) of exotic weeds in California, numbers of eradicated infestations, numbers of ongoing projects, and mean eradication effort for five infestation area categories.[a]

Initial infestation (ha)	<0.1	0.1–1	1.1–100	101–1000	>1000
No. of eradicated infestations	13	3	5	3	0
No. of on-going projects	2	4	9	10	4
Mean eradication effort per infestation (work hours)					
Eradicated	63	180	1496	1845	–
On-going	174	277	1577	17,194	42,751

[a]The data include 18 noxious weedy species (2 aquatic and 16 terrestrial) representing 53 separate infestations.
From Rejmánek & Pitcairn (2002).

weed infestations smaller than 1 ha is usually possible. In addition, about one third of infestations between 1 and 100 ha and a quarter of infestations between 101 and 1000 ha have been eradicated. However, the costs of eradication projects increase dramatically. With a realistic amount of resources, it is very unlikely that infestations larger than 1000 ha can be eradicated (Table 13.3).

Early detection of the presence of an invasive harmful taxon can make the difference between being able to employ offensive strategies (eradication) and the necessity of retreating to a defensive strategy that usually means an infinite financial commitment (Panetta *et al.* 2011). Nevertheless, depending on the potential impact of individual invaders, even infestations larger than 1000 ha should be targeted for eradication effort or, at least, substantial reduction and containment. If a non-native weed is already widespread, then species-specific biological control may be the only long-term effective method able to suppress its abundance over large areas (Van Driesche *et al.* 2008).

Finally, it is important to stress that many large-scale invasive plant management efforts have had only moderate restoration success. One of the major reasons has been only the limited focus on revegetation with natives after invasive control or eradication (Kettenring & Adams 2011).

Regardless of their environmental and/or economical effects, plant invasions provide unique chances to understand some basic ecological and evolutionary processes that are otherwise beyond the capacity or ethics of standard ecological experiments. We are just beginning to fully appreciate these opportunities and we still have a long way to go to achieve a more complete understanding and more rational decision making.

References

Aslan, C.A. & Rejmánek, M. (2010) Avian use of introduced plants: ornithologist records illuminate interspecific associations and research needs. *Ecological Applications* 20, 1005–1020.

Baker, H.G. (1965) Characteristics and modes of origin of weeds. In: *The Genetics of Colonizing Species* (eds H.G. Baker & G.L. Stebbins), pp. 147–172. Academic Press, New York, NY.

Bennett, A.E., Thomsen, M. & Strauss, S.Y. (2011) Multiple mechanisms enable invasive species to suppress native species. *American Journal of Botany* 98, 1086–1094.

Blair, A.C., Nissen, S.J., Brunk, G.R. & Hufbauer, R.A. (2006) A lack of evidence for ecological role of the putative allelochemical (±)-catechin in spotted knapweed invasion success. *Journal of Chemical Ecology* 32, 2327–2331.

Bruun, H.H. & Ejrnaes, R. (2006) Community-level birth rate: a missing link between ecology, evolution and diversity. *Oikos* 113, 185–191.

Burke, M.J.W. & Grime, J.P. (1996) An experimental study of plant community invasibility. *Ecology* 77, 776–790.

Burns, J.H. (2008) Demographic performance predicts invasiveness of species in the Comelinaceae under high-nutrient conditions. *Ecological Applications* 18, 335–346.

Cadenasso, M.L. & Pickett, S.T.A. (2001) Effect of edge structure on the flux of species into forest interiors. *Conservation Biology* 15, 91–97.

Callaway, R.M. (2011) Novel weapons hypothesis. In: *Encyclopedia of Biological Invasions* (eds D. Simberloff & M. Rejmánek), pp. 492–493. University of California Press, Berkeley, CA.

Callaway, R.M., Thelen, G.C., Rodriguez, A. & Holben, W.E. (2004) Soil biota and exotic plant invasion. *Nature* 427, 731–733.

Chesson, P. (1990) MacArthur's consumer-resource model. *Theoretical Population Biology* 37, 26–38.

Chesson, P. & Kuang, J.J. (2008) The interaction between predation and competition. *Nature* **456**, 235–238.

Chittka, L. & Schürkens, S. (2001) Successful invasion of a floral market. *Nature* **411**, 653.

Chun, Y.J., van Kleunen, M. & Dawson, W. (2010) The role of enemy release, tolerance and resistance in plant invasions: linking damage to performance. *Ecology Letters* **13**, 937–946.

Chytrý, M., Jarošík, V., Pyšek, P. *et al.* (2008) Separating habitat invasibility by alien plants from the actual level of invasion. *Ecology* **89**, 1541–1553.

Crawley, M.J., Brown, S.L., Heard, M.S. & Edwards, G.G. (1999) Invasion-resistance in experimental grassland communities: species richness or species identity? *Ecology Letters* **2**, 140–148.

Crutsinger, G.M., Souza, L. & Sanders, N.J. (2008) Intraspecific diversity and dominant genotypes resist plant invasions. *Ecology Letters* **11**, 16–23.

Daehler, C. (1998) The taxonomic distribution of invasive angiosperm plants: ecological insights and comparison to agricultural weeds. *Biological Conservation* **84**, 167–180.

Davidson, A.M., Jennions, M. & Nicotra, A.B. (2011) Do invasive species show higher phenotypic plasticity than native species and, if so, is it adaptive? A meta-analysis. *Ecology Letters*, **14**, 419–431.

Davies, K.F., Chesson, P., Harrison, S. *et al.* (2005) Spatial heterogeneity explains the scale dependence of the native-exotic diversity relationship. *Ecology* **86**, 1602–1610.

Davies, K.F., Cavender-Bares, J. & Deacon, N. (2011) Native communities determine the identity of exotic invaders even at scales at which communities are unsaturated. *Diversity and Distributions* **17**, 35–42.

Davis, M.A. & Pelsor, M. (2001) Experimental support for a resource-based mechanistic model of invasibility. *Ecology Letters* **4**, 421–428.

Davis, M.A., Grime, J.P. & Thompson, K. (2000) Fluctuating resources in plant communities: a general theory of invasibility. *Journal of Ecology* **88**, 528–534.

Dawson, W., Burslem, D.F.R.P. & Hulme, P.E. (2009) Factors explaining alien plant invasion success in a tropical ecosystem differ at each stage of invasion. *Journal of Ecology* **97**, 657–665.

Dukes, J.S., Chiariello, N.R., Loarie, S.R. & Field, C.B. (2011) Strong response of an invasive plant species (*Centaurea solstitialis* L.) to global environmental changes. *Ecological Applications* **21**, 1887–1894.

Duncan, R.P. (2011) Propagule pressure. In: *Encyclopedia of Biological Invasions* (eds D. Simberloff & M. Rejmánek), pp. 561–563. University of California Press, Berkeley, CA.

Ehrenfeld, J.G. (2010) Ecosystem consequences of biological invasions. *Annual Review of Ecology, Evolution and Systematics* **41**, 59–80.

Ellenberg, H., Weber, H.E., Düll, R. *et al.* (1992) Zeigerwerte von Pflanzen in Mitteleuropa. *Scripta Geobotanica* **18**, 1–248.

Emery, S.M. & Gross, K.L. (2007) Dominant species identity, not community evenness, regulates invasion in experimental grassland plant communities. *Ecology* **88**, 954–964.

Eschtruth, A.K. & Battles, J.J. (2011) The importance of quantifying propagule pressure to understand invasion: an examination of riparian forest invasibility. *Ecology* **92**, 1314–1322.

Fargione, J. & Tilman, D. (2005) Diversity decreases invasion via both sampling and complementarity effects. *Ecology Letters* **8**, 604–611.

Fargione, J., Brown, C.S. & Tilman, D. (2003) Community assembly and invasion: an experimental test of neutral versus niche processes. *Proceedings of the National Academy of Sciences of the United States of America* **100**, 8916–8920.

Foster, B.L. (2001) Constraints on colonization and species richness along a grassland productivity gradient: the role of propagule availability. *Ecology Letters* **4**, 530–535.

Foxcroft, L.C., Richardson, D.M., Rejmánek, M. & Pyšek, P. (2010) Alien plant invasions in tropical and sub-tropical savannas: patterns, processes and prospects. *Biological Invasions* **12**, 3913–3933.

Franklin, J. (2010) *Mapping Species Distributions: Spatial Inference and Prediction.* Cambridge University Press, Cambridge.

Fridley, J.D. (2011) Biodiversity as a bulwark against invasion: conceptual threads since Elton. In: *Fifty Years of Invasion Ecology: The Legacy of Charles Elton* (ed. D.M. Richardson), pp. 121–130. Wiley-Blackwell, Oxford.

Fridley, J.D., Stachowicz, J.J., Naem, S. *et al.* (2007) The invasion paradox: reconciling pattern and process in species invasions. *Ecology* **88**, 3–17.

Funk, J.L. & Vitousek, P.M. (2007) Resource-use efficiency and plant invasion in low-resource systems. *Nature* **446**, 1079–1081.

Gallien, L., Munkemuller, T., Albert, C.H., Boulangeat, I. & Thuiller, W. (2010) Predicting potential distributions of invasive species: where to go from here? *Diversity and Distributions* **16**, 331–342.

Gerhold, P., Pärtel, M., Tackenberg, O. *et al.* (2011) Phylogenetically poor plant communities receive more alien species, which more easily coexist with natives. *The American Naturalist* **177**, 668–680.

Gotelli, N.J. & McCabe, D.J. (2002) Species co-occurrence: a meta-analysis of J.M. Diamond's assembly rules model. *Ecology* **83**, 2091–2096.

Grabas, G.P. & Laverty, M. (1999) The effect of purple loosestrife (*Lythrum salicaria* L.; Lythraceae) on the pollination and reproductive success of sympatric co-flowering wetland plants. *EcoScience* **6**, 230–242.

Grime, J.P., Hodgson, J.G. & Hunt, R. (2007) *Comparative Plant Ecology*, 2nd edn. Castlepoint Press, Dalbeattie.

Grotkopp, E., Rejmánek, M. & Rost, T.L. (2002) Toward a causal explanation of plant invasiveness: seedling growth and life-history strategies of 29 pine (*Pinus*) species. *The American Naturalist* **159**, 396–419.

Grotkopp, E., Erskine-Ogden, J. & Rejmánek, M. (2010) Assessing potential invasiveness of woody horticultural plant species using seedling growth rate traits. *Journal of Applied Ecology* **47**, 1320–1328.

Hall, R.J. & Ayers, D.R. (2009) What can mathematical modeling tell us about hybrid invasions? *Biological Invasions* **11**, 1217–1224.

Hejda, M., Pyšek, P. & Jarošík, V. (2009a) Impact of invasive plants on the species richness, diversity and composition of invaded communities. *Journal of Ecology* **97**, 393–403.

Hejda, M., Pyšek, P., Pergl, J., Sádlo, J., Chytrý, M. & Jarošík, V. (2009b) Invasion success of alien plants: do habitat affinities in the native distribution range matter? *Global Ecology and Biogeography* **18**, 372–382.

Herben, T. (2007) General patterns in plant invasions: a family of quasi-neutral models. In: *Scaling Biodiversity* (eds D. Storch, P.A. Marquet & J.H. Brown), pp. 376–395. Cambridge University Press, Cambridge.

Higgins, S.I., Turpie, J.K., Costanza, R., Cowling, R.M., Le Maitre, D.C., Marais, C. & Midgley, G.F. (1997) An ecologically-economic simulation model of mountain fynbos ecosystems: dynamics, valuation and management. *Ecological Economics* **22**, 155–169.

Hobbs, R.J. & Huenneke, L.F. (1992) Disturbance, diversity and invasion: implications for conservation. *Conservation Biology* **6**, 324–337.

Hodkinson, D.J. & Thompson, K. (1997) Plant dispersal: the role of man. *Journal of Applied Ecology* **34**, 1484–1496.

Hooper, D.U. & Dukes, J.S. (2010) Functional composition controls invasion success in a California serpentine grassland. *Journal of Ecology* **98**, 764–777.

Hovick, S.M., Bunker, D.E., Peterson, C.J. & Carson, W.P. (2011) Purple loosestrife suppresses plant species colonization far more than broad-leaved cattail: experimental evidence with plant community implications. *Journal of Ecology* **99**, 225–234.

Hubbell, S.P., Ahumada, J.A., Condit, R. & Foster, R. (2001) Local neighborhood effects on long-term survival of individual trees in a Neotropical forest. *Ecological Research* **16**, 859–875.

Inderjit & van der Putten, W.H. (2010) Impact of soil microbial communities on exotic plant invasions. *Trends in Ecology and Evolution* **25**, 512–519.

Inderjit, Callaway, R.M. & Vivanco, J.M. (2006) Can plant biochemistry contribute to understanding of invasion ecology? *Trends in Plant Science* **11**, 574–580.

Kartesz, J.T. & Meacham, C.A. (1999) *Synthesis of the North American Flora*. CD-ROM Version 1.0. North Carolina Botanical Garden, Chapel Hill, NC.

Kennedy, T.A., Naeem, S., Howe, K.M. *et al.* (2002) Biodiversity as a barrier to ecological invasion. *Nature* **417**, 636–638.

Kettenring, K.M. & Adams, C.R. (2011) Lessons learned from invasive plant control experiments: a systematic review and meta-analysis. *Journal of Applied Ecology* **48**, 970–979.

Klironomos, J.N. (2002) Feedback with soil biota contributes to plant rarity and invasiveness in communities. *Nature* **417**, 67–70.

Koltunow, A.M., Okada, T. & Bicknell, R.A. (2011) Apomixis. In: *Encyclopedia of Biological Invasions* (eds D. Simberloff & M. Rejmánek), pp. 24–27. University of California Press, Berkeley, CA.

Lanta, V. & Lepš, J. (2008) Effect of plant species richness on invasibility of experimental plant communities. *Plant Ecology* **198**, 253–263.

Law, R. (1999) Theoretical aspects of community assembly. In: *Advanced Ecological Theory* (ed. J. McGlade), pp. 143–171. Blackwell Science, Oxford.

Levine, J.M., Adler, P.B. & Yelenik, S.G. (2004) A meta-analysis of biotic resistance to exotic plant invasions. *Ecology Letters* **7**, 975–989.

Lonsdale, W.M. (1994) Inviting trouble: introduced pasture species in northern Australia. *Australian Journal of Ecology* **19**, 345–354.

Lonsdale, W.M. (1999) Global patterns of plant invasions and the concept of invasibility. *Ecology* **80**, 1522–1536.

MacDougall, A.S., Gilbert, B. & Levine, J.M. (2009) Plant invasions and the niche. *Journal of Ecology* **97**, 609–615.

Mack, R.N. (2011) Cheatgrass. In: *Encyclopedia of Biological Invasions* (eds D. Simberloff & M. Rejmánek), pp. 108–113. University of California Press, Berkeley, CA.

Magee, T.K., Ringold, P.L., Bollman, M.A. & Ernst, T.L. (2010) Index of alien impact: a method for evaluating potential ecological impact of alien plant species. *Environmental Management* **45**, 759–778.

Maron, J.L. & Marler, M. (2008) Effects of native species diversity and resource additions on invader impact. *The American Naturalist* **172**, S18–S33.

Martin, P.H., Canham, C.D. & Marks, P.L. (2009) Why forests appear resistant to exotic plant invasions: intentional introductions, stand dynamics, and the role of shade tolerance. *Frontiers in Ecology and Environment* **7**, 142–149.

May, R.M. (1981) Patterns in multi-species communities. In: *Theoretical Ecology. Principles and Applications* (ed. R.M. May), pp. 197–227. Blackwell Scientific, Oxford.

Meiners, S.J., Pickett, S.T.A. & Cadenasso, M.L. (2001) Effects of plant invasions on the species richness of abandoned agricultural land. *Ecography* **24**, 633–644.

Meiners, S.J., Pickett, S.T.A. & Cadenasso, M.L. (2002) Exotic plant invasions over 40 years of old field succession: community patterns and associations. *Ecography* **25**, 215–223.

Mercure, M. & Bruneau, A. (2008) Hybridization between the escaped *Rosa rugosa* (Rosaceae) and native *R. blanda* in eastern North America. *American Journal of Botany* **95**, 597–607.

Moles, A.T., Flores-Moreno, H., Bonser, S.P. *et al.* (2012) Invasions: the trail behind, the path ahead, and a test of a disturbing idea. *Journal of Ecology* **100**, 116–127.

Morales, C.L. & Traveset, A. (2009) A meta-analysis of impacts of alien vs. native plants on pollinator visitation and reproductive success of co-flowering plants. *Ecology Letters* **12**, 716–728.

Novak, S.J. (2011) Geographic origins and introduction dynamics. In: *Encyclopedia of Biological Invasions* (eds D. Simberloff & M. Rejmánek), pp. 273–280. University of California Press, Berkeley, CA.

Nunez, M.A., Moretti, A. & Simberloff, D. (2011) Propagule pressure hypothesis not supported by an 80-year experiment on woody species invasion. *Oikos* **120**, 1311–1316.

Panetta, F.D., Cacho, O., Hestler, S., Sims-Chilton, N. & Brooks, S. (2011) Estimating and influencing the duration of weed eradication programmes. *Journal of Applied Ecology* **48**, 980–988.

Parker, J.D., Burkepile, D.E., Lajeunesse, M.J. & Lind, E.M. (2012) Phylogenetic isolation increases plant success despite increasing susceptibility to generalists. *Diversity and Distributions* **18**, 1–9.

Pemberton, R.W. & Liu, H. (2009) Marketing time predicts naturalization of horticultural plants. *Ecology* **90**, 69–80.

Petermann, J.S., Fergus, A.J.F., Roscher, C., Turnbull, L.A., Weigelt, A. & Schmid, B. (2010) Biology, chance, or history? The predictable reassembly of temperate grassland communities. *Ecology* **91**, 408–421.

Petipierre, B., Kueffer, C., Broennimann, O., Radin, C., Daehler, C. & Guisan, A. (2012) Climatic niche shifts are rare among terrestrial plant invaders. *Science* **335**, 1344–1348.

Powell, K.I., Chase, J.M. & Knight, T.M. (2011) A synthesis of plant invasion effects on biodiversity across spatial scales. *American Journal of Botany* 98, 539–548.

Proches, S., Wilson, J.R.U., Richardson, D.M. & Rejmánek, M. (2008) Searching for phylogenetic pattern in biological invasions. *Global Ecology and Biogeography* 17, 5–10.

Pyšek, P. (1998) Is there a taxonomic pattern to plant invasions? *Oikos* 92, 282–294.

Pyšek, P., Jarošík, V. & Kučera, T. (2002a) Patterns of invasion in temperate nature reserves. *Biological Conservation* 104, 13–24.

Pyšek, P., Sádlo, J. & Mandák, B. (2002b) Catalogue of alien plants of the Czech Republic. *Preslia* 74, 97–186.

Pyšek, P., Richardson, D.M., Rejmánek, M. *et al.* (2004) Alien plants in checklists and floras: towards better communication between taxonomists and ecologists. *Taxon* 53, 131–143.

Rejmánek, M. (1989) Invasibility of plant communities. In: *Biological Invasions. A Global Perspective* (eds J.A. Drake, H.A. Mooney, F. di Castri, R.H. Groves, F.J. Kruger, M. Rejmánek & M. Williamson), pp. 369–388. John Wiley & Sons, Ltd, Chichester.

Rejmánek, M. (1996) Species richness and resistance to invasions. In: *Diversity and Processes in Tropical Forest Ecosystems* (eds G.H. Orians, R. Dirzo & J.H. Cushman), pp. 153–72. Springer-Verlag, Berlin.

Rejmánek, M. (1999) Invasive plant species and invasible ecosystems. In: *Invasive Species and Biodiversity Management* (eds O.T. Sandlund, P.J. Schei & A. Viken), pp. 79–102. Kluwer Academic Publishers, Dordrecht.

Rejmánek, M. (2000) Invasive plants: approaches and predictions. *Austral Ecology* 25, 497–506.

Rejmánek, M. (2011) Invasiveness. In: *Encyclopedia of Biological Invasions* (eds D. Simberloff & M. Rejmánek), pp. 379–385. University of California Press, Berkeley, CA.

Rejmánek, M. & Pitcairn, M.J. (2002) When is eradication of exotic plant pests a realistic goal? In: *Turning the Tide: The Eradication of Invasive Species* (eds C.R. Veitch & M.N. Clout), pp. 249–253. IUCN, Gland, Switzerland and Cambridge, UK.

Rejmánek, M. & Richardson, D.M. (1996) What attributes make some plant species more invasive? *Ecology* 77, 1655–1661.

Rejmánek, M., Richardson, D.M., Higgins, S.I., Pitcairn, M.J. & Grotkopp, E. (2005) Ecology of invasive plants: state of the art. In: *Invasive Alien Species: Searching for Solutions* (eds H.A. Mooney, J.A. McNeelly, L. Neville, P.J. Schei & J. Waage), pp. 104–161. Island Press, Washington, DC.

Richards, J.H. & Janes, B.R. (2011) Tolerance limits, plants. In: *Encyclopedia of Biological Invasions* (eds D. Simberloff & M. Rejmánek), pp. 663–667. University of California Press, Berkeley, CA.

Richardson, D.M. & Rejmánek, M. (1998) *Metrosideros excelsa* takes off in the fynbos. *Veld & Flora* 85, 14–16.

Richardson, D.M. & Rejmánek, M. (2011) Trees and shrubs as invasive alien species – a global review. *Diversity and Distributions* 17, 788–809.

Richardson, D.M., Macdonald, I.A.W. & Forsyth, G.G. (1989) Reductions in plant species richness under stands of alien trees and shrubs in the fynbos biome. *South African Forestry Journal* 149, 1–8.

Richardson, D.M., Allsopp, N., D'Antonio, C.M., Milton, S.J. &. Rejmánek, M. (2000a) Plant invasions – the role of mutualisms. *Biological Reviews of the Cambridge Philosophical Society* 75, 65–93.

Richardson, D.M., Pyšek, P., Rejmánek, M. *et al.* (2000b) Naturalization and invasion of alien plants: concepts and definitions. *Diversity and Distributions* 6, 93–107.

Roscher, C., Bessler, H., Oelmann, Y. *et al.* (2009) Resources, recruitment limitation and invader species identity determine pattern of spontaneous invasion. *Journal of Ecology* 97, 32–47.

Sax, D.F., Brown, J.H. & Gaines, S.D. (2002) Species invasions exceed extinctions on islands world-wide: a comparative study of plants and birds. *The American Naturalist* 160, 766–783.

Schamp, B.S. & Aarssen, L.W. (2010) The role of plant species size in invasibility: a field experiment. *Oecologia* 162, 995–1004.

Schmidt, W., Dölle, M., Bernhardt-Römermann, M. & Parth, A. (2009) Neophyten in der Ackerbrachen-sukzession – Ergebnisse eines Dauerflächen-Versuchs. *Tuexenia* 29, 236–260.

Schutzenhofer, M.R. & Valone, T.J. (2006) Positive and negative effects of exotic *Erodium cicutarium* on an arid ecosystem. *Biological Conservation* 132, 376–381.

Shea, K. & Chesson, P. (2002) Community ecology theory as a framework for biological invasions. *Trends in Ecology & Evolution* **17**, 170–176.

Simberloff, D. & Von Holle, B. (1999) Positive interactions of nonindigenous species: invasional meltdown? *Biological Invasions* **1**, 21–32.

Simberloff, D. & 141 signatories (2011) Non-natives: 141 scientists object. *Nature* **475**, 36.

Smith, S.D., Huxman, T.E., Zitzer, S.F. *et al.* (2000) Elevated CO_2 increases productivity and invasive species success in an arid ecosystem. *Nature* **408**, 79–82.

Stohlgren, T.J., Binkley, D., Chong, G.W. *et al.* (1999) Exotic plant species invade hot spots of native plant diversity. *Ecological Monographs* **69**, 25–46.

Stohlgren, T.J., Jarnevich, C., Chong, G.W. & Evangelista, P.H. (2006) Scale and plant invasions: a theory of biotic acceptance. *Preslia* **78**, 405–426.

Stohlgren, T.J., Pyšek, P., Kartesz, J. *et al.* (2011) Widespread plant species: natives versus aliens in our changing word. *Biological Invasions* **13**, 1931–1944.

Symstad, A.J. (2000) A test of the effects of functional groups richness and composition of grassland invasibility. *Ecology* **81**, 99–109.

te Beest, M., Le Roux, J.J., Richardson, D.M. *et al.* (2012) The more the better? The role of polyploidy in facilitating plant invasions. *Annals of Botany* **109**, 19–45.

Thiele, J., Isermann, M., Otte, A. & Kollmann, J. (2010) Competitive displacement or biotic resistance? Disentangling relationships between community diversity and invasion success of tall herbs and shrubs. *Journal of Vegetation Science* **21**, 213–220.

Thiele, J., Isermann, M., Kollmann, J. & Otte, A. (2011) Impact scores of invasive plants are biased by disregard of environmental co-variation and non-linearity. *NeoBiota* **10**, 65–79.

Thomson, D.M. (2005) Matrix models as a tool for understanding invasive plant and native plant interactions. *Conservation Biology* **19**, 917–928.

Tilman, D. (1999) The ecological consequences of changes in biodiversity: a search for general principles. *Ecology* **80**, 1455–1474.

Tilman, D. (2004) Niche tradeoffs, neutrality, and community structure: a stochastic theory of resource competition, invasion, and community assembly. *Proceedings of the National Academy of Sciences of the United States of America* **101**, 10854–10861.

Tognetti, P.M., Chaneton, E.J., Omacini, M., Trebino, H.J. & León, J.C. (2010) Exotic vs. native plant dominance over 20 years of old-field succession on set-aside farmland in Argentina. *Biological Conservation* **143**, 2494–2503.

Trueman, M., Atkinson, R., Guézou, A. & Wurm, P. (2010) Residence time and human-mediated propagule pressure at work in the alien flora of Galapagos. *Biological Invasions* **12**, 3949–3960.

Tye A. (2006) Can we infer island introduction and naturalization rates from inventory data? Evidence from introduced plants in Galapagos. *Biological Invasions* **8**, 201–215.

Van Driesche, R., Hoddle, M. & Center, T. (eds.) (2008) *Control of Pests and Weeds by Natural Enemies.* Blackwell Publishing, Oxford.

van Kleunen, M., Weber, E. & Fischer, M. (2010) A meta-analysis of trait difference between invasive and non-invasive plant species. *Ecology Letters* **13**, 235–245.

Vilà, M. & Weiner, J. (2004) Are invasive plant species better competitors than native plant species? – Evidence from pair-wise experiments. *Oikos* **105**, 229–238.

Vilà, M., Espinar, J.L., Hejda, M. *et al.* (2011) Ecological impacts of invasive alien plants: a meta-analysis of their effects on species, communities and ecosystems. *Ecology Letters* **14**, 702–708.

Vitousek, P.M. & Walker, L.R. (1987) Colonization, succession and resource availability: ecosystem-level interactions. In: *Colonization, Succession and Stability* (eds A.J. Gray, M.J. Crawley & P.J. Edwards), pp. 207–223. Blackwell Science, Oxford.

Vitousek, P.M. & Walker, L.R. (1989) Biological invasion by *Myrica faya* in Hawai'i: plant demography, nitrogen fixation, ecosystem effects. *Ecological Monographs* **59**, 247–265.

Vitousek, P.M., D'Antonio, C.M., Loope, L.L., Rejmánek, M. & Westbrooks, R. (1997) Introduced species: a significant component of human-caused global change. *New Zealand Journal of Ecology* **21**, 1–16.

Von der Lippe, M. & Kovarik, I. (2007) Long-distance dispersal of plants by vehicles as a driver of plant invasion. *Conservation Biology* **21**, 986–996.

Von Holle, B., Delcourt, H.R. & Simberloff, D. (2003) The importance of biological inertia in plant community resistance to invasion. *Journal of Vegetation Science* **14**, 425–432.

Walker, S., Wilson, J.B. & Lee, W.G. (2005) Does fluctuating resource availability increase invasibility? Evidence from field experiments in New Zealand short tussock grassland. *Biological Invasions* 7, 195–211.

Williams, P.A., Nicol, E. & Newfield, M. (2001) Assessing the risk to indigenous biota of new plant taxa new to New Zealand. In: *Weed Risk Assessment* (eds R.H. Groves, F.D. Panetta & J.G. Virtue), pp. 100–116. CSIRO Publishing, Collingwood, Victoria.

Williamson, M. & Fitter, A. (1996) The varying success of invaders. *Ecology* 77, 1661–1666.

Wilson, J.R.U., Richardson, D.M., Rouget, M. *et al.* (2007) Residence time and potential range: crucial considerations in modelling plant invasions. *Diversity and Distributions* 13, 11–22.

14

Vegetation Conservation, Management and Restoration

Jan P. Bakker

University of Groningen, The Netherlands

14.1 Introduction

In the past few decades the importance of management, and more recently restoration, as a tool for nature conservation have increased considerably. This chapter will review the development from agricultural exploitation via maintenance management towards restoration management.

Conservation is carried out all over the world to maintain existing areas with nature conservation interests (e.g. Pickett *et al.* 1997). This does not imply preservation of biodiversity in general, but rather in terms of specific diversity of plants, mammals, birds, insects, etc. (Bakker *et al.* 2000). Conservation may imply the absence of human interference in case of a still existing natural system. It may also include human interference, management, as far as exploitation coincides with nature conservation interests. Management is mainly practised in industrialized countries in Europe (Westhoff 1983; Spellerberg *et al.* 1991), Australia (Lindenmayer *et al.* 2010) and New Zealand (Craig *et al.* 2000). Restoration is mainly practised in industrialized countries in Europe (e.g. Wheeler *et al.* 1995), North America, Australia and New Zealand (see many papers in the journals *Restoration Ecology*, *Applied Vegetation Science* and *Basic and Applied Ecology*, as well as the book *Restoration Ecology* (van Andel & Aronson 2012).

In the history of the exploitation of terrestrial ecosystems we may discern three periods: the 'natural' period, the 'semi-natural' period and the 'cultural' period. The natural period is characterized by the dominance of communities, landscapes and processes without any noticeable human influence (Bakker & Londo 1998). The major patterns in the landscape were largely determined by climatic and geomorphological factors; these were inserted upon the geological matrix. There was grazing and browsing by indigenous herbivores. Hence, the

Vegetation Ecology, Second Edition. Eddy van der Maarel and Janet Franklin.
© 2013 John Wiley & Sons, Ltd. Published 2013 by John Wiley & Sons, Ltd.

natural landscape can be defined by the species assemblage of the original flora, vegetation and fauna (Westhoff 1983). This can be forest or scrub under favourable abiotic conditions, or open landscapes with limiting harsh abiotic conditions such as bogs (wetness), salt marshes (salt), tundras (low temperature), short-grass prairies and savannas (drought).

In north-west Europe, the first agricultural immigration took place about 7000 BP; it was followed by a second one around 4600 BP, both from southeastern Europe and southern Russia. These people grew arable crops in a shifting cultivation system after the clearance of primeval forest. For the greater part, indigenous large herbivores were gradually replaced by livestock. In medieval times degradation and destruction of primeval forests continued and large oligotrophic bogs, mesotrophic fens and eutrophic reedbeds were drained, reclaimed and even completely removed for fuel (Wheeler *et al.* 1995). Not only the natural communities but also certain landscape-building processes disappeared, through the exclusion of the influence of the sea and rivers and through the regulation of hydrological conditions. Although the resulting open landscape was new, many species that invaded the emerged grasslands and heathlands were already present as elements in the understorey of open forest, in small glades, fringes along streams, fens and bogs, and in larger open areas along the coast. Hence, the definition of the semi-natural landscape includes the original flora and fauna but also a transformation of the original vegetation by humans (Westhoff 1983). These are the landscapes of the semi-natural period. They are known from Europe but also from other continents as indicated below.

In the Serengeti, pastoralism was practiced from 3500 BP. First the landscape was altered through the use of fire, and later through the domestication of livestock (Olff & Hopcraft 2008). Savanna systems in Kenya harbour temporary settlements. Around occupied settlements in savanna bushland and woodland, woody plant abundance tends to be reduced and large patches may become devoid of vegetation. Once abandoned again, these patches become very productive grasslands and can persist for decades. Settlement activity and succession after abandonment seem to be an important force creating the bush–grass mosaic and patch dynamics of the savannas (Muchiru *et al.* 2009).

In other parts of the world, various ways of exploitation to enhance the productivity of the soil by grazing and fire (Bowman *et al.* 2009) were applied tens of thousands of years prior to European settlement. This rendered large areas of semi-natural pastures and rangeland in North America, South America, Africa, Asia and Australia (Foley *et al.* 2005). The above implies that many open landscapes in different parts of the world are not natural, but belong to the semi-natural period. They need human activities such as fire, grazing or cutting to prevent them from transformation into scrub or forest by secondary succession after abandonment. For approaches in North America and Europe see Bakker & Londo (1998).

The character of human impact also changed. First only the biotic component was influenced by cutting trees and grazing livestock. Abiotic conditions were only influenced indirectly by, for instance, trampling and nutrient transport, and directly by superficial ploughing. Large areas of semi-natural landscapes such as heathland and grassland on infertile soils used for common grazing, were not enclosed in private fields but belonged to the local community – the 'commons'. The commons were predominantly found on the drier, sandy parts. Here the geological matrix remained more or less intact. As human impact was stable during many centuries, it became superimposed by a historical matrix. In the 'semi-natural period', regulation of hydrological conditions by drains and ditches in wet parts enabled direct influences on the abiotic conditions by reclamation, deep ploughing and soil levelling. These activities, as well as the division of the landscape into private properties, resulted in the enclosed semi-natural landscape, where fields became delimited by ditches and hedgerows. The geological matrix became severely disturbed.

The transition from the semi-natural to the cultural period in north-west Europe was triggered by the introduction of organic manure or waste from cities and artificial, inorganic fertilizers. The large-scale reclamation and subsequent eutrophication of common grassland and heathland occurred after 1920 when intensification in agriculture started. It resulted in the development of the cultivated landscape, in which not only the vegetation but also the flora and fauna became heavily influenced by people (Westhoff 1983). Indigenous species were eradicated by herbicides and non-indigenous species were introduced. These landscapes represent the cultural period in which we are living now, not only in Europe, but all over the world.

14.2 From agricultural exploitation to nature conservation

Since most semi-natural grasslands and heathlands are marginal from an agricultural point of view, these areas tend to be the first to be neglected or abandoned. This was common local practice earlier. In Europe this has recently been enforced by the European Union agricultural policy; this facilitated highly productive farms and led to the closing of less productive ones, on which so-called low-intensity farming was practised. In the 1990s, the total area of low-intensity farming was 56 million ha. The relative areas varied strongly among European countries. Such farming systems feature 82% of the agricultural area in Spain, 61% in Greece, 60% in Portugal, 35% in Ireland, 31% in Italy, 25% in France, 23% in Hungary, 14% in Poland, and 11% in the United Kingdom (Bignal & McCracken 1996). Low-intensity farming areas taken out of the agricultural system in, for instance, the Netherlands or Denmark may still be exploited in, for instance, Spain or France. On the other hand, artificially fertilized grasslands can be taken out of the agricultural system in the Netherlands with the aim of restoring species-rich grassland or heathland. Such plant communities are still widespread, although decreasing, in East European countries, notably Poland. Also in other parts of the world, large areas of previously exploited arable fields have been abandoned and turned into old fields. In the USA, the total area would be about 80 million ha (Cramer & Hobbs 2007). These open landscapes with low-intensity farming and semi-natural communities with high nature conservation interest, are increasingly threatened all over Europe, either by intensified, industrial agriculture or abandonment (Veen *et al.* 2009).

The degradation of fauna, flora and vegetation in natural and semi-natural landscapes has become a matter of great concern. The problem affects all continents, but particularly north-west Europe, and most of all the Netherlands and Belgium, due to the high population density and the advanced level of agriculture and technical development resulting in expanding urban and industrial areas, connected by a dense network of roads. Although rural areas outside of urbanized areas still have a predominantly agricultural land use, the intensity of agricultural exploitation leaves little room for species diversity.

Methods to counteract the degradation of flora and vegetation have developed. From the beginning of the 20th century onwards, areas have been acquired by private organizations for landscape and nature conservation purposes, and more recently also by governmental bodies. Most of these reserves in north-west European countries with intensive agriculture represent small fragments of areas with conservation interest. Other countries in Europe feature very large reserves, such as Białowieza (250 000 ha) in Poland and Belarus, Cevennes (913 000 ha) in France, Doñana (540 000 ha) and Montagüe, Extremadura (195 000 ha) in Spain, Gran Paradiso (70 000 ha) in Italy, the adjoining national parks Stora Sjöfallet (128 000 ha), Padjelanta (198 000 ha) and Sarek (197 000 ha) in Sweden. Some of these reserves have a largely semi-natural character, such as Stora Alvaret in Sweden (25 000 ha). Very large reserves are found in other continents such as Jasper in Canada (1 090 000 ha and Serengeti (3 000 000 ha) in east Africa.

Several forms of management have been developed aiming at nature conservation. In the UK, Wells (1980) distinguished between '*reclamation management*', carried out only once, and regular '*maintenance management*'. When the traditional agricultural use of these grasslands and heathlands is discontinued by cessation of grazing and mowing, coarse grasses, sedges and shrubs take over. When such an abandoned area still has potential as a nature reserve or/and amenity area, it has first to be reclaimed. In the UK and the USA restoration management is referred to as '*biological habitat reconstruction*' and '*restoration, reclamation and regeneration of degraded and destroyed ecosystems*'.

In the Netherlands two different situations occur. In the first situation conservation interest is still great and in need of '*nature management*' in a strict sense (Bakker & Londo 1998) (Fig. 14.1). This usually implies continuing or reintroducing management practices such as coppicing, haymaking, cutting sods and livestock grazing. In the case of semi-natural landscapes, management has to be carried out with a certain regularity, while little or nothing needs to be done for a near-natural landscape. Except for sod cutting, these practices, whether for agricultural exploitation or for present-day nature conservation, affect the structure of the vegetation.

In the second situation, nature conservation interest is low and '*new nature*' has to be developed – '*nature development*'. Usually the owners have one or more '*target communities*' in mind. Two phases are distinguished (Fig. 14.1). Phase 1, '*environmental restoration*', is necessary when the abiotic environment has been degraded – for example, after lowering the groundwater table, levelling the original relief or eutrophication. This may include removal of the eutrophicated topsoil down to 50 cm, restoration of the relief and raising the

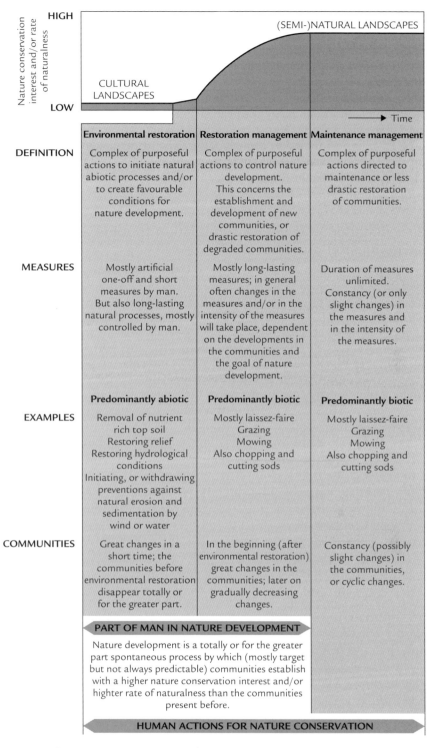

Fig. 14.1 Definitions, measures, examples and communities in relation to human activities for nature conservation in the framework of cultural and (semi-)natural landscapes. (After Bakker & Londo 1998.)

groundwater table. The perspectives of environmental restoration will depend on both the quality of the new environmental conditions and the availability of target species.

If the abiotic conditions have been changed after environmental restoration or the abiotic conditions have not been degraded, environmental restoration is not necessary and phase 2, '*restoration management*' can be implemented directly (Fig. 14.1). This may include a change in the existing management practices – for example, cessation of fertilizer application, followed by haymaking or grazing at a low stocking rate. In this way existing plant communities are turned into the target communities.

14.3   Vegetation management in relation to a hierarchy of environmental processes

Management of plant communities and plant species should take into account the various natural and human-influenced processes. The impacts of these processes can be considered in a hierarchical scheme according to C.G. van Leeuwen (see van der Maarel 1980), ranging – in order of impact – from atmosphere/climate, geology, geomorphology, (ground)water and soil to vegetation and fauna. Londo (1997) elaborated this scheme to indicate the position of environmental restoration and restoration management (Fig. 14.2).

When applying management measures on a lower level, one should be aware of the ecological processes occurring at a higher level that are beyond the influence of the local management. As an example, the mean atmospheric deposition of $40 \, kg\text{-}N \cdot ha^{-1} \cdot yr^{-1}$ in the Netherlands in the 1990s is about twice that of the critical nitrogen load for plant communities on mesotrophic and oligotrophic soils (Bobbink *et al.* 1998). A lowering of the groundwater table by 60 cm in a *Calthion palustris* fen meadow results in an even more alarming increase of nitrogen availability from 50 to $450 \, kg\text{-}N \cdot ha^{-1} \cdot yr^{-1}$. Deep drainage can result in an irreversible desiccation of the soil, the subsequent mineralization of organic matter and acidification because of the replacement of deep calcium-rich seepage water by shallow calcium-poor seepage water. Restoration by simply raising the groundwater table is then insufficient (Grootjans *et al.* 1996).

Vegetation is of course not only influenced by higher-level abiotic processes but also by the fauna (Fig. 14.2). Small herbivores can even act as keystone species in certain ecosystems (Mortimer *et al.* 1999; see also Chapter 8). Geese can destroy the vegetation by grubbing below-ground parts and can even degrade the soil (Jefferies & Rockwell 2002; see also Chapter 10). The establishment of the shrub *Prunus spinosa*, playing a major role in the shifting mosaic of grassland, scrub and forest, is controlled by rabbits rather than by cattle (Olff *et al.* 1999; see Chapter 8).

14.4   Laissez-faire and the wilderness concept

Some managers wish to restore abiotic conditions by removing the topsoil or rewetting, introducing large herbivores, thus reflecting the Pleistocene past, and

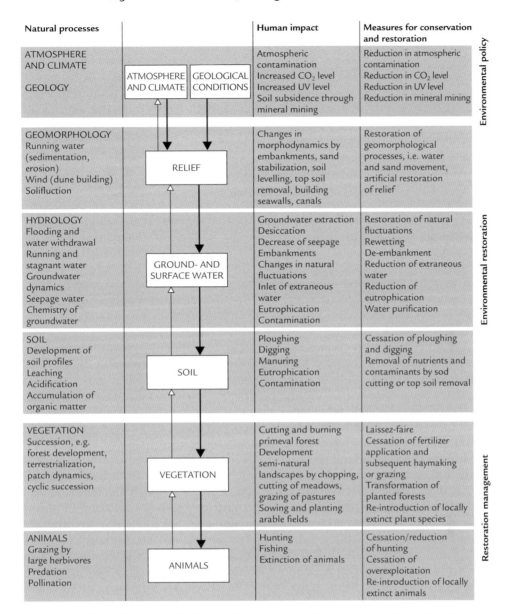

Fig. 14.2 Hierarchical model of levels influencing each other mutually (only influences between adjacent levels are indicated). Thickness of the arrows indicate the strength of the influence. At each level natural processes, human interferences and measurements of nature conservation and restoration are indicated. (After Londo 1997.)

leave the area alone as '*new nature*' or wilderness. Such schemes are adopted in, for example, New Zealand, Saudi Arabia, the Russian Far East and the Netherlands (Marris 2009). Here, giving way to natural processes becomes a goal in itself and not a means to reach a well-defined target This is a development which cannot be properly evaluated (Bakker *et al.* 2000). On the one hand very little systematic data or scientific papers have been published so far, but on the other hand it is difficult to have control areas on a scale of thousands of hectares. Lack of defined goals is incompatible with current management aims based on targets for species and habitats of conservation concern which are guided by Biodiversity Action Plans in the UK. Moreover, it is likely to be very difficult to impose the wilderness ideology on the busy cultural landscapes of Britain (Hodder *et al.* 2005).

The idea of wilderness 'creation' was promoted by Vera (2000) who suggested that species-rich open forest can be created or maintained after the introduction of large herbivores related to domestic livestock such as Konik horses (*Equus ferus ferus*), Heck cattle (*Bos domesticus*), and wild herbivores such as European bison (*Bison bonasus*), elk (*Alces alces*), red deer (*Cervus elaphus*), roe deer (*Capreolus capreolus*) and beaver (*Castor fiber*). Such a landscape is believed to have occurred in the Pleistocene before human intervention, and hence is regarded as a natural landscape. However, no indications exist that natural herbivores ever occurred in dense populations over large areas and that open forest occurred before agriculture started (Svenning 2002; Birks 2005; Szabo 2009). Only riverine landscapes subjected to natural dynamics may have been open with a shifting mosaic of grassland, tall forb communities, scrub and forest. Such landscapes occur at present along rivers under low stocking densities (Olff *et al.* 1999). Open vegetation may also have occurred on infertile soil.

Human-introduced land-use transitions generally go from natural ecosytems through stages of frontier clearings, subsistence agriculture to intensive agriculture with different velocities in various parts of the world (Foley *et al.* 2005). Fire emerged as likely potential key factor in creating open vegetation in northwest Europe and other continents, initially used as domestic fire, starting around 50 000 to 100 000 years ago, and as agricultural fire from 10 000 years ago (Bowman *et al.* 2009). Fire would probably also have been important in the maintenance of light-demanding or short-statured woody species within closed upland forests. Many plant species of calcareous grasslands in West and Central Europe have been found as macrofossils for the first time long after human impact in the landscape started, for example *Centaurea scabiosa* and *Primula veris* since the Roman period (Table 14.1) and a currently abundant species such as *Bromus erectus* only since medieval times (Poschlod & WallisDeVries 2002). This implies that many plant species and communities we wish to preserve depend on human activities in semi-natural landscapes. Hence, the cessation of human activities in semi-natural landscapes might eventually result in closed forest in most lowland biotopes, with almost certainly short-term losses of biodiversity, which may never be restored again. The development of mature natural forests will take centuries, that of fens and bogs even millennia.

Table 14.1   First appearance (+) from archaeological excavations of plant species characteristic of calcareous grasslands (Mesobromion) in the Lower Rhine Valley. Remarkable is that currently occurring characteristic grass species such as *Bromus erectus, Koeleria pyramidata, K. gracilis, Phleum phleoides, Festuca cinerea* and *F. rupicola* were not found before the Modern Age.

	Neolithic period	Bronze Age	Iron Age	Roman Empire	Middle Ages
Number of sites for sedges and herbs	66	11	?	>50	>80
Number of sites for grasses	34	6	34	27	26
Euphorbia cyparissias	+	+	+	+	+
Potentilla tabernaemontani	+	+	+	+	+
Scabiosa columbaria	+	+	+	+	+
Silene vulgaris	+	+	+	+	+
Ajuga genevensis	+	+	+	+	+
Campanula trachelium	+	+	+	+	+
Stachys recta	+	+	+	+	+
Festuca ovina	+	+	+	+	+
Brachypodium pinnatum	+	+	+	+	+
Stipa pennata	+				
Pimpinella saxifraga	+	+	+	+	
Carex caryophyllea	+	+	+		
Medicago lupulina	+	+	+		
Plantago media	+	+	+		
Avenochloa pratensis	+				
Campanula rapunculus	+	+			
Centaurea scabiosa		+	+		
Euphorbia seguierana	+		+		
Hippocrepis comosa			+		+
Peucedanum officinale		+		+	
Primula veris			+		+
Salvia pratensis			+		+
Sanguisorba minor			+		+
Silene vulgaris			+		+
Briza media				+	

After Poschlod & WallisDeVries (2002).

14.5   Management and restoration imply setting targets

Management and restoration of vegetation should have targets to follow. The Society for Ecological Restoration (2004) provided nine targets and target-related measures for the restored ecosystem:

1 it contains a characteristic set of the species that occur in the reference system;
2 it consists of indigenous species to the greatest practical extent;

3   all functional groups necessary for its continued development and/or stability
     are represented or have the potential to colonize;
4   its physical environment is capable of reproducing populations of the species
     necessary for its continued stability or development;
5   it functions normally for its ecological stage of development;
6   it is suitably integrated into a larger ecological matrix of landscape;
7   potential threats to its health and integrity from the surrounding landscape
     have been eliminated or reduced as much as possible;
8   it is sufficiently resilient to endure the normal periodic stress events in the
     local environment that serve to maintain the integrity of the ecosystem;
9   it is self-sustaining to the same degree as the reference ecosystem, and
     has the potential to persist indefinitely under existing environmental
     conditions.

Setting targets also implies gaining knowledge about the history of the land-scape: is it natural or semi-natural? When the history is not well known, the concept of laissez-faire (considered 'natural') in semi-natural landscapes may result in losses of the semi-natural landscapes that managers wish to protect. The heavy North American emphasis on 'naturalness' affects protection and restora-tion goals, determination of what is worthy of protection and restoration, and decisions about appropriate or inappropriate management tools (Wedin 1992). More recently, the idea that nowadays seemingly natural landscapes are very much the product of the history of anthropogenic activities, is taken into account (Foster *et al.* 2003). The nature conservation interest with respect to targets in Europe is acknowledged according to the European Nature Information System (EUNIS 2012) habitat classification. The EUNIS types can be considered as reference communities, which can be the subject of conservation, and when these communities are damaged or destroyed, their defined species composition will be the targets of restoration. These targets represent the final situation; it may take a very long time until these are all realized. Examples of how authori-ties in charge of nature management and restoration deal with targets are dis-cussed here.

Several strategies for the development of targets can be adopted (Bakker & Londo 1998). These are simple for the few natural landscapes left in lowland Europe and America, where human influence has always been modest. When geological processes such as sedimentation and erosion by water and wind are predominant, older successional stages can be eroded locally and young stages can emerge at other places. This may happen in coastal and inland dunes, salt marshes and along rivers.

In certain cultural and semi-natural landscapes a more natural landscape can be restored, first of all through hydrological measures, such as digging side-channels along rivers, building of dams to catch rainwater for bog development or de-embankment for salt-marsh restoration.

Restoration management in communal semi-natural landscapes with grass-land, heathland, scrub and/or wooded meadows, can be carried out by removing former borders in the enclosed landscape or allowing them to disappear. This can be accomplished by abiotic management such as neglecting drainage systems, giving way to eroding forces of wind and water, sod cutting (up to 10 cm) or topsoil removal. Biotic management implies fencing in large areas and grazing by large herbivores of different breeds such as Konik horse, Exmoor ponies, Scottish Highland cattle, Galloways, Schoonebeker or Mergelland sheep, but also heifers of dairy cattle.

Enclosed semi-natural landscapes are typically restored within their field borders including drainage systems. Restoration management deals with oligo-trophic or mesotrophic grassland communities by cutting, sometimes followed by grazing with high stocking rates. The process of restoration can be enhanced by topsoil removal, also for small isolated fields where heathland is the target. The effect of restoration management is thought to be enhanced by connecting nearby fields by corridors in which species can disperse. In cultural landscapes species-rich plant communities may be created in the margin of grasslands and arable fields.

Within semi-natural landscapes more tangible targets are needed. In the Netherlands a system of *'nature target types'* was developed, something between habitat and community types (Bal & Hoogeveen 1995). Each type includes lists of plant and animal species. These types also harbour many Red List species at the national level. It turned out that these target types may be useful to strive after in the long term. However, it should be taken into account that even in a small country such as the Netherlands. Some Red List species have a regional distribution, and do not occur all over the country. Moreover, the target types are not feasible for the short- and mid-term management practice to be fulfilled, because of dispersal constraints. It is now recognized that several pathways may lead to certain target types. Landscape matrices show the relation-ships of the target type with eutrophicated communities at the same substrate and successional relationships of open low canopy, closed low canopy, scrub and forest (Schaminée & Jansen 1998). Such a matrix for the target type of wet heathland, *Ericetum tetralicis*, is shown in Fig. 14.3. Species to be expected at several time intervals after the start of the restoration are listed. After sod cutting, *Molinia caerulea* may start colonization. Most target species are supposed to emerge immediately; this expectation is based on their presence in established vegetation, or in the soil seed bank for species with a high longev-ity index (Bekker *et al.* 1998). Starting from fertilized communities after sod cutting or topsoil removal, an initial establishment of species of fertilized habitats and a few target species that must have a long-term persistent seed bank will occur. Target species with a low longevity index are supposed to establish within 10 years (Fig. 14.3). The latter need dispersal from elsewhere. From 2001 onwards the Dutch government compensates the costs of management based on the fulfilment of targets set during 10 years. Clearly, authorities in charge of management have to set realistic mid-term targets instead of ideal long-term targets.

The framework of conservation targets should be the plant community system. The recent classification of plant communities in the Netherlands was based on *c.* 350 000 recent relevés from the period 1930–2000 (Hennekens & Schaminée 2001; see also Chapter 2). From this classification, based on synoptic tables, lists of target species for target plant communities are derived. As the geographical

(a)

Species	Number of years				Longevity index
	0–1	1–3	3–10	10–25	
Drosera intermedia					–
Juncus squarrosus					1
Rhynchospora alba					–
Rhynchospora fusca					–
Carex panicea					0.36
Calluna vulgaris					0.74
Drosera rotundifolia					–
Lycopodium inundatum					–
Erica tetralix					0.42
Molinia caerulea					0.30
Gentiana pneumonanthe					–
Scirpus cespitosus					–
Salix repens					0
Narthecium ossifragum					–

(b)

Species	Number of years					Longevity index
	0–1	1–3	3–10	10–25	>25	
Gnaphalium uliginosum						0.89
Erigeron canadensis						–
Rorippa sylvestris						–
Cirsium arvense						0.33
Rumex obtusifolius						0.62
Trifolium repens						0.38
Drosera intermedia						–
Ornithopus perpusillus						0
Leucanthemum vulgare						0.36
Juncus squarrosus						1
Calluna vulgaris						0.74
Rumex acetosella						0.71
Holcus lanatus						0.44
Erodium cicutarium						0.14
Carex panicea						0.36
Genista anglica						–
Filago minima						–
Gentiana pneumonanthe						–
Hypochaeris radicata						0.39
Erica tetralix						0.42
Molinia caerulea						0.30
Scirpus cespitosus						–
Narthecium ossifragum						–

Fig. 14.3 Restoration perspectives for wet heathlands on Pleistocene sandy soils in the Netherlands, indicated by the expected occurrence of plant species in various time periods. Thick lines indicate higher abundances. A. Starting from heathland overgrown by grasses, and the restoration measurement sod cutting. B. Starting from arable fields or grasslands applied with fertilizer, and the restoration measurements sod cutting and re-instalment of the original hydrological conditions. Seed bank data are derived from Thompson *et al.* (1997), longevity index values from Bekker *et al.* (1998). (After Schaminée & Jansen 1998.)

position of the relevés is known, lists of target species can be regionalized. These lists can be completed by records from a detailed floristic survey of the Netherlands on a 5×5 km basis, which started early in the 20th century. In this way, realistic sets of regional target species for restoration management can be achieved. Predictions for the establishment of species may be derived from data sets including life history traits such as seed longevity and dispersal characteristics (Kleyer *et al.* 2008). Data mining and working with such large data sets is an aspect of '*eco-informatics*' as pointed out in the Special Feature of *Journal of Vegetation Science* (Bekker *et al.* 2007).

The plant community concept assumes that plant species form more or less stable assemblages responding to the local environmental conditions. However, from a study of the above-mentioned phytosociological data from the Netherlands, it became clear that in most community types the floristic composition had changed since 1930, even if the physiognomy of the vegetation and the occurrence of many characteristic species has remained the same (Schaminée *et al.* 2002). Eutrophication through air and water pollution is probably responsible for most of the changes. Management authorities have to take into account such changes when planning the re-introduction of endangered plant species. Although characteristic species may still be present in communities subject to eutrophication, their populations seem to diminish in size by lack of rejuvenation. In such circumstances it is uncertain whether the introduction of seeds will be effective (Strykstra 2000). This may become even more questionable when the species has disappeared from the surroundings.

Plant communities include both common and rare species, which will have different environmental amplitudes. An analysis of 300 relevés and associated soil chemical properties from a range of heathlands and acidic grasslands across the Netherlands showed that of 12 measured soil parameters only the soil ammonium (NH_4) concentration and the ammonium/nitrate (NH_4/NO_3) ratio were significantly higher in sites with mainly common species compared to sites with rare species. The other parameters did not differ, but on average rare species had a significantly narrower ecological amplitude than common species (Kleijn *et al.* 2008). Apparently, rare species of heathlands and acidic grasslands are more susceptible to higher NH_4 concentrations than common species. Such higher concentrations may be due to atmospheric deposition (Bobbink *et al.* 1998).

14.6 Setting targets implies monitoring

The evaluation of restoration projects with targets evokes repeated vegetation monitoring. This can be carried out at different levels of resolution. On the basis of repeated aerial photographs, changes in the size of community patches and the extent of bare soil and structural types such as short vegetation, tall forb communities, scrub and forest can be detected (see Chapter 4), but no information on nature target types can be obtained in this way.

Information at the level of plant communities requires repeated vegetation mapping (see Chapter 16). Aerial photographs or other means of remote sensing, and/or field surveys enable stratified sampling of the elements to be discerned. The size of vegetation relevés may differ according to the structural class.

Changes in the presence/absence or the cover of individual species can only be monitored in permanent plots. Long-term permanent plots are important as they can help in separating trends and superimposed fluctuations (Huisman *et al.* 1993), and are needed to test ecological models that are often based on assumptions and not derived from solid field studies (Bakker *et al.* 1996). The study of long-term permanent plots has made it particularly clear that vegetation development in many ecosystems under restoration was different from the final state that was anticipated. This may generate new hypotheses (Klötzli & Grootjans 2001). Because a limited number of permanent plots will not cover all spatio-temporal changes in vegetation, it is better to establish permanent transects with adjacent grid cells varying in size depending on the vegetation structure (van der Maarel 1978; Olff *et al.* 1997) or even by large grid cells covering the entire study site (Verhagen *et al.* 2001).

Long-term recordings are needed to validate the effects of management measures. Experimental changes in salt-marsh management (Bos *et al.* 2002) and calcareous grassland management (Kahmen *et al.* 2002) revealed clear changes after 15–20 years, stressing the importance of long-term monitoring. Silvertown *et al.* (2010) made a plea for maintaining long-term ecological experimental projects in the UK, such as the Park Grass Experiment at Rothamsted Experimental Station and mention the Ecological Continuity Trust. Reviews on the success of restoration face the problem that very few studies have actually been published, and their information is often incomplete (lack of controls and documentation of failures) (Wolters *et al.* 2005; Klimkowska *et al.* 2007). Nature managers should be encouraged to publish their results in cooperation with researchers and knowledge platforms as evidence-based conservation (Sutherland *et al.* 2004).

14.7   Effects of management and restoration practices

14.7.1   Haymaking

Effects of restoration management (Fig. 14.1) and small-scale environmental restoration in enclosed previously fertilized or abandoned grasslands, and arable fields, will be discussed, especially regarding different mowing regimes. Targets are a decrease in yield and the establishment of more species-rich target communities.

Cessation of fertilizer application and an annual haymaking regime reduced the yield in mesotrophic *Mesobromion erecti* grassland on calcareous soil (Willems 2001) and oligotrophic *Nardo–Galion saxatilis* grassland on sand (Bakker *et al.* 2002). Moreover, the proportion of species indicating high and low soil fertility decreased and increased, respectively. Annual haymaking reduced the yield, but after 25 years the standing crop was still about twice the level of the target community, i.e. $200–300\,\mathrm{g\text{-}dw\,m^{-2}}$. The removal of nitrogen was gradually balanced by the input through atmospheric deposition. Two annual cuts and removal of

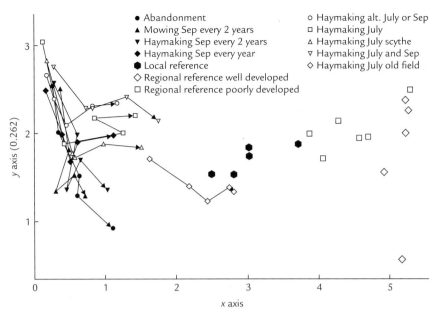

Fig. 14.4 Ordination by detrended correspondence analysis of all species with different management practices in the 'new' and 'old' field, the local reference and the poorly and well-developed regional references of *Nardo–Galion saxatilis* communities. (After Bakker *et al.* 2002.)

most nutrients, initially showed the highest species richness, but the species number decreased again after 15 years (Bakker *et al.* 2002). Annual haymaking regimes, involving removal of nutrients, brought the grassland closer to the target *Nardo–Galion* community than regimes where less nutrients were removed, such as mulching or haymaking every second year (Fig. 14.4). Still, the community composition resulting from the 'best' practice is far from a local reference site at *c.* 500 m (Bakker *et al.* 2002).

Attempts to restore a *Cirsio–Molinietum* wet fen meadow in an agriculturally improved pasture failed because of the very high yield of 1200 g-dw m^{-2} that did not decrease with annual haymaking. Target species from a nearby (500 m) reference community did not invade the meadow. Topsoil removal of 15–20 cm reduced yield by 50% and total soil phosphorus by 85%, and depleted plant P availability. Target species were planted as seedlings. Where the topsoil had not been removed, the vegetation became dominated by a few competitive species and although many of the planted target species were still present after 4 years, they were not abundant. Removal of the topsoil created suitable edaphic conditions for all planted target species to remain well established (Tallowin & Smith 2001).

The above-mentioned discrepancy between the effects of haymaking and mulching in restoration management was also found in grassland dominated by trivial species such as *Poa trivialis* (Oomes *et al.* 1996). A 25-year study on calcareous grassland revealed that the swards in the haymaking and mulching regimes resembled each other, but both regimes deviated from the control grazing regime (Kahmen *et al.* 2002). Similar conclusions were reached for an *Arrhenatherion elatioris* community in southern Germany (Moog *et al.* 2002). Apparently, cut but not removed biomass decomposed very fast with the high late-summer temperatures in southern Germany.

For an abandoned calcareous grassland overgrown by the tall perennial grass *Brachypodium pinnatum*, with a subsequent decrease of species richness, and for a mown grassland subject to atmospheric N deposition, haymaking in August, before the reallocation of nutrients to below-ground storage organs, turned out to be the right management practice to prevent *Brachypodium pinnatum* from becoming dominant (Bobbink & Willems 1991). For moist grasslands, the longer the period of fertilizer application, the more the soil seed bank is depleted (Bekker *et al.* 1997). The problem of the lack of diaspores in the soil seed bank can be overcome when fields are 'connected' by haymaking machinery, thus turning mowing machines into sowing machines (Strykstra *et al.* 1996).

14.7.2 Fire

Burning is a relatively cheap way to remove above-ground biomass and some nutrients. Initial results from calcareous grasslands in Germany seemed promising. However, after 25 years the grass *Brachypodium pinnatum* became dominant to the extent that the plant community resembled that of an abandoned field (Kahmen *et al.* 2002). Prescribed burning is used routinely in the management of upland heaths within the UK, where it is successfully applied in rotation. The aim is to remove the vegetation and allow the dominant species (*Calluna vulgaris*) to regenerate for the stem bases. Usually a 6- to 12-year rotation is followed (Gimingham 1992). It was found in North American tallgrass prairie that fire in the absence of grazers homogenizes the vegetation by uniformly removing the above-ground biomass and litter (Collins 1992).

14.7.3 Grazing

Grazing as a tool in restoration management and environmental restoration (Fig. 14.1) will be discussed for a communal semi-natural landscape with previously fertilized or abandoned grasslands and arable fields, especially for grazing regimes integrating different units. An area on sandy soil harboured 40 ha of open heathland with grasses and 20 ha of forest. Locally it was dominated by *Deschampsia flexuosa* and *Molinia caerulea* in 1983 when cattle grazing at a stocking rate of $0.2\,animal\cdot ha^{-1}$ started in the entire area, and some additional tree cutting was done in the heathland. Grazing did not reduce the cover of grasses, neither did it prevent grass invasion in the heathland. The soil contained a viable *Calluna vulgaris* seed bank. A positive correlation was found between the number of seedlings that emerged and seed density. Grazing by free-ranging cattle did not prevent encroachment by *Pinus sylvestris* and *Betula pubescens*. It also did not remove the high atmospheric nutrient input. Substantial amounts were redistributed from the grass lawns to the forest (Bokdam & Gleichman 2000). As a result of cattle grazing abandoned pastures on clay soil in Finland, the vegetation changed only slightly towards that of old pastures after 5 years (Pykälä 2003). Apparently restoration takes more time.

Livestock grazing is often practised on dry soils of previously fertilized pastures or arable fields to restore species-rich grasslands or heathland on oligotrophic soil. Changes from species indicating eutrophic to mesotrophic soil conditions are recorded, but succession does not proceed beyond stands dominated by *Holcus lanatus* and *Agrostis capillaris*. When wet or moist sites are included, the herbivores tend to avoid these sites and subsequently tall forb stands develop. Livestock grazing is also introduced to control tall-grass dominance in various ecosystems. Where extensive grazing occurs on areas with plant production exceeding herbivore use, vegetation compositional and structural patterns are produced at different scales (Bakker 1998) that cannot be mimicked by cutting. It was also found in North American tall-grass prairie that grazing by bison can generate small-scale heterogeneity (Veen *et al.* 2008).

14.7.4 Topsoil removal

Cutting and/or grazing do not always, or may but only very slowly, result in the targets set. Environmental restoration by topsoil removal may accelerate the process and render a closer approach of the target. To reduce the amount of nutrients from previously fertilized pastures and arable fields, the topsoil was removed to restore heathland and other plant communities characteristic of oligotrophic soil conditions. However, due to lack of money, the topsoil is not always removed from the site, but re-allocated within the site, thus creating depressions and mounds. This results in local topsoil removal and local accumulation of soil and nutrients with subsequent establishment of non-target species of eutrophic soil conditions. Where possible, previous pastures or arable fields are fenced in together with adjacent reference areas which still harbour the target communities; grazing livestock is then supposed to disperse propagules from the reference area into the target area. The majority of viable seeds found in dung includes species of eutrophic soil. Apparently the herbivores feed selectively on species of high forage quality (Mitlacher *et al.* 2002). The similarity between relevés of permanent plots and of reference relevés increased up to 30% for some communities (Verhagen *et al.* 2001). For the sites under study it was possible to collect data on the occurrence of target species in the surroundings of the sites derived from the Dutch 5×5 km grid data. Species occurring in the same grid cell as the study site or in the eight adjacent cells can be regarded as the local species pool (Zobel *et al.* 1998). Most of the target species were present in the local species pool, but very few target species appeared in most of the treated study sites. Part of these must have emerged from the soil seed bank as they are known for their longevity, such as the typical wet heathland species *Erica tetralix*, *Calluna vulgaris* and *Juncus squarrosus*. These species are still found in sites that have been reclaimed from heathland more than 70 years ago.

Bekker (2009) evaluated over 300 sites with topsoil removal in the Nether-lands. The abundances of plant species characteristic of nutrient-poor soil conditions were negatively related to the cover of the newly established vegeta-tion. The longer gaps are available in the sward after topsoil removal, the more chance these target species get to establish. The accuracy of topsoil removal also plays a role. The less nutrients (organic matter) that were left behind, the more successful the vegetation that developed. Yet, these sites need a vast amount of follow-up management to prevent tree species and common rush (*Juncus effusus*) from dominating the vegetation within a few years. Sites with $<200\,\mu mol\,P{\cdot}kg^{-1}$ and $<10\,\mu mol\,P{\cdot}kg^{-1}$ (P as P-Olsen) were identified as low-risk conditions for the spreading of *Juncus effusus*.

Suitable environments for rare species cannot always be provided by changing the hydrological conditions or by grazing. High NH_4 concentrations and high NH_4/NO_3 ratios can be reduced by turf cutting (Kleijn *et al.* 2008). However, this practice is only successful when combined with liming to increase the pH of the soil (Dorland *et al.* 2005).

14.7.5 Rewetting

Environmental restoration (Fig. 14.1) can also start with changing the hydrologi-cal conditions. An increase in the groundwater table by 30 cm in a field on peaty clay overlaying peat resulted in an increase of species of wet soils within 5 years. This was independent of the vegetation management including haymaking or mulching (Oomes *et al.* 1996).

Wet *Cirsio–Molinietum* meadows are annually cut for hay. In the Netherlands, these are all threatened by desiccation, acidification and eutrophication. Restora-tion in the high Pleistocene part of the country is feasible when the hydrological conditions are only slightly disturbed and dependent on local and regional groundwater systems, or when hydrological measures are carried out in combi-nation with sod cutting. Digging of shallow ditches may promote surface run-off of acid rainwater and upward seepage of base-rich groundwater. Restoration prospects in the low Holocene part of the country are small as they depend on very large-scale hydrological systems. Species that re-established seem to have emerged from a long-term persistent seed bank, whereas species with a short-term persistent or a transient seed bank were still locally present in the nature reserves under treatment (Jansen *et al.* 2000). Restoration of *Cirsio–Molinietum* by flooding during winter and spring, and additional sod cutting was not successful. Supposed limiting dispersal capacity of target species was encountered by introduction, but they did not survive after the second year of the experiment. It turned out that the flooding water was poor in base cations (van Duren *et al.* 1998).

Restoration of a cut-over bog on dried and shrunken *Eriophorum–Sphagnum* peat showed that after reflooding following blockage of surrounding ditches, ombrotrophic *Sphagnum* species failed. The very acid and nutrient-poor mire water reaching the surface had to be fertilized to promote any plant growth. After that, a floating mat was formed by minerotrophic *Sphagnum* species which eventually will develop a new bog. As a result of a large precipitation deficit in summer periods, enormous quantities of water have to be stored during winter periods by building high dams. The flooding water contains large amounts of nutrients, hence the mire resumes its functions as a nutrient sink. The vegetation includes very productive reeds and sedges. These stands may eventually transform into low-productive small-sedge communities, when a fen acrotelm is formed that can fix nutrients, thus offering Red List species a niche without the need for human intervention (Pfadenhauer & Grootjans 1999).

A review by Klimkowska *et al.* (2007) of over 90 restoration projects for wet meadow restoration, taking into account reference plant communities in various regions in Western Europe, revealed that rewetting alone had no measurable success. Topsoil removal was the key factor for success in restoration. Diaspore transfer was only successful when combined with topsoil removal. An overview of possibilities and constraints in the restoration of fen and wet meadow systems is given by Grootjans *et al.* (2002a) (Fig. 14.5).

In dune slacks, restoration is often successfully carried out by rewetting in combination with topsoil removal (Grootjans *et al.* 2002b). Restoration of salt marshes, i.e. building salt marshes by the interaction of vegetation and sedimentation, started recently in north-west Europe (Cooper *et al.* 2001). A review of

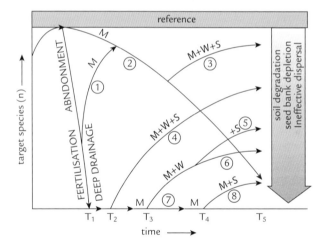

Fig. 14.5 Conceptual model of occurrence of target species (reference) in fen and wet meadows under restoration management. Continuation of the traditional management (mowing without fertilizer application) in meadow reserves cannot always prevent the extinction of many endangered species by negative influences from surrounding agricultural areas. Restoration measures (e.g. rewetting and sod cutting) are less effective after long-term exposure to these influences. Resuming a traditional management after a short period of abandonment is often successful with respect to re-establishment of target species. If restoration management is resumed shortly after cessation of agricultural exploitation, the restoration success is usually high compared to situations where long-term intensive fertilization has taken place. M, mowing with fertilizer applictaion; W, (re)wetting; S, sod cutting. (After Grootjans *et al.* 2002a.)

70 de-embanked sites in north-west Europe revealed that successful restoration of salt-marsh communities was positively related to elevation with respect to mean high tide, the input of sediment and the size of the site. Most sites younger than 20 years old contain more target species than older sites, especially when the latter are not grazed or mown (Wolters *et al.* 2005).

14.7.6   Re-introduction

To restore species-rich plant communities, low productivity levels are essential, but these cannot guarantee successful restoration (Berendse *et al.* 1992). Deliberately re-introducing disappeared species in the present fragmented landscape is an issue of restoration management that might be practised more in the future. Moreover, nowadays movements along the ecological infrastructure (i.e. machinery and herds of livestock; cf. Poschlod *et al.* 1996) in the former low-intensity farming system are lacking. Successful introduction experiments suggest that dispersal of propagules may indeed be a constraint. Diaspore transfer with plant material proved to be a very successful method in restoring species-rich flood meadows. After 4 years, 102 species had established in previously eutrophic flood meadows in Germany, among them many rare and endangered species. High-quality plant material and suitable site conditions with low competition (after topsoil removal) in early stages of the succession seem to be essential prerequisites (Hölzel & Otte 2003). However, other experiments reveal that common species do establish better than rare species (Tiikka *et al.* 2001). Failure to establish target plant species may be due to the lack of accompanying mycorrhizal fungal species (e.g. van der Heijden *et al.* 1998). Experiments including the introduction of soil from reference communities are practised.

In a review of species introduction projects in grasslands in central and north-western Europe, Kiehl *et al.* (2010) discuss the effects of seeding of site-specific seed mixtures, transfer of fresh seed-containing hay, vacuum harvesting of seeds and transfer of turves or seed-containing soil. In fact, seed limitation can be overcome successfully by most of these measures for species introduction. Sites with bare soil of former arable fields after tilling or topsoil removal, or raw substrate, for instance in mining areas, were most successful, whereas sites with species-poor grassland without soil disturbance and older arable fields with dense weed stands were less successful (Kiehl *et al.* 2010).

14.8   Constraints in management and restoration

It is clear from the studies discussed earlier that rewetting, haymaking or grazing alone will not result in restoring target communities on mesotrophic and oligotrophic soils. Because of atmospheric deposition and acidification, succession is unlikely to go beyond communities characteristic of mesotrophic soil. Because of a lack of long-term persistent seeds in the soil seed bank and poor dispersal in the present fragmented landscape, many rare and endangered Red List species are unlikely to establish. The combination of reference sites and areas to be restored within one fence in order to be grazed by livestock needs further study with respect to the role of herbivores in plant dispersal. Because of the preference for high-quality forage, dispersal may include transport of diaspores of non-target species. Only experiments including the introduction of target species, can reveal the causes of a standstill in an ongoing succession, be it abiotic conditions or lack of dispersal of propagules from elsewhere. Discussions on the genetic basis of introduced propagules can be overcome by introducing hay from nearby reference communities into the impoverished target area. Ecotypes become established that are adapted to local conditions.

Ozinga *et al.* (2009) studied species losses during the past century in Germany, the Netherlands and the UK, and tried to relate losses of species to their dispersal capacity. They found that dispersal traits make a large and significant contribution to explaining interspecific patterns of species losses, of the same order of magnitude as the effect of eutrophication. The results are consistent across the three countries. Species with a high potential for dispersal in the fur of large mammals or by running water are significantly more likely to decline than those using other dispersal vectors. Conversely, species with a high potential for dispersal by wind or birds are less likely to decline. The results also demonstrate that species with the ability to accumulate a persistent soil seed bank ('dispersal through time') perform relatively well (see also Chapter 6).

Herbivores can disperse seeds through their dung (Mouissie *et al.* 2005a) or in their fur (Mouissie *et al.* 2005b). Grazed sites often include a mosaic of eutrophic/mesotrophic sites to be restored and still-existing dry grassland and heathland communities on oligotrophic soils to allow dispersal. Unfortunately, herbivores may enrich the sites through deposition of dung, while the seeds they introduce in this way seem to originate mainly from species in the sites with the highest nutrient levels, which offer better forage quality for the herbivores (Mouissie *et al.* 2005a). Indeed, dispersal occurs, but 'in the wrong direction'.

Agricultural intensification has resulted in high nutrient levels in the soil. After the cessation of fertilizer application, the levels of nitrogen drop quickly. However, high phosphorus levels may be found even after removal of 50 cm of topsoil. Restoration and maintenance of soil phosphorus as the primary limiting nutrient is essential where there is a risk of nitrogen becoming non-limited through atmospheric input (Tallowin & Smith 2001). In this respect it is striking that very high numbers of species per 100 m^2 were found in 281 ancient meadows in five European countries when the P content did not exceed 5 mg·100 g^{-1} soil (Janssens *et al.* 1999).

Drainage has resulted in desiccation of many wetlands. For some plant communities, rewetting may be carried out at the scale of individual fields. However, fragmentation of landownership often causes these efforts to fail. In particular, the restoration of communities depending on deep seepage water requires that entire catchment areas be included so that their hydrological conditions are independent from those of neighbouring areas and can therefore be managed separately.

Taking into account the constraints in the restoration of ecological diversity (Bakker & Berendse 1999), it is not surprising that the results of restoration management may differ to some extent from the reference we have in mind

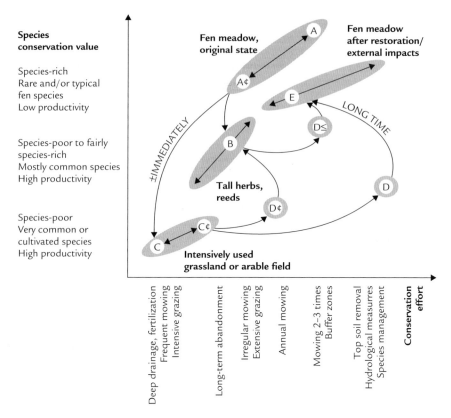

Fig. 14.6 Model relating the species conservation value of a fen meadow site to conservation effort. Each state A–D covers a certain range of effort and of value. This allows reversible changes in management. Beyond a certain threshold, sites can transform into a new state. Restoration to other states requires high conservation efforts and is only partially successful: the original species conservation value may no longer be reached, and the site to be 'restored' needs more efforts than previously for its maintenance. (After Güsewell 1997.)

(Fig. 14.6) (Güsewell 1997). It is clear that restoration is difficult, time consuming, expensive and perhaps not always possible in countries with intensive agricultural exploitation. Therefore, one of the best measures in nature conservation at a European level, is to use a regional approach, and maintain the small remnants of (near-) natural and semi-natural landscapes in densely populated parts of the continent. Remote regions with a low population density still present large areas of (near)natural landscape, which should be preserved. Finally, in regions with intermediate population densities, the still existing semi-natural landscapes with low-intensity farming systems should be maintained (Bignal & McCracken 1996).

14.9   Strategies in management and restoration

It will be clear from the above presentation that the best strategy in management and restoration depends on the intensity of agricultural practices in different countries. In Europe, this implies continuation of the still-existing low-intensity farming system in countries such as Greece, Spain, parts of France and the UK (Bignal & McCracken 1996) or eastern European countries (Veen *et al.* 2009). Here, agro-environment schemes can be extremely successful. In other European countries with intensive agriculture, ecosystems with nature conservation interest are often nature reserves. Here the management in charge of restoration has more or less taken over the former farming practices, and agro-environment schemes perform very poorly.

Sites with nature conservation interest are protected within the Natura 2000 framework (www.natura2000.org). This is a European network of protected reserves in the 27 European Union member states. It aims to protect habitats and individual species. The aim is to stop and restore the losses of plant and animal species by 2020. The Natura 2000 network includes reserves already protected within the EU Bird Directive of 1979 (http://ec.europa.eu/environment/nature/legislation/birdsdirective/index_en.htm) and the EU Habitats Directive of 1992 (http://ec.europa.eu/environment/nature/legislation/habitatsdirective/index_en.htm). Ultimately, the network will consist of 26000 protected areas covering over 850000 km², i.e. about 18% of the land of the countries involved. Data on species, habitats and sites compiled in the framework of Natura 2000 are collected in EUNIS (2012), the European Nature Information System. This is a comprehensive pan-European system to facilitate the harmonized description and collection of data across Europe through the use of criteria for habitat identification; it covers all types of habitats from natural to artificial, from terrestrial to freshwater and marine.

Nature reserves should be large, and connected to each other by corridors. In the Netherlands, this is carried out in the framework of the National Ecological Network. The importance of corridors connecting highly fragmented old-fields is also advocated in other parts of the world, for example Australia (Standish *et al.* 2007). In fact this is considered a general issue for restoration, in a broader landscape perspective and incorporating connectivity as a key characteristic to be maintained and restored. This also with a view to maintain or improve the potential for species movement in response to future climate changes (Harris *et al.* 2006).

A final point of discussion concerns the targets to be set in relation to existing or desired gradients of agricultural intensity within a country, and to take into account the overall differences in agricultural intensity between countries. The highest levels of fertilizer application may result in botanically poor grasslands, but these can cope with large amounts of winter and spring staging geese, for example in the Netherlands. Common meadow birds can cope with still relatively high levels of fertilizer application, and conventional agricultural practice can be combined with these nature conservation targets. On the other hand, like plants, more '*critical*' meadow birds such as ruff (*Philomachus pugnax*), redshank (*Tringa totanus*) and blacktailed godwit (*Limosa limosa*) can only cope with low levels of fertilizer application and little drainage (van Wieren & Bakker 1998). If farmers are willing to maintain grasslands with suboptimal production to achieve nature conservation targets, they need to be economically compensated. The effectiveness of these costly agro-environment schemes is under discussion (Kleijn *et al.* 2001). Botanical conservation interests are met to some extent in field margins along ditches outside the area of agricultural exploitation. It should be realized that the role of farmers for many nature conservation goals is very minor in countries with an overall high agricultural production level. Nevertheless they can be important with respect to the scenery, i.e. the open countryside in densely populated countries. In other countries, low-intensity farming is very important for many nature conservation goals.

For restoration purposes, it is important to know the spatial scales at which different groups of organisms operate or are being affected by environmental conditions. This knowledge makes it clear which constraints exist within a particular size of nature reserve. Cutting regimes act at the field scale. Large herbivores act at the field scale only when used as substitute for haymaking machinery. In small reserves, livestock are brought in with the aim of managing species of plants and animals that depend on a short sward (entomofauna, small mammals, birds). Here, the management of the populations of large herbivores is not important: they are covered by veterinary laws and need supplementary food in poor seasons. When management includes the restoration of viable populations of large herbivores, areas of 100–1000 km^2 are necessary (Table 14.2). In such large nature reserves in the Netherlands, mosaics of grassland, heathland, scrub and forest often occur. The control of numbers of large herbivores can be (i) 'bottom-up', including starvation in case of populations that are too large, or provision of supplementary food, or (ii) 'top-down', by predators such as wolves or culling by the management. Both solutions are heavily debated in populated cultural landscapes.

From the above, the idea emerges that very large areas, beyond the usual scale of cultural and semi-natural landscapes, would be necessary for the maintenance of viable populations of many plant and animal species, which contribute to a high biodiversity. What does this mean for the type of communities and their management? Implications for conservation and restoration can be derived from the pre-agricultural Holocene landscape. Closed forest, forest glades, pasture-woodland, meadows, grassland, scrub and heathland as well as their associated organisms would have a significant presence in north-west Europe under present natural conditions. Therefore, all these natural habitats should be considered of high conservation interest (Svenning 2002). Large herbivores and natural fires would be key agents creating and maintaining this diversity of habitats. Notably, open spaces within forests are important habitats in terms of diversity for many groups of woodland-associated organisms. Free-ranging large herbivores would also be important as dispersers of many herbaceous plants. Furthermore, large herbivores and fires would also provide the special microhabitats needed by a range of dung- and fire-dependent species, many of which have become rare or locally extinct in north-west Europe. Consequently it must be a conservation

Table 14.2 Relevant spatial dimensions important for the occurrence of various groups of organisms (biodiversity) and the spatial scale at which grazing and haymaking are effective (management).

	Dimensions					
	dm^2	m^2	ha	km^2	10^2–10^3 km^2	Continent
Management						
Grazing	Bite	Feeding station	Feeding site	Home range for wild ungulates	Viable population	
Cutting			Field size			
Biodiversity						
Plants	Germination, seed dispersal	Clonal growth, shoot growth	Viable population depending on rainwater	Viable population depending on seepage water		
Butterflies	Oviposition	Larval microhabitat	Local population	Viable (meta)-population		
Birds		Individual breeding territory	Viable population, seasonal range of migratory birds			Annual range of migratory birds

After Poschlod & WallisDeVries (2002).

priority to re-establish native large herbivores and natural fire regimes wherever possible, and mimic their effects by management such as grazing by domestic herbivores and the use of prescribed burning where this is not feasible. Finally, it is important to note that the most widespread natural vegetation type would be closed old-growth forest and that scattered old trees and dead wood probably would also be present in the more open vegetation. Many organisms dependent on old-growth forest, old trees, or dead wood have not persisted in the semi-natural and the cultural landscape. Thus, ancient forests should be protected and the presence of old trees and dead wood promoted in most habitats (Svenning 2002).

In other parts of the world, less emphasis is placed on semi-natural landscapes. More and more old-fields are abandoned in North and Central America and Australia (Cramer *et al.* 2008). They have been cleared from forest. The longer ago they were abandoned, and the more intensively they were exploited, the less successful the restoration efforts have been. Soil alterations, depletion of the soil seed bank, seed dispersal limitation and weed invasion are likely after intensive agriculture. This implies a huge and expensive effort to restore historical communities. Such old-fields remain in a persistent degraded state. Hence, restoration in the form of spontaneous succession may allows old-fields to contribute towards broader conservation goals and provide ecosystem services such as productivity or against erosion (Cramer *et al.* 2008). This might be a cost-effective alternative for increasing the natural value of a disturbed site (Prach & Hobbs 2008).

References

Bakker, J.P. (1998) The impact of grazing on plant communities. In: *Grazing and Conservation Management* (eds M.F. WallisDeVries, J.P. Bakker & S.E. van Wieren), pp. 137–184. Kluwer Academic Publishers, Dordrecht.

Bakker, J.P. & Berendse, F. (1999) Constraints in the restoration of ecological diversity in grassland and heathland communities. *Trends in Ecology & Evolution* 14, 63–68.

Bakker, J.P. & Londo, G. (1998) Grazing for conservation management in historical perspective. In: *Grazing and Conservation Management* (eds M.F. WallisDeVries, J.P. Bakker & S.E. van Wieren), pp. 21–54. Kluwer Academic Publishers, Dordrecht.

Bakker, J.P., Olff, H., Willems, J.H. & Zobel, M. (1996) Why do we need permanent plots in the study of long-term vegetation dynamics? *Journal of Vegetation Science* 7, 147–156.

Bakker, J.P., Grootjans, A.P., Hermy, M. & Poschlod, P. (2000) How to define targets for ecological restoration? *Applied Vegetation Science* 3, 3–6.

Bakker, J.P., Elzinga, J.A. & de Vries, Y. (2002) Effects of long-term cutting in a grassland system: possibilities for restoration of plant communities on nutrient-poor soils. *Applied Vegetation Science* 5, 107–120.

Bal, D. & Hoogeveen, Y. (eds) (1995) *Handboek Natuurdoeltypen in Nederland*. Report 11 IKC-Natuurbeheer, Wageningen. [In Dutch.]

Bekker, R.M. (2009). 20 jaar ontgronden voor natuur op zandgronden. *De Levende Natuur* 110, 9–15 [In Dutch with English summary.]

Bekker, R.M., Verweij, G.L., Smith, R.E.N. *et al.* (1997) Soil seed banks in European grasslands: does land use affect regeneration perspectives? *Journal of Applied Ecology* 34, 1293–1310.

Bekker, R.M., Bakker, J.P., Grandin, U., Kalamees, R., Milberg, P., Poschlod, P., Thompson, K. & Willems, J.H. (1998) Seed shape and vertical distribution in the soil: indicators for seed longevity. *Functional Ecology* 12, 834–842.

Bekker, R.M., Bruelheide, H. & Woods, K. (eds) (2007) Long-term datasets: from descriptive to predictive data using ecoinformatics. *Journal of Vegetation Science* 18, 457–570. Special Feature.

Berendse, F., Oomes, M.J.M., Altena, H.J. & Elberse, W.T. (1992) Experiments on the restoration of species-rich meadows in the Netherlands. *Biological Conservation* 62, 59–65.

Bignal, E.M. & McCracken, D.I. (1996) Low-intensity farming systems in the conservation of the countryside. *Journal of Applied Ecology* 33, 413–424.

Birks, H.J.B. (2005) Mind the gap: how open were European primeval forests? *Trends in Ecology & Evolution* 20, 154–156.

Bobbink, R. & Willems, J.H. (1991) Impact of different cutting regimes on the performance of *Brachypodium pinnatum* in Dutch chalk grassland. *Biological Conservation* 56, 1–21.

Bobbink, R., Hornung, M. & Roelofs, J.G.M. (1998) The effects of air-borne nitrogen pollutants on species diversity and semi-natural European vegetation. *Journal of Ecology* 86, 717–738.

Bokdam, J. & Gleichman, J.M. (2000) Effects of grazing by free-ranging cattle on vegetation dynamics in a continental north-west European heathland. *Journal of Applied Ecology* 37, 415–431.

Bos, D., Bakker, J.P., de Vries, Y. & van Lieshout, S. (2002) Vegetation changes in experimentally grazed and ungrazed back-barrier marshes in the Wadden Sea over a 25-year period. *Applied Vegetation Science* 5, 45–54.

Bowman, D.M.J.S., Balch, J.K., Artaxo, P. *et al.* (2009) Fire in the earth system. *Science* 324, 481–484.

Collins, S.L. (1992) Fire frequency and community heterogeneity in tallgrass prairie vegetation. *Ecology* 73, 2001–2006.

Cooper, N.J., Cooper, T. & Burd, F. (2001) 25 years of salt marsh erosion in Essex: implementatoin for coastal defence and nature conservation. *Journal of Coastal Conservation* 7, 31–40.

Craig, J., Anderson, S., Clout, M. *et al.* (2000). Conservation issues in New Zealand. *Annual Review of Ecology and Systematics* 31, 61–78.

Cramer V.A. & Hobbs, R.J. (eds) (2007) *Old Fields – Dynamics and Restoration of Abandoned Farmland*. Island Press, Washington, DC.

Cramer, V.A., Hobbs, R.J. & Standish, R.J. (2008). What's new about old fields? Land abandonment and ecosystem assembly. *Trends in Ecology & Evolution* 23, 104–112.

Dorland, E., Hart, M.A.C., Vermeer, M.L. & Bobbink, R. (2005) Assessing the success of wet heath restoration by combined sod cutting and liming. *Applied Vegetation Science* 8, 209–218.

EUNIS (2012) Habitat Classification of the European Environment Agency, Copenhagen. http://eunis.eea.europa.eu/habitats.jsp (accessed 25 May 2012).

Foley, J.A., DeFries, R., Asner, G.P. *et al.* (2005) Global consequences of land use. *Science* 309, 570–574.

Foster, D., Swanson, F., Aber, J. *et al.* (2003) The importance of land-use legacies to ecology and conservation. *Bioscience* 53, 77–88.

Gimingham, C.H. (1992). *The Lowland Heathland Management Book*. English Nature, Peterborough.

Grootjans, A.P., van Wirdum, G., Kemmers, R. & van Diggelen, R. (1996) Ecohydrology in The Netherlands: principles of an application-driven interdiscipline. *Acta Botanica Neerlandica* 45, 491–516.

Grootjans, A.P., Bakker, J.P., Janse, A.J.M. & Kemmers, R.H. (2002a) Restoration of brook valley meadows in the Netherlands. *Hydrobiologia* 478, 149–170.

Grootjans, A.P., Geelen, H.W.T., Jansen, A.J.M. & Lammerts, E.J. (2002b) Restoration of coastal dune slacks in the Netherlands. *Hydrobiologia* 478, 181–203.

Güsewell, S. (1997) Evaluation and management of fen meadows invaded by *Phragmites australis*. PhD Thesis, Federal Swiss Institute of Technology, Zürich.

Harris, J.A., Hobbs, R.J., Higgs, E. & Aronson, J. (2006). Ecological restoration and global climate change. *Restoration Ecology* 14, 170–176.

Hennekens, S.M. & Schaminée, J.H.J. 2001. TURBOVEG, a comprehensive data base management system for vegetation data. *Journal of Vegetation Science* 12, 589–591.

Hodder, K.H., Bullock, J.M., Buckland, P.C. & Kirby, K.J. (2005) *Large Herbivores in the Wildwood and Modern Naturalistic Grazing Systems*. English Nature Research Reports No. 648. English Nature, Peterborough.

Hölzel, N. & Otte, A. (2003) Restoration of a species-rich flood meadow by topsoil removal and diaspore transfer with plant material. *Applied Vegetation Science* 6, 131–140.

Huisman, J., Olff, H. & Fresco, L.F.M. (1993) A hierarchical set of models for species response analysis. *Journal of Vegetation Science* 4, 37–46.

Jansen, A.J.M., Grootjans, A.P. & Jalink, M.H. (2000) Hydrology of Dutch Cirsio-Molinietum meadows: Prospects for restoration. *Applied Vegetation Science* 3, 51–64.

Janssens, F., Peeters, A.A., Tallowin, J.R.B., Bakker, J.P., Bekker, R.M., Fillat, F. & Oomes, M.J.M. (1999) Relationship between soil chemical factors and grassland diversity. *Plant and Soil* 202, 279–298.

Jefferies, R.L. & Rockwell, R.F. (2002) Foraging geese, vegetation loss and soil degradation in an Arctic salt marsh. *Applied Vegetation Science* 5, 7–16.

Kahmen. S., Poschlod, P. & Schreiber, K.F. (2002) Management practice of calcareous grasslands. Changes in plant species composition and the response of plant functional traits during 24 years. *Biological Conservation* 104, 319–328.

Kiehl, K., Kirmer, A., Donath, T., Rasran, L. & Hölzel, N. (2010) Species introduction in restoration projects –evaluation of different techniques for the establishment of semi-natural grasslands in Central and Northwestern Europe. *Basic and Applied Ecology* 11, 285–299.

Kleijn, D., Berendse, F., Smit, R. & Gillissen, N. (2001) Agri-environment schemes do not effectively protect biodiversity in Dutch agricultural landscapes. *Nature* 413, 723–725.

Kleijn, D., Bekker, R.M., Bobbink, R., De Graaf, M.C.C. & Roelofs, J.G.M. (2008) In search for key geochemical factors affecting plant species persistence in heathland and acidic grasslands: a comparison of common and rare species. *Journal of Applied Ecology* 45, 680–687.

Kleyer, M., Bekker, R.M., Knevel, I.C. *et al.* (2008) The LEDA Traitbase: a database of life-history traits of the Northwest European flora. *Journal of Ecology* 96, 1266–1274.

Klimkowska, A., van Diggelen, R., Bakker, J.P. & Grootjans, A.P. 2007) Wet meadow restoration in Western Europe: a quantitative assessment of the effectiveness of several techniques. *Biological Conservation* 140, 318–328.

Klötzli, F. & Grootjans, A.P. (2001) Restoration of natural and semi-natural wetland systems in Central Europe: progress and predictability of developments. *Restoration Ecology* 9, 209–219.

Lindenmayer, D.B., Bennett, A.F. & Hobbs, R.J. (eds) (2010). *Temperate Woodland Conservation and Management*. CSIRO Publishing, Melbourne.

Londo, G. (1997) *Natuurontwikkeling*. Backhuys Publishers, Leiden. [In Dutch.]

Marris, E. (2009) Reflecting the past. *Nature* 462, 30–32.

Mitlacher, K., Poschlod, P., Rosén, E. & Bakker, J.P. (2002) Restoration of wooded meadows – comparative analysis along a chronosequence on Öland (Sweden). *Applied Vegetation Science* 5, 63–74.

Moog, D., Poschlod, P., Kahmen, S. & Schreiber, K.F. (2002) Comparison of species composition between different grassland management treatments after 25 years. *Applied Vegetation Science* 5, 99–106.

Mortimer, S.R., van der Putten, W.H. & Brown, V.K. (1999) Insect and nematode herbivory below ground: interactions and role in vegetation succession. In: *Herbivores: Between Plants and Predators* (eds H. Olff, V.K. Brown & R.H. Drent), pp. 205–238. Blackwell Science, Oxford.

Mouissie, A.M., van der Veen, C.E.J., Veen, G.F. & van Diggelen, R. (2005a) Ecological correlates of seed survival after ingestion by Fallow Deer. *Functional Ecology* 19, 284–290.

Mouissie, A.M., Lengkeek, W. & van Diggelen, R. (2005b) Estimating adhesive seed-dispersal distances: field experiments and correlated random walks. *Functional Ecology* 19, 478–486.

Muchiru, A.N., Western, D. & Reid, R.S. (2009) The impact of abandoned pastoral settlements on plant and nutrient succession in an African savanna ecosystem. *Journal of Arid Environments* 73, 322–331.

Olff, H., de Leeuw, J., Bakker, J.P. *et al.* (1997) Vegetation succession and herbivory in salt marsh: changes induced by sea level rise and silt deposition along an elevational gradient. *Journal of Ecology* 85, 799–814.

Olff, H. & Hopcraft, G.C. (2008) The resource basis of human–wildlife interaction. In: *Serengeti III: The Future of an Ecosystem*. (eds A.R.E. Sinclair, C. Packer, S.A.R. Mduma & J.M. Fryxell), pp. 95–133. University of Chicago Press, Chicago.

Olff, H., Vera, F.M., Bokdam, J. *et al.* (1999) Shifting mosaics in grazed woodlands driven by alternation of plant facilitation and competition. *Plant Biology* 1, 127–137.

Oomes, M.J.M., Olff, H. & Altena, H.J. (1996) Effects of vegetation management and raising the water table on nutrient dynamics and vegetation change in wet grassland. *Journal of Applied Ecology* 33, 576–588.

Ozinga, W.A., Römermann, C., Bekker, R.M. *et al.* (2009) Dispersal failure contributes to plant losses in NW Europe. *Ecology Letters* 11, 66–74.

Pfadenhauer, J. & Grootjans, A.P. (1999) Wetland restoration in Central Europe: aims and methods. *Applied Vegetation Science* 2, 95–106.

Pickett, S.T.A., Ostfeld, R.S., Shachak, M. & Likens, G.E. (1997) *The Ecological Basis of Conservation – Heterogeneity, Ecosystems, and Biodiversity*. Chapman and Hall, New York, NY.

Poschlod, P., Bakker, J.P., Bonn, S. & Fischer, S. (1996) Dispersal of plants in fragmented landscapes. In: *Species Survival in Fragmented Landscapes* (eds J. Settele, C. Margules, P. Poschlod & K. Henle), pp. 123–127. Kluwer Academic Publishers, Dordrecht.

Poschlod, P. & WallisDeVries, M.F. (2002) The historical and socioeconomic perspective of calcareous grasslands – lessons from the distant and recent past. *Biological Conservation* 104, 361–376.

Prach, K. & Hobbs, R.J. (2008). Spontaneous succession versus technical reclamation in the restoration of disturbed sites. *Restoration Ecology* 16, 363–366.

Pykälä, J. (2003) Efects of restoration with cattle grazing on plant species compostition and richness of semi-natural grasslands. *Biodiversity and Conservation* 12, 2211–2226.

Schaminée, J.H.J. & Jansen, A. (eds) (1998) *Wegen naar natuurdoeltypen*. Report 26 IKC-Natuurbeheer, Wageningen. [In Dutch.]

Schaminée, J.H.J., van Kley, J.E. & Ozinga, W.A. (2002) The analysis of long-term changes in plant communities: case studies from the Netherlands. *Phytocoenologia* 32, 317–335.

Silvertown, J, Tallowin, J., Stevens, C. *et al.* (2010) Environmental myopia: a diagnosis and a remedy. *Trends in Ecology end Evolution* 25, 556–561.

Society for Ecological Restoration Science and Policy Working Group (2004) *The SER Primer on Ecological Restoration*, version 2. www.ser.org/content/ecological_restoration_primer.asp (accessed 25 May 2012).

Spellerberg. I.F., Goldsmith, F.B. & Morris, M.G. (eds) (1991) *The Scientific Management of Temperate Communities for Conservation*. Blackwell Scientific Publications, Oxford.

Standish, R.J., Cramer, V.A., Wild, S.L. & Hobbs, R.J. (2007). Seed dispersal and recruitment limitation are barriers to native recolonization of old-fields in western Australia. *Journal of Applied Ecology* 44, 435–445.

Strykstra, R.J. (2000) Reintroduction of plant species: s(h)ifting settings. PhD Thesis, University of Groningen, Groningen.

Strykstra, R.J., Verweij, G.L. & Bakker, J.P. (1996) Seed dispersal by mowing machinery in a Dutch brook valley system. *Acta Botanica Neerlandica* 46, 387–401.

Sutherland, W.A., Pullin, A.S., Dolman, P.M. & Knight, T.M. (2004) The need for evidence-based conservation. *Trends in Ecology & Evolution* 19, 305–308.

Svenning, J.C. (2002) A review of natural vegetation openness in north-western Europe. *Biological Conservation* 104, 133–148.

Szabo, P. (2009) Open woodland in Europe in the mesolithic and in the Middle Ages: can there be a connection? *Forest Ecology and Management* 257, 2327–2330.

Tallowin, J.R.B. & Smith, R.E.N. (2001) Restoration of a Cirsio-Molinietum fen meadow on an agriculturally improved pasture. *Restoration Ecology* 9, 167–178.

Thompson, K., Bakker, J.P. & Bekker, R.M. (1997) *The Soil Seed Banks of North West Europe: Methodology, Density and Longevity*. Cambridge University Press, Cambridge.

Tiikka, P.M., Heikkilä, T., Heiskanen, M. & Kuitunen, M. (2001) The role of competition and rarity in the restoration of a dry grassland in Finland. *Applied Vegetation Science* 4, 139–146.

van Andel, J. & Aronson, J. (eds) (2012) *Restoration Ecology – The New Frontier*, 2nd edn. Blackwell Publishing, Oxford.

van der Heijden, M.G.A., Boller, T., Wiemken, A. & Sanders, I.S. (1998) Different arbuscular mycorrhizal fungal species are potential determinants of plant community structure. *Ecology* 79, 2082–2091.

van der Maarel, E. (1978) Experimental succession research in a coastal dune grassland, a preliminary report. *Vegetatio* 38, 21–28.

van der Maarel, E. (1980) Towards an ecological theory of nature management. *Verhandlungen der Gesellschaft für Ökologie* 8, 13–24.

van Duren, I.C., Strykstra, R.J., Grootjans, A.P., ter Heerdt, G.N.J. & Pegtel, D.M. (1998) A multidisciplinary evaluation of restoration measures in a degraded fen meadow (Cirsio-Molinietum). *Applied Vegetation Science* 1, 115–130.

van Wieren, S.E. & Bakker, J.P. (1998) Grazing for conservation in the twenty-first century. In: *Grazing and Conservation Management* (eds M.F. WallisDeVries, J.P. Bakker & S.E. van Wieren), pp. 349–363. Kluwer Academic Publishers, Dordrecht.

Veen, G.F., Blair, J.M., Smith, M.D. & Collins, S.L. (2008) Influence of grazing and fire frequency on small-scale plant community structure and resource variability in native tallgrass prairie. *Oikos* 117, 859–866.

Veen, P., Jefferson, R., de Smidt, J.T. & van der Straaten, J. (2009) *Grasslands in Europe of High Nature Value*. KNNV Publishing, Zeist.

Vera, F.W.M. (2000) *Grazing Ecology and Forest History*. CABI International, New York.

Verhagen, R., Klooker, J., Bakker, J.P. & van Diggelen, R. (2001) Restoration success of low-production plant communities on former agricultural soils after top-soil removal. *Applied Vegetation Science* 4, 75–82.

WallisDeVries, M.F. (2002). Options for the conservation of wet grasslands in relation to spatial scale and habitat quality. In: *Multifunctional Grasslands, Quality Forages, Animal Products and Landscapes* (eds J.L. Durand, J.C. Emile, C. Huyghe & G. Lemaire), pp. 883–892. European Grassland Federation, La Rochelle.

Wedin, D.A. (1992) Biodiversity conservation in Europe and North America: grasslands, a common challenge. *Restoration and Management Notes* 10, 137–143.

Wells, T.C.E. (1980) Management options for lowland grassland. In: *Amenity Grasland, an Ecological Perspective* (eds I.H. Rorison & R. Hunt), pp. 175–195. John Wiley & Sons, Ltd, Chichester.

Westhoff, V. (1983) Man's attitude towards vegetation. In: *Man's Impact on Vegetation* (eds W. Holzner, M.J.A. Werger & I. Ikusima), pp. 7–24. Junk, The Hague.

Wheeler, B.D., Shaw, S.C., Fojt, W.J. & Robertson, R.A. (eds) (1995) *Restoration of Temperate Wetlands*. John Wiley & Sons, Ltd, Chichester.

Willems, J.H. (2001) Problems, approaches, and results in restoration of Dutch calcareous grassland during the last 30 years. *Restoration Ecology* 9, 147–154.

Wolters, M., Garbutt, A. & Bakker, J.P. (2005) Salt-marsh restoration: evaluating the success of de-embankments in north-west Europe. *Biological Conservation* 123, 249–268.

Zobel, M., van der Maarel, E. & Dupré, C. (1998) Species pool: the concept, its determination and significance for community restoration. *Applied Vegetation Science* 1, 55–66.

15

Vegetation Types and Their Broad-scale Distribution

Elgene O. Box[1] and Kazue Fujiwara[2]
[1]University of Georgia, USA
[2]Yokohama City University, Japan

15.1 Introduction: vegetation and plant community

Vegetation is the aggregate of all the plants growing in an area and, as such, is the most conspicuous feature of most landscapes, already recognized by the early Greeks as a way of distinguishing one region from another. Vegetation involves populations of species of the local flora, which in turn involve different genetic, migration, historical or ecological elements. Vegetation is also composed of various plant forms and ecological plant types, reflecting various sizes, shapes and combinations. Finally, vegetation is shaped by physical and other environmental influences, including climate, substrate, soil microbes and disturbance regimes, which affect plant physiology. Plants and vegetation tend to integrate these environmental effects to produce vegetation structures adapted to and reflecting environmental conditions. In treating vegetation types and their broad-scale distribution, it is necessary to look at four main questions: what do we mean by vegetation types, what are the main ones at broad scale, how are they distributed geographically, and why? This last question may suggest how they may change in the future.

The first concepts of vegetation types developed from the great botanical voyages of the 1800s, especially by Alexander von Humboldt, sometimes called the father of plant geography. World vegetation types and regions were depicted on early world maps but with poorly known regions often represented by environmental surrogates, such as 'tropical forest' (see review and references in de Laubenfels 1975). Where the vegetation had been seen, sampled and described adequately, units were defined mainly by **physiognomy**, i.e. general appearance of the vegetation, first with simple distinctions between forest, grassland and desert.

The most recognizable large-area vegetation units correspond to major regional ecosystems, such as the Amazon rainforest, Great Plains grassland,

Vegetation Ecology, Second Edition. Eddy van der Maarel and Janet Franklin.
© 2013 John Wiley & Sons, Ltd. Published 2013 by John Wiley & Sons, Ltd.

Mediterranean maquis, or Siberian larch forest. These were called **formations** by Grisebach and are recognized by a relatively uniform physiognomy constituted by plants from the regional flora (see review and references in Grabherr & Kojima 1993). Similar formations occur in corresponding locations on different continents and are called vegetation **formation types**, such as tropical rainforest or boreal coniferous forest. More recently, the term **biome** has come to be used for a regional ecosystem roughly corresponding to a vegetation formation but understood also to consider the regional fauna.

The early mappers also recognized that natural vegetation regions are most closely related to climatic conditions, and regions were differentiated by obvious features such as wet versus dry climates or permanently warm versus cold-winter conditions. The occurrence or absence of frost was recognized early as perhaps the most important factor separating the lowland tropical and extra-tropical regions, but the upward decrease of temperatures in mountains complicates the zonation. Landform, soil types and other environmental factors were sometimes recognized, and early maps generally recognized mountain and lowland areas as distinct.

Vegetation types are often recognized as loosely equivalent to plant communities, which have been described, quite graphically, as what one would see as different parts of the pattern when looking out over a landscape (Kent & Coker 1992). A **community** is a group of species (populations) living together and includes animals as well as plants. Plant communities differ visibly by general physiognomy, related mainly to the different growth-forms of the plant species involved. If a community recurs with a consistent species composition, it may be recognized as an **association**.

A long-standing question in ecology has asked whether communities represent merely the intersection of overlapping ranges of species with similar requirements (individualistic concept) or more functionally integrated units (organismal concept, Clements 1916) (see Chapter 1). If vegetation does constitute an organized whole, it operates at a higher level of integration than the individual species and may possess emergent properties not found at the species level, such as symbiosis and interspecific competition (see Chapter 7). As such, vegetation provides not only the physical structure but also the functional framework of ecosystems (Chapter 10).

Plant geography and plant ecology were originally one field but began to diverge around 1900, with plant ecology focusing more on process and plant and vegetation geography proceeding mainly floristically and historically. These two main perspectives remain somewhat distinct: a historical-floristic perspective concerned with migration, dispersal and the historical development of regional floras; and an environmental perspective concerned with environmental constraints and ecological relations influencing distributions.

15.2   Form and function, in plants and vegetation

Plant and vegetation types are recognized by their form, but this is intimately related to how plants function. Photosynthesis increases roughly in proportion to green surface area, which increases in larger plants. Plants also, however, lose water in rough proportion to size and leaf-surface area, since water is lost when leaf pores are open to take in CO_2. As a result, large plants and dense vegetation such as forest are limited to climates (or landscapes) that provide sufficient water. In drier climates plants become smaller and vegetation becomes more contracted and sparse, unless the more extensive root systems of larger plants can meet the demand for more water. Water uptake, photosynthesis and other plant processes are slowed at lower temperatures, and plant tissues may be damaged or completely killed by internal ice formation. Plants can modify their vulnerability to drought, frost and other environmental stresses through form adaptations such as smaller or harder leaves, deeper root systems and the loss of leaves and other exposed surface areas during unfavourable periods. Broad-scale vegetation types, such as temperate deciduous forest, can thus be characterized by their **phenophysiognomy**, which combines structure and its seasonal changes. This follows, of course, from the form and seasonality of the larger plants. Form–function adaptations, however, involve functional trade-offs. For example, the 'softer' deciduous leaves that usually have higher photosynthetic rates do lose water faster, lose the first part of the growing season while they are growing out, and may require more energy and nutrients for their construction than do longer-living evergreen leaves.

One of the most basic environmental limitations on plants and vegetation involves limiting cold temperatures called **cardinal temperatures**, summarized in Table 15.1. Although some tropical plants cannot survive 'cold' temperatures well above freezing, most plants are not damaged until frost occurs. Compared with animals, plant tissues have relatively large 'empty' spaces between and within cells. As a result, most temperate-zone plants, even evergreens, can tolerate some ice formation in their intercellular fluids. Higher plants, however, do not survive significant ice formation in the fluids *inside* cells, because this results in mechanical damage to the cells themselves. Ice begins to form inside the cells of even relatively thick evergreen leaves at about $-15\,°C$, so this cardinal temperature, even over short durations, represents a limit for broad-leaved evergreen woody plants. (Lower-growing herbaceous evergreens may survive somewhat lower temperatures, since they are near the ground and perhaps insulated by fallen litter.) Deciduous trees and shrubs tolerate much colder conditions, but eventually ice may form inside the cells of their woody branches and

Table 15.1 Main cardinal temperature limits for plant types.

Upper limits:	40–45°C	most species
Lower limits:	5°C	many tropical species; also most un-reinforced malacophylls
	−2°C	most subtropical species (foliage)
	−15°C	temperate evergreen broad-leaved species (foliage)
	−40°C	ring-porous broad-leaved trees, etc. (woody parts)

Cardinal temperatures represent limits for potential damage to cells and tissues (e.g. freezing of intra-cellular fluids) as well as limitations to plant metabolism (e.g. collapse of the respiratory mechanism at high temperatures). Limitation may result from single events, such as extreme overnight low temperature, for which local absolute minimum temperature is often a useful index. The values shown were first suggested by Sakai (1971), Larcher (1973), George *et al.* (1974) and others (cf. Woodward 1987).

trunks, depending on wood structure. Trees with ring-porous wood, i.e. with most of the active water-conducting structure concentrated in the outer wood from the most recent growing seasons, are vulnerable to lethal ice formation beginning at around −40 °C (George *et al.* 1974). On the other hand, trees with diffuse-porous wood, such as birches and other 'boreal' deciduous trees, can tolerate even lower temperatures, as can boreal conifers, due to their quite different wood structure (see summary by Archibold 1995).

Leaf shape can have subtle effects on a plant's energy and water budgets. Compound leaves with smaller leaflets, for example, can be deployed rapidly and may provide greater ventilation, reducing overheating. Compact shapes, such as needles and scales, have higher volume-to-area ratios and tend to have 'harder' surfaces, providing greater control over water loss. Linear leaves, including the flat 'needles' of *Abies* and some other conifers, may be less compact but still restrict water loss more than broader leaves, especially in warm climates, and may permit higher photosynthetic rates than round needles. More significant for plants, and ultimately for vegetation dynamics, is the **consistency** ('hardness') of the photosynthetic surface, which may include not only leaves but also highly lignified but photosynthetic phyllodes, succulent surfaces and woody but green stems. The main functional types of leaves are summarized in Table 15.2, in terms of hardness and shade tolerance (see also Chapter 12).

The fundamental trade-off between photosynthetic rate and potential water loss can sometimes be circumvented by other adaptations, such as **seasonality** patterns. Plants in seasonally dry or cold climates often produce soft leaves with high photosynthesis rates and simply drop them in the unfavourable season, greatly reducing water loss as well as respiration demands. Seasonality, however, may also follow more subtle environmental factors, such as the short dry periods in otherwise non-seasonal equatorial climates that tend to synchronize plant functions. Plants in non-seasonal tropical rainforests may be not just evergreen but actually evergrowing, in the sense that they have very little 'down time' and may perform the normal functions of foliation, blooming, fruiting and defoliation continuously, even simultaneously on different branches of the same tree. Most evergreens, on the other hand, can be described as seasonal or leaf-changing evergreens, which drop old leaves all at once, normally in springtime, just as new leaves are produced.

Concepts of basic plant types facilitate the study of the relationships between form and function. Vegetation at a site will be composed of plants with particular combinations of form characters that permit the plant to function successfully in that environment. When similar morphological or physiognomic responses occur in unrelated taxa in similar but widely separated environments, they may be called convergent characteristics, such as the occurrence of sclerophyll shrubs in the world's five regions of mediterranean-type climate. The first classification of plant forms is credited by J.J. Barkman to Theophrastos (371–286 BC). Plant types defined entirely by their structure, such as broad-leaved trees, stem-succulents or graminoids, are called **growth-forms**, a concept derived from the *Hauptformen* of von Humboldt (1806), the 54 types of Grisebach (1872), and the 55 forms of Drude (1896), which he eventually called *Wuchsformen* (see Barkman 1988).

Table 15.2 Main leaf types, as defined by hardness and shade tolerance, with examples.

Light Requirement (and color) Hardness ↓	Light-demanding, shade-intolerant (light, sometimes yellowish green)	Intermediate (light/medium green)	Shade-tolerant (dark/darker green)
Soft and thin (malacophyllous)	Pioneers, e.g. *Prunus pensylvanica* *Populus tremula*, *Robinia pseudoacacia*	M A L A C O P H Y L L O U S Summergreen *Fraxinus*, *Celtis*; raingreen *Macaranga*, *Tectona*	*Acer saccharum*, *Carpinus*
Thin but somewhat reinforced (e.g. chartaceous or thin-coriaceous)	Pioneers, some on drier sites, e.g. *Quercus laevis*, *Liquidambar*	Deciduous *Quercus*, evergreen/deciduous *Nothofagus*	**L** TRF trees; *Castanopsis*, **A** *Laurus*; deciduous **U** *Fagus*
Leathery and thicker but pliable (coriaceous)	C O R I A C E O U S *Pinus taeda*, etc.; some *Smilax*	*Ficus microcarpa*, *Arbutus*, *Pittosporum*, *Corynocarpus*	**R** some *Ilex*, **O** *Nothofagus menziesii*, **P** some *Castanopsis* **H**
Hard and at least somewhat brittle (sclerophyllous)	S C L E R O P H Y L L O U S e.g. *Eucalyptus*, *Coccoloba uvifera*; many taller palms	*Olea europaea*, *Quercus geminata*	**Y** *Magnolia grandiflora*, **L** *Ilex opaca*; some TRF **L** understorey

TRF = tropical rainforest.

When structure is interpreted as an ecologically significant adaptation to environmental conditions, the forms involved are called **life-forms** (*Lebensformen*, from Warming 1895; see summary by Fekete & Szujko Lacza 1970). Life-forms may be interpreted as basic ecological types, grouping taxa with similar form and ecological requirements, resulting from similar morphological responses to similar environmental conditions. A more specific life-form concept was offered by Raunkiær (1934), who defined types based on the location of the renewal buds that must survive the cold winter. Growth-forms and life-forms both provide a convenient way of describing vegetation structure without having to treat large numbers of species individually. Another convenient concept is the **synusia**, which groups species of similar size that occupy a particular vegetation layer and are subject to similar micro-environmental conditions, such as ground-layer herbs subject to low light levels on a forest floor.

Functional similarity is also recognized in the concept of vicariance, referring to closely related and ecologically similar species that occur in different, usually distant regions. For example, the evergreen, sclerophyllous 'live oaks' of the warm-temperate south-eastern US coastal plain (*Quercus virginiana, Q. geminata*) and of mediterranean southern California (*Q. agrifolia, Q. wislizenii*) are quite similar in structure and general ecology. These species evolved separately after the rise of the Rocky Mountains effectively separated the eastern and western parts of the range of their ancestors. Vicariance has been hypothesized based on floristic similarities between eastern Asia and eastern North America (first recognized by Gray 1846) and between eastern North America and nemoral Europe (e.g. Manthey & Box 2007).

Global systems of plant types were developed partly by extension of temperate-zone systems to the tropics, as by Lebrun (1966), Schnell (1970–1977) and Vareschi (1980 and earlier). A very comprehensive, global classification of plant life-forms is found as Appendix A in Mueller-Dombois & Ellenberg (1974). The first global classification of plant types for modelling purposes was the structurally defined but functionally interpreted 'ecophysiognomic' types of Box (1981), which explicitly hypothesized relationships between form and function. This use of basic plant types was adopted by subsequent global modelling efforts (cf. Cramer & Leemans 1993; Foley 1995) and led to today's commonly used term 'plant function type' (PFT; see Chapter 12), which grew out of an emphasis on biosphere response to increasing atmospheric CO_2 levels and higher temperatures (see review by Smith *et al.* 1993). PFTs have been defined as 'groups of species that use the same resources and respond to the environment in a similar way' (Pausas & Austin 2001). PFTs were conceived as purely functional plant types, but so far, no general, world-applicable set of purely 'functional' types has been presented. On the other hand, world classifications of plant *forms* have been presented and used for modelling (e.g. Box 1981; cf. Cramer & Leemans 1993). At first it seems that at least 100 types are needed to capture the main features of plant form and function at a global scale (see the ongoing compilation of types in Table 15.3). A list of 15 generalized plant forms for use in global modelling was also presented, as well as a list of important functions that do not have form manifestations (Box 1996; see also Körner 1991).

Many physical and ecological processes seem to follow familiar global geographical patterns, four of which are shown in Fig. 15.1 (from Box & Meentemeyer 1991). Solar radiation and many aspects of temperature, including evaporative power, follow a *thermal* pattern, with highs in the low latitudes and lows near the poles. Annual precipitation and some measures of available moisture follow a *moisture* pattern, with highs near the equator and on windward coastlines (and lows in subtropical and continental deserts). Actual evapotranspiration, primary production, detrital decomposition and many other processes require simultaneous availability of both energy and moisture, represent biophysical 'work' and describe a *throughput* pattern, with the highest values in the humid tropics. Storages, such as soil carbon or standing biomass, represent balances between production and losses; these follow a less distinct *accumulation* pattern, with highs in cool humid regions (e.g. temperate rainforests and boreal climates, but also in wetlands). These patterns transcend vegetation or biome types, but most biomes can be characterized by high, medium or low levels for each pattern.

Table 15.3 Main terrestrial plant life-forms. This listing summarizes a classification of 146 plant forms, represented by over 550 taxa (ongoing, expanded from 1st edition of this book). Physiognomic definitions, including phenology and type of photosynthetic surface, are shown in the left column; subtypes (often with further subdivisions) are listed at the right.

Trees

Evergreen	Laurophyll	4	Tropical-rainforest, tropical-microphyll, warm-perhumid, cool-maritime
	Coriaceous	4	Tropical, mediterranean, warm-temperate, cool-maritime
– Semi-EG (tropical)		3	coriaceous, microphyll, sclerophyll
	Sclerophyll	5	Tropical, warm-temp., mediterranean, tall-humid, lauro-sclerophyll
	Malacophyll	1	Cool-maritime
Deciduous	Raingreen	3	Monsoon incl. montane, xero-microphyll, bottle trees
	Summergreen	3	Mesophytic, xerophytic, short-summer notophyll
Conifer	Platyphyll	4	Lauro-linear, broad-oxyphyll, linear, phylloclade
(evergr.)	Feather-leaved	1	Cool-oceanic
	Needle-leaved	5	Heliophilic, submediterr., awn-needle, temperate, boreal/subalpine
	Scale-leaved	3	Xeric, mediterranean, hygrophilic
(summergreen)		2	Feather-leaved, boreal

Small trees

Evergreen	Laurophyll	4	Rainforest, cloud-forest, subtropical/warm-temperate, cool-maritime
	Microphyll	1	Tropical coriaceous
	Sclerophyll	1	Savanna-sclerophyll
Deciduous		2	Savanna-raingreen, summergreen
Conifer		1	Needle/scale-leaved

Tuft-trees and treelets — 5 — Palms, bottle palms, palmettos/pandans, understorey, trunk-cycads

Arborescents

Evergreen	Laurophyll	2	Tropical, subtropical/warm-temperate
	Sclerophyll	1	Tropical/subtropical
Deciduous		2	Raingreen, summergreen
Stemgreen		2	Microphyll, xeric/leafless

Tuft-arborescents — 4 — Tree-fern, tropical-alpine, coriaceous, sclerophyll/succulent

Krummholz (needle-leaved) — 2 — Evergreen, summergreen

Shrubs

Evergreen	Laurophyll	3	Tropical, subtropical/warm-temperate, ericad
	Sclerophyll	2	Mediterranean, hot-desert

(Continued)

Table 15.3 (*Continued*)

	Succulent	1	Leaf-succulent
	Needle-leaved	2	Xeric, temperate-oceanic
Semi-evergreen		1	Temperate-xerophytic
Summergreen		2	Mesophytic, xeromorphic
Dwarf-shrubs			
Evergreen		4	Mediterranean, temperate, maritime-heath, boreal/tundra
Summergreen		1	Boreal/tundra
Xeromorphic		2	Leptophyll, stemgreen
Cushion-form		2	Mesophytic-evergreen, xerophytic-compact
Rosette-shrubs		2	Trunkless palm, xeric leaf-succulent
Stem-succulents		5	Columnar, branched-arborescent, frutescent, compact, cryptic
Semi-shrubs		3	Xylopodial, mesic-caducous, xeric-caducous
Graminoids			
Bambusoid		2	Arborescent, dwarf
Tall		3	Cane-graminoid, typical-tall, tall-tussock
Short		5	Spreading, bunch, short-tussock, sclerophyll, desert-grass
Forbs			
Evergreen		4	Arborescent, tropical-frutescent, temperate-understorey, rosette
Deciduous	Raingreen	1	Understorey
	Summergreen	3	Tall, understorey, taproot-perennial
Geophyte		3	Raingreen, spring-ephemeral, desert ephemeroid
Dwarf-xerophytic		2	Succulent, polar/alpine cushion
Ruderal		2	Evergreen-perennial, typical
Ephemeral		2	Desert-ephemeral, polar/alpine
Ferns		4	Mesic-understorey, semi-evergreen, resurrecting summergreen
Vines/Lianas		2	Tropical, summergreen
Evergreen vines		3	Understorey laurophyll, sprawling, temperate climbing
Deciduous vines		2	Raingreen, summergreen
Epiphytes	Rosette	2	Large-tropical, stenophyll
	Succulent	2	Erect stem-succulent, climbing stem-succulent
	Shrublet	1	Wintergreen mistletoe
	Herbs	3	Tropical-forb, fern, small-fern
	Vines	2	Understorey, leafless sprawling
Cryptogams		3	Mat-forming, lichenoid, algae

1. a **Thermal** pattern, representing
  energy availability:
  -insolation
  -temperature
  -potential evapotranspiration
  -respiration rate

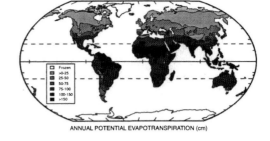

ANNUAL POTENTIAL EVAPOTRANSPIRATION (cm)

2. a **Moisture** pattern, representing
  moisture availability:
  -precipitation
  -climatic moisture balance
  -rooting depth
  -shoot-root ratios

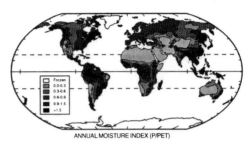

ANNUAL MOISTURE INDEX (P/PET)

3. a **Throughput** pattern, representing
  productive functions/work
  (simultaneous availability of
  energy and moisture):
  -soil texture, water holding, etc.
  -photosynthetic productivity
  -decompostion rates
  -biodiversity

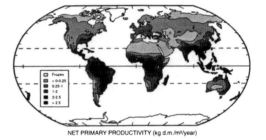

NET PRIMARY PRODUCTIVITY (kg d.m./m²/year)

4. an **Accumulation** pattern, representing
  storage at equilibrium
  (accumulations of net throughput):
  -soil carbon storage
  -standing live biomass
  -detrital accumulations

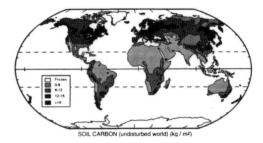

SOIL CARBON (undisturbed world) (kg / m2)

Fig. 15.1 Basic geographical patterns of ecological processes. Basic ecological and biophysical functions transcend biome or vegetation types, but their global geographic patterns may be fairly consistent, with extremes often located in particular biomes. Examples: highest throughput in tropical rainforests (equatorial climate); lowest throughput in arid or cold deserts; highest energy inputs in subtropical deserts (subtropical arid climate); lowest energy inputs in polar regions (polar climate); highest potential biotic accumulations in cool, humid biomes: biomass in temperate rainforests (marine west-coast climate), soil carbon in boreal landscapes (boreal climate). (From Box & Meentemeyer 1991.)

15.3　Vegetation types

Vegetation types are recognized most commonly by species composition or general appearance. Local plant communities can be defined by species presence, with a formal description and classification based on complete floristic composition and relative species abundances. The best known and most universally accepted methodology for such floristic field sampling (description) and classification is **phytosociology**, introduced by Braun-Blanquet (1928) and described more fully in Chapters 1 and 2 (see also Westhoff & van der Maarel 1974; Kent & Coker 1992).

At broader scales, vegetation can be described most simply by physiognomy (see Beard 1973), using obvious descriptors such as height, density and dominant plant types. Height criteria vary considerably, since 'tall' in boreal regions or in a desert may be quite different from 'tall' in the humid tropics. Density can be described somewhat more clearly, using the term 'closed' if crowns in the highest vegetation layer are touching (as in a forest) or 'open' otherwise. Even so, dense grassland, for example, can be viewed as closed (dense sod) or open (taller plants, if any, very widely spaced). **Dominant** plant types (or taxa) are usually the largest, and are able to outcompete other plants by controlling the availability of light, water and other resources. As vegetation science has developed, physiognomic concepts such as forest, shrubland, grassland and desert were relatively straightforward, but new categories were needed for mixes of plant types, such as savanna or scrub (see Eiten 1968 for a good classification). Generally accepted concepts of vegetation physiognomic classes, with the necessary plant structural types for each, are summarized in Table 15.4. Physiognomic classes can be broken down further by reference to leaf type and seasonal habit, and geographic descriptors can be added to complete the names of major vegetation formation types, such as tropical rainforest, boreal coniferous forest, or temperate grasslands. Where multiple structural types are co-dominant, the vegetation is often referred to as 'mixed', most commonly for mixtures of broad-leaved and coniferous trees in montane or sub-boreal 'mixed forest'. A very complete classification of vegetation structural types was offered by Mueller-Dombois & Ellenberg (1974: Appendix B).

A major complication in the recognition and classification of vegetation types is the fact that vegetation is dynamic (Chapter 4). Vegetation development through successional stages and the stability of 'climax' vegetation were described and formalized by Clements (1916, 1936), although the basic ideas had already been around for some time. Since climate was seen as the overriding control, the final stage of vegetation succession would be a stable 'climatic climax'. Vegetation that becomes stable before reaching its climatic potential, due to precluding non-climatic constraints, might be called an edaphic climax (unusual soil conditions), a fire climax (naturally recurring fire), or a topogenic climax (such as terrestrial wetlands). Another way of incorporating dynamics into concepts of vegetation type was presented by Daubenmire (1968), who classified and mapped 'habitat types' in the north-western USA based on their successional sequences and the ultimate potential dominant tree species. Of course the concepts of succession and climax do not apply everywhere. In humid tropical forests, for example, vegetation regenerates and develops through 'gap-phase dynamics' following

Table 15.4 Physiognomic vegetation structures and their main plant types.

Vegetation structure	Main/dominant Plant forms	Stature/Closure	Other features
Forest	**Trees**	Tall and closed	
– Rainforest		– Even taller	Evergreen, with closed evergreen understorey
Woodland	**Trees**	Short or tall and open	
– Parkland	(+grass)		With regular openings
Scrub	**Woody forms** (mixed)	Open or closed but not tall	Multiple woody forms, no one dominant
– Thicket		– Very dense	Usually localized
– Dwarf-scrub		– Very short	Usually in extreme habitats
Shrubland	**Shrubs**	Open or closed	(not sparse)
– Shrub-steppe		– Quite open	Shrubs regularly spaced, as in semi-desert
Savanna	**Grass + trees**	Grass closed or nearly so	Trees widely scattered, not in groves
– Grove-savanna			Trees in scattered groves
– Savanna-woodland		Trees more dense	Trees and grass layer almost equal in importance
– Shrub savanna	Grass + shrubs	Grass more open?	Shrubs scattered, no groups
Grassland	**Grasses**	Tall or short, closed or open	Spreading or bunch grasses, essentially no woody plants
– Steppe		Short and open	
Meadow	**Forbs + graminoids**	Tall or short, closed or open	Essentially no woody forms, forbs important
Tundra	Graminoids, forbs and dwarf-shrubs	Very short, closed or open	Micro-mosaics dependent on topography and local relief
Semi-Desert	Almost any forms	Mostly short, very open	Desert-like but with vegetation
– Cold-desert	Mosses and lichens, few small herbs	Very short, open	Desert-like, in coldest climates
Desert	Cryptogams, small herbs, if any	Extremely sparse	No or almost no vegetation at all

disturbance (see Chapter 4). In drier areas, where closed canopies are not possible, vegetation dynamics tends to involve spurts of re-growth after catastrophic distur-bance, as in frequently burned grasslands and mediterranean scrub.

In the 1800s, what European explorers saw outside semi-natural north-west and central Europe could often be interpreted as 'natural' vegetation, and this is what was depicted on the first global vegetation maps. In Europe, on the other hand, mapping often focused on the actual vegetation, its dynamics (e.g. Faliński 1991; Pedrotti 2004), and concepts of 'naturalness', developed especially by H. Sukopp (see reviews by Dierschke 1984 and Westphal *et al.* 2004). The concept of **potential natural vegetation** (PNV) was formalized by Tüxen (1956) as the vegetation type that would arise on an area if all outside influences were removed. PNV is not necessarily the 'original' vegetation, since the physical environment may have been altered, as by the massive soil erosion that occurred around the Mediterranean Sea after the Romans cut the oak forests to build their ships. PNV is a useful geographical concept, however, because vegetation integrates many environmental factors and thus provides a useful index of environmental or land-scape potential. On the other hand, whatever vegetation exists on an area now is called **actual vegetation**, which may be the PNV or some unstable kind of **substi-tute vegetation** such as crops, pasture, tree plantations or roadside weeds, main-tained by some human land-use regime. Any vegetation that is naturally stable, i.e. can resist invasion by other vegetation types, can be called **permanent vegetation** (*Dauergesellschaft*), whether it is the recognized 'climax' vegetation or not.

15.4   Distribution of the main world vegetation types

At a broad scale, vegetation distribution has been treated in two ways, by geo-graphic region and by biome type. Regional treatments of (mainly) natural vegeta-tion were presented in the 'Vegetation der einzelnen Großräume' series (H. Walter, editor) and various others. The term 'biome' came into more general use as the organizational basis for the projects of the International Biological Program (1964–1974) and for the encyclopaedic 'Ecosystems of the World' series of eco-system descriptions (D. Goodall, editor). Comprehensive modern treatments of world vegetation by biome type have been presented, in particular, by Walter (1968, 1973; summarized in English, Walter 1985), by Eyre (1968, with conti-nental maps), by Schmithüsen (1968), and in the encyclopaedic textbook by Archibold (1995), with many black and white photographs from all regions.

Biome types may occur in multiple large regions, in a somewhat regular global geographic pattern that was apparent from even the earliest maps and verbal treatments of world vegetation (cf. Grisebach 1872; Schimper 1898). Rübel's (1930) *Pflanzengesellschaften der Erde* was probably the first attempt to quantify the climatic limits of vegetation regions (formation types), thus demonstrating the unity of the global system. The rough north–south symmetry but also asym-metry in the world system of biomes was demonstrated quite graphically by the map of an 'average continent' (*Durchschnittskontinent*) by Troll (1948).

The geographic regularity of vegetation distribution arises, of course, from the geographic regularity of Earth's main climatic regions, driven by the global circu-lation pattern of the Earth's atmosphere. This circulation system consists of a

zone of low pressure and frequent precipitation near the equator (the Intertropi-
cal Convergence zone, or ITC); a zone of high pressure and dry conditions in the
subtropics of each hemisphere (stronger on the west side of continents); and
prevailing winds in each hemisphere that blow from the high-pressure belts
toward the ITC (trade winds, or tropical easterlies) and toward the poles (quickly
deflected by the Coriolis effect into west-to-east flows called the westerlies). Since
these pressure and wind belts migrate north–south with the seasons (trailing solar
declination by roughly one month), many latitudes experience seasonal winds and
precipitation. The resulting east–west latitudinal bands of similar conditions are
called **zones** and give rise to the idea of bioclimatic **zonation**, which represents
the fundamental geographic framework for the locations of biome types and
many other earth features. Climates that correspond to the expected, dominant
pattern within each zone are called **zonal** climates. Soil and vegetation types that
correspond to zonal climatic conditions are also called zonal, while aberrant types
are called **azonal**, as in wetlands, coastal dunes, and areas of serpentine soil.

The most widely used system of climate types based on this global zonation is
that of Walter (1977; but see also Troll & Pfaffen 1964). The Köppen climate
classification, used in most atlases, separates climates by quantitative indices and
lends itself readily to mapping, but is otherwise less flexible. Zonal systems, on the
other hand, are not quantitative but focus rather on the mechanisms that generate
distinct climate types and regions, thus focusing tacitly on core areas rather than
boundaries. Recognizing that Walter's 'warm-temperate' climate (type V) occurs
on the east and west sides of continents for very different reasons (the only mechanistic inconsistency in Walter's system), the first author of this chapter separated
type V into a 'warm-temperate' climate on continental east sides and a 'marine
west-coast' climate (an existing term) on west sides. The relative locations of these
Walter climates on an 'ideal continent' are shown in Fig. 15.2 (Box 2002; from
classroom lectures since 1980). Simple subtypes with one-letter notations were
also added, including arid (a), more continental (c), and maritime (m). For
example, the boreal climate (VIII) has an ultra-continental subtype (VIIIc) in the
coldest interior parts of north-eastern Siberia, where evergreen boreal conifers
such as *Picea obovata* are largely replaced by deciduous *Larix*; it also has a maritime subtype (VIIIm) in places like Iceland and coastal Alaska, where conifers are
replaced by deciduous broad-leaved trees such as *Betula*. At a global scale, the most
important subtypes are probably the temperate arid climate (VIIa) of the mid-
latitude cold-winter deserts and the ultra-continental boreal type (VIIIc).

Due to their good representation of fundamentally different climatic situations and potential limitation mechanisms, zonal climates correspond well to the
locations of the world's main biomes. For each climatic zone one can readily
identify up to three zonal biomes that occur as natural landscapes (at least in
lowlands) in that zone and essentially nowhere else. For example, the zonal
vegetation of the equatorial climate (I) is tropical rainforest, occurring where
there is no extended dry period, temperatures never fall below about 10 °C, and
plants are essentially evergrowing. On the other hand, the tropical summer-rain
climate (II), with distinct wet and dry seasons, includes three zonal vegetation
types that occupy different portions of the total range of annual precipitation:
tropical moist deciduous forest (wet season longer than dry season), dry deciduous forest and woodland in the mid-range (e.g. miombo woodland in

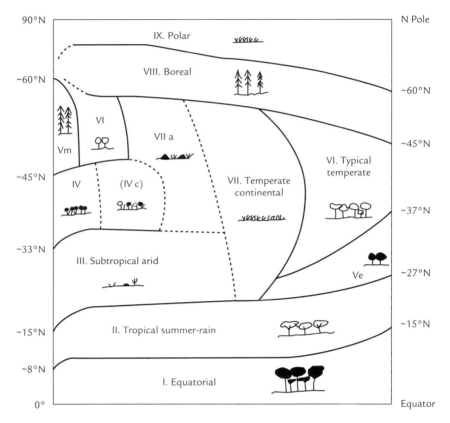

Fig. 15.2 Climatic regions on an ideal continent. The climates are the genetic climate types of Walter (1977; cf 1968, 1973, 1985), modified by splitting the original Walter V climate into marine west-coast (Vm) and warm-temperate east-coast (Ve) types, due to the quite different atmospheric mechanisms that produce them.

Recognizable subtypes include the following:

a = arid:	Ia	= Dry Equatorial (e.g. East Africa)
	VIIa	= Temperate Arid (interior Eurasia, western North America)
c = continental:	IVc	= Continental Mediterranean (e.g. interior Middle East)
	Vc	= Dry Warm-Temperate (e.g. central Texas, south Australia)
	VIc	= Monsoonal Humid Temperate (north and north-east China)
	VIIIc	= Ultra-Continental Boreal (eastern Siberia)
m = maritime:	IIm	= Windward Monsoonal (e.g. Kerala, Bangladesh)
	IIIm	= Coastal Fog Deserts (e.g. Namib, Atacama)
	VIm	= Maritime Nemoral (e.g. British Isles)
	VIIm	= Maritime Dry-Temperate (e.g. Patagonian & NZ grasslands)
	VIIIm	= Maritime Boreal (e.g. Iceland, Kamchatka)
	IXm	= Maritime Polar (e.g. subantarctic islands)
x = dry-maritime:	VIIx	= Maritime Arid-Temperate (Patagonian semi-desert)

south-central Africa), and raingreen thorn-scrub or savanna at the dry end (depending also on topography or other substrate conditions). The entire global system of zonal vegetation types is formalized in Table 15.5, which shows the main terrestrial biomes, the corresponding Walter climate type, the main climatic and non-climatic (e.g. edaphic) variants, and the expected vertical zonation in mountains, where the different vegetation levels are called **belts** rather than zones. Although mountain vegetation belts may be quite distinct, there is often considerable vertical integration in mountain ecosystems, to the extent that mountain ranges were called 'orobiomes' by Walter (1976).

Although seasonal temperature levels and precipitation amounts correlate with the world distribution of biomes, the lengths of the warm, wet and dry periods are more decisive, as first shown by Lauer (1952) and formalized by Lauer & Rafiqpoor (2002). An attempt to summarize the relationship of period lengths and some aspects of temperature for the main zonal biomes is shown in Table 15.6. One can recognize climate types that, at least on long-term average, are wet all year (no dry period longer than perhaps one month), dry all year (no significant wet season), have distinct wet and dry seasons, or are too cold for taller plants. Most climate zones have a single zonal vegetation type, such as tundra in the polar (IX), but climates with both wet and dry seasons generally have several zonal vegetation types. The Mediterranean climate (IV) also has three zonal vegetation types: sclerophyll forest at the wet end, maquis (chaparral, matorral) in the middle, and more open dwarf scrub (e.g. garrigue, phrygana, coastal sage) at the dry end. In both climates II and IV, the vegetation types toward the drier end may also occur as degradation forms substituting for potentially more robust zonal types toward the wetter end; for example, the widespread garrigue of southern Europe, where rainfall would permit forest but soil erosion has left too little water-holding capacity to support forest.

15.5 Regional vegetation

Even zonal vegetation types will differ on different continents, and vegetation patterns may also depend on topography, unusual substrates, fire, or other factors. Land-mass shapes may even play a role, as in the unusual tropical winter-rain region in south-eastern India caused by north-easterly monsoonal winter winds crossing the ocean before hitting land again. The most detailed description of world vegetation within a zonal framework was given by Walter (1968, 1973). Even more detail was provided by Archibold (1995), in a more flexible biome framework. The most detailed classification by vegetation type, rather than zones, with coordinated mapping of regional vegetation formations, was by Schmithüsen (1968, maps in Schmithüsen 1976). A compilation of regional occurrences of the main global-scale vegetation formation types, within the framework of bioclimatic zonation, is available from the first author.

For conservation purposes, greater regional and local accuracy is needed, including recognition of concrete regional and local ecosystems. The first global system of regional ecosystems was probably that of Udvardy (1975), developed for the International Union for Conservation of Nature and based on

Table 15.5 Main zonal biomes of the World, zonal climates, associated climatic and other vegetation variants, and vertical vegetation zonation in mountains.

Biome	Climate	Climatic variants	Other, azonal vegetation	Montane zonation
Tropical rainforest	I	Semi-evergreen forests, dry EG forests/woods, dry scrub (Ia)	Derived woodlands, scrub, savannas	Páramo Cloud forest Montane rainforest
Tropical deciduous forest and woodlands	II	Sclerophyll woodlands, thorn-scrub, savannas	Seasonal wetlands	Moist puna grassland EG montane forest Semi-EG forest
– Tropical savanna	II–III		(thorn-scrub)	
Warm deserts	III	Fog desert (IIIm) Xeric shrub-steppe	Oases, wadis, playas	Dry puna grassland Montane scrub
Mediterranean forests, woodlands, shrublands	IV	Deciduous scrub	Degraded scrub (stable), sclerophyll savanna-woodland	Dry cushion-scrub Montane conifer forest (foothill woodlands)
Temperate rainforest (evergreen broad-leaved)	Vm	Coniferous rainforest, Giant forests (IV–Vm)	Deciduous forest, heaths, wetlands	Wet alpine 'tundra' Wet montane forest
Evergreen broad-Leaved ('laurel') forest	Ve	Evergreen mixed forest, Sclerophyll woodlands	Pine forest/woodland, summergreen forest, swamps/other wetlands	Alpine 'tundra' Subalpine conifer forest Montane mixed forest
Temperate deciduous (summergreen) forest	VI	Cool-summer and warm-temperate decid. forest	Pine forest, grasslands, temperate wetlands	Alpine 'tundra' Subalpine/mixed forests
Temperate grassland (prairie, steppe)	VII	Oceanic tussock grasslands	Grove belts, riparian forests, degraded steppes, wetlands	Dry alpine 'tundra' Subalpine/dry conifer forests
– Temperate desert	VIIa	Oceanic cushion-steppe	Riparian woods, salt/rock veg.	(similar)
Boreal forest (evergreen coniferous)	VIII	Larch forest (VIIIc) Deciduous BL forest (VIIIm)	Bog forests, open conifer woods, forest-tundra, bogs	Alpine tundra
Polar tundra	IX	Maritime 'tundra' Moss-lichen cold-desert	Riparian/other wetlands	–

BL = broad-leaved, EG = evergreen.

Table 15.6 World biome types and their typical climatic limits in lowlands.

Natural biomes	Climate	Number of months Wet	Dry	≥10°	T_{min}	T_{abmin}
Wet (year-round)						
Tropical rainforests	I		≤1	all	≥18°	>10°
Laurel/EG-BL forests	Ve		≤1	>8	≥1°	>−15°
Temperate rainforests	Vm		≤2	4–10	>−1°	>−20°
Summergreen forests	VI		≤1	>4	<10°	<−10°
Boreal forests	VIII		≤1	1–3	<−1°	<−10°
– Boreal Larch woods	VIIIc		≤2	1–3	<−25°	<−40°
Dry (year-round)						
Warm deserts	III	≤1		most	>>0°	>−10°
Cold-winter deserts	VIIa	≤1		some	<5°	<−10°
– Maritime cold-winter	VIIx	≤1		some	<10°	<0°
Wet and dry seasons						
Raingreen woods, savanna, thorn-scrub	II	≥3	≥3	most	>10°	>−15°
Mediterranean woods, scrub and shrublands	IV	≥3	≥3	many	>1°	>−15°
Temperate grasslands	VII	≤3		some	<10°	
Cold-summer						
Polar tundra	IX			none	<0°	<−10°
– Maritime tundra	IXm	all		none	<10°	

Numbers of months indicated must be consecutive; T_{min} = mean temperature of coldest month, T_{abmin} = absolute minimum temperature (lowest ever measured at location)
Climate types are those of Walter, with the following sub-type modifications:
a = arid   c = continental   m = maritime   x = converse, e.g. maritime dry in rain shadow.

combinations of bioclimatic regions, surface physiography, vegetation associations, and local plant and animal ranges. More detail was added in second-generation regionalizations that recognize what have come to be called bioregions or **ecoregions** (Bailey 1983). Fairly detailed world maps with up to about 200 such ecoregions (Olson & Dinerstein 1998) are now used as a basis for conservation planning.

Understanding constraints on vegetation distributions has also been strengthened by results from large-area vegetation surveys (see Chapter 2), some of which were specifically for inventory and description (e.g. Miyawaki *et al.* 1994), while others were intended from the beginning to produce formal classifications. Among the latter was the exhaustive 10-year inventory that produced a complete, unified phytosociological classification of Japanese vegetation and communities, with formal phytosociological tables and many maps (Miyawaki 1980–1989). Vegetation compositional data were gathered from remnants of natural or nearly natural vegetation, especially from traditionally protected forest remnants around shrines and temples, and used to infer the general pattern of the potential natural vegetation. These data, along with topographic maps, were used as the basis for mapping potential and current actual vegetation, using known relationships between vegetation and topography. The largest such project, though, was the European Vegetation Survey, which attempted to integrate traditional national approaches into a single classification system for all of Europe and to produce a unified map (see Bohn & Neuhäusl 2003). The recent US National Vegetation Classification (Grossman *et al.* 1998) is a useful model, since it is also designed to integrate concepts from various earlier classification approaches over a large, diverse region (Chapter 2). In both the European and US projects, classification is based, at the highest level(s), on pheno-physiognomy and then on concrete plant associations at lower levels in the hierarchy. This permits the larger units in the systems to be identified while more data are being gathered to fill out the smaller, more local units in more floristic detail. One wonders whether this approach would work in the tropics, where biodiversity is so much greater, phenology can be less consistent and endpoints of landscape development are less clear.

For mapping vegetation over large regions, resolution of different approaches and development of an appropriate classification become the indispensable first step. The main vegetation map of China (Hou *et al.* 1979) seems to be based on the classification by Wu *et al.* (1980), with types usually defined by two dominant species per unit. The European Vegetation Survey, on the other hand, was able to use associations as the basis, within a physiognomic framework. Floristic approaches cannot be applied to the whole globe, however, because of the large floristic differences between the northern and southern hemispheres (despite some convergence in adaptations, cf. Box 2002). Potential vegetation of large regions has also been mapped over (*inter alia*) the USA (Küchler 1964), the Mediterranean region (UNESCO/FAO 1968), South America (Hueck & Seibert 1972, UNESCO 1980–81), Western Australia (Beard 1974–1981), the former Soviet Union (Isachenko & Gribovoy 1977), Africa (White 1983), the circumpolar Arctic (CAVM Team 2003), and present-day Russia (Yurkovskaya *et al.* 2006) (see also Chapter 16).

15.6   Vegetation modelling and mapping at broad scales

Modelling provides a way to put concepts into action, to estimate unknowns, to simulate or predict behaviour and to map results. In particular, this provides a way to test hypotheses of the controls of distributions and other patterns, and to gain new insights. These possibilities became attractive with the advent of computers and urgent with the prospect of a disrupted biosphere responding to broad-scale climatic and land-use changes. Models can be anything from regressions based on field measurements to so-called 'process models', i.e. complex, dynamic system simulations of physical and physiological processes. Most vegetation modelling also involves mapping. Although satellite data (Chapter 16) are often used now, vegetation modelling was based originally on climate, not only because it is the overriding control on broad-scale vegetation patterns but also because, prior to satellites, it represented the only global database available.

Useful reviews of the development of dynamic global vegetation modelling are given by Foley (1995), Peng (2000), and Prentice *et al*. (2007). Possibly the first model-driven global computer map was the 'Miami Model' map of net primary productivity (NPP) predicted from annual temperature and precipitation, based on data from about 50 NPP measurement sites representing most world biomes (Box *et al*. 1971; see also Lieth 1975). A world regression model for gross primary production (GPP) was also produced (without the highest GPP estimates later considered invalid), and both maps were quantified, giving the first systematic, globally representative estimates of the carbon balance of the land vegetation cover (Box 1978). The interplay of CO_2 drawdown by NPP and release by respiration and decomposition, producing seasonal biospheric CO_2 source and sink regions, was simulated by model MONTHLYC (Box 1988), and its sensitivity to the accuracy of respiration estimates was explored (Box 2004). The climatic equilibrium accumulation of standing biomass on the world's land areas was first presented in 1980–1981, but only for woody vegetation; once additional litter-fall data became available for herbaceous vegetation, a complete global map of potential biomass was produced (mid-1990s, see Miyawaki & Box 2006), yielding a global estimate of about 1700 gT of potential (but probably unattainable) carbon sequestration.

For the distribution of vegetation types, on the other hand, the first quantitative treatment was by Rübel (1930). The first complete global system of climatically predictable types was the so-called 'life zones' of Holdridge (1947), which did not specify seasonality but were mapped and used widely for land planning in tropical and subtropical countries. Both of these represent early concepts of what came to be called **climatic envelopes**, which were developed in a global-modelling context in the 1970s (Box 1981). A climatic envelope is a set of estimated upper and lower limiting values for climatic variables that represent what are thought to be the main factors that control the geographic range of the target biotic entity, which could be a species, a more general plant type, or a vegetation type – or even an animal (see Chapter 16). Biomes were predicted and mapped first as a basis for mapping energy fixation and photosynthetic efficiency (Box 1979).

At broad scales, modelling of vegetation or ecosystem distributions must be based mainly on climate, since climate is the overriding control. Complexities are revealed at more local scales, where the range of climatic variation is reduced and other factors, such as substrate, topography and history, become more important in determining vegetation patterns. Anomalies occur, especially on unusually young or nutrient-poor substrates and in marginal environments where disturbance and stochastic processes may determine which of several possible vegetation types becomes stable. Even so, on the unusual substrates of Florida, climatic envelopes could predict the ranges of the main woody species surprisingly well (Box *et al*. 1993).

Recognition of additional local vegetation patterns was improved, of course, by the advent of satellite-based imagery. This revolutionized vegetation study because it brought the means not only to monitor large areas but also to recognize cover types and related characteristics of real, increasingly disturbed if not completely human-altered landscapes over much of the world (Chapter

16). Complete global coverage by satellite imagery, albeit at coarse resolution, became available in the early 1980s, with the Advanced Very High-Resolution Radiometer (AVHRR) on polar-orbiting NOAA satellites. A greenness-enhancing ratio of two spectral bands, called the NDVI, became the most widely used index for vegetation monitoring, and the NDVI ratio has been continued with newer MODIS and other satellite data.

Most maps of large-area vegetation patterns rely on pheno-physiognomy, i.e. vegetation types defined by structure (forest, grassland, etc.) and seasonal activity (evergreen vs. deciduous). This approach has the advantages that: (1) the types can be identified from field data; (2) their geographic occurrence, at least as the potential dominant vegetation (PDV) of the area, can be predicted with considerable accuracy from climate data (Box 1995b); and (3) the types can also be recognized spectrally by satellites and perhaps eventually also by (airborne) LIDAR. A world classification of pheno-physiognomic PDV types, designed to represent world vegetation with as few types as possible, is shown in Table 15.7. This set was derived by a sort of 'geographic regression' (trial and error) that posed types, predicted their occurrence at 1600 climatic sites worldwide (by climatic envelopes, see Fig. 15.3), and then added or modified types until all sites were adequately described.

The resulting 50 PDV types can be grouped into 15 more strictly phenophysiognomic types, which should be readily distinguishable by multispectral satellite data. These 15 types represent world vegetation significantly better than the usual approximately 10 'types' of the spectrally inspired 2×2 model (evergreen/deciduous $\times$ broad-leaved/needle-leaved) used in most global models. With the increased availability of geographic information systems, maps of predicted potential vegetation (e.g. Prentice *et al.* 1992) have increasingly been used in modelling efforts involving biosphere–atmosphere interactions. A predictive mapping of the PDV types of Table 15.7, from climatic envelopes, is shown in Plate 15.1. This is superior to other maps because its vegetation units are rigorously defined by structure, predictions are improved in some problem areas, and mapped results show no four-way boundaries (that betray less rigorous valuation of limiting values).

One purpose of vegetation modelling is to suggest biosphere behaviour under global climate change. The first attempt to model vegetation composition at global scale and how it may change involved 90 plant growth-forms defined by structural type (e.g. tree), leaf form and consistency (e.g. broad-leaved malacophyll), relative plant and leaf size, and seasonal activity (e.g. summergreen) (Box 1981). A climatic envelope was constructed for each form, based on estimated physiological limits and actual geographic ranges, and the occurrence and proximity to the closest limit of these growth-forms were then predicted worldwide. The results were compared with vegetation descriptions for a geographically representative set of locations, representing the first, and one of very few, attempts to validate a global ecological model (Peters 1991). The results also provide some suggestion of relative abundance and importance within vegetation stands.

A major goal in vegetation modelling since the latter 1980s has been development of 'dynamic global vegetation models' (DGVM) that can simulate changes in vegetation patterns and be used with global atmospheric models (see also

Table 15.7   Main world pheno-physiognomic vegetation types.

Physiognomic class	Individual types
1. Tropical rainforests	– Tropical rainforest (lowland)
	– Tropical montane rainforest
	– Tropical subalpine (cloud) forest
	– Subtropical rainforest
2. Tropical seasonal woodlands/savanna	– Tropical semi-evergreen forest
	– Raingreen forest
	– Raingreen scrub (incl. montane)
	– Tropical dry evergreen forest
3. Evergreen broad-leaved forests	– Evergreen broad-leaved forest
	– Mediterranean evergreen forest
	– Cool-temperate evergreen BL forest
	– Subpolar evergreen BL forest
4. Temperate rainforest	(Evergreen BL/mixed/NL)
5. Summergreen BL forests and woods	– Summergreen broad-leaved forest
	– Summergreen broad-leaved woodland
	– Subpolar summergreen BL forest
6. Needle-neaved evergreen forests/ woods	– Dry conifer forest
	– Mediterranean conifer forest
	– Boreal conifer forest (EG, incl. dry)
	– Subpolar/subalpine conifer woodland
7. Summergreen needle-leaved (Larch) forests/woods	
8. Subhumid woodlands/scrub	– Semi-evergreen dry woodland/scrub
	– Mediterranean woodland/scrub
9. Shrubland/Krummholz (seasonal/evergreen)	– Cool-evergreen BL scrub/Krummholz
	– Subhumid shrubland/low scrub
10. Grasslands	– Tropical savanna
	– Temperate grassland
	– Cool-maritime grassland
11. Tropical alpine vegetation	– Páramo (scrub/shrubland)
	– Puna (grassland/steppe)
12. Tundra and related Krummholz/dwarf-scrub	– Cool-summergreen BL Scrub/krummholz
	– Subalpine conifer krummholz
	– Polar/alpine tundra
	– Maritime tundra
	– Cold desert/semi-desert
13. Semi-desert scrub	
14. Deserts (extreme)	– Arid desert
	– Polar/subnival cold-desert
	– Ice desert

(Derived by 'geographic regression', see Box 1995b).

1. **Climatic Envelopes:**

	BT	TMAX	TMIN	DTY	MI	PMIN	PMTMAX
Tropical Rainforest							
Maximum	30	35	30	8	****	***	***
Minimum	18	21	18	0	1.00	20	20
Tropical Deciduous Forest							
Maximum	31	35	30	18	****	25	***
Minimum	16	17	13	0	0.58	0	30

.
.   (about 40 types)
.

2. **Check Inclusion/Exclusion and Distance to Closest Limit:**

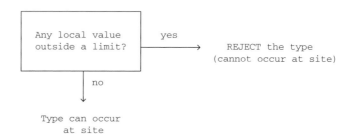

3. **Initial Result** for the particular site:

	Limiting factor	Distance to closest limit
Example: Harbin (China)		
1. Summergreen broad-leaved woodland	MI	0.15˙
2. Temperate grassland	MI	0.08
3. Summergreen broad-leaved forest	MI	0.02

4. Apply **Competition Model** for Potential Dominant Type:

   Forest > Woodland > Grassland   --->   Forest

      BUT: forest is very close to a climatic limit at Harbin,
             not so strong (competitive)

   Woodland is farthest from any limit, so the PDV type is

      --->   **Summergreen broad-leaved woodland**

Fig. 15.3 Modelling global potential dominant vegetation (PDV) types. This procedure identifies vegetation types that are climatically possible at sites. Essentially the same procedure was used to calibrate the limits of the climatic envelopes, based on comparison of thus predicted results with the distribution patterns of natural vegetation types and relevant bioclimatic isolines.

BT   = biotemperature (sum of monthly mean temperatures > 0°C, divided by 12)
Tmax, Tmin = maximum and minimum mean monthly temperatures (°C)
DTY   = annual range of mean monthly temperature (=Tmax – Tmin) (°C)
MI   = annual moisture index (precipitation divided by potential evapotranspiration)
Pmin   = average precipitation of the driest month (mm)
PmTmax   = average precipitation of the warmest month (mm)

Chapter 17). The first DGVM was produced by Foley *et al.* (1996), and various others have been developed since. Even so, 'there has been little effort to compare simulated vegetation patterns with observed vegetation distribution' (Peng 2000). Different models also use quite different estimators for potential evapotranspiration, yielding different PET geographies and model results.

The first climate-based global vegetation model (Box 1981) deliberately used variables that could be valuated worldwide from readily available climatic data alone. Though often described as 'correlative', envelope models can be based on physiologically significant variables and limiting values (cf. Box 1995a). Better identification of the limitation mechanisms and availability of more climatic data now permit better modelling. A first step was inclusion of extreme minimum temperatures as limits for some plant types (cf. Table 15.1), first estimated by Sakai (1971) and Larcher (1973). One can appreciate this limitation by observing that most of the interior south-eastern USA has deciduous forest, despite much higher mean winter temperatures – but lower extremes – than corresponding latitudes in East Asia, which have evergreen broad-leaved forest (cf. Box 1997). Inclusion of absolute minimum temperature greatly improved model predictions in the mid-latitudes as well as tropical and subtropical mountains.

More insightful vegetation modelling, however, requires consideration of both limitation mechanisms and positive plant physiological requirements. IBP work had provided data for understanding and estimating the climatic relations of photosynthesis and productivity, in particular the strong temperature dependence of autotrophic respiration and thus of the production-respiration balance, which is critical not only to plant survival and growth but also to biosphere–atmosphere CO_2 exchange (e.g. Cooper 1975; Shidei & Kira 1977). Climatic limitations include mainly low temperature, since plants can evade moisture shortage through deciduousness, by collapsing completely (caducousness), by dormancy, or by some combination. Climate-related requirements and limitations are summarized in Table 15.8, along with variables that could be used to represent them.

Development of better satellite-based methodologies requires going beyond the widely used but not really analytical statistical-signature approach, which makes many mistakes. For example, the southern hemisphere has nothing similar to the boreal coniferous forest, even if spectral signatures (based largely on greenness duration or sun angle) are similar in some imagery; more local maps also show large mistakes. Ancillary climatic data could be used (but rarely are) as a control on interpretation of the satellite imagery. Despite identifiable errors, the early global map of actual land cover by Tateishi & Kajiwara (1991) probably remains one of the best. A more insightful use of satellite data in vegetation modelling and mapping has involved the development of 'metrics' that analyse the shape of annual curves of NDVI or other signals to identify factors and timings thought to control vegetation functions, such as the timing of rapid warm-up in spring and overall growing-season length. This approach was pioneered by Malingreau (1986) and has been pursued subsequently by (*inter alia*) Reed *et al.* (1994), DeFries *et al.* (1995) and Hansen *et al.* (2002).

At broader scales, all models of vegetation distributions, processes and dynamics are inherently geographic models and should be based on data representing

Table 15.8 Climatic factors limiting taxon distributions.

Climatic factor	Mechanism of limitation	Potential variable(s)
Temperature levels:		
1. low temperature levels (T sums, winter extremes, etc.)	NPP $\to$ <<0 (GPP $\to$ 0?)	T_{min} lower limit T_{max} lower limit
2. high temperature levels (mainly summer)	R $\to$ >GPP (not desiccation)	T_{max} upper limit
3. high winter temperatures	lack of vernalization	T_{min} upper limit
Extreme Temperatures:		
1. high temperature extremes	GPP shutdown, enzyme damage, etc.	T_{abmax} upper limit
2. low temperature extremes	frost/cold damage to leaves/buds	T_{abmin} lower limit (T_{min}, T_{mmin}?)
	freezing death of whole plants	T_{abmin} lower limit
Water availability:		
1. drought	desiccation (PET >> tissue water)	lower limits of MI, Py, P_{min}
2. physiological drought a. from extreme heat b. from frozen soil	desiccation – PET >> tissue water – water uptake inhibited	lower MI and/or – upper T_{max}, T_{abmax} – lower T_{min}, T_{abmax}
Water Excess (flooding, saturated soil)	lack of aeration	P_{min} upper limit

Abbreviations: GPP = gross primary production, R = respiration, NPP = net primary production, PET = potential evapotranspiration.

all geographically important types. Resulting models must also be validated geographically, i.e. by testing with data from the different types and situations (Box & Meentemeyer 1991). In ecology, the term 'verification' is generally understood to mean confirmation that a model reproduces observed system behaviour, while 'validation' requires testing with independent data, not used in model development, in order to permit confident application to other situations (Odum 1983, p. 579; Rykiel 1996). Validation can be done more formally at local scales (e.g. Box *et al.* 1993), but geographic evaluation was also attempted early at a global scale, for NDVI imagery as an estimator of NPP, biomass and net CO_2 flux (Box *et al.* 1989). Validation of global-change models is especially problematic (Rastetter 1996) and may, strictly speaking, be impossible. One good approach is to predict reciprocally between the present and the past, such as during or just after the last glacial period (e.g. Martínez-Meyer *et al.* 2004). Validation at sites is more rigorous than comparison of pixel values (areal averages). A good recent step in this direction, for global 'process' models, has involved comparison of MODIS-based NPP and GPP estimates (i.e. pixel values) at 1-km resolution against eddy-flux estimates at sites in nine different biomes (Turner *et al.* 2006).

15.7   Vegetation and global change

Vegetation has responded to changing climatic conditions throughout much of Earth's history, with migrations of taxa and vegetation zones as well as the disappearance of communities and the recombination of taxa into new vegetation types. The continuously changing nature of natural landscapes provides much of the basis for their high levels of biodiversity. The prospect of global warming provides a new challenge, however, not only to vegetation and ecosystem function, but also to landscape stability and familiar concepts of vegetation. Some hypothetical patterns of the response of vegetation zones to global warming are summarized in Table 15.9 and include physiological stresses; changes in metabolic balance, potential biomass and fitness; poleward/upslope shifts of climate spaces and perhaps actual ranges by at least some species; 'weedification' (Ehrlich & Mooney 1983) of some landscapes as invasive secondary (and some alien) species migrate faster than less vagile but potentially stabilizing larger species; and perhaps long-term instability of landscapes, especially on the equatorward side. These responses mean that, with warming, the zonal potentials shown in Fig. 15.3 will shift towards the poles (and upwards in mountains). Many taxa, though, may be quite idiosyncratic, with specific advantages or disadvantages in changing environments. Potentially stabilizing larger plants will decline, due to their longer development periods, unless their propagules can be dispersed widely and effectively, as by birds (e.g. tree Lauraceae).

With the very first productivity models it was possible to predict that land NPP might change by about 5% with each 1 °C change in average global temperature (Lieth 1976). Early inventories of actual land biomass and its carbon equivalent were made, especially by Olson *et al.* (1983). Envelope models suggest the sensitivities of plant types to warming and have been used to project consequent shifting climate spaces of plant species (e.g. Box *et al.* 1999, Iverson *et al.* 1999). An especially disturbing aspect of such range shifts involves the potential for lost integrity and even break-up of familiar plant communities and landscapes if the ranges of their main structuring elements diverge (cf. Crumpacker *et al.* 2001). Missing until recently from most global-change modelling were questions such as *which* taxa will migrate, *how* will they get to their new, preferred locations, and *how long* will it take? These questions require more than just envelope models or sophisticated 'process models'. Projecting migration is finally being addressed by several newer modelling approaches (e.g. Iverson *et al.* 2004; Neilson *et al.* 2005; Engler & Guisan 2009).

Global change is not just climatic warming or drying; it also involves the biosphere response and that *other* driver of change, namely human activities, especially overpopulation, land-use changes and the effects of globalization (see also Chapter 17). Perhaps most threatening is the sheer magnitude of the land-use changes and landscape fragmentation. Effects on vegetation include

Table 15.9 Hypothetical effects of climatic warming on the poleward and equatorward margins of zonal biomes.

Phenomenon	Poleward margin	Equatorward margin
Water availability	less effect?	potential drought stress, especially in summer
Plant metabolism:		
– Photosynthesis/ respiration balance	increased photosynthesis but still higher respiration (greater T sensitivity), both summer and winter	
– Net productivity	increased unless drier	(increase or decrease)
→ **Physiological fitness**	**reduced some**	**reduced more, esp. larger species**
Biomass (standing)	less biomass supportable under higher respiration; big trees die off slowly	
	may be replaced by others of same species	not replaced; slow immigration of new potential dominants
Decomposition	possible large increase	increased unless much drier; fire increases C releases
→ **Net carbon balance**	**D maybe > NPP: CO$_2$ source**	**turnover maybe > NPP increase: CO$_2$ source**
Species migrations	migration poleward, newcomers mainly from within biome	massive invasion by weedy species from warmer areas
Competition	diverging competitive abilities	stressed local species eliminated by invaders
Dominance	(little change?)	local dominants stressed, replaced by invaders
Dynamics/Succession	proceeds further as new species enter	succession stops with invading species; most new stabilizers arrive much more slowly
→ **Overall effect**	colonization poleward, **reorganization** with more species	decline, long-term **instability**, as disturbance continues and new stabilizers slow to arrive

Abbreviations: D = decomposition   NPP = net primary production   T = temperature.

degradation, fragmentation and complete destruction of natural, or at least stable, self-maintaining, quasi-natural vegetation. The resulting surfaces are covered by unstable substitute vegetation or by none at all, both situations requiring costly management efforts by humans. Potential natural vegetation provides a graphic basis for land-use and other environmental planning, and serves as an ever more important 'benchmark' (Box 1995b) as the world's vegetation cover is destroyed more and more completely. The biosphere will respond only slowly to climate change, which itself may be happening somewhat faster than many predicted. As fast as the climate may change, though, the human-induced changes are happening much faster and will dominate our efforts to maintain a stable global vegetation cover and its services.

References

Archibold, O.W. (1995) *Ecology of World Vegetation*. Chapman and Hall, London.

Bailey, R.G. (1983) Delineation of ecosystem regions. *Environmental Management* 7, 365–373.

Barkman, J.J. (1988) New systems of plant growth forms and phenological plant types. In: *Plant Form and Vegetation Structure* (eds M.J.A. Werger, P.J.M. van der Aart, H.J. During & J.T.A. Verhoeven), pp. 9–44. SPB Academic Publishers, The Hague.

Beard, J.S. (1973) The physiognomic approach. In: *Ordination and Classification of Communities* (ed. R.H. Whittaker), pp. 355–386. Dr. W. Junk, The Hague.

Beard, J.S. (1974–1981) Vegetation Survey of Western Australia. 1:1,000,000 Series. University of Western Australia Press, Nedlands, WA. 7 sheets + Explanatory Notes (1981) cf 1:250,000 Series, Vegmap Publications.

Bohn, U. & R. Neuhäusl (2003) *Karte der Natürlichen Vegetation Europas* [Map of the Natural Vegetation of Europe]. Explanatory Text. + map (1:2,500,000). BfN–Schriftenvertrieb, Münster.

Box, E.O. (1978) Geographical dimensions of terrestrial net and gross primary productivity. *Radiation and Environmental Biophysics* 15, 305–322.

Box, E.O. (1979) Use of synagraphic computer mapping in geoecology. In: *Computer Mapping in Education, Research, and Medicine*, Harvard Library of Computer Mapping 5, 11–27. Harvard University, Cambridge, MA.

Box, E.O. (1981) *Macroclimate and Plant Forms: An Introduction to Predictive Modeling in Phytogeography*. Tasks for Vegetation Science, Vol. 1. Dr. W. Junk, The Hague.

Box, E.O. (1988) Estimating the seasonal carbon source-sink geography of a natural steady-state terrestrial biosphere. *Journal of Applied Meteorology* 27, 1109–1124.

Box, E.O. (1995a) Factors determining distributions of tree species and plant functional types. *Vegetatio* 121, 101–116.

Box, E.O. (1995b) Global potential natural vegetation: dynamic benchmark in the era of disruption. In: *Toward Global Planning of Sustainable Use of the Earth – Development of Global Eco-engineering* (ed Sh. Murai), pp. 77–95. Elsevier, Amsterdam.

Box, E.O. (1996) Plant functional types and climate at the global scale. *Journal of Vegetation Science* 7, 309–320.

Box, E.O. (1997) Bioclimatic position of evergreen broad-leaved forests. In: *Island and High-Mountain Vegetation: Biodiversity, Bioclimate and Conservation*, pp. 17–38. Proceedings IAVS meeting, April 1993. Universidad de La Laguna, Servicio de Publicaciones, Tenerife.

Box, E.O. (2002) Vegetation analogs and differences in the Northern and Southern Hemispheres: a global comparison. *Plant Ecology*, 163, 139–154 [appendix missing: ask author].

Box, E.O. (2004) Gross production, respiration and biosphere CO_2 fluxes under global warming. *Tropical Ecology* 45, 13–29.

Box, E.O. & Meentemeyer, V. (1991) Geographic modeling and modern ecology. In: *Modern Ecology: Basic and Applied Aspects* (eds G. Esser & D. Overdieck), pp. 773–804. Elsevier, Amsterdam.

Box, E.O., Lieth, H. & Wolaver, T. (1971) Miami model: primary productivity predicted from temperature and precipitation averages. World map published by Lieth (1973) in *Human Ecology* 1, 303–332.

Box, E.O., Holben, B.N. & Kalb, V. (1989) Accuracy of the AVHRR Vegetation Index as a predictor of biomass, primary productivity, and net CO_2 flux. *Vegetatio* 80, 71–89.

Box, E.O., Crumpacker, D.W. & Hardin, E.D. (1993) A climatic model for location of plant species in Florida, USA. *Journal of Biogeography* 20, 629–644.

Box, E.O., Crumpacker, D.W. & Hardin, E.D. (1999) Predicted effects of climatic change on distribution of ecologically important native tree and shrub species in Florida. *Climatic Change* 41, 213–248.

Braun-Blanquet, J. (1928) *Pflanzensoziologie: Grundzüge der Vegetationskunde*, Springer, Berlin. (2nd edn 1951, 3rd edn 1964, Wien; English 3rd edn 1965: *Plant Sociology*, Hafner, New York, NY.)

CAVM Team (2003) *Circumpolar Arctic Vegetation Map*. 1:7,500,000. Conservation of Arctic Flora and Fauna (CAFF) Map No. 1. US Fish and Wildlife Service, Anchorage, AK.

Clements, F.E. (1916.) *Plant Succession, an Analysis of the Development of Vegetation*. Publication No. 242. Carnegie Institute of Washington, Washington, DC.

Clements, F.E. (1936) Nature and structure of the climax. *Journal of Ecology* **24**, 252–284.

Cooper, J.P. (ed.) (1975) *Photosynthesis and Productivity in Different Environments*. IBP series, Vol. 3. Cambridge University Press, Cambridge and London.

Cramer, W.P. & Leemans, R. (1993) Assessing impacts of climate change on vegetation using climate classification systems. In: *Vegetation Dynamics and Global Change* (eds A.M. Solomon & H.H. Shugart), pp. 190–217. Chapman & Hall, New York, NY.

Crumpacker, D.W., Box, E.O. & Hardin, E.D. (2001) Potential breakup of Florida plant communities as a result of climatic warming. *Florida Scientist* **64**, 29–43.

Daubenmire, R.F. (1968) *Plant Communities*. Harper & Row, New York, NY.

DeFries, R., Hansen, M. & Townshend, J. (1995) Global discrimination of land cover types from metrics derived from AVHRR Pathfinder data. *Remote Sensing of Environment* **54**, 209–222.

de Laubenfels, D.J. (1975) *Mapping the World's Vegetation*. Syracuse University Press, Syracuse, NY.

Dierschke, H. (1984) Natürlichkeitsgrade von Pflanzengesellschaften unter besonderer Berücksichtigung der Vegetation Mitteleuropas. *Phytocoenologia* **12**, 173–184.

Drude, O. (1896) *Deutschlands Pflanzengeographie*. Engelhorn, Stuttgart.

Ehrlich, P.R. & Mooney, H.A. (1983) Extinction, substitution and ecosystem services. *BioScience* **33**, 248–254.

Eiten, G. (1968) Vegetation forms: a classification of vegetation based on structure, growth form of the components, and vegetative periodicity. *Boletim do Instituto de Botânica* (São Paulo) **4**, 88 pp.

Engler, R. & Guisan, A. (2009). MigClim: predicting plant distribution and dispersal in a changing climate. *Diversity and Distributions* **15**, 590–601.

Eyre, S.R. (1968) *Vegetation and Soils: A World Picture*, 2nd edn. Arnold, London.

Faliński, J.B. (ed.) (1991) *Vegetation Processes as Subject of Geobotanical Maps*. Proceedings, 33rd Symposium International Association for Vegetation Science. *Phytocoenosis* (Warszawa-Bialowieza), 3(2), Supplementum Cartographiae Geobotanicae + map volume.

Fekete, G. & Szujko Lacza, J. (1970) A survey of plant life-form systems and the respective approaches. *Annales Historico-Naturales Musei Nationalis Hungarici, Pars Botanica* **62**, 115–127.

Foley, J.A. (1995) Numerical models of the terrestrial biosphere. *Journal of Biogeography* **22**, 837–842.

Foley, J.A., Prentice, I.C., Ramankutty, N. *et al.* (1996) An integrated biosphere model of land surface processes, terrestrial carbon balance, and vegetation dynamics. *Global Biogeochemistry Cycles* **10**, 603–628.

George, M.F., Burke, M.J., Pellett, H.M. & Johson, A.G. (1974) Low-temperature exotherms and woody plant distribution. *HortScience* **9**, 519–522.

Grabherr, G. & Kojima, S. (1993) Vegetation diversity and classification systems. In: *Vegetation Dynamics and Global Change* (eds A.M. Solomon & H.H. Shugart), pp. 218–232. Chapman & Hall, New York, NY.

Gray, A. (1846) Analogy between the flora of Japan and that of the United States. *American Journal of Science and Arts* **II**(2), 135–136.

Grisebach, A.R.H. (1872) *Die Vegetation der Erde nach ihrer klimatischen Anordnung. Ein Abriss der Vergleichenden Geographie der Pflanzen*, Vol. 2. W. Engelmann, Leipzig.

Grossman, D.H., Faber-Langendoen, D., Weakley, A.S. *et al.* (1998) *The National Vegetation Classification System*. Vol. 1. In: *International Classification of Ecological Communities*. The Nature Conservancy, Arlington, VA.

Hansen, M.C., DeFries, R.S., Townshend, J.R.G. *et al.* (2002) Toward an operational MODIS continuous field of percent-tree-cover algorithm: examples using AVHRR and MODIS data. *Remote Sensing of Environment* **83**, 303–319.

Holdridge, L.R. (1947) Determination of world plant formations from simple climatic data. *Science* **105**, 367–368.

Hou, X.-Yu (editor/main author) *et al.* (1979) [Vegetation Map of China.] Institute of Botany, Academia Sinica. Map Press, Beijing. Map (1:4,000,000, in Chinese) + manuals [Chinese and English].

Hueck, K. & Seibert, P. (1972) *Vegetationskarte von Südamerika*. Vegetation der einzelnen Großräume series, Vol. **IIa**. Gustav Fischer Verlag, Stuttgart.

Isachenko, T.I. & Gribovoy, S.A. (1977) Rastitel'nost' SSSR [Vegetation of USSR]. 1:25,000,000. In: *Bol'shaya Sovyetskaya Entsiklopediya* [Soviet Encyclopedia], Vol. 24-II. NRKCh GUGK, Moskva [in Russian].

Iverson, L., Prasad, A.M., Hale, B.J. & Sutherland, E.K. (1999) *Atlas of Current and Potential Future Distributions of Common Trees of the Eastern United States*. US Forest Service report NE-265. US Department of Agriculture, Washington, DC.

Iverson, L.R., Schwartz, R.W. & Prasad, A.M. (2004) How fast and far might tree species migrate in the eastern United States due to climate change? *Global Ecology and Biogeography* 13, 209–219.

Kent, M. & Coker, P. (1992) *Vegetation Description and Analysis: A Practical Approach*. John Wiley & Sons, Ltd, Chichester. [With diskette.]

Körner, Ch. (1991) Some often overlooked plant characteristics as determinants of plant growth: a reconsideration. *Functional Ecology* 5, 162–173.

Küchler, A.W. (1964) *The Potential Natural Vegetation of the Conterminous United States*, Special Research Publication 36 (map + manual). American Geographical Society, New York, NY.

Larcher, W. (1973) Limiting temperatures for life functions in plants. In: *Temperature and Life* (eds H. Precht, J. Christophersen, H. Hensel & W. Larcher), pp. 195–231. Springer-Verlag, Berlin and New York.

Lauer, W. (1952) Humide und aride Jahreszeiten in Afrika und Südamerika und ihre Beziehung zu den Vegetationsgürteln. *Bonner Geographische Abhandlungen* 9, 15–98.

Lauer, W. & Rafiqpoor, D. (2002) *Die Klimate der Erde: Eine Klassifikation auf der Grundlage der ökologischen Merkmale der realen Vegetation*. Franz-Steiner-Verlag, Stuttgart.

Lebrun, J. (1966) Les formes biologiques dans la végétation tropicale. *Mémoires de la Société Botanique de France* 1966, 166–177.

Lieth, H. (1975) Modeling the primary productivity of the world. In: *Primary Productivity of the Biosphere* (eds H. Lieth & R.H. Whittaker), pp. 237–263. Springer, New York, NY.

Lieth, H. (1976) Possible effects of climate change on natural vegetation. In: *Atmospheric Quality and Climatic Change* (ed. R.J. Kopec), pp. 150–159. University of North Carolina, Chapel Hill, NC.

Malingreau, J.-P. (1986) Global vegetation dynamics: satellite observations over Asia. *International Journal of Remote Sensing* 7, 1121–1146.

Manthey, M. & Box, E.O. (2007) Realized climatic niches of deciduous trees: comparing western Eurasia and eastern North America. *Journal of Biogeography* 34, 1028–1040.

Martínez-Meyer, E., Peterson, A.T. & Hargrove, W.W. (2004) Ecological niches as stable distributional constraints on mammal species, with implications for Pleistocene extinctions and climate-change projections for biodiversity. *Global Ecology and Biogeography* 13, 305–314.

Miyawaki, A. (ed.) (1980–1989) *Nippon Shokusei-Shi* [Vegetation of Japan]. 10 vols, plus vegetation tables and colour maps [in Japanese, with German or English summary]. Shibundô, Tokyo.

Miyawaki, A. & Box, E.O. (2006) *The Healing Power of Forests*. Kōsei Publishing Co., Tokyo.

Miyawaki, A., Iwatsuki, K. & Grandtner, M.M. (eds.) (1994) *Vegetation in Eastern North America*. University of Tokyo Press, Tokyo.

Mueller-Dombois, D. & Ellenberg, H. (1974) *Aims and Methods of Vegetation Ecology*. John Wiley & Sons, Ltd, New York, NY.

Neilson, R.P., Pitelka, L.F., Solomon, A.M. et al. (2005) Forecasting regional to global plant migration in response to climate change. *BioScience* 55, 749–759.

Odum, H.T. (1983) *Systems Ecology: An Introduction*. John Wiley & Sons, Ltd, New York, NY.

Olson, D.M. &. Dinerstein, E. (1998) The Global 200: A representation approach to conserving the earth's most biologically valuable ecoregions. *Conservation Biology* 3, 502–515.

Olson, J.S., Watts, J.A. & Allison, L.J. (1983) *Carbon in Live Vegetation of Major World Ecosystems*. Report DOE/NBB-0037. US Department of Energy, Washington, DC. [With map.]

Pausas, J.G. & Austin, M.P. (2001) Patterns of plant species richness in relation to different environments: an appraisal. *Journal of Vegetation Science* 12, 153–166.

Pedrotti, F. (2004) *Cartografia Geobotanica*. Pitagora Editrice, Bologna.

Peng, Ch.-H. (2000) From static biogeographical model to dynamic global vegetation model: a global perspective on modeling vegetation dynamics. *Ecological Modeling* 135, 33–54.

Peters, R.H. (1991) *A Critique for Ecology*. Cambridge University Press, Cambridge.

Prentice, I.C., Bondeau, A., Cramer, W. *et al.* (2007) Dynamic global vegetation modeling: quantifying terrestrial ecosystem responses to large-scale environmental change. In: *Terrestrial Ecosystems in a Changing World* (eds J. Canadell, D. Pataki & L. Pitelka), pp. 175–192. Springer-Verlag, Heidelberg.

Prentice, I.C., Cramer, W., Harrison, S.P. *et al.* (1992). A global biome model based on plant physiology and dominance, soil properties and climate. *Journal of Biogeography*, **19**, 117–134.

Rastetter, E.B. (1996) Validating models of ecosystem response to global change. *BioScience* **46**, 190–198.

Raunkiær, C. (1934) *The Life Forms of Plants and Statistical Plant Geography*. Clarendon, Oxford.

Reed, B.D., Brown, J.F., van der Zee, D. *et al.* (1994) Measuring phenological variability from satellite imagery. *Journal of Vegetation Science* **5**, 703–714.

Rübel, E. (1930) *Pflanzengesellschaften der Erde*. Huber, Berlin.

Rykiel, E.J. (1996) Testing ecological models: the meaning of validation. *Ecological Modeling* **90**, 229–244.

Sakai, A. (1971) Freezing resistance of relicts from the Arcto-Tertiary flora. *New Phytologist* **70**, 1199–1205.

Schimper, A.F.W. (1898) *Pflanzengeographie auf physiologischer Grundlage*. Gustav Fischer Verlag, Jena (3rd edn 1935, with F.C. von Faber); English translation 1903, by W.R. Fisher. Oxford Press, Oxford.

Schmithüsen, J. (1968) *Vegetationsgeographie*, 3rd edn. Walter de Gruyter, Berlin.

Schmithüsen (1976) *Atlas zur Biogeographie*. Meyers Großer Physischer Weltatlas. Bibliographisches Institut, Mannheim.

Schnell, R. (1970–1977) *Introduction à la Phytogéographie des Pays Tropicaux*. 4 vols. (see Vol. II: chapter 3). Gauthier-Villars, Paris.

Shidei, T. & Kira, T. (eds.) (1977) *Primary Productivity of Japanese Forests*. JIBP Synthesis series, Vol. 16. Tokyo University Press, Tokyo.

Smith, T.M., Shugart, H.H., Woodward, F.I. & Burton, P.J. (1993) Plant functional types. In: *Vegetation Dynamics and Global Change* (eds A.M. Solomon & H.H. Shugart), pp. 272–292. Chapman & Hall, New York, NY.

Tateishi, R. & Kajiwara, K. (1991) Global land-cover classification by NOAA GVI data: thirteen land-cover types by cluster analysis. In: *Applications of Remote Sensing in Asia and Oceania* (ed. Sh. Murai), pp. 9–14. Asian Association for Remote Sensing, Tokyo University, Tokyo.

Troll, C. (1948) Der asymmetrische Aufbau der Vegetationszonen und Vegetationsstufen auf der Nord- und Südhalbkugel. *Jahresbericht des Geobotanischen Instituts Rübel* **1947**, 46–83.

Troll, C. & Pfaffen, K.H. (1964) Karte der Jahreszeitenklimate der Erde. *Erdkunde* **18**, 5–28.

Turner, D.P., Ritts, W.D., Cohen, W.B. *et al.* (2006) Evaluation of MODIS NPP and GPP products across multiple biomes. *Remote Sensing of Environment* **102**, 282–292.

Tüxen, R. (1956) Die heutige potentielle natürliche Vegetation als Gegenstand der Vegetationskartierung. *Angewandte Pflanzensoziologie* (Stolzenau) **13**, 5–42.

Udvardy, M.D.F. (1975) *A Classification of the Biogeographical Provinces of the World*. Occasional Paper 18. IUCN, Morges.

UNESCO/FAO (1968) *Vegetation Map of the Mediterranean Region*. 2 sheets, 1:5,000,000, with explanatory handbook. UNESCO, Paris.

UNESCO (1980–1981) *Vegetation Map of South America*. 2 sheets, 1:5,000,000, with explanatory notes. UNESCO, Paris.

Vareschi, V. (1980) *Vegetationsökologie der Tropen*. Ulmer Verlag, Stuttgart.

von Humboldt, A. (1806) *Ideen zu einer Physiognomik der Gewächse*. F.G. Cotta, Tübingen. Reprinted 1957 by Akademische Verlagsgesellschaft, Leipzig.

Walter, H. (1968, 1973) *Die Vegetation der Erde in öko-physiologischer Betrachtung*. Vol. 1 (3rd edn); Vol. II. Fischer Verlag, Stuttgart. [English edition 1985.]

Walter, H. (1976) *Die ökologischen Systeme der Kontinente (Biogeosphäre)*. Gustav Fischer Verlag, Stuttgart. [See English summary in *Vegetatio* **32**, 72–81.]

Walter, H. (1977) *Vegetationszonen und Klima*, 3rd edn. Eugen Ulmer Verlag, Stuttgart.

Walter, H. (1985) *Vegetation of the Earth and Ecological Systems of the Geobiosphere*, 3rd edn. Springer, New York.

Warming, E. (1895) *Plantesamfund: Gruntraek af de økologiske Plantegeografi*. København. English version 1909: *Oecology of Plants*, Humphrey Milford & Oxford University Press, Oxford.

Westhoff, V. & van der Maarel, E. (1974) The Braun-Blanquet approach. In: *Ordination and Classification of Communities* (ed. R.H. Whittaker), pp. 617–726. Dr. W. Junk, The Hague.

Westphal, C., Härdtle, W. & von Oheimb, G. (2004) Forest history, continuity and dynamic naturalness. In: *Forest Biodiversity: Lessons from History for Conservation* (eds O. Honnay, K. Verheyen, B. Bossuyt & M. Hermy), pp. 205–220. CAB International, Wallingford.

White, F. (1983) *The Vegetation of Africa*. Descriptive Memoir to UNESCO/AETFAT/UNSO *Vegetation Map of Africa*. UNESCO, Paris.

Woodward, F.I. (1987) *Climate and Plant Distribution*. Cambridge University Press, Cambridge.

Wu, Zh.-Y. and committee (eds) (1980) *Zhongguo Zhibei* [Vegetation of China]. Science Press, Beijing. [With 339 black and white photos (in Chinese); 2nd edn 1995.]

Yurkovskaya, T.K., Iljina, I.S. & Safronova, I.N. (2006) [Vegetation map of Russia]. Scale 1:15,000,000. In: [*National Atlas of Russia*], Vol. 1. Moskva [in Russian].

16

Mapping Vegetation from Landscape to Regional Scales

Janet Franklin

Arizona State University, USA

16.1 Introduction

This is not your grandmother's vegetation map.

As someone who was interested, at an early age, not just in plants (that was odd enough), but in patterns of vegetation on the landscape, the discovery of Küchler's pronouncement that the plant community is 'the only tangible, integrated expression of the entire ecosystem' (Beard 1975; Küchler 1984) justified my unusual interests. Mapping plant communities provides spatial information about the entire ecosystem, and is part and parcel with understanding what environmental factors control their distributions. Vegetation mapping efforts have grown since the mid 20th century because they form a basis for natural resources inventory and land-use planning, and provide a baseline against which to measure future landscape change.

DeMers has often been cited (see Franklin 1995; Millington & Alexander 2000) for distinguishing the communication versus analytical perspectives in vegetation mapping (DeMers 1991). Historically, vegetation maps aimed to communicate, synthetically, the geographic patterns of the vegetation classification employed in the mapping, and paper maps were the only available way of storing and presenting this information (as noted by Brzeziecki *et al.* 1993). Accordingly, textbooks on methods in vegetation ecology emphasized issues such as the use of colour and symbols in crafting effective vegetation maps (chapter 14.6 in Mueller-Dombois & Ellenberg 1974), and other cartographic considerations (chapters 8–13 in Küchler & Zonneveld 1988). With the advent of **geographic information systems** (GIS), maps are stored as digital data, and can more easily have multiple attributes (if vector data) or attributes can be separated into different map layers (raster data). This gives the vegetation mapper flexibility,

Vegetation Ecology, Second Edition. Eddy van der Maarel and Janet Franklin.
© 2013 John Wiley & Sons, Ltd. Published 2013 by John Wiley & Sons, Ltd.

because complex attributes do not necessarily have to be synthesized into a single thematic (choropleth) map, but also responsibility to understand the nature of spatial and categorical generalization used in traditional maps that are ingested by digitizing, and of the geospatial data that are available for contemporary vegetation mapping efforts.

In many realms, vegetation maps have effectively been replaced with multi-attribute vegetation databases (e.g. Franklin *et al.* 2000; Ohmann *et al.* 2011). Traditional (analogue) vegetation maps have frequently been scanned or digitized and treated as georeferenced digital data. This was demonstrated for a small area (large cartographic scale) vegetation map by DeMers (1991). At coarse, global scales, early models of the biosphere were driven by digitized global vegetation maps to establish initial conditions (reviewed by Nemani & Running 1996; see Chapter 15). However, as Goodchild (1994) cautioned, maps are far from straightforward, and GISs are more than just containers for maps. A geographic data model provides a framework for understanding how various types of spatial vegetation data can be represented and what they represent (Section 16.2).

Recent developments (over the past 50 years) in both source data and interpretation methods for mapping vegetation are often presented as if they followed an orderly progression of methods, data and scale from (a) field survey and visual interpretation of aerial photos for small-area, detailed mapping, to (b) sophisticated image processing methods applied to multidimensional remotely sensed imagery for mapping larger regions. However, while it would be convenient if there were a simple relationship between data, method and scale, these boundaries have become quite blurred. Very large area vegetation maps have been developed by visual interpretation and manual delineation of multidimensional satellite imagery (Table 16.1) while sophisticated image processing methods have been applied to hyperspectral, high-resolution imagery to create fine-scale maps of individual tree crowns (Section 16.3).

The general purpose of vegetation maps is to depict and help understand the causes of vegetation patterns. This has not really changed during this time of tremendous improvements in mapping data and methods. The tradition of developing vegetation maps for agricultural land use planning and land management dates back to the early decades of the 20th century in some countries. However, it is becoming much easier to use vegetation maps, and analyse those patterns, in conjunction with other spatial data in a GIS (reviewed by Franklin 1995; Millington & Alexander 2000). Vegetation maps are now widely used as data for natural resource (land) management, conservation planning and decision support for environmental policymaking (Scott *et al.* 1993; Margules & Pressey 2000), topics discussed in Chapter 14. Vegetation maps at various scales are also used to drive spatially explicit models of ecosystem processes and community dynamics (for example Mladenoff 2004; Scheller *et al.* 2007), and to predict the impacts of global change (Cramer *et al.* 2001; Lenihan *et al.* 2008) as discussed in Chapters 15 and 17.

The vegetation **attributes** that are mapped for these myriad purposes usually include (a) vegetation class or type defined on the basis of physiognomy, species composition and/or structure (Section 16.2), but can also include (b) biophysical properties of vegetation such as leaf area index (LAI), biomass, net primary

Table 16.1 Hierarchical framework for vegetation mapping with examples.

Vegetation attribute[a]	Grain[b] (MMU)	Extent[c]	Environmental drivers[d]	Data[e]	Published examples
Plant formation; biome; biophysical attributes	1 – 2,500 km² (up to 5° × 5°)	Global	climate	Coarse scale multispectral satellite imagery (AVHRR, MODIS); interpolated climate data	Circumpolar Arctic Vegetation map (Walker et al. 2005); Vegetation map of tropical continental Asia (Blasco et al. 1996); Vegetation map of Europe (Neuhäusl 1991)
Land Cover (Anderson I)	10 ha – 10(–100) km²	Continental – Global	Climate, topography, land use	Coarse- to moderate-scale (Landsat TM) multispectral satellite imagery	MODIS Global Land Cover (Friedl et al. 2002)
Vegetation types defined by physiognomy and dominant species (Anderson II)	1 ha – 1(–10) km²	Regional	Topography, soil type	Moderate-resolution airborne and spaceborne imagery	Land cover of Wyoming (Driese et al. 1997)
Communities defined by species composition, structure class[f] (Anderson III)	0.1 – 10 ha	Landscape	Topography, soil properties, biotic interactions, disturbance history	Moderate- to high resolution airborne and spaceborne imagery; aerial photographs; field survey	US National Forest maps (Franklin et al. 2000)
Crown delineation and tree species identification (patch map)	100 m² – 1 ha	Local	Biotic interactions	High-resolution airborne imagery	Identify native, non-native species (Pu & Gong 2000; Asner et al. 2008)

AVHRR – Advanced Very High Resolution Radiometer on NOAA series satellites, 1-km GSD; MODIS – Moderate Resolution Imaging Spectroradiometer, on the NASA Terra and Aquasatellites; Landsat Thematic Mapper (TM), and Enhanced Thematic Mapper (ETM+).

[a]Hierarchy of vegetation attributes synthesized from Urban et al. (1987) and Franklin & Woodcock (1997). 'Anderson' refers to the hierarchical level as defined in the Anderson land cover classification system of the US Geological Survey (Anderson et al. 1976).

[b]MMU, minimum mapping unit or the area of the smallest polygon in the output map.

[c]Based on Pearson & Dawson (2003).

[d]Based on Franklin (1995), Guisan & Zimmermann (2000), Mackey & Lindenmayer (2001) and Pearson & Dawson (2003).

[e]Based on Millington & Alexander (2000) and chapter 5 in Franklin (2010a).

[f]Vegetation structure attributes could include cover, canopy height and biomass.

productivity (NPP), canopy height and canopy closure; (c) community properties (especially measures of species diversity); and even (d) the distribution of individual plants (stem maps). Maps of individual plant canopies are usually used for basic research rather than resource management, but with potential to be scaled up using remote sensing. Further, attributes of the site or environment can also be interpreted from the vegetation, for example non-native species, fire effects, and other types of disturbance.

In this chapter, the data (Section 16.3) and methods (Section 16.4) currently used to map vegetation at various scales (Section 16.2) will be reviewed, with an emphasis on landscape- to regional-scale mapping (Table 16.1). Global scale vegetation classification and mapping is discussed in more detail in Chapter 15. The history of vegetation mapping has been reviewed elsewhere (Küchler 1967).

16.2 Scale and vegetation mapping

Vegetation mapping is based on a conceptual model of vegetation as a geographical phenomenon, consisting of **gradients** and **patches**. Vegetation attributes usually have to be estimated over some discrete area, and are defined everywhere in the map area, consistent with the mental model of geography as a multivariate field (Goodchild 1994). Vegetation classification systems are typically hierarchical (reviewed by Franklin & Woodcock 1997), with finer divisions nested within broader categories. These hierarchies can either be taxonomic (species, community, cover type, formation), or ecological. Ecological hierarchies are based on differences in population (individual, population, community) or ecosystem (functional component, ecosystem, biosphere) process rates (O'Neill *et al.* 1986). Taxonomic or process hierarchies may not be spatially nested (Allen & Hoekstra 1990), but rather a particular class of phenomena, such as a vegetation type or cover class, will be distributed discontinuously across the landscape. This is in contrast with a land system or ecoregion approach to mapping, in which hierarchical units are spatially nested (e.g. Bailey 2004).

The components of spatial scale are grain and extent (Turner *et al.* 2001; reviewed in Skidmore *et al.* 2011). **Grain** is the size of the smallest observation or map unit, and can be expressed in terms of resolution (pixel size) or the minimum mapping unit (MMU), while **extent** describes the total size of the mapped area. For vegetation mapping, the nested hierarchy of spatial units on the landscape (Fig. 16.1) has been described as: plant/gap, stand, community, province (cover type), formation (biome) (Urban *et al.* 1987). A **vegetation stand** is defined as a contiguous area of similar species composition, physiognomy and structure (Mueller-Dombois & Ellenberg 1974; Franklin & Woodcock 1997; Jennings *et al.* 2009), where structure describes the horizontal and vertical patterns of vegetation, distribution of sizes of individuals, biomass, and so forth (chapter 9 in Küchler 1967). A stand is typically the smallest spatial entity represented in landscape-scale community type maps (Table 16.1).

Landscape- to regional-scale vegetation maps often represent the community, cover type and formation level in the taxonomic hierarchy (Franklin *et al.* 2003), corresponding to Levels III (species dominance), II (cover type) and I (land

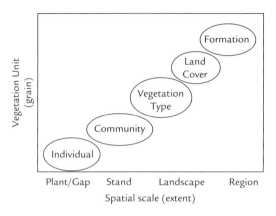

Fig. 16.1 Typical hierarchy of vegetation units (see Urban *et al.* 1987), arrayed in terms of spatial grain and extent. See text for discussion of terms. (Modified from Franklin & Woodcock 1997.)

cover) in the Anderson *et al.* (1976) land-cover classification (discussed in detail in Franklin 2001). See Web Resource 16.1 for an example of these three levels. In Table 16.1, this spatial and categorical hierarchy of vegetation mapping is linked to typical mapping scales, environmental drivers of vegetation pattern at those scales, and data sources.

Phinn and colleagues (Phinn 1998; Phinn *et al.* 2003) presented a general framework, based on hierarchy theory and remote-sensing scene models (Strahler *et al.* 1986; see Section 16.4.2), for selecting appropriately scaled remotely sensed data (Section 16.3) and analysis techniques (Section 16.4) to address a particular environmental monitoring need. In this framework, the information needed, such as vegetation mapping, is matched to the spatial and temporal scales, and the spectral and radiometric resolution, of available imagery. This framework makes explicit those decisions that are often implicitly made in a remote-sensing-based mapping study, and should become increasingly useful as the number of remote-sensing systems and types of data available for vegetation mapping grows (Jones & Vaughan 2010). Remote-sensing-based vegetation mapping should be carried out in the framework of well-defined vegetation classification and mapping standards (Chapter 2 and Jennings *et al.* 2009) so that the remote sensing supports vegetation classification goals, rather than developing *ad hoc* categorizations based on what can be discriminated by a particular type of imagery.

16.3   Data for vegetation mapping

Vegetation mapping typically involves classifying vegetation into categories based on data from quantitative vegetation surveys (Chapter 2), and then labelling map units according to those categories. The map units can be derived or delineated using a variety of data sources and methods of inference described in this section.

Data from vegetation surveys, collected for the purpose of vegetation classification, can be used either to 'train' or calibrate supervised classification of remotely sensed data, or guide visual interpretation and manual delineation of air photos or imagery. Often these efforts – vegetation survey and classification and vegetation mapping – have been carried out quite separately, but ideally they are more closely integrated. The following subsections describe the data sources used to map vegetation units.

16.3.1   Field observations

Prior to the availability of aerial photography, field observation was the primary means of mapping vegetation patterns on the landscape (see chapter 14 in Mueller-Dombois & Ellenberg 1974). Although it might be thought that field observation is only practical for exhaustively delineating boundaries of vegetation types over limited extents, there are some spectacular examples of large-area field-based species and vegetation mapping and classification. For example, plant species were mapped in the Netherlands starting around 1900. The data were collected in the field on a 5 × 5 km grid basis, and the survey was repeated in the 1950s to 1960s on a 1 × 1 km grid and digitized. Since that time there have been other well-known efforts to develop species atlases based on systematic field observations and a variety of other data sources including natural history collections, for example the *Atlas of the British Flora* (Perring & Walters 1962). That project was also updated recently using contemporary methods and digital maps (Preston *et al.* 2002).

In another example, the Wieslander Vegetation Type Mapping (VTM) Project for California, USA (Wieslander 1935) was conducted in the early 1930s by field crews who sketched vegetation boundaries, viewed from ridges, peaks and vantage points, onto 1 : 62 500 scale topographic maps, with a minimum mapping unit of about 16 ha (Table 16.2). At the same time, vegetation plot surveys were conducted and voucher photographs were taken. The maps cover 16 million ha (Kelly *et al.* 2005), and the University of California has archived these historic vegetation maps and data (http://vtm.berkeley.edu/).

Because precise surveying would have been impractical, the attributes of the VTM maps may be more reliable than the map unit boundaries (Davis *et al.* 1995). However, more recently, small-area vegetation mapping has taken advantage of widely available, low-cost geographical positioning systems (GPS), to precisely delimit vegetation units in the field, and even characterize their 'overlap' in the form of ecotones or gradual transitions in composition and structure (Steers *et al.* 2008). Wyatt (2000) pointed out that vegetation mapping from the ground, air and space can be complementary approaches, used fruitfully together.

16.3.2   Aerial photographs

Aerial photography (air photos) was the primary source of data for mapping vegetation and land systems from the 1930s to the 1970s. Related to vegetation mapping, land systems or landscape units defined on the basis of both vegetation and physical environment were mapped in early integrated land resources surveys

Table 16.2 Examples of vegetation mapping efforts in California, USA.[a]

Mapping Project	Extent	Resolution	Dates	Vegetation attributes	Method
Vegetation type map (VTM) survey (Wieslander 1935; Kelly et al. 2005)	16 m ha (40% of the state)	16 ha mmu	1928–1940	220 vegetation types, species composition	Hand-drawn on topographic base maps in the field
CALVEG US Forest Service	Entire state	160–320 ha mmu (average 15,400 ha)	1979–1991	Series-level species groups	Photointerpretation of Landsat MSS CIR prints
CALVEG US Forest Service 2nd generation (Woodcock et al. 1994; Franklin et al. 2000)	Entire state	2.5 ha mmu	1988–present	220 series-level vegetation and land-use types, forest canopy cover and crown size class	Advanced digital image processing of Landsat TM and predictive modelling using terrain data
California gap analysis (multi-attribute vegetation database) (Stoms et al. 1992; Davis et al. 1995)	Entire state; 21,000 polygons	1 km² mmu	1990–1996	Up to 3 overstory species and vegetation types	Boundaries photo-interpreted from Landsat TM; labels from air photos, field survey and VTM
Integrated vegetation mapping and classification	20% of the state (National Parks, other)	0.25–0.5 ha mmu	1996–present	Detailed plant communities, association-level (NVCS)	Photointerpretation of digital or analogue air photos; sampling and classification of vegetation

[a]Described in detail in Keeler-Wolf (2007) and Franklin et al. (2000), illustrating the variety of methods and data sources that have been applied to map vegetation communities with high categorical detail regarding life-form and floristic composition (Anderson Level III or greater) for a very large area (California's entire land area is 403,933 km²).

CALVEG, California Vegetation classification system and map; CIR, color infrared; MSS, multi-spectral scanner; mmu, minimum mapping unit; NVCS, National Vegetation Classification System of the United States (Grossman et al. 1998); TM, thematic mapper.

In the US NVCS, association is the finest level of floristic classification, defined on the bases of all diagnostic species; series is a broader level of classification, based on the dominant species.

(reviewed by Franklin & Woodcock 1997) conducted in order to evaluate large inaccessible areas of Africa, Australia and elsewhere for their potential for commercial agricultural development (Christian & Stewart 1968; Astle *et al.* 1969; Beard 1975). The growing availability of aerial photography for civilian applications following the Second World War made this type of mapping possible.

The film-based technology of analogue cameras has largely been replaced by digital photography in recent years, blurring the distinction between air photos and other types of remotely sensed imagery from airborne and spaceborne platforms. Google Earth™ users are viewing something like a photographic image, although it may have been acquired from a satellite. National archives of analogue air photos are being scanned and digitized (for example the National Aerial Photography Program of the US Geological Survey). Digital image processing methods (Section 16.4.2) have been applied to scanned and digitized analogue air photos to derive map units for high-resolution vegetation mapping (Carmel & Kadmon 1998; Mullerova 2004).

Orthophotos are air photos that are geometrically corrected for camera tilt and topographic displacement so that they are in the correct orthographic position or 'orthorectified' (Paine & Kiser 2003). Orthophotos that are georeferenced (established location in terms of a coordinate system or map projection) are equivalent to planimetric maps – maps that accurately portray the horizontal positions of features. Analogue methods of orthorectification have largely been replaced by digital methods ('soft copy' photogrammetry), and digital orthoimagery is becoming widely available and used for vegetation mapping.

The use of analogue air photos for vegetation mapping has now largely been replaced by remotely sensed imagery (or digital orthophotography), a trend since the mid 1980s (Millington & Alexander 2000: figure 18.1). Perhaps in the future, the source of the imagery for delineating vegetation units (airborne or spaceborne, digital camera or multispectral instrument) will be a less important distinction than the resolution and the interpretation method – visual interpretation, traditionally developed for air photos (Section 16.4.1), versus automated pattern recognition algorithms typically applied to multivariate remotely sensed imagery (Section 16.4.2), or some combination.

16.3.3   Airborne and satellite remote sensing

Multispectral (with a few, broad wavelength bands for detecting electromagnetic radiation) and hyperspectral (many narrow wavebands) remotely sensed data typically discriminate biophysical attributes of the vegetation better than they discriminate species composition (Kerr & Ostrovsky 2003). Plant species do not all have unique spectral signatures and remotely sensed data typically do not identify communities (assemblages of species) directly. Communities may have diagnostic spectral, spatial and temporal characteristics that can be identified in certain types of imagery, for example a characteristic mixture of different leaf types (broad-leaf, needle-leaf), a texture pattern related to the arrangement of crown shapes and sizes (stand structure), or a certain timing of annual green-up. Therefore, airborne and satellite imagery are often best used for discriminating physiognomic and structural classes of vegetation corresponding to formations, land cover and general vegetation types (Franklin 1995; Franklin & Woodcock 1997; Millington & Alexander 2000).

A recent textbook on remote sensing of vegetation described the types of imagery from spaceborne and airborne sensors that are typically used to map vegetation from landscape to global scales (Jones & Vaughan 2010, chapter 5.8, p. 118–125). However, even the specialized text demurred that there are so many sources of remotely sensed data, and they change so quickly, that it is difficult to give an up-to-date and comprehensive list (but see their appendix 3 which is quite useful). The features of different sensors, and the characteristics of the imagery used for vegetation mapping was also reviewed by Xie *et al.* (2008; see their table 1). Remotely sensed imagery is usually described in terms of its **spatial** (grain and extent), **spectral**, and **radiometric** resolution, and the **temporal** frequency of data acquisition (Jensen 2000). In the following paragraphs an overview is given of the types of remotely sensed data that have recently been used in major vegetation mapping efforts, grouped according to their spatial resolution (pixel size or **ground sample distance**, GSD).

Building on the foundations of early efforts (Goward *et al.* 1985; Hansen *et al.* 2000) low-resolution (e.g. AVHRR) and medium-resolution (e.g. MODIS, Landsat) systems have been used as the basis for regional to global and continental-scale land-cover maps (for an example see Web Resource 16.2) which are periodically updated (Section 16.6) and therefore referred to as 'data products.' These include the MODIS land-cover maps (500-m resolution, global), the CORINE land-cover project of the European Union, and the Global Land Cover 2000 program of the European Commission. These data products, developed to monitor global environmental change, contain information about vegetation composition and physiognomy (structure) at various levels of detail. For example, the land-cover map for Africa for the year 2000 (see Web Resource 16.3) depicts 27 categories of vegetation and land cover at 1-km resolution derived from imagery from the VEGETATION sensor (1-km GSD) on the SPOT-4 satellite (Mayaux *et al.* 2004; Cabral *et al.* 2006). CORINE land-cover maps, produced for 29 countries in Europe using roughly a 25-ha minimum mapping unit, describe 28 classes for vegetated land cover (Feranec *et al.* 2007), for example conifer forest, broad-leaved forest, heathland. The 1-km resolution land-cover database of North America 2000, also derived from imagery from the VEGETA-TION sensor (Latifovic *et al.* 2002), uses a life-form classification system based on characteristics of the overstorey layer of vegetation as a basis for regional land-cover classes. In this database and map, 202 land-cover and vegetation categories are defined based on species dominance, for example Anderson level III (for example, Ponderosa pine forest).

Low- to moderate-resolution satellite data often have moderate to high temporal resolution, from daily to every few weeks. Building on early work that focused on broadscale (global) vegetation (Justice *et al.* 1985), multidate imagery has been used to help discriminate vegetation types based on temporal profiles of vegetation cover or leaf phenology (Fuller *et al.* 1994; Mayaux *et al.* 2004; Cabral *et al.* 2006).

Moderate resolution data from the Landsat satellites (especially Thematic Mapper and Enhanced Thematic Mapper sensors, or TM and ETM+) have been used extensively for mapping vegetation types distinguished by community composition, structure, and other attributes, e.g. for the land-cover map of Great Britain (Fuller *et al.* 1994). While its moderate GSD (30 m), spectral resolution (seven shortwave and thermal infrared wavebands) and relatively infrequent acquisition (every 16 days) may limit its usefulness for detailed plant community mapping (Xie *et al.* 2008), there are numerous examples of the effective use of data from these sensors, often in combination with other types of mapped environmental data, dating from the 1980s to the present (Franklin *et al.* 1986; Skidmore 1989; Franklin *et al.* 2000; Franklin & Wulder 2002; Sesnie *et al.* 2010; Frohn *et al.* 2011). Continuity of the Landsat mission is very important for vegetation mapping and other types of earth system monitoring, but it is far from secure (Wulder *et al.* 2008).

High-resolution **hyperspectral** imagery such as from the AVIRIS airborne sensor, are primarily acquired from sensors on aircraft, and therefore there are not widely available globally. Recent decades have seen increasing use of hyperspectral imagery for delineating individual tree canopies and identifying tree species (Culvenor 2003; Clark *et al.* 2005; Asner *et al.* 2008), as well as for vegetation type mapping (Kokaly *et al.* 2003).

Hyperspatial resolution imagery, with submeter to 4 m GSD, are available from satellite sensors such as from IKONOS™ and Quickbird™, as well as from airborne sensors such as ADAR (Airborne Data Acquisition and Registration) and CASI (Franklin 1994; Stow *et al.* 1996; Lefsky *et al.* 2001). These data have spatial resolution approaching that of digital orthoimagery but have multiple spectral bands with similar wavelength ranges to Landsat or SPOT, allowing for multispectral image processing, as well as change detection of the images are radiometrically corrected or matched (Coulter & Stow 2009). Data from these sensors are being used for local scale vegetation mapping (Coulter *et al.* 2000; Treitz & Howarth 2000).

16.4   Methods for vegetation mapping

The interpretation methods applied to photographic and multispectral imagery range from visual interpretation for delineating map units and assigning them a vegetation attribute (labeling them), to automated computer-based image processing for object delineation and labeling (classification).

16.4.1   Interpretation of aerial photography

In air photo interpretation, objects or scene elements (see Section 16.4.2) are identified based on their tone, texture, pattern, size, shape, shadow and context. The basic principles of air photo interpretation are discussed in detail in classic texts (for example, Arnold 1997; Paine & Kiser 2003). Objects such as vegetation stands or even tree crowns are delineated based on visual interpretation of boundaries between the mapping units (for example see Web Resource 16.4). Planimetric maps are developed using well-established photogrammetric techniques (formerly analogue, now largely replaced by digital photogrammetry).

Vegetation mapping based on air photo interpretation has been driven by the information needs of resource managers in forestry and rangeland management (Paine & Kiser 2003).

The growing availability of georeferenced photo-like imagery, e.g. digital orthoimagery, and tools for on-screen digitizing using GIS, have lead air photo interpretation for vegetation mapping back to the future. Object identification, especially for complex objects such as vegetation stands, is something the human eye does very well and computer pattern recognition algorithms are challenged by, which is why object-based image classification is such an active area of research (e.g. Yu *et al.* 2006). Contemporary vegetation mapping efforts aimed at identifying communities at high levels of categorical detail often combine phytosociological survey (of plant community composition), vegetation classification, and mapping based on visual interpretation and manual delineation of vegetation map units from air photos (for example Keeler-Wolf 2007, and see the vegetation mapping protocols established by the U.S. National Park Service, http://science.nature.nps.gov/im/inventory/veg/). Photointerpreters have always used landscape context as a source of information for delineation and identification of map units. Nowadays photointerpreters working in a GIS environment can bring in environmental data layers such as geology, topography, soils, fire history and climate maps, and vegetation field data, in order to help delineate and identify vegetation units based on correlations of vegetation with these environmental variables. In fact, in a contemporary approach that harkens back to land systems mapping, GIS-assisted air photo interpretation uses vegetation polygons as the base unit of land analysis wherein vegetation attributes can be coupled with other landscape attributes such as type and level of human disturbance (road density, invasive exotics, land use, erosion). This allows the vegetation map to support environmental analysis and land use planning objectives such as site quality ranking and habitat corridor planning (Keeler-Wolf 2007).

16.4.2   Pattern recognition or image classification

Before reviewing the methods used to develop a thematic map of vegetation types from remotely sensed imagery and other digital environmental maps, collectively termed **pattern recognition** or {image} classification (reviewed in Franklin *et al.* 2003), I will introduce the **scene model**. The scene model explicitly considers the spatial, temporal, spectral and radiometric properties of the scene which is defined as some vegetated portion of the Earth surface viewed at a specific scale (Strahler *et al.* 1986). Different forms of information extraction rely on the assumption that image pixels (ground resolution elements) area either larger (low or 'L' resolution) or smaller (high or 'H' resolution) than the target to be mapped. Image classification is a H-resolution method because the categorical attribute of interest, vegetation type, occurs over spatial extents (stands) that are larger than the spatial resolution of the sensor. L-resolution approaches, such as spectral mixture analysis or regression, have also been used to produce maps of a continuous vegetation attribute, such as cover, biomass or leaf area index.

In the H-resolution classification approach to vegetation mapping, the multivariate (multispectral) vectors representing pixels are grouped according to similarity, and assigned to information classes using either an unsupervised (clustering) or supervised (classification) approach (reviewed in Franklin *et al.* 2003). In the unsupervised case, patterns (clusters) of observations in measurement space are first described using some type of clustering algorithm, and then assigned to information classes. In the supervised case, a sample of observations of the categorical response variable, vegetation type, is used to train a classifier, that is, estimate coefficients or quantitative rules that can be applied to new observations in order to assign them to a vegetation class. Parametric maximum likelihood and non-parametric k-nearest neighbor classification have been widely used for thematic mapping (reviewed in Franklin *et al.* 2003). Both supervised and unsupervised approaches have long been used for vegetation mapping (e.g. Franklin *et al.* 1986; Stow *et al.* 2000).

Although often the vegetation categories are defined independent of the mapping effort using vegetation survey and classification (Chapter 2), in some cases phytosociological survey has been integrated with digital image processing of satellite imagery (Zak & Cabido 2002), e.g. by the U.S. National Park Service.

16.4.3   Predictive vegetation mapping

There have been tremendous developments in statistics (modern regression) and machine learning methods in recent decades, and many of these methods can be considered to be forms of statistical learning or supervised classification (Breiman 2001; Hastie *et al.* 2001) when they are applied to categorical response variables such as vegetation type. These supervised classifiers, including artificial neural networks (Foody *et al.* 1995; Carpenter *et al.* 1999), decision trees (Friedl *et al.* 1999; Franklin *et al.* 2001) and support vector machines (Sesnie *et al.* 2010) have been applied to remotely sensed imagery for vegetation mapping.

One of the most useful aspects of these new approaches is that they can be very flexible about incorporating different types of data, and in particular they are useful for combining remotely-sensed with other environmental predictors for vegetation mapping. Even in the earliest uses of remote sensing for vegetation mapping, digital environmental maps (of terrain, geology, soils and so forth) were called 'ancillary' or 'collateral' data and were used in various ways with imagery to develop regional forest maps (reviewed in Franklin 1995). This was because, while classification of Landsat imagery provided information on vegetation structure, plant communities (defined by species composition) could almost never be reliably discriminated. When combined even using fairly simple methods (decision rules, map algebra), multispectral satellite imagery and other digital mapped environmental data were better able to map plant species assemblages (Strahler 1981; Hutchinson 1982). This complementarity of data and modelling approaches when mapping vegetation structure and land cover versus vegetation composition was described by Lees & Ritman (1991), and is shown diagrammatically in Fig. 16.2.

The mapped environmental variables that are useful for vegetation mapping are those that describe the **primary environmental regimes** of light, heat, water

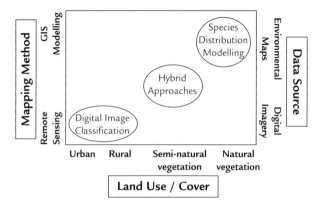

Fig. 16.2 A conceptual model linking land use and land cover (human-altered to natural), mapping methods and typical data sources. Species distribution modelling is sometimes called predictive vegetation modelling when applied to vegetation mapping (Franklin 1995). 'Environmental maps' refers to digital maps of environmental variables related to the primary environmental regimes, including climate, substrate and terrain (as described in the text). (Based on Lees & Ritman 1991.)

and mineral nutrients that influence plant distributions, e.g. climate, substrate, and terrain (Mackey 1993; Franklin 1995; Guisan & Zimmermann 2000), as shown in Table 16.1. Those factors have been described in terms of a spatial hierarchy outlining their scales of influence (Mackey & Lindenmayer 2001; Pearson & Dawson 2003), and the types and sources of GIS data that have been used to represent them are reviewed by Franklin (2010a, Chapter 5). Chapter 15 also discusses the broadscale drivers of vegetation patterns (primarily determined by climate).

Statistical and machine learning approaches have, in recent decades, been extensively applied to the problem of **species distribution modelling**, also called environmental niche modelling. While SDM typically focuses on the spatial prediction of the occurrence, abundance or other attribute of a single species of plant, animal, or other type of organism, it shares similarities with vegetation mapping in terms of the modelling methods and source data. The main difference is that the response variable for vegetation mapping is an attribute of the vegetation or plant community, such as community type, vegetation composition or structure, as noted in Section 16.1.

SDM has been described in detail elsewhere (Elith & Leathwick 2009; Franklin 2010a). I will limit my discussion to those approaches that have been developed for spatial modelling of plant communities in order to map vegetation. Ferrier & Guisan (2006) in their comprehensive review of this topic describe three strategies for community mapping based on predictive modelling. (1) Models can be derived for pre-defined community types or other community attributes using some form of supervised classification based on modern regression and machine learning classifiers (Lees & Ritman 1991; Brzeziecki *et al.* 1993; Hilbert & Ostendorf 2001). (2) Alternatively, models can be developed for plant species and those separate spatial predictions for species can be combined in some way to create vegetation maps (Austin 1998; Zimmermann & Kienast 1999; Leathwick 2001; Ferrier *et al.* 2002). (3) Finally, if complete phytosociological data are available, the communities can be defined and their distribution predicted at the same time using multiresponse modelling methods (e.g. De'ath 2002; Leathwick *et al.* 2005), constrained classification or ordination (Guisan *et al.* 1999; Ohmann & Gregory 2002), or compositional dissimilarity modelling (Ferrier *et al.* 2007). For example, the nearest neighbour imputation method of constrained ordination (Ohmann *et al.* 2011) was used to develop detailed forest attribute maps for the state of Oregon, USA (see Web Resource 16.5).

This third approach, classifying and predicting the distribution of plant communities at the same time, holds a lot of potential for vegetation mapping. It requires extensive vegetation survey data, but when these are available they typically include species lists and abundances for all plants in sample areas, and support combined classification and mapping of plant communities using multivariate approaches not unfamiliar to plant community ecologists.

The use of SDM for vegetation mapping can be thought of as a kind of **interpolation** in multivariate environmental predictor space (and typically in geographical space as well) where, in the terminology of supervised classification, training data are associated with a set of predictors, and rules for classifying new observations are derived from the training data using some modelling method. These rules are then applied to locations where the response variable is unknown (but the predictors are) to 'fill in the gaps' between surveyed or observed locations (Franklin 2010a). SDM is also becoming very widely used for projecting biotic distributions under environmental change scenarios, for example forecasting the impacts of climate change on species distributions, or the potential distribution of non-native (invasive) species (Peterson & Vieglais 2001; Iverson *et al.* 2008). This usually involves **extrapolating** to novel or non-analogue environmental conditions, or projection outside the range of predictor values used in training. This is something your statistics instructor probably told you never to do (e.g. with a regression model), and the limitations of this approach, as well as potential solutions, have been extensively written about (e.g. Hijmans & Graham 2006; Araújo & Luoto 2007; Dormann 2007; Elith & Leathwick 2009; Franklin 2010b).

16.4.4   Mapping plant diversity using remote sensing

In addition to vegetation *types*, measures of community diversity (e.g. species richness) have been estimated and mapped using remote sensing from landscape to global scales (Stoms & Estes 1993; Nagendra 2001; Turner *et al.* 2003). A number of recent studies have linked biochemical diversity measured from hyperspectral AVIRIS imagery (Carlson *et al.* 2007), and structural diversity and measured from multispectral ETM+ imagery (Gillespie 2005) to tree species richness. At even coarser scales, reflectance, surface temperature and NDVI summaries from MODIS and AVHRR have been used to measure spatial and temporal patterns of primary productivity, and related to continental-scale patterns of tree species richness (Waring *et al.* 2006; Saatchi *et al.* 2008). Is it worth noting that remote sensing does not sense diversity directly, but can identify the spectral-spatial-temporal signatures of vegetation communities that are more or less diverse, based on the same attributes used for remote-sensing-based community mapping – leaf type, leaf phenology, canopy chemistry, structure and arrangement, and so forth. Although there are a growing number of studies mapping biodiversity metrics from remote sensing, this is still an active research area, rather than a field that has progressed to the point of producing standardized data products.

16.4.5   Vegetation map accuracy

Quantitative descriptions of accuracy should be associated with all maps, and in the case of vegetation maps the most important measure may be its thematic or categorical accuracy – the correct assignment of vegetation type labels to locations. The subject of thematic map accuracy has been extensively developed in the literature (Fielding & Bell 1997; Stehman & Czaplewski 1998; Foody 2002; Pontius 2002) and draws heavily from methods used in the statistical analysis of categorical data. Although it is beyond the scope of this chapter to address this important topic, any producer or user of a vegetation map should have a basic understanding of the methods used to characterize map accuracy, including the sampling requirements for a statistically rigorous assessment, and the statistics used to describe accuracy or agreement between mapped and observed classes.

16.5   Examples of recent vegetation maps illustrating their different uses

Beginning in the 1990s, a number of very large area (small cartographic map scale) vegetation maps were developed and published, particularly in *Journal of Vegetation Science* (a journal established in 1990 by the International Association of Vegetation Science), complementing earlier maps compiled at similar scales in the 1970s and 1980s for Africa, South America and island south-east Asia (Malesia) under the auspices of the United Nations Educational, Scientific and Cultural Organization (UNESCO).

For example, a vegetation map was compiled for Europe at 1:2.5 million scale and an accompanying book published describing it (Neuhäusl 1991). This map was the collaborative effort of over 50 vegetation scientists from 24 European countries, and depicted the natural vegetation that exists today or would exist on a site in the absence of human influence. It was compiled from numerous finer-scaled efforts using a variety of methods and data sources and the mapping units correspond to the association-level in the Braun-Blanquet system. In another example, a 1:5 million scale map of vegetation formations for tropical continental Asia was compiled from published and unpublished maps onto a common base map, and the boundaries of each mapping unit were updated based on manual delineation of hundreds of Landsat Multi-spectral Scanner (MSS) scenes (Blasco *et al.* 1996).

The vegetation map of South Africa, Lesotho and Swaziland is another notable contemporary example of a map produced for a large area with significant categorical (floristic) detail (Mucina & Rutherford 2006). It was developed for national-level conservation and land management planning, encompasses 1.3 million km², and was developed between 1995 and 2005 at a nominal scale of 1:250000. The basic mapping unit is the vegetation unit, defined as vegetation community complexes that occupy habitat complexes at the landscape scale (Mucina *et al.* 2006). Vegetation units are grouped hierarchically into vegetation groups and biomes.

The South African map was developed using a complex hybrid of approaches ranging from supervised classification of satellite images (for forests) to manual delineation of boundaries and labelling of map units in a GIS informed by digital maps of climate, geology, topography, soil type and other vegetation maps. For example, in the arid biomes, multivariate analysis of vegetation data with environmental variables lead to the use of mapped land type boundaries to define vegetation unit boundaries. In other biomes, digital elevation models were used to derive slope angle and aspect, which were important in delineating some vegetation units. Map units are described in terms of growth-form composition, and also include lists of important, biogeographically important, and endemic plant species as attributes. Species lists are based on extensive records from the national herbarium database (Pretoria Computerized Information System, PRECIS). Thus, this is a multi-attribute vegetation database. Over 100 people were directly involved in mapping and describing the more than 17000 vegetation map units. Subregions were mapped separately by teams of experts and then GIS procedures were used to edit the final map.

16.6   Dynamic vegetation mapping

A notable feature of contemporary large-area vegetation mapping programs, especially those conducted by government agencies, is that it is explicitly acknowledged that vegetation and land cover are dynamic, and that the maps need to be periodically updated, either by developing an entirely new map, or through map updating, in order for those maps to remain useful. Dynamic vegetation maps, produced by any number of means, can and are being used to detect pathways of succession, study the impact of land use changes, trace the effects of global warming, and support spatially explicit models of ecosystem processes and community dynamics.

A very large research field of land-use change mapping using remotely sensed and other data has been developed in support of these operational mapping programs. Because Landsat image data have been acquired and archived for a reasonably long time (since the early 1980s), they have been used to map vegetation change in a number of regions (reviewed by Xie *et al.* 2008). For example, in the state of California, USA, remote-sensing change detection methods, based on enhancing spectral differences in multidate Landsat TM imagery and relating those differences to land-cover change (Rogan *et al.* 2003), are used to produce periodic updates to land-cover maps (provided by the California Department of Forestry and Fire Protection).

The concept of repeated vegetation mapping in order to study spatial-temporal vegetation dynamics is not new and is not restricted to large area, coarse-scale, remote-sensing-based mapping. For example, in the fine-scale (small area) study by van Dorp *et al.* (1985) on the dynamics of the vegetation of a 150-ha dune area at Voorne, the Netherlands, vegetation development over 45 years after cessation of overgrazing was followed using two detailed vegetation maps and additional air photo interpretations. In this study multiple succession pathways were detected, meaning that the next phase in the succession of each vegetation type was dependent on the surrounding vegetation (Chapter 4).

16.7   Future of vegetation mapping research and practice

Vegetation data and mapping procedures must be matched to the information needs of the scientists, agencies, policymakers or resource managers who will use these maps. We have seen several examples of traditional methods, such as air photo interpretation and manual delineation of map units, being applied to novel data sources, from digital orthoimagery to coarse-scale satellite data (Table 16.1).

This can be illustrated using the recent history of vegetation mapping in California, USA (Keeler-Wolf 2007), summarized in Table 16.2. Field-based mapping (described in Section 16.3.1) carried out in the 1930s VTM programme yielded maps that have a surprisingly detailed level of information about species composition in forests and shrublands, but developed in the pre-GIS era these maps had large mapping units (polygons) whose boundaries locations were not precisely georeferenced. Vegetation mapping efforts by the Forest Service in the 1980s to 1990s relied on Landsat imagery and yielded somewhat spatially and categorically generalized vegetation maps. The GAP analysis vegetation database produced in the 1990s used a hybrid approach to mapping and yielded a map that was spatially generalized (large map units), but with a fair amount of categorical detail in the vegetation attributes. Ongoing efforts to integrate vegetation classification and mapping rely on GIS-supported photointerpretation of map units and extensive survey and classification of plant communities, resulting in fine-scale, detailed vegetation maps that cover very large areas. These maps are being integrated with additional attributes for the purpose of conservation analysis (see table 1.10 in Keeler-Wolf 2007).

While it is beyond the scope of this chapter to exhaustively review all recent landscape to regional-scale vegetation mapping efforts, examples were chosen to illustrate that contemporary vegetation mapping creatively integrates old and new data and methods including field survey, high resolution aerial imagery, satellite remote sensing, photointerpretation, image processing and GIS modelling (e.g. see Web Resource 16.6) to develop vegetation maps that are useful for environmental analysis and planning.

Acknowledgements

I thank T. Keeler-Wolf, E. van der Maarel, J. Ripplinger and A. H. Strahler for greatly improving this chapter with their comments on earlier drafts.

References

Allen, T.F.H. & Hoekstra, T.W. (1990) The confusion between scale-defined levels and conventional levels of organization in ecology. *Journal of Vegetation Science* 1, 5–12.

Anderson, J.R., Hardy, E.E., Roach, J.T. & Witmer, R.E. (1976) *A land use and land cover classification system for use with remote sensor data. US Geological Survey Professional Paper* 964. US Geologic Survey, Arlington, VA.

Araújo, M.B. & Luoto, M. (2007) The importance of biotic interactions for modelling species distributions under climate change. *Global Ecology and Biogeography* 16, 743–753.

Arnold, R.H. (1997) *Interpretation of Airphotos and Remotely Sensed Imagery*. Prentice Hall, Upper Saddle River, NJ.

Asner, G.P., Jones, M.O., Martin, R.E., Knapp, D.E. & Hughes, R.F. (2008) Remote sensing of native and invasive species in Hawaiian forests. *Remote Sensing of Environment* 112, 1912–1926.

Astle, W.L., Webster, R. & Lawrance, C.J. (1969) Land classification for management planning in the Luangwa Valley of Zambia. *Journal of Applied Ecology* 6, 143–169.

Austin, M.P. (1998) An ecological perspective on biodiversity investigations: examples from Australian eucalypt forests. *Annals of the Missouri Botanical Garden* 85, 2–17.

Bailey, R.G. (2004) Identifying ecoregion boundaries. *Environmental Management* 34, S14–S26.

Beard, J.S. (1975) The vegetation survey of Western Australia. *Vegetatio* 30, 179–187.

Blasco, F., Bellan, M.F. & Aizpuru, M. (1996) A vegetation map of tropical continental Asia at scale 1:5 million. *Journal of Vegetation Science* 7, 623–634.

Breiman, L. (2001) Random forests. *Machine Learning* 45, 15–32.

Brzeziecki, B., Kienast, F. & Wildi, O. (1993) A simulated map of the potential natural forest vegetation of Switzerland. *Journal of Vegetation Science* 4, 499–508.

Cabral, A.I.R., Vasconcelos, M.J.P., Pereira, J.M.C., Martins, E. & Bartholome, E. (2006) A land cover map of southern hemisphere Africa based on SPOT-4 Vegetation data. *International Journal of Remote Sensing* 27, 1053–1074.

Carlson, K.M., Asner, G.P., Hughes, R.F., Ostertag, R. & Martin, R.E. (2007) Hyperspectral remote sensing of canopy biodiversity in Hawaiian lowland rainforests. *Ecosystems* 10, 536–549.

Carmel, Y. & Kadmon, R. (1998) Computerized classification of Mediterranean vegetation using pan-chromatic aerial photographs. *Journal of Vegetation Science* 9, 445–454.

Carpenter, G.A., Gopal, S., Macomber, S. *et al.* (1999) A neural network method for efficient vegetation mapping. *Remote Sensing of Environment* 70, 326–338.

Christian, C.S. & Stewart, G.A. (1968) Methodogy of integrated surveys. In: *Aerial Surveys and Integrated Studies. Proceedings Toulouse Conference*, pp. 233–280. UNESCO, Paris.

Clark, M.L., Roberts, D.A. & Clark, D.B. (2005) Hyperspectral discrimination of tropical rain forest tree species at leaf to crown scales. *Remote Sensing of Environment* 96, 375–398.

Coulter, L.L. & Stow, D.A. (2009) Monitoring habitat preserves in southern California using high spatial resolution multispectral imagery. *Environmental Monitoring and Assessment* 152, 343–356.

Coulter, L., Stow, D., Hope, A. *et al.* (2000) Comparison of high spatial resolution imagery for efficient generation of GIS vegetation layers. *Photogrammetric Engineering and Remote Sensing* 66, 1329–1335.

Cramer, W., Bondeau, A., Woodward, F.I. *et al.* (2001) Global response of terrestrial ecosystem structure and function to CO_2 and climate change: results from six dynamic global vegetation models. *Global Change Biology* 7, 357–373.

Culvenor, D.S. (2003) Extracting individual tree information: a survey of tehcniques for high spatial resolution imagery. In: *Remote Sensing of Forest Environments: Concepts and Case Studies* (eds M.A. Wulder & S.E. Franklin), pp. 255–277. Kluwer Academic Publishers, Boston, MA.

Davis, F.W., Stine, P.A., Stoms, D.M., Borchert, M.I. & Hollander, A. (1995) Gap analysis of the actual vegetation of California: 1. The southwestern region. *Madroño* **42**, 40–78.

De'ath, G. (2002) Multivariate regression trees: a new technique for modeling species–environment relationships. *Ecology* **83**, 1105–1117.

DeMers, M. (1991). Classification and purpose in automated vegetation maps. *Geographical Review* **81**, 267–280.

Dormann, C.F. (2007). Promising the future? Global change projections of species distributions. *Basic and Applied Ecology* **8**, 387–397.

Driese, K.L., Reiners, W.A., Merrill, E.H. & Gerow, K.G. (1997) A digital land cover map of Wyoming, USA: a tool for vegetation analysis. *Journal of Vegetation Science* **8**, 133–146.

Elith, J. & Leathwick, J.R. (2009) Species distribution models: ecological explanation and prediction across space and time. *Annual Review of Ecology, Evolution and Systematics* **40**, 677–697.

Feranec, J., Hazeu, G., Christensen, S. & Jaffrain, G. (2007) Corine land cover change detection in Europe (case studies of the Netherlands and Slovakia). *Land Use Policy* **24**, 234–247.

Ferrier, S. & Guisan, A. (2006) Spatial modelling of biodiversity at the community level. *Journal of Applied Ecology* **43**, 393–404.

Ferrier, S., Drielsma, M., Manion, G. & Watson, G. (2002) Extended statistical approaches to modelling spatial pattern in biodiversity in northeast New South Wales. II. Community-level modeling. *Biodiversity and Conservation* **11**, 2309–2338.

Ferrier, S., Manion, G., Elith, J. & Richardson, K. (2007) Using generalized dissimilarity modelling to analyse and predict patterns of beta diversity in regional biodiversity assessment. *Diversity and Distributions* **13**, 252–264.

Fielding, A. & Bell, J. (1997) A review of methods for the assessment of prediction errors in conservation presence/absence models. *Environmental Conservation* **24**, 38–49.

Foody, G.M. (2002) Status of land cover classification accuracy assessment. *Remote Sensing of Environment* **80**, 185–201.

Foody, G.M., McCulloch, M.B. & Yates, W.B. (1995) Classification of remotely sensed data by an artificial neural network: issues related to training data characteristics. *Photogrammetric Engineering and Remote Sensing* **61**, 391–401.

Franklin, J. (1995) Predictive vegetation mapping: geographic modeling of biospatial patterns in relation to environmental gradients. *Progress in Physical Geography* **19**, 474–499.

Franklin, J. (2010a) *Mapping Species Distributions: Spatial Inference and Prediction*. Cambridge University Press, Cambridge.

Franklin, J. (2010b) Moving beyond static species distribution models in support of conservation biogeography. *Diversity and Distributions* **16**, 321–330.

Franklin, J. & Woodcock, C.E. (1997) Multiscale vegetation data for the mountains of Southern California: spatial and categorical resolution. In: *Scale in Remote Sensing and GIS* (eds D.A. Quattrochi & M.F. Goodchild), pp. 141–168. CRC/Lewis Publishers Inc., Boca Raton, FL.

Franklin, J., Logan, T., Woodcock, C.E. & Strahler, A.H. (1986) Coniferous forest classification and inventory using Landsat and digital terrain data. *IEEE Transactions on Geoscience and Remote Sensing* **GE-24**, 139–149.

Franklin, J., Woodcock, C.E. & Warbington, R. (2000) Multi-attribute vegetation maps of Forest Service lands in California supporting resource management decisions. *Photogrammetric Engineering and Remote Sensing* **66**, 1209–1217.

Franklin, J., Phinn, S.R., Woodcock, C.E. & Rogan, J. (2003) Rationale and conceptual framework for classification approaches to assess forest resources and properties. In: *Remote Sensing of Forest Environments: Concepts and Case Studies* (eds M.A. Wulder & S.E. Franklin), pp. 279–300. Kluwer Academic Publishers, Boston, MA.

Franklin, S.E. (1994) Discrimination of sub-alpine forest and canoy density using digital ACSI, SPOT PLA and Landsat TM data. *Photogrammetric Engineering and Remote Sensing* **60**, 1233–1241.

Franklin, S.E. (2001) *Remote Sensing for Sustainable Forest Management*. Lewis Publishers, Boca Raton, FL.

Franklin, S.E. & Wulder, M.A. (2002) Remote sensing methods in medium spatial resolution satellite data land cover classification of large areas. *Progress in Physical Geography* **26**, 173–205.

Franklin, S.E., Stenhouse, G.B., Hansen, M.J. *et al.* (2001) An Integrated Decision Tree Approach (IDTA) to mapping landcover using satellite remote sensing in support of grizzly bear habitat analysis in the Alberta yellowhead ecosystem. *Canadian Journal of Remote Sensing* **27**, 579–592.

Friedl, M., Brodley, C. & Strahler, A. (1999) Maximizing land cover classification accuracies produced by decision trees at continental to global scales. *IEEE Transactions on Geoscience and Remote Sensing* **37**, 969–977.

Friedl, M.A., McIver, D.K., Hodges, J.C.F. *et al* (2002) Global land cover mapping from MODIS: algorithms and early results. *Remote Sensing of Environment* **83**, 287–302.

Frohn, R.C., Autrey, B.C., Lane, C.R. & Reif, M. (2011) Segmentation and object-oriented classification of wetlands in a karst Florida landscape using multi-season Landsat-7 ETM+ imagery. *International Journal of Remote Sensing* **32**, 1471–1489.

Fuller, R.M., Groom, G.B. & Jones, A.R. (1994) The land-cover map of Great Britain – an automated classification of Landsat Thematic Mapper data. *Photogrammetric Engineering and Remote Sensing* **60**, 553–562.

Gillespie, T.W. (2005) Predicting woody-plant species richness in tropical dry forests: A case study from south Florida, USA. *Ecological Applications* **15**, 27–37.

Goodchild, M.F. (1994) Integrating GIS and remote sensing for vegetation analysis and modeling: methodological issues. *Journal of Vegetation Science* **5**, 615–626.

Goward, S.N., Tucker, C.J. & Dye, D.G. (1985) North American vegetation patterns observed with NOAA-7 Advanced Very High Resolution Radiometer. *Vegetatio* **64**, 3–14.

Grossman, D.H., Faber-Langendoen, D., Weakley, A.S. *et al.* (1998) *International Classification of Ecological Communities: Terrestrial Vegetation of the United States* Vol. 1. *The National Vegetation Classification System: Development, Status and Applications*. The Nature Conservancy, Washington, DC.

Guisan, A. & Zimmermann, N.E. (2000) Predictive habitat distribution models in ecology. *Ecological Modelling* **135**, 147–186.

Guisan, A., Weiss, S. & Weiss, A. (1999) GLM versus CCA spatial modeling of plant species distributions. *Plant Ecology* **143**, 107–122.

Hansen, M.C., Defries, R.S., Townshend, J.R.G. & Sohlberg, R. (2000) Global land cover classification at 1km spatial resolution using a classification tree approach. *International Journal of Remote Sensing* **21**, 1331–1364.

Hastie, T., Tibshirani, R. & Friedman, J. (2001) *The Elements of Statistical Learning: Data Mining, Inference and Prediction*. Springer-Verlag, New York.

Hijmans, R.J. & Graham, C.H. (2006) The ability of climate envelope models to predict the effect of climate change on species distributions. *Global Change Biology* **12**, 2272–2281.

Hilbert, D.W. & Ostendorf, B. (2001) The utility of artificial neural networks for modelling the distribution of vegetation in past, present and future climates. *Ecological Modelling* **146**, 311–327.

Hutchinson, C.F. (1982) Techniques for combining Landsat and ancillary data for digital classification improvement. *Photogrammetric Engineering and Remote Sensing* **48**, 123–130.

Iverson, L.R., Prasad, A.M., Matthews, S.N. & Peters, M. (2008) Estimating potential habitat for 134 eastern US tree species under six climate scenarios. *Forest Ecology and Management* **254**, 390–406.

Jennings, M.D., Faber-Langendoen, D., Loucks, O.L., Peet, R.K. & Roverts, D. (2009) U.S. plant community classificaition. *Ecological Monographs* **79**, 173–199.

Jensen, J.R. (2000) *Remote Sensing of the Environment: An Earth Resource Perspective*. Prentice Hall, Upper Saddle River, NJ.

Jones, H.G. & Vaughan, R.A. (2010) *Remote Sensing of Vegetation*. Oxford University Press, Oxford, UK.

Justice, C.O., Townshend, J.R.G., Holben, B.N. & Tucker, C.J. (1985. Analysis of the phenology of global vegetation using meteorological satellite data. *International Journal of Remote Sensing* **6**, 1271–1318.

Keeler-Wolf, T. (2007) The history of vegetation classification and mapping in California. In: *Terrestrial Vegetation of California* (eds M.G. Barbour, T. Keeler-Wolf, & A.A. Schoenherr), pp. 1–42. University of California Press, Berkeley, CA.

Kelly, M., Allen-Diaz, B. & Kobzina, N. (2005) Digitization of a historic dataset: the Wieslander California vegetation type mapping project. *Madroño* **52**, 191–201.

Kerr, J.T. & Ostrovsky, M. (2003) From space to species: ecological applications for remote sensing. *Trends in Ecology & Evolution* **16**, 299–305.

Kokaly, R.F., Despain, D.G., Clark, R.N. & Livo, K.E. (2003) Mapping vegetation in Yellowstone National Park using spectral feature analysis of AVIRIS data. *Remote Sensing of Environment* **84**, 437–456.

Küchler, A.W. (1967) *Vegetation Mapping*. Ronald Press Co., New York, NY.

Küchler, A.W. (1984) Ecological vegetation maps. *Vegetatio* **55**, 3–10.

Küchler, A.W. & Zonneveld, I.S. (eds.) (1988) *Vegetation Mapping*, Kluwer Academic Publishers, Boston.

Latifovic, R., Zhu, Z.-L., Cihlar, J. & Giri, C. (2002) *Land Cover of North America 2000*. Natural Resources Canada, Canada Center for Remote Sensing, US Geological Survey, EROS Date Center.

Leathwick, J.R. (2001) New Zealand's potential forest pattern as predicted from current species–environment relationships. *New Zealand Journal of Botany* **39**, 447–464.

Leathwick, J.R., Rowe, D., Richardson, J., Elith, J. & Hastie, T. (2005) Using multivariate adaptive regression splines to predict the distributions of New Zealand's freshwater diadromous fish. *Freshwater Biology* **50**, 2034–2052.

Lees, B.G. & Ritman, K. (1991) Decision-tree and rule-induction approach to integration of remotely sensed and GIS data in mapping vegetation in disturbed or hilly environments. *Environmental Management* **15**, 823–831.

Lefsky, M.A., Cohen, W.B. & Spies, T.A. (2001) An evaluation of alternate remote sensing products for forest inventory, monitoring, and mapping of Douglas-fir forests in western Oregon. *Canadian Journal of Forest Research – Revue Canadienne De Recherche Forestière* **31**, 78–87.

Lenihan, J.M., Bachelet, D., Neilson, R.P. & Drapek, R. (2008) Simulated response of conterminous United States ecosystems to climate change at different levels of fire suppression, CO_2 emission rate, and growth response to CO_2. *Global and Planetary Change* **64**, 16–25.

Mackey, B.G. (1993) Predicting the potential distribution of rain-forest structural characteristics. *Journal of Vegetation Science* **4**, 43–54.

Mackey, B.G. & Lindenmayer, D.B. (2001) Towards a hierarchical framework for modeling the spatial distribution of animals. *Journal of Biogeography* **28**, 1147–1166.

Margules, C. & Pressey, R. (2000) Systematic conservation planning. *Nature* **405**, 243–253.

Mayaux, P., Bartholome, E., Fritz, S. & Belward, A. (2004) A new land-cover map of Africa for the year 2000. *Journal of Biogeography* **31**, 861–877.

Millington, A.C. & Alexander, R. (2000) Vegetation mapping in the last three decades of the twentieth century. In: *Vegetation Mapping: From Patch to Planet* (eds R. Alexander & A.C. Millington), pp. 321–331. John Wiley & Sons, Ltd, Chichester.

Mladenoff, D.J. (2004) LANDIS and forest landscape models. *Ecological Modelling* **180**, 7–19.

Mucina, L. & Rutherford, M.C. (2006) *The Vegetation of South Africa, Lesotho and Swaziland*. South African Biodiversity Institute, Pretoria.

Mucina, L., Rutherford, M.C. & Powrie, L.W. (2006) The logic of the map: approaches and procedures. In: *The Vegetation of South Africa, Lesotho and Swaziland* (eds L. Mucina & M.C. Rutherford), pp. 13–29. South African Biodiversity Institute, Pretoria.

Mueller-Dombois, D. & Ellenberg, H. (1974) *Aims and Methods of Vegetation Ecology*. The Blackburn Press, Caldwell, NJ.

Mullerova, J. (2004) Use of digital aerial photography for sub-alpine vegetation mapping: a case study from the Krkonose Mts., Czech Republic. *Plant Ecology* **175**, 259–272.

Nagendra, H. (2001) Using remote sensing to assess biodiversity. *International Journal of Remote Sensing* **22**, 2377–2400.

Nemani, R. & Running, S.W. (1996) Implementation of a hierarchical global vegetation classification in ecosystem function models. *Journal of Vegetation Science* **7**, 337–346.

Neuhäusl, R. (1991) Vegetation map of Europe – first results and current state. *Journal of Vegetation Science* **2**, 131–134.

O'Neill, R.V., DeAngelis, D.L., Waide, J.B. & Allen, T.H.F. (1986) *A Hierarchical Concept of Ecosystems*. Princeton University Press, Princeton, NJ.

Ohmann, J.L. & Gregory, M.J. (2002) Predictive mapping of forest composition and structure with direct gradient analysis and nearest neighbor imputation in coastal Oregon, U.S.A. *Canadian Journal of Forest Research* **32**, 725–741.

Ohmann, J.L., Gregory, M.J., Henderson, E.B. & Roberts, H.M. (2011) Mapping gradients of community composition with nearest-neighbor imputation: extending plot data for landscape analysis. *Journal of Vegetation Science* **22**, 660–676.

Paine, D.P. & Kiser, J.D. (2003) *Aerial Photography and Image Interpretation*. John Wiley & Sons, Ltd, Hoboken, NJ.

Pearson, R.G. & Dawson, T.P. (2003) Predicting the impacts of climate change on the distribution of species: are bioclimatic envelope models useful? *Global Ecology & Biogeography* **12**, 361–371.

Perring, F.H. & Walters, S.M. (1962) *Atlas of the British Flora*. Botanical Society of the British Isles.

Peterson, A.T. & Vieglais, D.A. (2001) Predicting species invasions using ecological niche modeling: new approaches from bioinformatics attack a pressing problem. *Bioscience* **51**, 363–371.

Phinn, S.R. (1998) A framework for selecting appropriate remotely sensed data dimensions for environmental monitoring and management. *International Journal of Remote Sensing* **19**, 3457–3463.

Phinn, S.R., Stow, D.A., Franklin, J., Mertes, L.A.K. & Michaelsen, J. (2003) Remotely sensed data for ecosystem analyses: combining hierarchy theory and scene models. *Environmental Management* **31**, 429–441.

Pontius, R.G. (2002) Statistical methods to partition effects of quantity and location during comparison of categorical maps at multiple resolutions. *Photogrammetric Engineering and Remote Sensing* **68**, 1041–1049.

Preston, C.D., Pearman, D.A. & Dines, T.D. (eds) (2002) *New Atlas of the British and Irish Flora*. Oxford University Press, Oxford.

Pu, R. & Gong, P. (2000) Band selection from hyperspectral data for conifer species identification. *Geographic Information Science* **6**, 137–142.

Rogan, J., Miller, J., Stow, D. *et al.* (2003) Land cover change monitoring in southern California using multitemporal Landsat TM and ancillary data. *Photogrammetric Engineering and Remote Sensing* **69**, 793–804.

Saatchi, S., Buermann, W., Ter Steege, H., Mori, S. & Smith, T.B. (2008) Modeling distribution of Amazonian tree species and diversity using remote sensing measurements. *Remote Sensing of Environment* **112**, 2000–2017.

Scheller, R.M., Domingo, J.B., Sturtevant, B.R. *et al.* (2007) Design, development, and application of LANDIS-II, a spatial landscape simulation model with flexible temporal and spatial resolution. *Ecological Modelling* **201**, 409–419.

Scott, J.M., Davis, F., Csuti, B. *et al.* (1993) Gap analysis: a geographical approach to protection of biological diversity. *Wildlife Monographs* **123**, 1–41.

Sesnie, S.E., Finegan, B., Gessler, P.E. *et al.* (2010) The multispectral separability of Costa Rican rainforest types with support vector machines and Random Forest decision trees. *International Journal of Remote Sensing* **31**, 2885–2909.

Skidmore, A.K. (1989) An expert system classifies eucalypt forest types using Thematic Mapper data and a digital terrain model. *Photogrammetric Engineering and Remote Sensing* **55**, 1449–1464.

Skidmore, A.K., Franklin, J., Dawson, T.P. & Pilesjo, P. (2011) Geospatial tools address emerging issues in spatial ecology: a review and commentary on the Special Issue. *International Journal of Geographical Information Science* **25**, 337–365.

Steers, R.J., Curto, M. & Holland, V.L. (2008) Local scale vegetation mapping and ecotone analysis in the Southern Coast Range, California. *Madroño* **55**, 26–40.

Stehman, S.V. & Czaplewski, R.L. (1998) Design and analysis for thematic map accuracy assessment: fundamental principles. *Remote Sensing of Environment* **64**, 331–344.

Stoms, D. & Estes, J.E. (1993) A remote sensing research agenda for mapping and monitoring biodiversity. *International Journal of Remote Sensing* **14**, 1839–1860.

Stoms, D.M., Davis, F.W. & Cogan, C.B. (1992) Sensitivity of wildlife habitat models to uncertainties in GIS data. *Photogrammetric Engineering and Remote Sensing* **58**, 843–850.

Stow, D., Hope, A., Nguyen, A.T., Phinn, S. & Benkelman, C.A. (1996). Monitoring detailed land surface changes using an airborne multispectral digital camera system. *Ieee Transactions on Geoscience and Remote Sensing* **34**, 1191–1203.

Stow, D., Daeschner, S., Boynton, W. & Hope, A. (2000) Arctic tundra functional types by classification of single-date and AVHRR bi-weekly NDVI composite datasets. *International Journal of Remote Sensing* **21**, 1773–1779.

Strahler, A.H. (1981) Stratification of natural vegetation for forest and rangeland inventory using Landsat digital imagery and collateral data. *International Journal of Remote Sensing* **2**, 15–41.

Strahler, A.H., Woodcock, C.E. & Smith, J.A. (1986) On the nature of models in remote sensing. *Remote Sensing of Environment* **20**, 121–139.

Treitz, P. & Howarth, P. (2000) Integrating spectral, spatial, and terrain variables for forest ecosystem classification. *Photogrammetric Engineering and Remote Sensing* 66, 305–317.

Turner, M.G., Gardner, R.H. & O'Neill, R.V. (2001) *Landscape Ecology in Theory and Practice*. Springer-Verlag, New York, NY.

Turner, W., Spector, S., Gardiner, N. *et al.* (2003) Remote sensing for biodiversity science and conservation. *Trends in Ecology & Evolution* 18, 306–314.

Urban, D.L., O'Neill, R.V. & Shugart Jr., H.H. (1987) Landscape ecology. *Bioscience* 37, 119–127.

van Dorp, D., Boot, R. & van der Maarel, E. (1985) Vegetation succession on the dunes near Oostvoorne, The Netherlands, since 1934, interpreted from air photographs and vegetation maps. *Vegetatio* 58, 123–136.

Walker, D.A., Raynolds, M.K., Daniels, F.J.A. *et al.* (2005) The Circumpolar Arctic vegetation map. *Journal of Vegetation Science* 16, 267–282.

Waring, R.H., Coops, N.C., Fan, W. & Nightingale, J.M. (2006) MODIS enhanced vegetation index predicts tree species richness across forested ecoregions in the contiguous USA. *Remote Sensing of Environment* 103, 218–226.

Wieslander, A.E. (1935) A vegetation type map of California. *Madroño* 3, 140–144.

Woodcock, C.D., Collins, J., Gopal, S. *et al.* (1994) Mapping forest vegetation using Landsat TM imagery and a canopy reflectance model. *Remote Sensing of Environment* 50, 240–254.

Wulder, M.A., White, J.C., Goward, S.N. *et al.* (2008) Landsat continuity: issues and opportunities for land cover monitoring. *Remote Sensing of Environment* 112, 955–969.

Wyatt, B.K. (2000) Vegetation mapping from ground, air and space – competitive or complimentary techniques? In: *Vegetation Mapping* (eds R. Alexander & A.C. Millington), pp. 3–15. John Wiley & Sons, Ltd, Chichester.

Xie, Y., Sha, Z. & Yu, M. (2008) Remote sensing imagery in vegetation mapping: a review. *Journal of Plant Ecology* 1, 9–23.

Yu, Q., Gong, P., Clinton, N. *et al.* (2006) Object-based detailed vegetation classification. with airborne high spatial resolution remote sensing imagery. *Photogrammetric Engineering and Remote Sensing* 72, 799–811.

Zak, M.R. & Cabido, M. (2002) Spatial patterns of the Chaco vegetation of central Argentina: integration of remote sensing and phytosociology. *Applied Vegetation Science* 5, 213–226.

Zimmermann, N. & Kienast, F. (1999) Predictive mapping of alpine grasslands in Switzerland: species versus community approach. *Journal of Vegetation Science* 10, 469–482.

17

Vegetation Ecology and Global Change

Brian Huntley and Robert Baxter
University of Durham, UK

17.1 Introduction

The term 'global change' refers to the changes that are currently taking place in various aspects of the global environment as a consequence of human activities. These changes can be grouped into two broad categories. In the first category are changes to components of the earth system that are inherently global in their extent and impact, principally because they are changes to 'well-mixed' earth system components, notably the atmosphere. Amongst these we include:

1 climatic change;
2 increasing atmospheric concentrations of CO_2;
3 increased fluxes of ultraviolet-B (UV-B) as a consequence of 'thinning' of the stratospheric ozone (O_3) layer;
4 increasing rates of deposition of nitrogen compounds (NO_x and NH_3) from the atmosphere; and
5 increasing tropospheric concentrations of various pollutants, notably SO_2, NO_x and O_3.

In the second category are changes that individually are of local to regional extent, but that, because they recur worldwide, have a global impact upon biodiversity and/or upon earth system processes. The latter will especially arise where the changes taking place alter land surface qualities, and hence impact upon transfers of energy and materials between the land surface and the overlying atmosphere. In this category we include:

Vegetation Ecology, Second Edition. Eddy van der Maarel and Janet Franklin.
© 2013 John Wiley & Sons, Ltd. Published 2013 by John Wiley & Sons, Ltd.

6 land-cover changes as a consequence of human land use, often resulting in habitat loss and fragmentation, as well as in changes of land surface qualities;
7 selective pressures upon ecosystem components as a result of human activities (selective felling, hunting, persecution of carnivores) that, in addition to their biodiversity impacts, may result in changes in ecosystem structure and/ or function with consequent impacts upon the participation of ecosystems in global geochemical cycles; and
8 both deliberate and accidental introductions of 'alien' species that, as a result of the consequent changes in ecosystem composition, also may result in changes in ecosystem structure and/or function, with similar ultimate impacts.

In this chapter we will focus upon the first five of these changes that are truly global in their extent. Of these, we will also give greatest attention to climatic change, because, as we shall discuss, this has a qualitatively different and more far-reaching impact than the other four.

In the sections that follow we will discuss first the impacts upon vegetation of climatic change, and then secondly the confounding impacts of increases in CO_2 concentration, UV-B flux, nitrogen deposition and concentrations of atmospheric pollutants in the troposphere. We will focus upon the general principles underlying the observed or expected responses. In the final section we will then consider some of the important conclusions that emerge. We will also consider briefly some of the further factors and phenomena that will confound and limit our efforts to predict the consequences of global changes that might be expected over the next 100 years or so.

17.2 Vegetation and climatic change

17.2.1 Responses of species and vegetation

Fundamental to the response of vegetation to climatic change is the individualism of species' responses to their environment (see Chapter 3). Historically, vegetation scientists debated the extent to which plant communities could be considered 'organismal' in character, responding in their entirety to the environment, as opposed to being composed of component species that were independently and 'individualistically' responding to their environment. The individualistic view receives support not only from the observations of continuous spatial variation in present plant communities, but also from the Quaternary palaeoecological record, which shows that plant species assemblages change continuously through time (Huntley 1990b, 1991), as well as from experimental studies (Chapin & Shaver 1985). The principal responses of vegetation to climatic changes can be considered to be:

1 quantitative changes of composition, structure and/or function;
2 qualitative changes in composition and/or structure; and
3 adaptive responses of the component species.

Although our discussion will focus upon the responses of vegetation to climatic change, it is important to realize that the vegetation cover of land surfaces has fundamental influences upon their interaction with the surrounding atmosphere and hence upon climate. Vegetation should be considered as part of the global climatic system, actively participating in changes in the system rather than passively responding to such changes. This active participation is primarily through important feedback mechanisms and has been demonstrated by climatic modelling experiments (Street-Perrott *et al.* 1990; Mylne & Rowntree 1992; Foley *et al.* 1994). Amongst these feedbacks, the best documented relate to the effects of vegetation cover on albedo (e.g. boreal forest versus tundra, especially during spring months when trees mask snow cover) and the 're-cycling' of moisture to the atmosphere by transpiration and evaporation (e.g. large-scale clearance of rainforest). The 'greening' of much of the Sahara during the early Holocene is likely to have altered regional climate through both mechanisms (e.g. see Claussen & Gayler 1997; Patricola & Cook 2007).

Spatial and temporal scale are important considerations for the interactions of vegetation and climatic change. The responses that we expect in principle, and indeed that we observe in practice, differ according to the scale that we consider (Fig. 17.1). In the following discussion of the principal responses of vegetation to climatic change, we shall consider explicitly the scales at which each type of response is relevant.

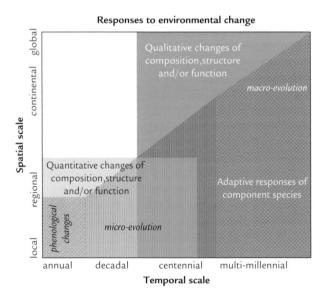

Fig. 17.1 Schematic illustration of the relationships between spatial and temporal scale and both the principal responses of vegetation and the adaptive responses of its component species to environmental change. (Shaded rectangles represent vegetation responses and hatching patterns represent the adaptive responses of component species.)

17.2.2　Quantitative changes

These are changes that involve only alterations in the abundance, whether measured as number of individuals, biomass, etc., of the component species of a plant community. Such alterations may result in changes in the structure of the vegetation, although these are unlikely to be major changes; a climatic change that results in a substantial change in vegetation structure (e.g. savanna to closed woodland), is likely also to lead to qualitative changes in species composition, and thus belongs in the next category. Quantitative changes in abundance of component species may also lead to changes in ecosystem function, for example changes in net primary productivity.

Quantitative changes occur within vegetation stands and hence are local in spatial scale. They are also potentially rapid in terms of temporal scale, being limited in their response time only by the inherent growth and life-cycle characteristics of the species comprising the plant community. Thus a community dominated by annual plant species may exhibit very marked inter-annual changes in the relative abundance of the component species in response to inter-annual changes in climatic conditions that differentially affect the germination, establishment and survival of different species. Even in communities where long-lived perennials dominate, such as chalk grasslands, extreme climatic events can result in marked shifts in the relative abundance of species (Hopkins 1978), albeit that such changes usually are transient in these cases.

Communities of long-lived perennial herbaceous species can also exhibit inter-annual variations in biomass or productivity amongst species in response to inter-annual climatic variability (Willis *et al.* 1995, see Fig. 17.2). Quantitative changes in the numbers of individuals, however, will occur more slowly, in response to decadal scale shifts in mean climatic conditions (Watt 1981). At centennial scales, even communities of woody species will show quantitative changes in composition as fluctuations in long-term mean climatic conditions differentially alter both survival and recruitment rates of species.

Quantitative responses thus can provide resilience to the climatic variability that occurs on inter-annual to centennial time scales, but that characteristically does not involve persistent changes in mean conditions that are maintained over centennial time scales. These responses, however, do not enable vegetation to respond either to rapid climatic changes of large magnitude, or to persistent changes in mean climatic conditions, whether these changes occur rapidly or relatively slowly.

17.2.3　Qualitative changes

Qualitative changes involve losses and gains of species from the plant community and can result not only in changes in composition but also, potentially, in profound changes in the structure and function of the plant community. In contrast to quantitative changes, qualitative changes may occur across a wide range of spatial scales, from local to continental, and across a correspondingly wide range of temporal scales, from decadal to multi-millennial.

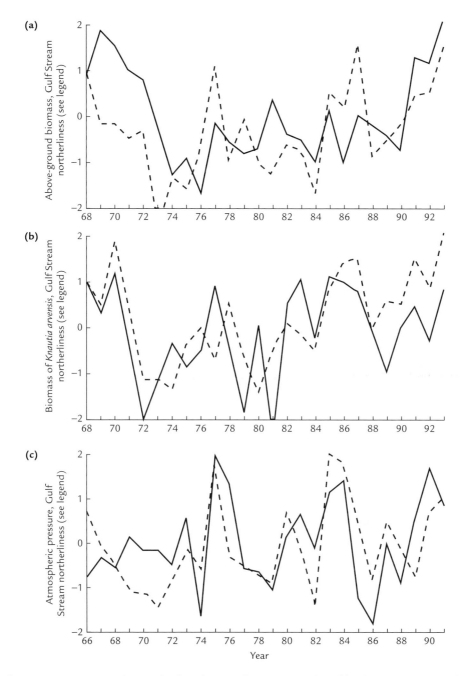

Fig. 17.2 Inter-annual quantitative changes in a community of herbaceous perennials on road verges at Bibury, Gloucestershire (Willis *et al.* 1995). Standardized time series (zero mean, unit variance). (a) Mean above-ground biomass (g·m^{-2}) of vegetation in the Bibury plots (solid line) and Gulf Stream northerliness index in the next-to-previous spring and summer months (March to August, broken line), $r = 0.489$. (b) Mean July biomass (g·m^{-2}) of *Knautia arvensis* (Field scabious) in the Bibury plots (solid line) and Gulf Stream northerliness index in the previous autumn and winter months (September to February, broken line), $r = 0.496$. (c) Log mean atmospheric pressure at sea level (mbar, solid line) and Gulf Stream northerliness index in the same August (broken line), $r = 0.551$.

At the scale of the individual stand, the time scale for such changes is often dependent upon the frequency of disturbance and/or the characteristic longevity of the component species of the present vegetation. In gap-regenerating communities of perennials, the death of individuals provides the opportunity for species favoured by the changed environmental conditions to become established. In such systems the incoming species take many years completely to replace species that are no longer favoured by the prevailing climatic conditions. As a consequence, a transient effect of the climatic change may be an increase in within-stand species diversity comparable to the enhanced diversity often observed spatially in relation to ecotones between adjacent stands of different communities. The transient plant community can thus perhaps best be considered the temporal equivalent of such ecotonal communities.

In communities that regenerate episodically following landscape-scale disturbance events (e.g. forest fires, windstorms, defoliating insect outbreaks), only the characteristic spatial and temporal scales of the process differ from those in gap-regenerating communities. Spatially, entire stands of such communities may be transformed following the disturbance event, with prevailing climatic conditions favouring the establishment and growth of a qualitatively different community to that which was present before the disturbance (Bradshaw & Zackrisson 1990). Temporally, the replacement of the previous community occurs within the time scale for stand regeneration, typically decades to a century. However, the inertia of the pre-disturbance community seems to enable it to persist in a progressively greater degree of disequilibrium with the changing climate until it is disturbed, so that the rate at which the qualitative change occurs across the landscape as a whole is determined by the characteristic return time for disturbance events. Thus, in boreal forest systems, where fire return times may typically be 200–500 years (Foster 1983; Segerström *et al.* 1996), it may require many centuries to pass before the entire landscape is transformed. In consequence, the period of transition is again characterized by a transient peak in diversity, in this case at the landscape scale and among stands. Once again, this temporal phenomenon can be considered equivalent to the spatial heterogeneity and associated biodiversity peak typically found when ecotonal zones are viewed at a landscape scale.

At more extensive spatial scales, and over longer time scales, qualitative changes in vegetation composition are the resultant of the 'migration' of species, i.e. their shifts in geographical distribution, as a response to persistent changes in climatic conditions at regional to continental scales. Such migrations of plant species, including long-lived woody taxa, are well documented by Quaternary palaeoecological data (Davis 1983; Huntley & Birks 1983; Webb 1987), and are the predominant response of species to persistent shifts in climatic conditions (Huntley & Webb 1989). Thus, as climate shifted from glacial to interglacial conditions around 11 400 years ago, temperate forest trees in Europe, North America and eastern Asia 'migrated' at rates of $0\cdot2$–$2\,km\cdot yr^{-1}$, shifting their geographical ranges in response to the changing climate (Fig. 17.3). Such range changes have continued throughout the Holocene: in Europe species such as *Fagus sylvatica* and *Picea abies* continued to exhibit changes of range and of prevalence in forests at the landscape scale during recent millennia (Björse & Bradshaw 1998). The qualitative changes seen at local and landscape scales are

Fig. 17.3 Isopoll map sequence for *Picea abies* (spruce) in Europe during the Holocene (Huntley 1988) showing how its pattern of distribution and abundance changed as climate changed during this period.

indeed generally an expression of the more extensive migrations of species; conversely, these extensive migrations are achieved as a result of, or at least facilitated by, the more local scale qualitative changes. All such qualitative changes are thus best viewed as part of one overall process of spatial response of species to climatic change.

At the geological time scales of glacial–interglacial cycles and beyond, a further factor can contribute to qualitative changes in the composition of vegetation, namely extinction. Trivially, the qualitative change in composition of a stand that leads to the loss of a species can be considered to be a local extinction (extirpation), and the same can be argued at landscape or regional scales. However, the persistent impact of an extinction only comes to be felt when the taxon becomes regionally extirpated, for example, *Tsuga* in Europe during the late Pleistocene, or globally extinct, for example, *Picea critchfieldii* in eastern North America at the last glacial–interglacial transition about 11–12 thousand years ago (Jackson & Weng 1999). Such extinctions most probably reflect the loss from the region where the taxon occurs on any area of the climatic conditions favouring that taxon (Huntley 1999). Subsequently conditions may once again be favourable for the taxon, but its extinction results in qualitatively different vegetation now developing under those climatic conditions. In this context it is also relevant to note that extinctions of herbivorous animals potentially can lead to both quantitative and qualitative changes in the character of plant communities. Whereas the possible impact on the native vegetation of the human-driven extinction of the nine species of moa, huge flightless birds (formerly found in New Zealand) has been the subject of speculation and debate (Atkinson & Greenwood 1989; Batcheler 1989; Bond *et al.* 2004; Lee *et al.* 2010), the potential impact upon European forest vegetation of the late-Pleistocene extinction of several forest-dwelling mammalian mega-herbivores, including the straight-tusked elephant (*Palaeoloxodon antiquus*) (Stuart 2005) and narrow-nosed rhinoceros (*Dicerorhinus hemitoechus*), has to date received little attention from ecologists.

17.2.4   Adaptive responses

Plant species may also exhibit adaptive responses (Fig. 17.1) when the climate changes, enabling them to persist in communities that may, as a result, exhibit neither qualitative nor quantitative changes. Such adaptive responses are in principle possible at spatial scales ranging from local to continental and across time scales from decadal to the multi-millennial time scales of geological time. In practice, such responses are primarily polarized to the two extremes of this range of time scales because they occur as a result of processes with very different temporal characteristics.

At local to regional spatial scales, and over relatively short time scales, the predominant processes are those that involve either phenotypic plasticity or selection amongst the many genotypes that arise each generation as a result of the recombination of alleles during sexual reproduction. Such adaptation, whether resulting from phenotypic plasticity or 'micro-evolution', does not alter the species' overall climatic range, in terms of its tolerances or requirements, but does enable a local or regional population of the species to adapt to changing climatic conditions. The extent of such an adaptive response, however, is limited by the species' overall climatic adaptability; in many cases it may be even more limited in scope if elements of the overall genetic variability exhibited by the species in relation to climate are limited in their distribution to only a fraction of the species' overall geographical distribution.

At somewhat longer time scales the latter limitation may be overcome by the 'migration' of genotypes as a result of pollen/spore as well as seed/fruit dispersal. This will be relatively more rapid and more extensive in the case of anemophilous species, in which pollen transport may facilitate the rapid and widespread migration of newly favoured alleles. At the longest time scales, species may evolve new capabilities with respect to climate as the result of the chance mutation of one or more genes to produce alleles that provide an extension of the range of climatic conditions under which the species may persist. Such mutations will arise infrequently, however, and, as a result, such 'macro-evolution' is only able to play any significant role at longer geological time scales.

The most obvious and widespread response of plants and ecosystems to recent climatic warming has been a phenological response, achieved principally as a result of phenotypic plasticity. Many species are coming into leaf or flowering earlier then they did a few decades ago (Sparks *et al.* 2000; Parmesan & Yohe 2003) and there is evidence of a general advance of spring greening in the northern hemisphere temperate and boreal zones (Schwartz *et al.* 2006). A species' phenological behaviour, however, is likely also to be under genetic control. Bennie *et al.* (2010) present evidence that selection favours different phenological strategies under different climatic regimes, and show that, as a result, local populations of a species may be unable to optimize their phenological response if the required genotypes are locally absent.

In an elegant experimental study, Franks *et al.* (2007) demonstrated how an annual plant species not only showed differing adaptations in relation to the onset of flowering between populations from a mesic site and from a dry site, but that exposure to a recent multi-year drought had resulted in an adaptive shift to earlier flowering in both populations, and that this was a result of genetic adaptation. As Huntley (2007) subsequently pointed out, however, even after the strong selection pressure exerted by the drought, flowering time of the mesic population under experimental dry conditions was significantly later than in the dry site population before the drought. This strongly suggests that, as Bradshaw & McNeilly (1991) argued on the basis of evidence from their studies of the evolution of heavy metal tolerance in various plant species, the extent to which a local population may be able to adapt to climatic change is unlikely to reflect the capabilities of the species as a whole. Rather, that adaptability is limited by the genetic variability, or capital, of the local population that in most cases is unlikely to encompass the whole of the species' genetic variability.

Although, as Good (1931) argued long ago, the rate and magnitude of recent and projected future climatic changes is expected to result in shifts in species' geographical distributions, the rate at which plant species are able to realize such distribution changes is limited both by their dispersal abilities and by the need, in many cases, for disturbance events to facilitate establishment (Bradshaw & Zackrisson, 1990). Thus there is little if any convincing evidence as yet of plant species' distribution changes caused by recent climatic changes. Such changes are, however, convincingly reported for more mobile taxa, notably butterflies (Parmesan *et al.* 1999) and birds (Devictor *et al.* 2008), although even amongst these groups, as the latter study reports, the rates of distribution shift are failing to track the rate of climatic change. Modelling studies have provided indications of the potential magnitude of plant species' distribution changes required to track projected climatic changes (Huntley *et al.* 1995; Sykes 1997; Thuiller 2004) and of the dynamic responses of vegetation to climatic change (Sykes & Prentice 1996). Models have also been developed that are able to simulate the dynamics of species' distribution changes (Carey 1996; Collingham *et al.* 1996). Simulating how plant species and vegetation will respond to climatic change requires, however, the integration of these three, to-date usually separate, modelling approaches to develop models that are able to simulate the dynamics of dispersal, population processes and vegetation. Strategies for the development of such models are currently emerging (Williams *et al.* 2008; Huntley *et al.* 2010; Midgley *et al.* 2010) but it is likely to be some years before the proposed fully integrated models are available.

17.3   Confounding effects of other aspects of global change

17.3.1   Increasing atmospheric concentrations of CO_2

Increased atmospheric CO_2 concentration not only leads to warming of the lower atmosphere, but also increases availability of the primary substrate for photosynthesis. In principle, higher atmospheric concentrations of CO_2 should stimulate photosynthesis and lead to faster growth. This generally is the case for species using the C_3 photosynthetic pathway (*c.* 95% of the world's higher plants). In these species, increased atmospheric partial pressure of CO_2 leads to reduced photorespiration and hence a net gain in carbon fixation, by an average of *c.* 30% (range −10 to +80%) for a doubling of CO_2 concentration. In addition, these species also benefit from physiological gains, notably a reduction in stomatal opening, as well as a potential reduction in stomatal density, that leads to increased water-use-efficiency as a result of reduced transpirational water loss. In contrast, species using the C_4 photosynthetic pathway (only *c.* 5% of the world's higher plants, but including *c.* 50% of grass species) experience only a very modest net gain in C fixation, estimated at *c.* 7%, for the same doubling of atmospheric CO_2 concentration. This substantial difference in photosynthetic stimulation reflects the evolutionary adaptations of physiology and morphology in C_4 species. Physiological adaptations include saturation of the key CO_2 fixation enzyme of C_4 photosynthesis (PEP carboxylase – phosphoenolpyruvate carboxylase) at much lower CO_2 concentrations than are required to saturate its C_3 counterpart (RuBP carboxylase – ribulose bisphosphate carboxylase). There are also morphological adaptations in the so-called Kranz anatomy, in which initial CO_2 fixation is spatially separated from the remainder of the photosynthetic pathway, and C_4 acids formed when CO_2 is fixed in the mesophyll are shunted to bundle sheath cells where CO_2 is released to enter the 'normal' C_3 pathway, but at higher concentrations than in the surrounding atmosphere. The benefits of this CO_2 concentrating mechanism are seen when the $CO_2 : O_2$ ratio is low and include (i) minimal photorespiration and (ii) an inherently higher water-use-efficiency. The latter is a consequence of a significantly lower stomatal conductance, C_4 species typically having lower stomatal densities and opening their stomata for shorter periods of the day. When the $CO_2 : O_2$ ratio increases, however, C_4 species are at a disadvantage relative to their C_3 counterparts because they expend greater amounts of energy concentrating the CO_2 in the bundle sheath cells.

The range in CO_2 response exhibited by C_3 species results, at least in part, from the phenomenon of 'acclimation'. When plants are exposed to elevated concentrations of atmospheric CO_2 there is, through time, a strong tendency for them to adapt physiologically and morphologically to their new environment. As a consequence, initial increases in C-fixation rates are often not maintained through time. Accurate predictions of future vegetation changes resulting directly from increasing concentrations of atmospheric CO_2 thus must be made with care. This is especially true when considering communities that may be impacted by the differential responses of slow-growing, long-lived perennial species, as opposed to fast-growing perennials. Whereas the former, often characteristic of more diverse but less productive communities, are relatively unresponsive, the latter, often dominant in less diverse communities of productive habitats, show a strong response to elevated CO_2 (Hunt *et al.* 1993).

C_4 species have significantly higher temperature optima for photosynthesis than C_3 species. This is associated with a strong latitudinal trend in the distribution and relative abundance of the former. C_4 species are increasingly predominant in the hotter and drier environments at lower latitudes where their high water-use-efficiency also contributes to their ability to outcompete C_3 species. Given the important differences in their physiological and morphological adaptations, a number of key questions remain concerning the direct effects of increased atmospheric CO_2 concentration upon vegetation. In particular, will the increasing atmospheric CO_2 concentration, through its direct effects, be the primary determinant of changes in the composition of plant communities, especially where C_3 and C_4 species co-occur, or will the altered climate, itself an indirect effect of increased atmospheric CO_2 concentration, have a greater impact, masking or even overriding any direct effects? Studies of Quaternary palaeovegetation and palaeoenvironments can help answer this question. The partial pressure of atmospheric CO_2 was markedly lower during the last glacial stage, at around 190 ppmv, than during either the preceding interglacial stage or the pre-industrial post-glacial, during both of which it was around 280–300 ppmv. Studies of glacial–interglacial variations in relative abundance of C_3 versus C_4 plants, however, suggest that regional climates exerted the strongest influence upon their relative abundance, and that in the absence of favourable moisture and temperature conditions, a low partial pressure of atmospheric CO_2 alone was insufficient to result in an increased proportion of C_4 plants in the vegetation (Huang *et al.* 2001). Thus, climate seems likely to be more important than the direct effects of CO_2 concentration in determining the presence or absence of C_4 species in a given environment, and thus is likely to have the greater role in determining future changes, quantitative and qualitative, in the relative contributions of C_3 and C_4 species to plant communities.

Many succulent plants employ a third mode of CO_2 carboxylation, termed crassulacean acid metabolism (CAM). CAM plants comprise *c.* 10% of the world's higher plant flora. Like C_4 species, they have the PEP carboxylase enzyme in addition to the enzymes of the C_3 Calvin cycle. In contrast to C_4 plants, however, the carboxylation process is temporally rather than spatially separated from the remainder of the photosynthetic pathway. CAM plants minimize transpirational water losses by opening their stomata during the cooler conditions at night, rather than during the heat of the day. CO_2 fixation by PEP carboxylase can occur in the dark, when cooler conditions reduce the amount of water lost for a given amount of CO_2 uptake, and the CO_2 is then released from the C acids during daylight hours and used by the 'normal' photosynthetic pathway.

In addition to those species that use only CAM (e.g. most Cactaceae, many Bromeliaceae and Orchidaceae) there are facultative CAM species that use CAM under conditions of stress resulting from drought or salinity (e.g. members of the Aizoaceae). This ability of facultative CAM species (some of which employ diverse variants of the pathway) to 'switch' metabolism gives them a plasticity that enables them to occupy otherwise sparsely inhabited ecological niches. The extent of any response by such species to increased atmospheric CO_2 concentration will depend upon the impacts of both prevailing temperature and water availability in the particular environments that they occupy.

It must also be noted that the productive investment of carbon fixed by photosynthesis requires adequate availability of nutrients, especially nitrogen. In part that N requirement may be met by reallocation within the plant, for example through increased nitrogen-use-efficiency as a result of reduced investment of N in enzymes of the photosynthetic pathway. Resultant changes in tissue chemistry, however, may translate into altered litter quality that in turn might impact the decomposition rates of such material and hence biogeochemical cycling. When grown in elevated CO_2 atmospheres, a wide range of species exhibits reduced leaf tissue nitrogen content, along with increases in lignin and cellulose, compared to the same species grown at present-day CO_2 concentrations (Fig. 17.4).

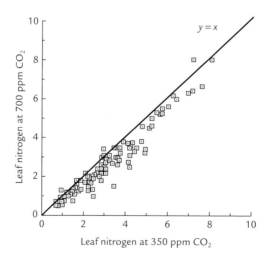

Fig. 17.4 Decline in leaf nitrogen concentration when plants are grown in a high CO_2 environment (from Bazzaz 1996).

A higher lignin content makes leaves more difficult to decompose. Altered litter chemistry, including higher $C:N$ ratios, thus is likely to slow down nutrient cycling rates in ecosystems, in turn reducing their responsiveness to elevated CO_2 concentrations (Bazzaz 1996).

Leaf tissue chemistry also has profound implications with respect to herbivory. Plant survival of herbivory may be altered as a result of changes in allocation of C and N to secondary metabolites that act as anti-herbivore defences, whilst increased $C:N$ ratios will reduce the quality of foliage as food for herbivores. Lower nitrogen content per unit mass of foliar tissue also requires the herbivore to consume more tissue to gain the same amount of nitrogen, increasing the impact of herbivory upon the plant. Such an increase in tissue predation levels is likely to have an impact upon competitive interactions amongst plant species, with resulting changes in plant community structure and composition. Evidence from studies of experimental plant communities exposed to herbivory suggests that species that do relatively well in these circumstances are characterized by their general competitive ability rather than by the extent of their CO_2 responsiveness (Bazzaz *et al.* 1995).

The potential consequences of increased atmospheric CO_2 concentration and increased temperatures, outlined above, ultimately may interact to alter the flux of carbon between the atmosphere and the soil–plant continuum. Any potential resulting shift in the balance between the carbon in the atmosphere and that sequestered in biomass, litter and soil is highly relevant to the discussion of future potential climates. The possibility of a net flux of carbon to the atmosphere as a result of increased rates of decomposition of biotic materials (as CO_2 from aerobic and as CH_4 from anaerobic, e.g. wetland, environments) generating a positive feedback through the consequent increased radiative forcing, as well as the direct effects upon vegetation of increased CO_2 concentration, is a key issue. Recent studies have demonstrated that carbon cycling in boreal and Arctic wetlands strongly influences the global climate (Panikov 1999). Plant productivity was found to exert very important biological controls on CH_4 flux both through stimulation of methanogenesis, by increasing C-substrate availability (input of organic substances to soil through root exudation and litter production), and through by-passing of potential sites of CH_4 oxidation in the upper layers of the soil (Christensen *et al.* 2003), as a result of enhanced gas transport from the soil to the atmosphere via root aerenchyma.

Many of the ecosystems on earth are currently net sinks for carbon, fixing more than they release. A shift in this balance could have significant impacts through positive feedbacks, resulting in faster and more extensive climatic changes than have hitherto been recorded. For example, sequestered carbon could be released into the atmosphere if climate change leads to melting permafrost or to more active disturbance regimes (fires or storms). Dynamic Global Vegetation Models (DGVMs; Prentice *et al.* 2007) provide a framework for exploring these feedbacks. DGVMs simulate ecosystem processes and vegetation dynamics (including competition) driven by a time series of climate data (e.g. solar radiation, temperature and precipitation), given constraints of latitude, topography and soil characteristics (reviewed in Skidmore *et al.* 2011). These models can be tailored to individual plant species, or they can use a simplified vegetation classification based upon plant functional types (PFTs; Chapter 12) (Hickler *et al.* 2009), or on global biomes (Haxeltine & Prentice 1996), as described in Chapter 15.

17.3.2   Increased fluxes of UV-B as a consequence of 'thinning' of the stratospheric ozone layer

Anthropogenic alterations to the trace-gas composition of the atmosphere have not been restricted to increases in naturally occurring greenhouse gases – CO_2, CH_4 and N_2O – but have included, especially during the latter half of the 20th Century, the introduction of increasing amounts of chlorofluorocarbons (CFCs) to the atmosphere. CFCs may reside in the atmosphere for many decades, accumulating in the stratosphere where, as a result of photolysis reactions, they release reactive halogens and halogen compounds, including chlorine and chlorine oxide, implicated in the breakdown of stratospheric O_3. Decreases in total-column ozone are now observed over large parts of the globe, permitting increased penetration of solar UV-B radiation to the Earth's surface. Depletion of O_3 in the atmospheric column is not uniform around the globe, but is more intense at higher latitudes and especially in polar regions. Since the 1970s, UV radiation reaching the surface during winter and spring has increased by *c.* 4–7% in northern and southern hemisphere mid-latitudes. Over the same period, UV reaching the surface during spring in the Antarctic and Arctic has increased by 130% and 22% respectively. Although it is at present difficult to predict longer-term future UV-B levels, current best estimates indicate that a slow return to pre-ozone depletion levels may occur within *c.* 50 years. However, confounding influences that remain poorly understood at the present time, especially the interactions of CFCs with other greenhouse gases and the atmospheric chemistry of CFC substitutes, render these predictions subject to considerable uncertainty.

Most of the UV-B radiation penetrating plant cells is absorbed, potentially causing acute injuries as a result of its high quantum energy. In addition to its photo-oxidative action, UV radiation causes photolesions, particularly in biomembranes. Although absorption of UV radiation by epicuticular waxes and by flavonoids dissolved in the cell sap provides higher plant cells with considerable protection from radiation injuries, some damage does occur to DNA, membranes, photosystem II of photosynthesis and photosynthetic pigments. Recent research has shown the importance of the dynamic balance between damage and protection/repair mechanisms (e.g. DNA excision repair, scavenging of radicals formed by the absorption of UV-B photons), and of the great variation between species with respect to this balance. Whereas, for example, some species have a high capacity for repair of DNA damaged by UV-B irradiation, others have a much weaker capacity. A growing body of evidence suggests that the effects of UV-B irradiation may be exerted primarily through altered patterns of gene activity rather than through physical damage (e.g. alteration of life-cycle timing, altered morphology, altered production of secondary metabolites leading to changes in palatability and in plant–herbivore interactions) (see e.g. Singh *et al.* 2010). Evidence to date strongly indicates that the primary responses of vegetation to increased UV-B levels result principally from shifts in the balance of competition between individual higher plant species in a community rather than from negative impacts upon the performance of individual species. However, it is currently difficult to predict the sign of such UV-mediated changes in species' interactions. It should also be noted that the responses of plants exposed to UV-B are modulated strongly by other environmental factors, such as the concentration of atmospheric CO_2, water availability, temperature and nutrient availability, all of which, as we have already seen, are also changing.

17.3.3   Increasing tropospheric concentrations and deposition of various pollutants – SO_2, NO_x, NH_3 and O_3

Across much of the developed world, emission of the gas SO_2 and its subsequent deposition from the atmosphere has historically been of great significance. This gas modifies plant growth responses through either acute, or more often chronic, toxic action, although with few, if any, visible symptoms. The phytotoxic effects of SO_2 gas and of its solution products have been studied extensively, particularly in relation to those taxa most susceptible to gaseous pollutants, such as the bryophytes and lichens that lack the protection of a cuticle.

Over recent decades, the shift from coal-burning to gas- and oil-fired power stations, coupled with large increases in road traffic volumes, has led to decreasing SO_2 emissions but increasing emissions of oxides of nitrogen (NO_x). In addition, there has been a substantial increase in emission of reduced nitrogen compounds, predominantly arising from intensive agricultural activities (e.g. ammonia (NH_3) from intensive animal rearing) (Cape *et al.* 2009a). A further complexity arising in the case of NO_2 emissions to the atmosphere is the sequence of complex photochemical interactions that ensue leading to the tropospheric formation of O_3, itself a highly phytotoxic gas.

Compared to the greenhouse gases, the residence times of SO_2, NO_x and O_3 in the atmosphere are short because of their highly reactive nature and relatively rapid deposition back onto the Earth's surface. Nevertheless, atmospheric monitoring networks have provided clear evidence of measurable deposition of N-containing and acidic compounds remote from their sources. The high latitudes of the Arctic regions are an excellent example, where local sources are negligible.

The duration of the passage of pollutant species through the atmosphere, following emission, has an important bearing upon the state in which they are deposited upon, or interact with, plants. They may interact with plant tissues whilst still in the gaseous state, or following deposition in aqueous droplets derived either from rain or from the fine droplets of mist or cloud – so-called dry, wet and 'occult' deposition respectively. In addition, aqueous deposition may be in either an undissociated (e.g. H_2SO_4 (aq), HNO_3 (aq)), or a dissociated (e.g. H^+, SO_4^{2-}, NO_3^-, NH_4^+) state. Furthermore, in many situations concurrent deposition of two or more of these pollutants will occur. The impacts of deposition of these pollutants must therefore be considered not only in terms of acidity, toxicity and nutrient ion content, but also in terms of potential antagonistic or synergistic effects arising from combined deposition.

Of present-day atmospheric pollutants, increased deposition rates of nitrogen compounds and significant tropospheric concentrations of ozone represent serious threats to vegetation in a wide range of terrestrial ecosystems, although their impacts vary greatly between species. For example, the varied impacts upon plant growth of atmospheric deposition of N compounds reflect great differences between species for their N requirements. Many plant species in natural and semi-natural ecosystems are adapted to grow in oligotrophic (low nutrient) environments. Such species often compete successfully with other species only in such environments. An increase in nitrogen supply may lead to quantitative or even qualitative changes in the composition of the vegetation as a result of competitive displacement of species adapted to grow under low nutrient conditions. This is the case, for example, in Dutch heathlands, where increased inputs of nitrogen (mainly as NH_3) and increased soil acidity have been associated with the decline of various heathland species, the increased abundance of invading species and the accelerated replacement of dwarf-shrub heath by grass heath as part of a succession progressing towards woodland (Bobbink 1991; Houdijk *et al.* 1993). The overall result is a reduction in both species richness and species diversity. Such qualitative and quantitative changes in plant community composition and structure may have positive feedbacks to rates of N cycling as a result of, for example, altered rates of soil nitrogen mineralization and nitrification (Table 17.1). However, high rates of deposition of N compounds will not always promote vegetation growth and development; in many plant communities other nutrient species, especially P, will be limiting. Increasing N deposition in such cases is more likely to result in increased N saturation of the soil and consequent increased leaching of N into drainage waters (Wilson *et al.* 1995).

Much recent research has been directed towards the determination of 'critical loads' and 'critical levels' of pollutants. A critical load can be defined as the maximum amount of deposition (flux) of a given compound which will not cause long-term harmful effects on ecosystem structure and function according to

Table 17.1 Changes in ecosystem properties associated with vegetation change in Dutch heathlands.

Vegetation	Net primary production $(g·m^{-2}·yr^{-1})$	Net N mineralization $(g-N·m^{-2}·yr^{-1})$	Per cent nitrified
Calluna (original)	730	6·2	4·8
Molinia (invader)	2050	10·9	33·0
Deschampsia (invader)	430	12·6	42·9

Note: As species composition has changed, the potential for nitrogen loss, as nitrate, has increased. Nitrogen mineralization is a measure of the rate at which nitrogen is made available to plants; the per cent nitrified is the percentage of the mineralized nitrogen that is converted to nitrate, which has the potential for being leached from the soil. Measurements were made in areas dominated by *Calluna vulgaris*, *Molinia caerulea* and *Deschampsia flexuosa*, respectively. After vanVuuren *et al.* (1992).

present knowledge; a 'critical level' referring to a concentration or dose of a gaseous pollutant. They are in both cases, therefore, a threshold which ecosystems can tolerate without damage (Sanders *et al.* 1995; Cape *et al.* 2009b, and references therein). This critical loads/levels approach (including direct measures of deposition in the field) is proving vital to future policy decisions on the abatement of atmospheric pollutants around the world.

17.3.4   Interactive effects of pollutants, their deposition products and human land use

No one tropospheric pollutant is found in isolation, nor does it operate in isolation. As mentioned earlier, antagonistic and/or synergistic interactions are often seen, with interactions between pollutants leading either to less damage when present together or, conversely, to greater damage when present in combination than when either is present alone. The character of the soil, including moisture status, temperature and nutrient status, is also of key importance in determining the impact of pollutants. These soil characteristics in turn may be a function of other pressures operating upon the ecosystem, such as management practices associated with particular land use. For example, in a recent modelling study of the carbon dynamics of northern hardwood forests, Ollinger *et al.* (2002) have shown that historical increases of atmospheric concentration of CO_2 and of N deposition (over the period 1700–2000) have stimulated forest growth and carbon uptake. However, the degree of stimulation differs depending upon the intensity of human land management because this alters soil C and N pools and hence also alters the degree of growth limitation by C versus N. When other components of atmospheric pollution (e.g. tropospheric ozone) are factored into the model, this substantially offsets the increases in growth and C uptake resulting from CO_2 and N deposition. Thus, for modelled intact temperate forests, at least, there is little evidence of altered growth since before the Industrial Revolution, despite substantial changes in the chemical environment experienced by these forests. Whether this result will be borne out by the field measurement and manipulation studies underway in temperate forest stands remains to be seen. Findings to date, however, from both experiments and models, highlight the potential importance of interactions between confounding factors that are operating at the present time. They also highlight the need to understand these interactions if we are to be able accurately to predict their likely future impacts upon vegetation. It should finally be noted that there also remain many uncertainties associated with the range of experimental protocols used and hence with the conclusions reached.

17.4   Conclusions

Amongst the various aspects of global environmental change, the impacts of climatic change are the most important. Climatic change elicits large-scale spatial responses by species that in turn lead to qualitative changes in the composition, as well as structural changes in many cases, of vegetation. The palaeoecological record shows the potential rates and magnitudes of these spatial responses, and also reveals their individualistic character (Huntley 1991). This individualism leads to the development of plant communities without a present-day analogue (Huntley 1990a; Overpeck *et al.* 1992); such no-analogue communities result primarily from the response of species to combinations of environmental conditions, including climate, that lack a present analogue (Williams & Jackson 2007). One of the few certainties about the next century is that the combinations of carbon dioxide concentration, UV-B flux and climate that will develop lack current analogues. We can thus predict with reasonable confidence that plant communities lacking a present-day analogue, in terms of their composition and/or structure, will arise during the next century in many parts of the world.

In contrast to climatic change, the other aspects of global change that we have considered – CO_2, UV-B, N deposition, tropospheric pollutants (SO_2, NO_x, NH_3, O_3) – are secondary in the magnitude of their impacts, many of which are likely to be quantitative. Nonetheless, the impacts of these secondary factors are often likely to become visible more quickly than the impacts of climatic change. This rapid visibility of their impacts must not be allowed to distract the attention of vegetation scientists from the longer-term but very profound impacts of climatic change.

As we have noted above, species' individualism, coupled to the emergence of no-analogue environments (Williams & Jackson 2007; Williams *et al.* 2007), will result in the emergence of new plant communities in the future, just as it did in the past. This affect is likely to be amplified further by two phenomena. Firstly, as noted above, transient communities and landscapes of greater species diversity may emerge in some cases during the period of transition from present to future communities. Secondly, because of the relatively rapid rate at which climate is expected to change over the next century compared to past changes (Jansen *et al.* 2007), the migration capabilities of species are likely to be exceeded. As a result, species' migrations may lag behind the changing climate. This is likely to affect the relatively slower growing and longer-lived species (S strategists sensu Grime 1978) much more than it will affect the rapidly growing species with short life cycles (R and C-R strategists). Because many of the latter species are also characterized by the production of numerous, small, widely-dispersed propagules, they are likely to experience a short-term benefit, because they will be able to migrate rapidly to exploit newly available areas. Many of them also will benefit from the increased availability of N in areas with high rates of deposition of N compounds from the atmosphere, and some are also likely to benefit from the increased concentration of atmospheric CO_2. As a result, many plant communities beyond present climatic ecotones may be replaced initially by communities dominated by shorter-lived, opportunist species of a relatively 'weedy' or ruderal character. Such communities may be analogous to early-successional communities within that present climatic boundary, although it is more likely that they will be without such analogues, because they will persist and develop without the influence of the late-successional species that would today come to dominate such stands in place of the early-successional taxa.

The potential for micro-evolution to select 'cryptic' genotypes with climatic tolerances or requirements not exhibited by a species' present population will also arise. This will result from selection of genotypes favoured by newly available no-analogue conditions, but that are destined always to die under present conditions. The extent to which this may occur is extremely difficult to assess; it is clear, however, that species were able to adapt to no-analogue conditions during the late-Quaternary. If it does occur to any significant extent, then selection of such 'cryptic' genotypes will render the prediction of species, and hence vegetation, responses to climatic change fraught with even greater difficulties. Such potentially unpredictable adaptive changes may lead to 'surprise' responses of species and vegetation to climatic changes, especially when they occur in tandem with the many other ongoing changes to the global environment.

References

Atkinson, I.A.E. & Greenwood, R.M. (1989) Relationships between moas and plants. *New Zealand Journal of Ecology* **12**, 67–96.

Batcheler, C.L. (1989) Moa browsing and vegetation formations, with particular reference to deciduous and poisonous plants. *New Zealand Journal of Ecology* **12**, 57–65.

Bazzaz, F.A. (1996) *Plants in Changing Environments. Linking Physiological, Population and Community Ecology.* Cambridge University Press, Cambridge.

Bazzaz, F.A., Miao, S.L. & Wayne, P.M. (1995) Microevolutionary responses in experimental populations of plants to CO_2-enriched environments: parallel results from two model systems. *Proceedings of the National Academy of Science, USA* **92**, 8161–8165.

Bennie, J., Kubin, E., Wiltshire, A., Huntley, B. & Baxter, R. (2010) Predicting spatial and temporal patterns of bud-burst and spring frost risk in north-west Europe: the implications of local adaptation to climate. *Global Change Biology* **16**, 1503–1514.

Björse, G. & Bradshaw, R. (1998) 2000 years of forest dynamics in southern Sweden: suggestions for forest management. *Forest Ecology and Management* **104**, 15–26.

Bobbink, R. (1991) Effects of nutrient enrichment in Dutch chalk grassland. *Journal of Applied Ecology* **28**, 28–41.

Bond, W.J., Lee, W.G. & Craine, J.M. (2004) Plant structural defences against browsing birds: a legacy of New Zealand's extinct moas. *Oikos* **104**, 500–508.

Bradshaw, A.D. & McNeilly, T. (1991) Evolutionary response to global climatic change. *Annals of Botany* **67**, 5–14.

Bradshaw, R.H.W. & Zackrisson, O. (1990) A two thousand year record of a northern Swedish boreal forest stand. *Journal of Vegetation Science* **1**, 519–528.

Cape, J.N., van der Eerden, L.J., Sheppard, L.J., Leith, I.D. & Sutton, M.A. (2009a) Evidence for changing the critical level for ammonia. *Environmental Pollution* **157**, 1033–1037.

Cape, J.N., van der Eerden, L., Fangmeier, A. *et al.* (2009b) Critical levels for ammonia. In: *Atmospheric Ammonia: Detecting Emission Changes and Environmental Impacts. Results of an Expert Workshop under the Convention on Long–range Transboundary Air Pollution* (eds M.A. Sutton, S. Reis & S.M.H. Baker), pp. 375–382. Springer, Heidelberg.

Carey, P.D. (1996) DISPERSE: a cellular automaton for predicting the distribution of species in a changed climate. *Global Ecology and Biogeography Letters* **5**, 217–226.

Chapin, F.S., III & Shaver, G.R. (1985) Individualistic growth response of tundra plant species to environmental manipulations in the field. *Ecology* **66**, 564–575.

Christensen, T.R., Panikov, N., Mastepanov, M. *et al.* (2003) Biotic controls on CO_2 and CH_4 exchange in wetlands – a closed environment study. *Biogeochemistry* **64**, 337–354.

Claussen, M. & Gayler, V. (1997) The greening of the Sahara during the mid-Holocene: results of an interactive atmosphere-biome model. *Global Ecology and Biogeography Letters* **6**, 369–377.

Collingham, Y.C., Hill, M.O. & Huntley, B. (1996) The migration of sessile organisms: a simulation model with measurable parameters. *Journal of Vegetation Science* 7, 831–846.

Davis, M.B. (1983) Holocene vegetational history of the eastern United States. In: *The Holocene* (ed. H.E. Wright, Jr.). Late-Quaternary Environments of the United States Vol. 2, pp. 166–181. University of Minnesota Press, Minneapolis, MN.

Devictor, V., Julliard, R., Couvet, D. & Jiguet, F. (2008) Birds are tracking climate warming, but not fast enough. *Proceedings of the Royal Society B Biological Sciences* 275, 2743–2748.

Foley, J.A., Kutzbach, J.E., Coe, M.T. & Levis, S. (1994) Feedbacks between climate and boreal forests during the Holocene epoch. *Nature* 371, 52–54.

Foster, D.R. (1983) The history and pattern of fire in the boreal forest of southeastern Labrador. *Canadian Journal of Botany* 61, 2459–2471.

Franks, S.J., Sim, S. & Weis, A.E. (2007) Rapid evolution of flowering time by an annual plant in response to a climatic fluctuation. *Proceedings of the National Academy of Sciences of the United States of America* 104, 1278–1282.

Good, R. (1931) A theory of plant geography. *New Phytologist* 30, 11–171.

Grime, J.P. (1978) *Plant Strategies and Vegetation Processes.* John Wiley & Sons, Ltd, New York, NY.

Haxeltine, A. & Prentice, I.C. 1996. BIOME3: an equilibrium terrestrial biosphere model based on ecophysiological constraints, resource availability, and competition among plant functional types. *Global Biogeochemical Cycles* 10, 693–709.

Hickler, T., Fronzek, S., Araújo, M.B. *et al.* (2009) An ecosystem model-based estimate of changes in water availability differs from water proxies that are commonly used in species distribution models. *Global Ecology and Biogeography* 18, 304–313.

Hopkins, B. (1978) The effects of the 1976 drought on chalk grassland in Sussex, England. *Biological Conservation* 14, 1–12.

Houdijk, A., Verbeek, P., van Dijk, H.F.G. & Roelofs, J. (1993) Distribution and decline of endangered herbaceous heathland species in relation to the chemical composition of the soil. *Plant and Soil* 148, 137–143.

Huang, Y., Street-Perrott, F.A., Metcalfe, S.E. *et al.* (2001) Climate change as the dominant control on glacial-interglacial variations in C3 and C4 plant abundance. *Science* 293, 1647–1651.

Hunt, R., Hand, D.W., Hannah, M.A. & Neal, A.M. (1993) Further responses to CO_2 enrichment in British herbaceous species. *Functional Ecology* 7, 661–668.

Huntley, B. (1988) Glacial and Holocene vegetation history: Europe. In: *Vegetation History* (eds. B. Huntley & T. Webb, III), pp. 341–383. Kluwer Academic Publishers, Dordrecht.

Huntley, B. (1990a) Dissimilarity mapping between fossil and contemporary pollen spectra in Europe for the past 13,000 years. *Quaternary Research* 33, 360–376.

Huntley, B. (1990b) European post-glacial forests: compositional changes in response to climatic change. *Journal of Vegetation Science* 1, 507–518.

Huntley, B. (1991) How plants respond to climate change: migration rates, individualism and the consequences for plant communities. *Annals of Botany* 67, 15–22.

Huntley, B. (1999) The dynamic response of plants to environmental change and the resulting risks of extinction. In: *Conservation in a Changing World* (eds. G.M. Mace, A. Balmford & J.R. Ginsberg), pp. 69–85. Cambridge University Press, Cambridge.

Huntley, B. (2007) Evolutionary response to climatic change? *Heredity* 98, 247–248.

Huntley, B., Barnard, P., Altwegg, R. *et al.* (2010) Beyond bioclimatic envelopes: dynamic species' range and abundance modelling in the context of climatic change. *Ecography* 33, 621–626.

Huntley, B., Berry, P.M., Cramer, W.P. & McDonald, A.P. (1995) Modelling present and potential future ranges of some European higher plants using climate response surfaces. *Journal of Biogeography* 22, 967–1001.

Huntley, B. & Birks, H.J.B. (1983) *An Atlas of Past and Present Pollen Maps for Europe: 0–13000 B.P.* Cambridge University Press, Cambridge.

Huntley, B. & Webb, T., III (1989) Migration: species' response to climatic variations caused by changes in the earth's orbit. *Journal of Biogeography* 16, 5–19.

Jackson, S.T. & Weng, C. (1999) Late Quaternary extinction of a tree species in eastern North America. *Proceedings of the National Academy of Science* 96, 13847–13852.

Jansen, E., Overpeck, J., Keith R.B. *et al.* (2007) Paleoclimate. *Climate Change 2007: The physical science basis. Contribution of Working Group I to the Fourth Assessment Report of the Intergovernmental*

Panel on Climate Change (eds S. Solomon, D. Qin, M. Manning, Z. *et al.*), pp. 433–497. Cambridge University Press, Cambridge, UK & New York, NY.

Lee, W.G., Wood, J.R. & Rogers, G.M. (2010) Legacy of avian-dominated plant–herbivore systems in New Zealand. *New Zealand Journal of Ecology* 34, 28–47.

Midgley, G.F., Davies, I.D., Albert, C.H. *et al.* (2010) BioMove – an integrated platform simulating the dynamic response of species to environmental change. *Ecography* 33, 612–616.

Mylne, M.F. & Rowntree, P.R. (1992) Modeling the effects of albedo change associated with tropical deforestation. *Climatic Change* 21, 317–343.

Ollinger, S.V., Aber, J.D., Reich, P.B. & Freuder, R.J. (2002) Interactive effects of nitrogen deposition, tropospheric ozone, elevated CO_2 and land use history on the carbon dynamics of northern hardwood forests. *Global Change Biology* 8, 545–562.

Overpeck, J.T., Webb, R.S. & Webb, T., III (1992) Mapping eastern North American vegetation change of the past 18 ka: no-analogs and the future. *Geology* 20, 1071–1074.

Panikov, N.S. (1999) Fluxes of CO_2 and CH_4 in high latitude wetlands: measuring, modelling and predicting response to climate change. *Polar Research* 18, 237–244.

Parmesan, C., Ryholm, N., Stefanescu, C. *et al.* (1999) Poleward shifts in geographical ranges of butterfly species associated with regional warming. *Nature* 399, 579–583.

Parmesan, C. & Yohe, G. (2003) A globally coherent fingerprint of climate change impacts across natural systems. *Nature* 421, 37–42.

Patricola, C.M. & Cook, K.H. (2007) Dynamics of the West African monsoon under mid-Holocene precessional forcing: regional climate model simulations. *Journal of Climate* 20, 694–716.

Prentice, I.C., Bondeau, A., Cramer, W. *et al.* 2007. Dynamic global vegetation modelling: quantifying terrestrial ecosystem responses to large-scale environmental change. In: *Terrestrial Ecosystems in a Changing World* (eds J. Canadell, D.E. Pataki & L.F. Pitelka), pp. 175–192. Springer, Berlin.

Sanders, G.E., Skärby, L., Ashmore, M.R. & Fuhrer, J. (1995) Establishing critical levels for the effects of air pollution on vegetation. *Water, Air & Soil Pollution* 85, 189–200.

Schwartz, M.D., Ahas, R. & Aasa, A. (2006) Onset of spring starting earlier across the Northern Hemisphere. *Global Change Biology* 12, 343–351.

Segerström, U., Hornberg, G. & Bradshaw, R. (1996) The 9000-year history of vegetation development and disturbance patterns of a swamp-forest in Dalarna, northern Sweden. *Holocene* 6, 37–48.

Singh, S., Roy, S., Choudhury, S. & Sengupta, D. (2010) DNA repair and recombination in higher plants: insights from comparative genomics of arabidopsis and rice. *BMC Genomics* 11, 443.

Skidmore, A., Franklin, J., Dawson, T. & Pilesjö, P. (2011) Geospatial tool address merging issues in spatial ecology: a review and commentary on the Special Issue. *International Journal of GIScience* 25, 337–365.

Sparks, T.H., Jeffree, E.P. & Jeffree, C.E. (2000) An examination of the relationship between flowering times and temperature at the national scale using long-term phenological records from the UK. *International Journal of Biometeorology* 44, 82–87.

Street-Perrott, F.A., Mitchell, J.F.B., Marchand, D.S. & Brunner, J.S. (1990) Milankovitch and albedo forcing of the tropical monsoons: a comparison of geological evidence and numerical simulations for 9000 yBP. *Transactions of the Royal Society of Edinburgh* 81, 407–427.

Stuart, A.J. (2005) The extinction of woolly mammoth (*Mammuthus primigenius*) and straight-tusked elephant (*Palaeoloxodon antiquus*) in Europe. *Quaternary International* 126–128, 171–177.

Sykes, M.T. (1997) The biogeographic consequences of forecast changes in the global environment: Individual species' potential range changes. In: *Past and Future Rapid Environmental Changes: The Spatial and Evolutionary Responses of Terrestrial Biota* (eds. B. Huntley, W. Cramer, A.V. Morgan, H.C. Prentice & J.R.M. Allen). *NATO ASI Series I: Global Environmental Change* 47, pp. 427–440. Springer-Verlag, Berlin.

Sykes, M.T. & Prentice, I.C. (1996) Climate change, tree species distributions and forest dynamics: A case study in the mixed conifer northern hardwoods zone of northern Europe. *Climatic Change* 34, 161–177.

Thuiller, W. (2004) Patterns and uncertainties of species' range shifts under climate change. *Global Change Biology* 10, 2020–2027.

vanVuuren, M.M.I., Aerts, R., Berendse, F. & de Visser, W. (1992) Nitrogen mineralisation in heathland ecosystems dominated by different plant species. *Biogeochemistry* 16, 151–166.

Watt, A.S. (1981) Further observations on the effects of excluding rabbits from grassland A in East Anglian Breckland: the pattern of change and factors affecting it (1936–73). *Journal of Ecology* **69**, 509–536.

Webb, T. III (1987) The appearance and disappearance of major vegetational assemblages: long-term vegetational dynamics in eastern North America. *Vegetatio* **69**, 177–187.

Williams, J.W. & Jackson, S.T. (2007) Novel climates, no-analog communities, and ecological surprises. *Frontiers in Ecology and the Environment* **5**, 475–482.

Williams, J.W., Jackson, S.T. & Kutzbach, J.E. (2007) Projected distributions of novel and disappearing climates by 2100 AD. *Proceedings of the National Academy of Sciences of the United States of America* **104**, 5738–5742.

Williams, S.E., Shoo, L.P., Isaac, J.L., Hoffmann, A.A. & Langham, G. (2008) Towards an integrated framework for assessing the vulnerability of species to climate change. *PLoS Biol* **6**, e325.

Willis, A.J., Dunnett, N.P., Hunt, R. & Grime, J.P. (1995) Does Gulf Stream position affect vegetation dynamics in Western Europe? *Oikos* **73**, 408–410.

Wilson, E.J., Wells, T.C.E. & Sparks, T.H. (1995) Are calcareous grasslands in the UK under threat from nitrogen deposition? – An experimental determination of a critical load. *Journal of Ecology* **83**, 823–832.

Index

Page references to figures are in *italic*, to tables are in **bold**
Lists of ecologists, (dis)similarity indices (coefficients), multivariate computer programmes, growth-forms and life-forms, geographic names, vascular plant taxa, other taxa, phytosociological units and vegetation types are presented at the end of this Index.

adaptation 16, 18, 113, 166–7, 184,
    186, 242, 248, 266–7, 347–8,
    351, 353, 356, 358, 365, 369,
    407, 409, 457–9, 472, 516–9,
    524, 527
aerial photography *see* vegetation
    mapping
agriculture *see* human use
alien species *see* species invasion
alliance *see* syntaxonomy
Anglo-American approaches (schools) 1,
    4, 7–8, 37
apomict, apomixis *144*, 147, 156, 408
archaeology 174, 433
archaeophyte **391–2**, 393, 433
assembly rule *see* community
association *see* classification, syntaxonomy
Atlas of the British Flora 491
atmosphere 24, 95, 113, 248, 285–6,
    289, 297, 299, *300–1*, *303*,
    375, 430, 445, *468*, 474,
    509–11
  CO_2 24, 460, 467, 477, 509, *513*,
    518–20, *520*, 522, 523–6

light 9, 111, 112, *112*, 120, 127–8,
    130, 133, 149, *150*, 151–2,
    153, 153–6, 166, 179, 180,
    182–3, 186–7, *189*, 203, 205,
    207–8, 210, 216, 244–5, 255–6,
    296, 317, 322–3, **352–3**, 355–7,
    366, *365*, 368–9, 376, 395,
    408, 412, 416, 432, **459**, 464,
    497
nitrogen 248, 261–2, 266–7, 276,
    437–8, 440, 444, 523
red:far red 180, 182

biodiversity: species diversity, species
    richness 1, 64, 166, 210, 308–9
diversity
  and animals (herbivores) 16, 250–2,
    254, 257
  and biotic interactions 216, 226,
    266, *323*, 332, *336*, 399
  and dispersal 166, 174, 189, 203,
    336
  and disturbance 29, 226–7, 322–4,
    323, *365 see also* disturbance

Vegetation Ecology, Second Edition. Eddy van der Maarel and Janet Franklin.
© 2013 John Wiley & Sons, Ltd. Published 2013 by John Wiley & Sons, Ltd.

List of ecologists (selection)

List of (dis)similarity indices (coefficients)

List of vascular plant taxa

List of other taxa (vascular plants listed above)